Lecture Notes in Networks and Systems

Volume 153

The series “Lecture Notes in Networks and Systems” publishes the latest developments in Networks and Systems—quickly, informally and with high quality. Original research reported in proceedings and post-proceedings represents the core of LNNS.

Volumes published in LNNS embrace all aspects and subfields of, as well as new challenges in, Networks and Systems.

The series contains proceedings and edited volumes in systems and networks, spanning the areas of Cyber-Physical Systems, Autonomous Systems, Sensor Networks, Control Systems, Energy Systems, Automotive Systems, Biological Systems, Vehicular Networking and Connected Vehicles, Aerospace Systems, Automation, Manufacturing, Smart Grids, Nonlinear Systems, Power Systems, Robotics, Social Systems, Economic Systems and other. Of particular value to both the contributors and the readership are the short publication timeframe and the world-wide distribution and exposure which enable both a wide and rapid dissemination of research output.

The series covers the theory, applications, and perspectives on the state of the art and future developments relevant to systems and networks, decision making, control, complex processes and related areas, as embedded in the fields of interdisciplinary and applied sciences, engineering, computer science, physics, economics, social, and life sciences, as well as the paradigms and methodologies behind them.

**** Indexing: The books of this series are submitted to ISI Proceedings, SCOPUS, Google Scholar and Springerlink ****

More information about this series at http://www.springer.com/series/15179

Nenad Mitrovic · Goran Mladenovic ·
Aleksandra Mitrovic
Editors

Experimental and Computational Investigations in Engineering

Proceedings of the International Conference of Experimental and Numerical Investigations and New Technologies, CNNTech 2020

Editors
Nenad Mitrovic
Faculty of Mechanical Engineering,
Department for Process Engineering
and Environmental Protection
University of Belgrade
Belgrade, Serbia

Goran Mladenovic
Faculty of Mechanical Engineering,
Department for Production Engineering
University of Belgrade
Belgrade, Serbia

Aleksandra Mitrovic
Faculty of Information Technology
and Engineering
University Union - Nikola Tesla
Belgrade, Serbia

ISSN 2367-3370 ISSN 2367-3389 (electronic)
Lecture Notes in Networks and Systems
ISBN 978-3-030-58361-3 ISBN 978-3-030-58362-0 (eBook)
https://doi.org/10.1007/978-3-030-58362-0

This Springer imprint is published by the registered company Springer Nature Switzerland AG
The registered company address is: Gewerbestrasse 11, 6330 Cham, Switzerland

Preface

The book is a collection of high-quality peer-reviewed research papers presented at the International Conference of Experimental and Numerical Investigations and New Technologies (CNNTech 2020) held at Zlatibor, Serbia, from 29 June to 02 July 2020. The conference is organized by the Innovation Center of the Faculty of Mechanical Engineering, Faculty of Mechanical Engineering at the University of Belgrade and Center for Business Trainings. Over 50 delegates were attending the CNNTech 2020—academicians, practitioners and scientists from 13 countries—presenting and authoring more than 70 papers. The conference programme included four keynote lectures, two mini symposia, three sessions (oral and poster), one workshop and B2B meetings. Twenty-six selected full papers went through the double-blind reviewing process.

The main goal of the conference is to make positive atmosphere for the discussion on a wide variety of industrial, engineering and scientific applications of the engineering techniques. Participation of a number of domestic and international authors, as well as the diversity of topics, has justified our efforts to organize this conference and contribute to exchange of knowledge, research results and experience of industry experts, research institutions and faculties which all share a common interest in the field in experimental and numerical investigations.

The CNNTech 2020 was focused on the following topics:

- Mechanical engineering,
- Engineering materials,
- Chemical and process engineering,
- Experimental techniques,
- Numerical methods,
- New technologies,
- Clear sky,
- Dental materials and structures,
- Sustainable design and new technologies,
- Industry and sustainable development: contemporary management perspectives.

We express our gratitude to all people involved in conference planning, preparation and realization, especially to:

- All authors, specially keynote speakers and invited speakers, who have contributed to the high scientific and professional level of the conference,
- All members of the Organizing Committee,
- All members of the International Scientific Committee for reviewing the papers and Chairing the Conference Sessions,
- Ministry of Education, Science and Technological Development of Republic of Serbia for supporting of the conference.

Organization

Scientific Committee

Name	Affiliation
Miloš Milošević (Chairman)	University of Belgrade, Faculty of Mechanical Engineering, Serbia
Nenad Mitrović (Co-chairman)	University of Belgrade, Faculty of Mechanical Engineering, Serbia
Aleksandar Sedmak	University of Belgrade, Faculty of Mechanical Engineering, Serbia
Hloch Sergej	Technical University of Košice, Faculty of Manufacturing Technologies, Slovakia
Dražan Kozak	University of Osijek, Faculty of Mechanical Engineering in Slavonski Brod, Croatia
Nenad Gubeljak	University of Maribor, Faculty of Mechanical Engineering, Slovenia
Monka Peter	Technical University of Kosice, Faculty of Manufacturing Technologies, Slovakia
Snežana Kirin	University of Belgrade Innovation Center of Faculty of Mechanical Engineering, Serbia
Ivan Samardžić	University of Osijek, Faculty of Mechanical Engineering in Slavonski Brod, Croatia
Martina Balać	University of Belgrade, Faculty of Mechanical Engineering, Serbia
Ludmila Mládková	University of Economics Prague, Czech Republic
Johanyák Zsolt Csaba	Athéné University, Faculty of Engineering and Computer Science, Hungary
Igor Svetel	University of Belgrade, Innovation centre of Faculty of Mechanical Engineering, Serbia
Aleksandra Mitrović	University of Belgrade, Faculty of Technology and Metallurgy, Serbia

Contents

Examination of Laminations in the Base Material on a Section of High Pressure Gas Supply Pipeline

Viktor Stojmanovski[1](✉), Vladimir Stojmanovski[2], and Blagoja Stavrov[3]

[1] Faculty of Mechanical Engineering, Ss. Cyril and Methodius University in Skopje, Karpos 2 BB, PO Box 464, 1000 Skopje, Macedonia
viktor.stojmanovski@mf.edu.mk

[2] Centre for Research Development and Continual Education CIRKO DOOEL Skopje, Rudjer Boskovic 18, IT-027, 1000 Skopje, Macedonia

[3] RAPID BILD DOO, Nikola Tesla 160, 1300 Kumanovo, Macedonia

Abstract. During the installation works of high pressure gas supply pipeline (working pressure 54 bar, pipe diameter Ø508 mm, wall thickness 6,4/7,9/9,53 mm, material X60M according to API5L, PSL2) at the performance of the in-field welding, appeared indications that in the base material of the pipe exist laminations. They are one of the most severe defects that cannot be tolerated in this kind of structures. That fact imposed the need for additional investigations in order to confirm and evaluate the severity of these defects before performing the 81 bar hydro test. Considering the fact that the entire section of the pipeline (cca 36 km) was already installed and buried under the ground, UT inspection of the section had to be done with special equipment that was moving inside the pipeline filled with water. Only by this manner it can be achieved to have inspection of the entire internal surface of the pipeline, in its full length of cca 36 km. With this inspection all the defects in the pipe wall were discovered at their exact location. On particular critical locations, to find out the exact dimensions and characteristics of the imperfections, additional examinations were performed with NDT methods. These findings in the critical locations of the entire 36 km pipeline obliged to be performed reparations. This paper covers the categorization of the defects on the critical locations (discovered by the UT examination and additional NDT testing with classical methods), the specifications and causes of their appearance. The proposed reparations on the entire gas supply section, due to these defects, are presented too.

Keywords: Laminations · UT inspection · NDT inspection · Gas supply pipeline inspection · Imperfections · Defects · Anomalies

1 Introduction

During the welding works on the high pressure gas supply pipeline (working pressure 54 bar, pipe diameter Ø508 mm, wall thickness 6,4/7,9/9,53 mm, material X60M according to API5L, PSL2), at the performance of the butt-weld, the welders noticed unusual

N. Mitrovic et al. (Eds.): CNNTech 2020, LNNS 153, pp. 1–19, 2021.
https://doi.org/10.1007/978-3-030-58362-0_1

occurrence during the melting of the metal. Specifically, during the melting of the metal at the welding, occurred trace that looks like crack. Due to this occurrence, the welding immediately was stopped. The weld with the defect which already was done was removed and detailed visual and ultrasonic inspections were performed at the affected location (the end of the pipe and the surface of non protected part of it). The inspection found out existence of defect with dimensions presented on Fig. 1 (the pipe of the left side). Considering the fact that pipes are manufactured according to API5L and each pipe has its own production number, further was investigated the next pipe from the production line. It was discovered that the defect continues on the next pipe with almost the same dimensions (Presented on Fig. 1 - the pipe of the right side).

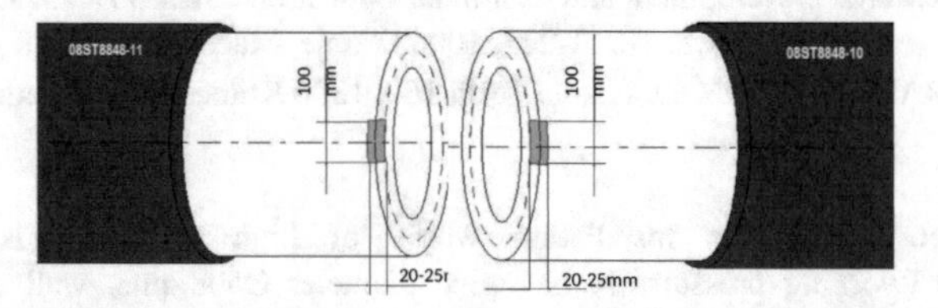

Fig. 1. Location of the defect and its overall dimensions

Having in mind that two neighboring pipes from the production line were affected, after the cutting of the pipes, it was suspected that in the sheet metal existed defect with dimensions of 100×100 mm. The character of the defect imposed the doubt that it is lamination.

2 Preliminary Examinations for Confirming the Indications for Lamination

In order to investigate the indications for existence of double layer laminations detailed examinations of affected locations were performed. These investigations besides visual control included metallographic examination and examination of mechanical properties. Methodology of examination is widely known and used as methodology in another similar application [1–4] and [8–10].

2.1 Metallographic Examinations

From the locations of the pipe where suspected defect are found, are taken samples for metallographic pictures. The defects in the transversal direction of the pipe axis from metallographic examination are presented on Fig. 2.

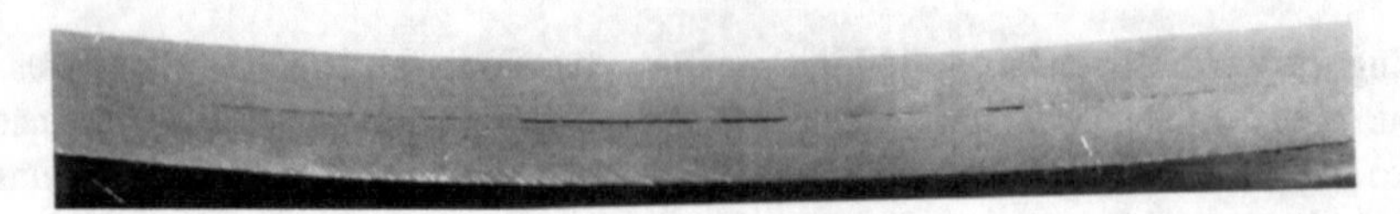

Fig. 2. Image of the defect

Wall thickness of the pipe is 6,4 mm. From visual inspection can be seen that the indication for lamination is in the middle of the wall.

2.2 Examination of Mechanical Properties

In order to confirm the character of the defect, and the sensitivity of the sheet metal, at the occurrence of double layered lamination, mechanical examination tests are performed on the location with and without defect. The sample taken from the location of the defect is presented on Fig. 3.

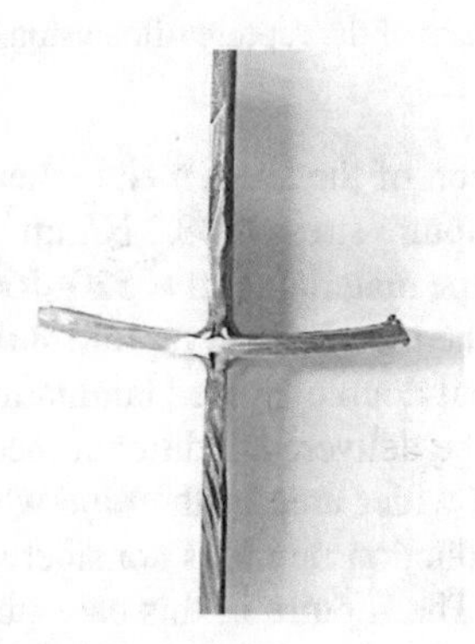

Fig. 3. Sample taken from the location of the defect

Images from tensile testing of the defect, before and after rupture, are presented on Fig. 6 (Fig. 4).

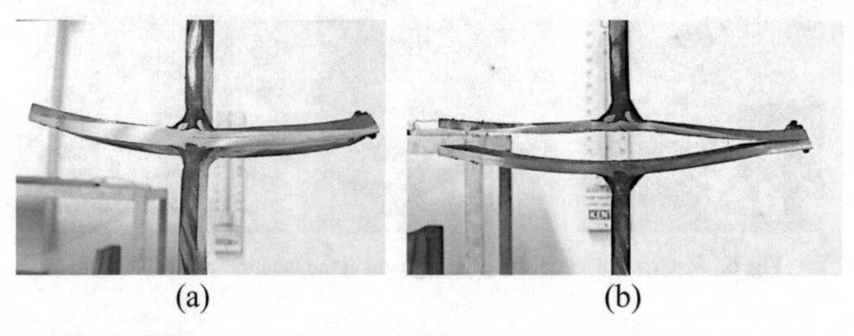

Fig. 4. a) Rupture test of the defect. b) Rupture occurred inside the pipe wall

The rupture of the specimen and total separation of the sheet metal in the location of the defect occurred at 154000 N. The image of the rupture surface is presented on Fig. 5.

Fig. 5. The surface of the rupture (dimensions 100 × 40 mm)

The area of the cross section of the sheet metal where separation occurred is A = 100 × 40 = 4000 mm^2. The rupture stress is 38,5 N/mm^2, and it is significantly smaller than the rupture stress of the pipe material (Rm = 520–760 N/mm^2 for X60M material). The geometry of the rupture, the image of the rupture surface and the small intensity of the stress suggest that it is about double layered lamination.

For definite conclusion to be delivered additional mechanical tests were performed with specimens selected from wider area of the pipe where this type of defect didn't existed (neighboring pipe production numbers considered in API). Images from these tests are presented on Fig. 5. The rupture in this case occurred at stress usual for this material and location of the rupture is presented on the right side image of Fig. 6.

Fig. 6. Mechanical test of the pipe segment at the location without defect.

From the performed examination, it is confirmed without doubt, that there is lamination on the suspected location. That fact imposed need for examination on the entire pipeline in order to check the existence of other laminations. The pipeline at the time was in the phase of building, already mounted and buried, waiting for hydro test. The overall length of the section of the pipeline that was about to be examined is 36 km.

In order to confirm the existence of other laminations in the transversal direction of the wall thickness, on the overall pipeline, the only possible way is to perform ultrasonic inspection of the entire pipeline.

3 Preparations for the UT Inspection of the Pipeline

Ultrasonic inspection of the entire pipeline is done by special equipment (tool) that moves in the pipe propelled by water and records the defect of the entire pipeline. Therefore, for successful test, it is crucial to provide proper movement by providing appropriate water flow. Providing sufficient amount of water, implied construction of installation for filling the pipeline with water and preserving the pressure of 30–38 bar during the examination. This is in order to enable continuous movement of the testing tool in the pipe. Previous to this operation, pipeline has to be clean and calibrated (tested that there is no violation in the ovality).

Considering the configuration of the terrain and available water, the pipeline of 36 km length was divided in 3 Sections:

- Section 1 - length 9,5 km,
- Section 2 - length 13,5 km,
- Section 3 - length 12,9 km.

The sections of the pipeline, vertical elevation, the difference in the hydrostatic pressure, the location of hydraulic equipment and the initial position of the in-line tools (cleaning, calibration and inspection tools) are presented on Fig. 7. The same scheme is used for the hydro test of the pipeline.

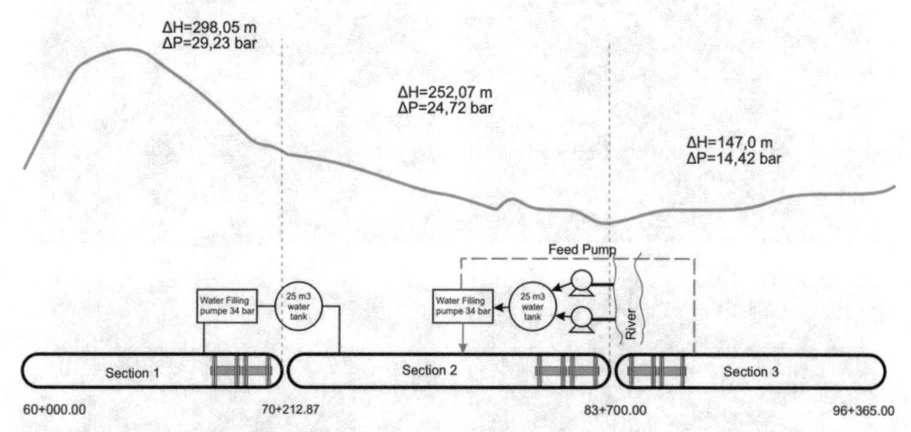

Fig. 7. Vertical section of the pipeline, elevations, position of hydraulic equipment and scheme for the hydro test.

The cleaning of the pipeline was done with special in-line tool (Fig. 9). Cleaning was done repeatedly until it was proved that the internal surface of the pipeline is totally clean and the debris from the internal side of the pipeline was removed. Images from cleaning process are presented on Fig. 8.

Fig. 8. Cleaning of the pipeline

After cleaning, the pipeline was calibrated with special in-line tool containing calibration plate. Diameter of the calibration plate is appropriate to internal diameter of the pipe. Figure 9 presents the calibration tool and contains images taken from the calibration of the pipeline. Criteria for acceptance of calibration is: when exiting from the pipeline the calibration plate not to be damaged. From figure can be seen non damaged calibration plate.

Fig. 9. Calibration of the pipeline

After confirming that the pipeline is clean and there is no violation of the ovality the next step may be proceeded: the ultrasonic inspection of the pipeline.

4 Ultrasonic Inspection of the Pipeline and Categorization of the Defects

Ultrasonic (UT) inspection of the pipeline was done with intelligent in-line tool containing 72 installed ultrasonic heads. The tool was put in the pipes of each section of the tested pipeline. The motion is enabled by the flow of the water maintaining continual pressure of 30 bar in the pipeline. The intelligent tool is presented on Fig. 10.

Position of the tool inside the pipeline and pressure inside the pipeline during the test are presented on Fig. 11.

Fig. 10. Intelligent in-line tool for UT inspection of the pipeline

Fig. 11. Position of the tool and pressure inside the pipeline during the examination

4.1 Results from Ultrasonic Inspection

UT inspection discovered the locations of all possible indications for defects in the pipeline in all of its length. These indications are:

- metal loss
- indications for laminations
- features that are not categorized as metal loss or lamination.

Indications for Metal Loss

Locations with loss of the metal on the wall thickness along the pipeline, of the Sect. 2, are presented on the diagram on Fig. 12. In this paper, Sect. 2 is taken as representative because it was section with most severe indications. Other sections didn't had indications with such severity.

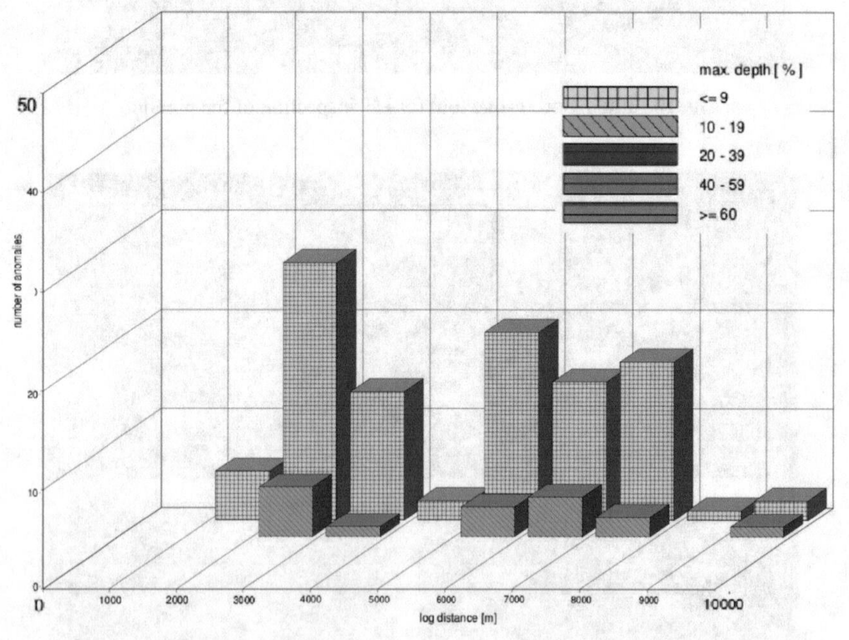

Fig. 12. Number of findings of metal loss of the wall thickness and categorization along the Section 2 of the pipeline

Comparing these findings with the allowable values, it may be concluded that there is no metal loss higher than allowed according to [2].

The location of the most severe metal loss (within the allowable range) is presented below. This defect will further be investigated for final evaluation. It is located on 4402,61 m from the start of the Sect. 2. Dimensions of the feature and its exact location are presented below.

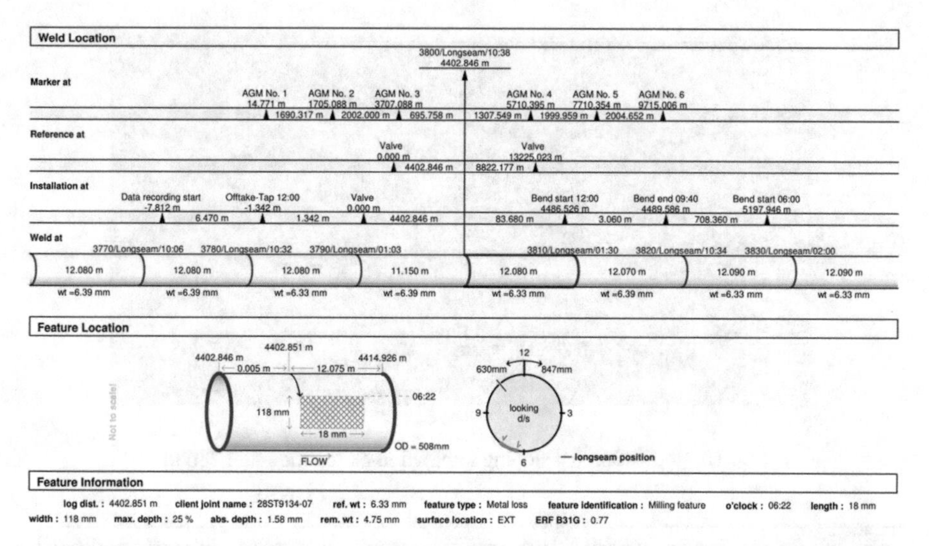

This feature, in order to be evaluated more precisely, is further analyzed with classical NDT methods.

Indications for Laminations

One location with most characteristic sloping lamination, found from the UT inspection, is presented below. Dimensions of the feature are provided as well as its position on the pipe cross section.

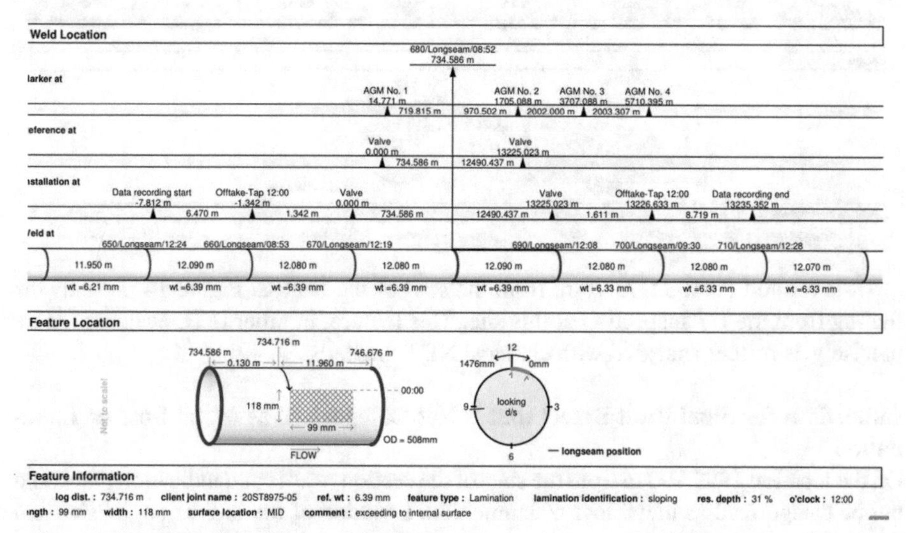

It is located on 734,716 m from the start of the Sect. 2. Figure 13 presents the finding from the UT inspection at this site. This feature, in order to be evaluated more precisely, is further analyzed with classical NDT methods.

Below is presented another indication for sloping lamination found in UT examination of the pipeline. Dimensions of the feature are provided as well as its position on the pipe cross section.

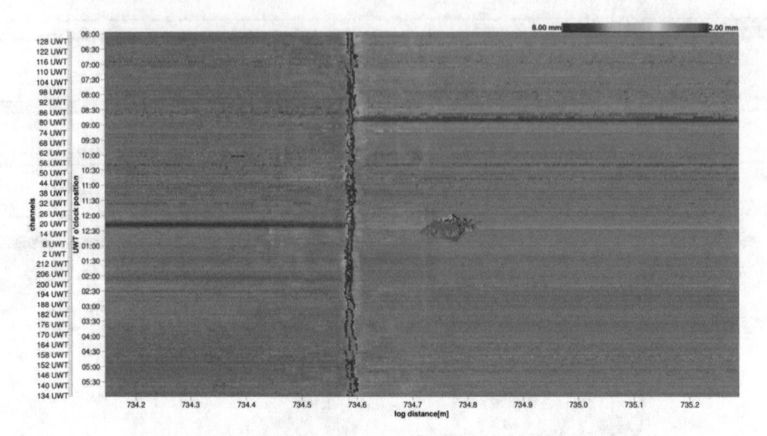

Fig. 13. Indication for sloping lamination on location 734,786 m

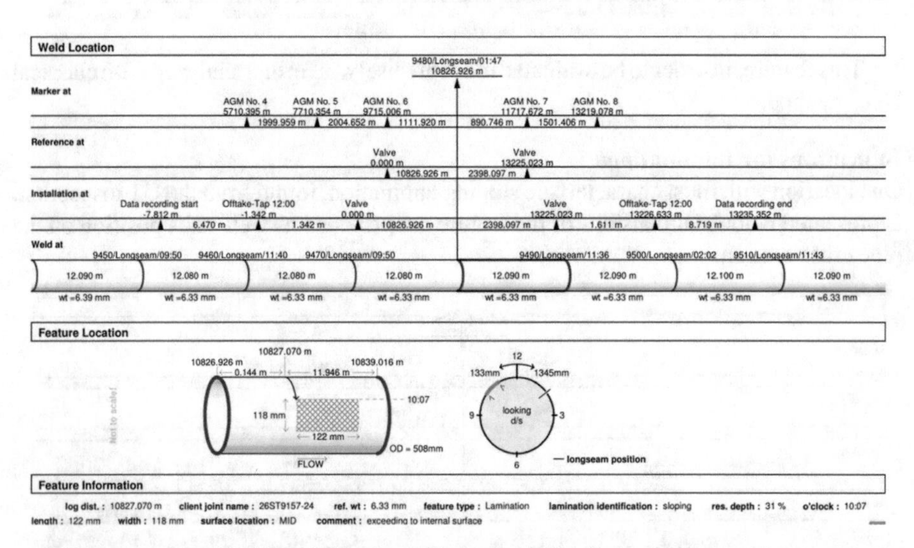

It is located on 108277,070 m from the start of the Sect. 2. Figure 14 presents the finding from the UT inspection at this site. This feature, in order to be evaluated more precisely, is further analyzed with classical NDT methods

Indications for Significant Defect that is Not Categorized as Metal Loss or Lamination

On the location 1999,700 m from the start of the section, was found indication that could not be categorized as metal loss or lamination. Location of the indication, its size and the position on the pipe cross section, are presented below.

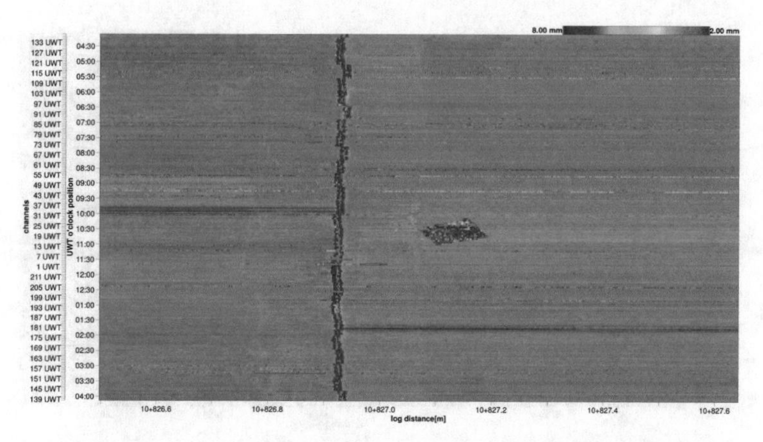

Fig. 14. Indication for sloping lamination on location 10827,070 m

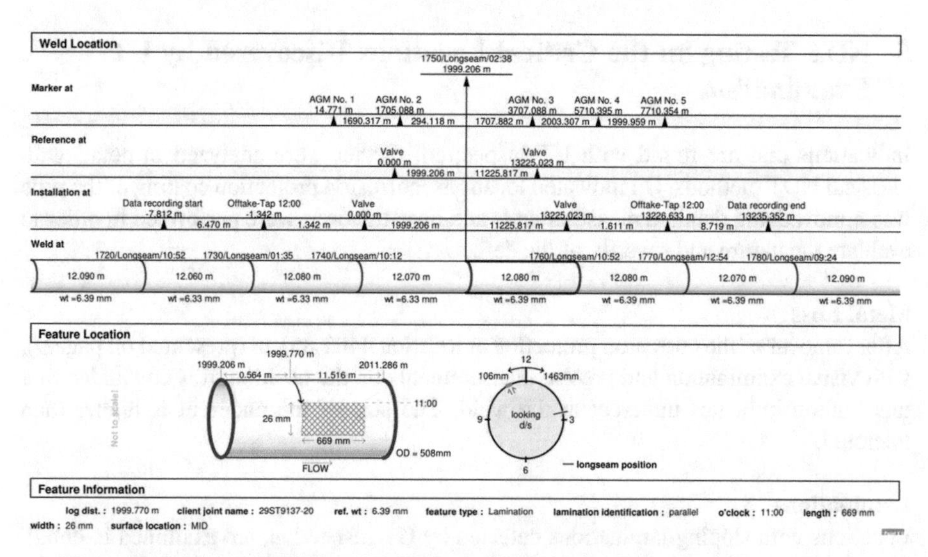

The record from the UT testing of this indication is presented on Fig. 15.

This indication requires special attention, and it was further analyzed with direct classical NDT examination presented in Sect. 5 of this paper.

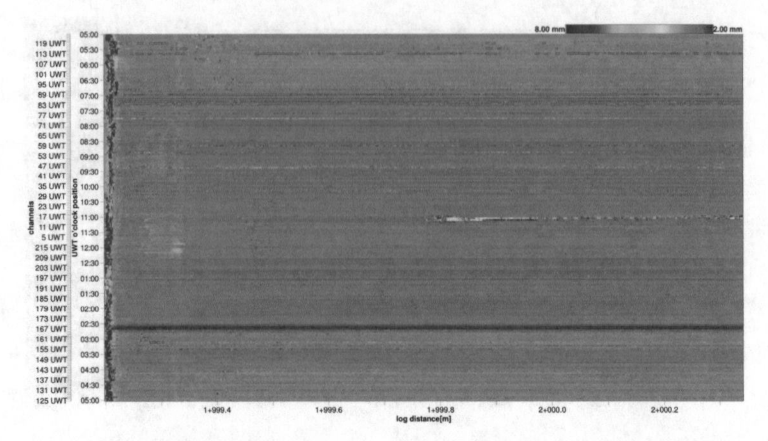

Fig. 15. Indication found on location 1999,206 m

5 NDT Testing on the Critical Locations Discovered by UT Examination

Indications that are found with UT inspection, further were analyzed in detail with classical NDT methods. On indicated locations, corrosion protection coating of the pipe was removed and detailed measurements and investigations were performed in order to evaluate the nature and severity of the defect.

Metal Loss

After removal of the corrosion protection at location 4402.851 m (presented on page 9), with visual examination and precise measurements on the position, it is concluded that this finding indicates undercut at the weld. The size of the undercut is higher than tolerated.

Laminations

Locations with sloping laminations detected by UT inspection, are examined in detail. The pipeline on those locations was excavated, parts of the pipes with indications for sloping laminations were removed and precise measurements with ultrasonic thickness meter was performed.

Indications for laminations discovered by UT were not correct. The indications were on the place where the manufacturer of the pipe placed labeling stickers on the internal side of the pipe. The measurements with the intelligent tool were so precise that they recorded the labeling stickers and the signal, having same blueprint, was false considered as sloping laminations. Findings on site confirmed that there are no laminations and labeling stickers give false positive indication for sloping lamination. Examples of these indications are those presented on Fig. 13 (location 734,716 m) and Fig. 14 (108277,070 m) from the Sect. 4 of this paper. Images taken from on site examination of the location 734.71 m are presented on Fig. 16.

Examinations of the location 108277,070 m and on-site findings are presented on Fig. 17.

Fig. 16. Examination on the location 734.71 m. Indicated lamination was found to be sticker from the manufacturer of the pipe

Fig. 17. Examination on the location 108277,070 m Indicated lamination was found to be sticker from the manufacturer of the pipe.

Significant Defect that is Not Categorized as Metal Loss or Lamination

On location No. 1999,206 m in-line UT inspection discovered indication for anomaly that can not be categorized as metal loss or lamination. Record of this finding is presented on Fig. 15. On this location excavation of the pipe was done and sample was taken for further analysis. The analysis included:

– Visual examination from the internal side of the pipe,
– Radiographic examination,
– Metallographic examination,
– Mechanical tests.

The defect is presented on Fig. 18.

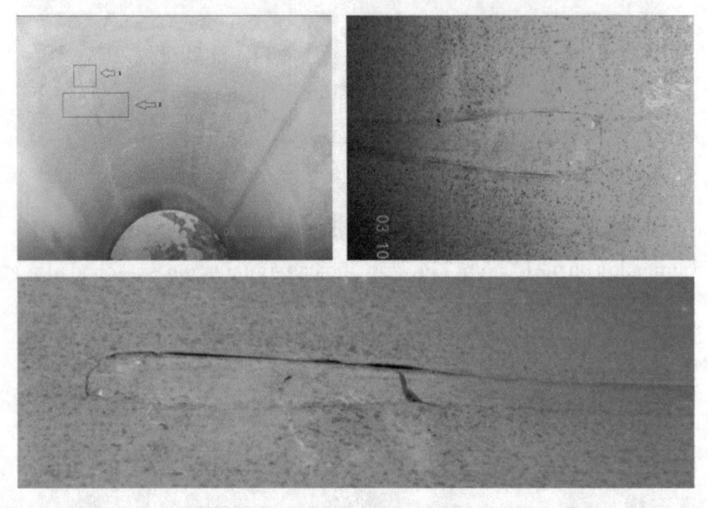

Fig. 18. Defect on location 1999.206 m. It is on the internal side of the pipe and cannot be categorized as metal loss or lamination.

On Fig. 19 is presented image from radiographic control of the defect.

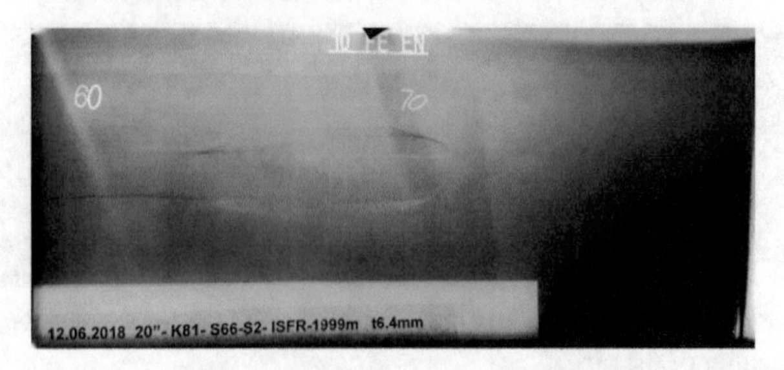

Fig. 19. Image from radiographic control

Figure 20 presents the nature of the defect, its shape and size related to the wall of the pipe. The defect is outside of the allowable limits and cannot be tolerated. Removal on this part of the pipe and replacement with new segment must be done.

Fig. 20. Image of the defect

6 Recommendation for Repair at the Locations Discovered by NDT Examination

Considering the fact that the pipeline was in the phase of building, locations where non tolerable defects were found, were removed by cutting off the pipe segment where the defect existed, and replacing it with new pipe segment (Fig. 21).

Fig. 21. Reparation of the pipeline by replacing defected segments

7 Leakage and Strength Test (Hydro-Test)

For ensuring that the pipeline will have proper integrity, the pipeline was tested for leakage and strength, by performing hydro-test. The scheme for the hydro-test is presented on Fig. 7. Each section of the pipeline was filled with water and special attention was put on proper deaeration of the pipeline.

After filling, the pressure was increased, until it reached the maximum working pressure of 81 bar. In this process, special attention was put on the configuration of the terrain of the pipeline, with particular attention on the hydrostatic pressure. Criteria for pressure, considering the terrain, is that minimum pressure (at the highest point of the section) not to be under 70,2 bar, and maximum pressure (at the lowest point of the section) not to exceed value that will produce stress greater than 95% of the yielding stress of the pipe material. Only in those conditions, the pipeline can be tested properly not violating the structural integrity. During the inspection of the Sect. 1, at the lowest altitude point, the testing pressure achieved value very close to produce critical stress (95% of the yielding point). Considering that there are imperfections of the pipe dimensions, and to be sure that the critical stress during the testing is not achieved at the lowest pipeline altitude the stress was monitored with strain gauges (see Fig. 7). The position of the strain gauges on the cross section is presented on Fig. 22. Other possibility is to monitor the strain state by DIC (Digital Image correlation) method [5–7].

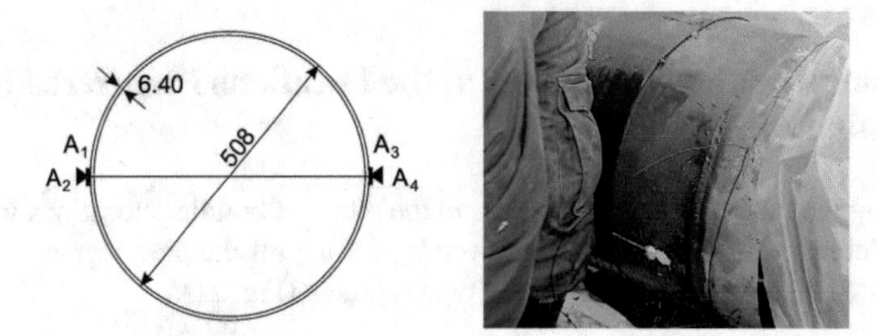

Fig. 22. Position of the strain gauges on the cross section at the lowest point of the pipeline

Monitoring of the stress achieved values during the test are presented on Fig. 23. The values were permanently monitored and if they reached the critical value the test would be stopped.

Progressively, continuously increasing, the testing pressure of 81 bar for strength test was achieved. The level of the achieved pressure previously was maintained on appropriate time required for pressure stabilization. Figure 24 presents monitoring of the pressure during the examination.

After the needed period for stabilization, the testing pressure was maintained for 8 h. After 8 h, gradually the pressure was decreased until reached the level of maximum working pressure. This pressure in fact represents the testing pressure for the leakage. This pressure (working) was maintained in the pipeline for the time of 24 h. It was done in order to ensure there is no leakage of the pipeline subjected to maximum working pressure for 24 h. Figure 25 presents the monitoring of the process of the leakage test.

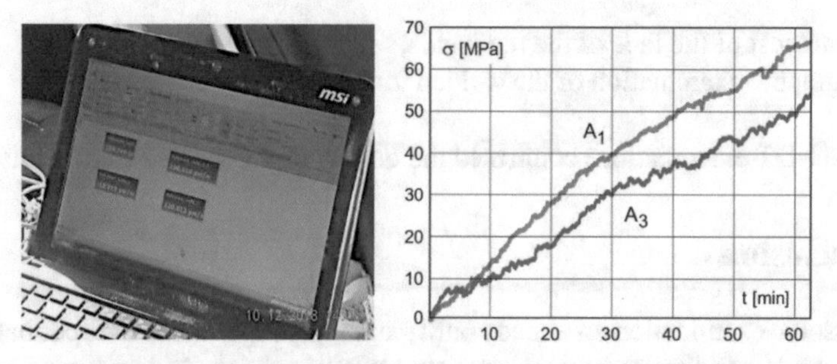

Fig. 23. Monitoring of the stress and achieved values of the stress during the test

Fig. 24. Monitoring of the pressure of the pipe during the strength test

Fig. 25. Monitoring of the pressure of the pipe during the leakage test

The process was monitored with calibrated manometers. Simultaneously during the test, the temperature of the surrounding air and the temperature of the wall of the pipe were measured on various locations of the pipeline section. During the strength and leakage test there wasn't noticed any leakage or deformation.

Welded joints that connect the tested Sects. 1, 2 and 3, were made after the hydro-test and they are not examined on strength and leakage. This kind of welds usually are noted as "red welds". In order to assure the quality of those red welds, additional NDT test are performed including:

– Penetrant test of the root of the red welds,

- Penetrant test of the face of the red welds,
- Radiographic examination of the welded joints.

These NDT examinations confirmed the quality of the welded (red) joints.

8 Conclusion

1. Ultrasonic examination on already built gas supply pipelines is exceptionally expensive and demanding process. For proper propulsion of the in-line tool to be met, the pipeline have to be filled with water under specific pressure and with continuous water supply, as result of difference of the pressure, the tool is moving inside the pipe.
2. With the ultrasonic sensors of the in-line inspection tool is inspected the internal surface of the entire tested pipeline. Only with this kind of test existed laminations can be detected on the entire pipeline.
3. On the locations where there are indications for existence of laminations and other defects, the pipe is excavated, protection coating is removed and detailed NDT inspection is performed with classical methods. Only by this way the character and the real dimensions of the defect can be precisely determined.
4. On the locations where the anomalies are outside of the tolerated values, the pipeline have to be repaired by cutting the damaged pipe and removing it with new segment.
5. These examinations are of significant importance for the overall safety and integrity of the pipeline.
6. With other kind of examination it is not possible to detect the defects analyzed in this paper, especially when the pipeline is already built and buried under ground.

References

1. Stojmanovski, Vl., Stojmanovski, V., Bogatinovski, Z.: Behavior of Butt-Welded Joints with Imperfections. LAP Lambert Academic Publishing, Saarbrucken (2015)
2. ANSI/ASME B31G: Manual for Determining the Remaining Strength of Corroded Pipelines. American Society of Mechanical Engineers (1991)
3. Stojmanovski, Vl., Bogatinovski, Z.: Research on Behavior of Butt-Welded Joints with Imperfections. Scientific Cooperations, IWME-2014, pp. 255–263, 2914
4. Stojmanovski, V., et al.: MFL and NDT examinations for integrity assessment of high pressure gas pipeline during exploitation. CIRKO Skopje (2017)
5. Milosevic, M., Milosevic, N., Sedmak, S., Tatic, U., Mitrovic, N., Hloch, S., Jovicic, R.: Digital image correlation in analysis of stiffness in local zones of welded joints. Tech. Gaz. **23**, 19–24 (2016)
6. Mitrovic, N., Petrovic, A., Milosevic, M.: Strain measurement of pressure equipment components using 3D DIC method. Struct. Integr. Procedia **13**, 1605–1608 (2018). 22nd European Conference on Fracture - ECF22
7. Milosevic, M., Mitrovic, N., Jovicic, R., Sedmak, A., Maneski, T., Petrovic, A., Aburuga, T.: Measurement of local tensile properties of welded joint using Digital Image Correlation method. Chem. Listy **106**, 485–488 (2012)

8. Černý, I., Mikulová, D., Sis, J.: Examples of actual defects in high pressure pipelines and probabilistic assessment of residual life for different types of pipeline steels. Procedia Struct. Integr. **7**, 431–437 (2017)
9. Goedecke, H.: Ultrasonic or MFL inspection: which technology is better for you? Pipeline Gas J. **230**, 34–41 (2003)
10. Pipeline Operator Forum: Specifications and requirements for intelligent pig inspection of pipelines, Version 3.2, January 2005

Digital Dissemination of Environmental and Social Initiatives. Investigation of Reporting Practices in German Shipping Industry

Slavica Cicvarić Kostić[1(✉)], Marko Mihić[1], Slavenko Djokić[2], and Dana Stojiljković[3]

[1] University of Belgrade, Jove Ilića 154, 11000 Belgrade, Serbia
slavica.cicvaric.kostic@fon.bg.ac.rs
[2] Monitoring and Evaluation Expert, Zrmanjska 3, 11000 Belgrade, Serbia
[3] Volkswagen Group Services GmbH, Niedersachsenstraße 2, 26723 Emden, Germany

Abstract. Both sustainable development and corporate social responsibility (CSR) have received particular attention recently, hence becoming increasingly important in the business world as a model in building equilibrium between economic benefits and growth on the one side, and the society and the environment on the other side. CSR is a strategic business concern that indicates responsiveness towards economic and environmental issues striving for a favorable influence on the development of society, and consequently sustainable development. Along with the trend of incorporation CSR initiatives into every-day business operations, the demand for more transparency on companies' activities in these areas has been expanding. As companies appear to increasingly engage in disclosure of information related to their environmental and social impacts, this study explores voluntary disclosure of data related to these two groups of concerns, particularly on companies' websites. The goal of the paper is to contribute to the understanding the reporting and disclosure of sustainability data. The study particular investigates the extent to which German freight shipping companies use websites as channels for presenting their social and environmental information to the public. The research included 82 freight shipping companies, members of the German ship-owners' association. For each company, the study measured the level of compatibility with the Global Reporting Initiative environmental and social reporting standards. The results indicate that companies seated in Hamburg, companies with group affiliation and companies with large fleets outperform those located in other German provinces, nonaffiliated shipping companies and modest-sized fleets.

Keywords: Environmental and social responsibility · Global reporting initiative standards · Digital disclosure · Freight shipping companies

1 Introduction

Sustainable Development and Corporate Social Responsibility (CSR) are closely related topics that have significantly influenced businesses in the recent period. Many studies have confirmed the undisputed interconnection between responsible business conduct

N. Mitrovic et al. (Eds.): CNNTech 2020, LNNS 153, pp. 20–34, 2021.
https://doi.org/10.1007/978-3-030-58362-0_2

and sustainable development (Ait Sidhoum and Serra 2018; Kolk and Van Tulder 2010; Moon 2007). According to these studies, the socially responsible business operations involve balancing the economic benefits with both environmental and societal responsibility, thus promoting and advancing sustainable development. Having in mind many challenges, such as the growing world population, extensive consumption of resources and other environmental concerns the only viable solution is the adoption of the sustainable development models and practices across the world. Leading companies from the developed economies have been showing a growing interest in this issue making substantial contributions to the sustainable development, by employing socially responsible business practices.

Both Sustainable Development and CSR have received particular attention recently, hence becoming increasingly important in the business world as a model in building equilibrium between economic benefits and growth on the one side, and the society and the environment on the other side (Crowther and Seifi 2018; Bansal and Song 2017). Thus, CSR as a new paradigm of the contemporary corporate practice contributes to the better understanding of companies' performance in the context of sustainable development.

Grounding on the Stakeholder theory and its core premise that companies are obligated to take into consideration the interests of all stakeholders groups affected by their actions (Freeman 1984), nowadays' business operations are analyzed continuously on their energy efficiency, environmental friendliness, community engagement, and other social issues, and not just pure business performance and profit (Djokic 2019; Mihic et al. 2014). More and more companies consider socially responsible business behavior and sustainability as a fundamental guiding principal embedded in their long-term growth strategies (Mihic et al. 2019; Vlastelica et al. 2015).

Along with the trend of incorporation CSR initiatives into every-day business operations, the demand for more transparency on companies' activities in these areas has been expanding (Taylor et al. 2018). As companies appear to increasingly engage in disclosure of information related to their environmental and social impacts (KPMG 2013), this study explores voluntary disclosure of data related to these two issues.

The goal of the paper is to contribute to the understanding the ever-growing CSR concerns, and to highlight the relevance of reporting and digital dissemination of these initiatives in order to better understand this relevant issue. The study particular investigates the extent to which German freight shipping companies use websites as channels for presenting their social and environmental information to the public.

The remainder of the paper is organized as follows: the first two sections define the conceptual framework and discusses sustainable development and corporate social responsibility, CSR related reporting practices, as well as the importance of digital disclosure of CSR initiatives; Sect. 3 illustrates the research method and results, and Sect. 4 presents the conclusions and contributions of the study.

2 Conceptual Framework: Sustainability and Socially Responsible Business Initiatives

2.1 The Evolution of the Sustainable Development Concept

A notion of importance of sustainability has evolved in response to growing global challenges resulting from industrialization and subsequently very aggressive exploitation of natural resources. Nevertheless, the concept itself was originated back in the 18th century as a strict control related to the number of harvested trees in the German forestry; and even the term derived from "nachhaltiger ertrag," which means "sustained yield" in German language. Hans Carl von Carlowitz is seen as the founding father of the modern sustainable development idea (Enders and Remig 2015), being the first to say that we have to find the right balance between resource use and the regeneration of natural capital. In his book Sylvicultura economica from 1713, von Carlowitz argued that the natural resources must be used in a way that nature can reproduce in a given period of time (Vogt and Weber 2019).

In the modern sense, the term "sustainable" is linked to the Club of Rome, and The Limits to Growth, a 1972 report that described the way in which the world population can reach to global equilibrium and balance. The Club of Rome triggered an international discussion in response to this report engaging the World Commission on Environment and Development under the patronage of the Secretary-General of the UN (Meadows et al. 1972). Five years later, the publication Our Common Future established a recommended path for sustainable development on a global level bringing the concept of sustainability into the focus of the international community, and providing the popular definition of Sustainable Development, as follows: "Sustainable development is a development that meets the needs of the present without compromising the ability of future generations to meet their own needs" (Brundtland Report, World Commission on Environment and Development 1987). The definition encompasses the following two essential elements: (1) needs of the world's poor who should be given ultimate priority; and (2) limits in terms of what environment can provide to meet present without compromising future needs given the state of technology and society overall. This definition resulted in some controversy, specifically the part about solving the global poverty problems as an ultimate priority. Many parties involved in this issue argued that environmental sustainability should be given ultimate priority instead, whereas the unresolved environmental problems would eventually lead to complete collapse of all aspects of human existence on the plant.

By September 2000, the UN had introduced a more coherent and consolidated approach towards the economic, social, and environmental dimensions of international development, and consequently adopted the UN Millennium Declaration, hence stipulating a new global partnership to reach the fifteen-year development targets known as the Millennium Development Goals (Millennium Development Goals 2019; Georgeson and Maslin 2018). Building on the momentum generated by the Millennium Development Goals, UN conducted a series of global consultations with civil society organizations, scientists, citizens, academics and private sector stakeholders from around the world to further strengthen convergence of the development agenda, and create a new, people-centered, development agenda that includes an expended series of time-bound targets known as the Sustainable Development Goals or the 2030 Agenda (Kumar et al. 2016). The 2030 Agenda, with 17 Sustainable Development Goals and 169 specific targets, amplifies the critical dimensions of sustainable development – economic growth, social inclusion, and environmental protection, as well as the rule of law and peace through institutional development and cooperation (Sustainable Development Goals – Knowledge Platform 2019; Georgeson and Maslin 2018).

Why is sustainable development important? In addition to many challenges, perhaps the most critical one is the fact that the world population is growing at a staggering rate; namely, there were 5.3 billion people in 1990, 7.6 billion in 2017, and the forecasts suggest that the world population will reach 8.6 billion in 2030 (UN Department of Economic and Social Affairs, Population Division 2017). According to the 2008 Living Plant Report, humans are consuming 30% more resources than the Earth can regenerate each year; if the resources continue to be used at the current rate, two planets will be insufficient to house the world population by mid-2030s (WWF 2008). Clearly, our Planet cannot provide sufficient resources at the current rate to support the growing population needs, hence the only viable solution is the adoption of the sustainable development models and practices across the world. Global, technologically advanced companies that emerged from the strongest economies have been showing enormous interest in this issue making tangible contributions to the sustainable development agenda worldwide. These companies paved the way forward toward sustainability making it possible through innovations, yet balancing their economic growth with community involvement and environmental actions.

2.2 Corporate Social Responsibility as a Strategic Business Concern

Corporate social responsibility is a strategic business concern that indicates responsiveness towards economic and environmental issues striving for a favorable influence on the development of society, and consequently sustainable development. Whereas the companies have a significant impact on a society as a whole, CSR emerged as a result of increased interest for appropriate role of companies in a society with regard to being responsive to the communities and environment (Albuquerque et al. 2018; Vlastelica et al. 2015; Dalshrud 2008; European Commission 2003). Being one of the first, Bowen (1953, p. 6) defined CSR as "an obligation to pursue policies to make decisions and to follow lines of action which are compatible with the objectives and values of society".

The conceptualization of CSR comprises the contribution of various academic attempts, as well as international organizations which institutionalize, promote and monitor socially responsible business operations (Aguinis and Glavas 2019; Wang et al. 2016; ISO 26000 2010; Dahlsrud 2008; Garriga and Mele 2004; European Commission 2003; Carrol 1991). CSR has been discussed with the objective of improving the understanding of the performance of companies in the context of sustainable development. Whereas the companies have a legal right to utilize the existing resources, they are therefore responsible for their use in a sustainable manner being responsive to the communities and environment, as well as customers and shareholders.

Carroll (1991) identified four areas that create a corporate social responsibility pyramid: legal, economic, ethical and philanthropic. This pyramid has become widely used to elaborate on the main responsibilities of business. In legal terms, companies will protect customers by obeying regulations. The customers rely upon the truthfulness of the companies about the products and services they sell. Being profitable and provide a return on investment to owners and shareholders, while generating employment in the communities are the primary economic responsibilities. It also considers streamlining production processes to find the most efficient ways of the business operations that will lead to increased revenues. Company's ethical responsibilities include managing waste, recycling and rational consumption of necessary resources to help preserve the nature for future generations. Other ethical duties come in the form of truthful labeling and advertisement of the products as to avoid any intentional misleading of the customers, and adequate treatment of employees, e.g., companies can provide more than minimum benefits package for their employees, and it can further invest resources in creating a work environment where employees would feel happy to come every day. Successful companies should also engage in philanthropy to spread goodwill by giving to those in need and promote the welfare of humans (Carroll 1991).

Elkington (1997) has expanded on these concerns pointing out that companies should focus on environmental and social matters as much as they focus on increasing their value through maximizing profit and outputs. Elkington was the first one to formulate the phrase "the triple bottom line" (TBL) while explaining that companies should be considering three different bottom lines in their business operations. TBL is the framework for reporting and tracking business performance, which includes three main indicators, economic (profit), social (people) and the environment (planet). By using this framework, the paper will focus on social and environmental sustainability. These dimensions incorporate various topics, related to the workplace – such as health and safety at a workplace, protection of human rights, equal treatment in employment and opportunities for all employees, staff development, respecting work-life balance; related to the community – such as investment program for local community, philanthropy, donations and humanitarian activities, employee volunteering; and related to the environment – such as responsible use of natural resources, reduced consumption of natural resources, preservation of existing resources, investing in the creation of new resources, preserving biodiversity, reducing $CO2$ emission (Vlastelica et al. 2018).

The growing importance of sustainability and socially responsible business practices is consequently recognized in the stock market. In 1999, the most prominent stock market index S&P Dow Jones Indices along with Robeco SAM jointly introduced the Dow Jones Sustainability Indices (DJSI) as a family of sustainability stock market benchmarks for investors who build their investment portfolios while considering sustainable business practices as critical in generating long-term shareholder value. DJSI is designed to track the stock performance of the world's leading companies in terms of economic, environmental and social criteria; it includes the top ten percent of the biggest stocks in the Dow Jones Global Indexes, covering more than 2,500 companies, and measuring their business activities in the context of sustainability. Companies represented in the index are assessed through a sophisticated grading system that considers social and environmental metrics, in addition to economic performance and other industry-specific indicators. Whereas the investors have been increasingly interested in socially responsible investments, the index has been widely recognized by many private wealth managers as a benchmark, with billions of assets under the management. The enterprises that are members of the index consider it as an opportunity to increase the shareholder awareness of environmental efforts, publicizing their index membership as an essential business accomplishment (López et al. 2007).

Academics have not reached a consensus on a unique and overall model or theory of a business case for CSR (Carrol and Shabana 2010). There are researches of the impact of certain aspects of CSR on certain business results, or the benefits of CSR for the different stakeholders. In terms of business performance, Weber (2008) explores and summarizes the benefits of CSR as follows: a positive impact on corporate reputation as well as on recruitment, motivation and retention of employees; reduction of operating costs and risks, and also increase in sales, market share and revenues. The academic literature is dominated by empirical studies of the positive impact of CSR on financial performance of companies (Vlastelica et al. 2015; Weber 2008; Mackey et al. 2007; Margolis and Walsh 2003). Certain studies also have been focusing on the impact of CSR on consumer reaction and consumer behavior (Aguilera et al. 2007; Ellen et al. 2006). Hence, for most consumers, CSR is still not among significant criteria when deciding to purchase (Vlastelica et al. 2015). One of the recent studies (Rivera et al. 2019) finds a direct, positive influence of CSR associations on loyalty, and also an indirect influence on brand awareness and consumer satisfaction. To summarize. Vlastelica et al. (2018) listed the researches about the main benefits of implementing CSR, ranging from cost reduction, over other performance indicators, benefits for different stakeholders, to identifying CSR as one of the main drivers of corporate reputation. In addition, a successful socially responsible business formula aims for a positive effect on bottom-line economic results without taking advantage of nature or people in doing so. Therefore, this paper is focused on the relevance of social and environmental concerns.

2.3 Reporting and Digital Disclosure of CSR Initiatives

Over the past decades, more and more companies have worked to monitor and manage their impacts and make environmental and social responsibility a core part of their corporate missions (European Commission 2019). Consequently, many different instruments to manage, measure, communicate and reward CSR performance have been developed to motivate business sector to assume responsibility for the social, ecological and economic consequences of their activities, such as the Organization for Economic Cooperation and Development (OECD) Guidelines for Multinational Enterprises, the Environmental Management System standard ISO 14001, the Global Reporting Initiative (GRI) Sustainability Reporting Guidelines, the United Nations Global Compact (UNGC), and the social responsibility guidance standard ISO 26000. These international guidelines have brought considerable changes in the developing CSR field (Zinenko et al. 2015), while OECD Guidelines, UNGC and ISO 26000 along with ILO Conventions are often referred to as the core set of internationally recognized principles and guidelines regarding CSR (Theuws and van Huijstee 2013).

The European Commission believes that CSR is essential for the sustainability, competitiveness, and innovation of the EU enterprises and the EU economy; it brings benefits for risk management, cost savings, access to capital, customer relationships, and human resource management. Accordingly, the EU has pursued an active role in promoting and fostering the business model inclusive of social responsibility. From very beginning, European Commission has been advocating for the CSR instruments that include critical guidance and benchmarks for sustainability performance (European Commission 2004); helping organizations to manage the quality of their processes, systems and impacts, and encouraging their best practice, while fostering the effective promotion of CSR.

Moving forward, EU has been actively encouraging the member states to promote the use of standards such as ISO 26000 Guidance on social responsibility to all organizations regardless of their size, type, activity, and location (European Commission 2019). Although it cannot be certified to, unlike ISO 14001 some other well-known ISO standards, ISO 26000 can provide guidance to the organizations striving for sustainable development and to encourage them to go beyond legal compliance, recognizing that an essential duty of any organization and a fundamental social responsibility is compliance with relevant laws (ISO Online Browsing Platform). However, according to the Marrakech Agreement that set foundations of the World Trade Organization, ISO 26000 cannot be referred to as an international standard, guideline or recommendation; hence it cannot be used as a basis for any legal action as such, whereas other two ISO standards are also linked to social responsibility and sustainable development – ISO 14000 and ISO 9000, dealing with environmental protection and quality management respectively. Furthermore, some recent studies have shown that the firms certified on ISO 14000 or ISO 9000, may consider ISO 26000 less valuable due of the lack of certification and external audits (Zinenko et al. 2015).

Among these various CSR instruments and guidance, the UNGC is the most widespread in terms of adopters (Ortas et al. 2015; Rasche et al. 2013), with more than 13,500 participants including 9,913 companies in 159 countries and 62,002 reports (United Nations Global Compact 2019). Launched by the UN Secretary-General at the 1999 World Economic Forum in Switzerland, the UNGC is based on 10 universal principles on human rights, labor, environment and anti-corruption derived from the Universal Declaration of Human Rights, the ILO's Declaration on Fundamental Principles and Rights at Work, the Rio Declaration on Environment and Development, and the UN Convention against Corruption. The UNGC guides firms in adopting standards and initiatives that support CSR, helping these companies operate in a sustainable way. To join the UNGC and obtain an active status, firms are required to prepare a letter of commitment, expressing adherence to the UNGC's 10 principles. To maintain this status, participants must annually provide a report called "Communication on Progress;" and, two consecutive failures in submitting the report causes the firm to be delisted (Orzes et al. 2018). In September 2015, the UN also launched its SDG initiative, which could be seen as a complement to the UNGC (Mhlanga et al. 2018).

Perhaps the most critical aspect of CSR is reporting, whereas the companies are able to examine their sustainability footprint in a transparent way. Even though the European Union issued a directive to large firms requiring public disclosure of data related to social, environmental, and governance issues (Deloitte 2014), reporting is a mostly voluntary activity that provides better understanding of companies' impacts on all three bottom lines of TBL framework (Thijssens et al. 2016; Bonsón and Bednárová

2015), and disclosure of these impacts enables companies to be more transparent about the risks and opportunities they face (https://www.globalreporting.org/).

Regardless of whether it is mandatory or voluntary initiative of the company in order to make its business more transparent, reporting requires the existence of a reliable and efficient system for internal data collection, processing and dissemination. When it comes to compulsory reports, the structure of these documents is usually predefined and standardized, usually depending on the purpose of the report and the institution that requires the report.

If a company voluntarily disclose a report on CSR initiatives, as it is mentioned in a paragraph above, there are various forms and standards that can serve as guidelines, such as GRI, ISO 26000, CSR Sustainability Monitor (Sethi et al. 2017), or UNGC Principles, OECD Guidelines for Multinational Enterprises, AA1000, ISO 14001, SA88000 (Bonsón and Bednárová 2015).

Reporting and disclosure of sustainability data can help organizations to understand and communicate their economic, environmental, social, and governance performances, and then set goals, and manage change more effectively. A sustainability report is the key platform for communicating sustainability performance and impacts – whether positive or negative (Global Reporting Initiative 2019). The GRI Sustainability Reporting Standards are the first and most widely adopted global standards for sustainability reporting (Michelon et al. 2015), also perceived trusted and unbiased due to their dual governance structure. Since GRI's inception in 1997, it has been transformed from a niche practice to one now adopted by a growing majority of organizations. In fact, 93% of the world's largest 250 corporations report on their sustainability performance through GRI (Global Reporting Initiative 2019). According to some recent studies, the GRI's guidelines to voluntary disclosure of sustainability data appear to be more robust and rigorous compared to the guidelines offered by ISO 26000 (Sethi et al. 2017).

Increasing the transparency of companies through the dissemination of company's data via internet has been also imposed by the European Union in the recent years (Bonsón and Escobar 2006). The authors emphasize that in the technology-driven environment, digital disclosure of information is a part of a modern corporate identity that drives the image and reputation as well. Coluccia et al. (2016) reported the increase of digital dissemination of CSR initiatives during recent years, even though it is still in infancy, it is a common practice for a growing number of companies. This research focused on the voluntary dissemination of CSR initiatives on companies' websites.

3 Research Method and Results

3.1 Research Framework

An empirical study examined the digital disclosure of GRI topics, particularly social and environmental topics, by an analysis of the 82 freight shipping companies, members of the German ship-owners' association (VDR - Der Verband Deutscher Reeder). The German ship-owners' association consists of 149 regular members located in five of the sixteen German provinces. The focus of our research was on 82 freight shipping companies. The remaining 67 companies were excluded for the following reasons: 19 of them had no website, 18 were in restructuring process, another 18 conducted passenger transport only, while the remaining 12 were providing marine towing services.

The companies' websites were primary source of information while exploring digital disclosure of GRI topics, presented in the Table 1. The topics were disaggregated in two groups: Environmental and Social. The presence of a disclosure related to a specific topic counted as one point on the scale of maximum 28 GRI covering points.

Table 1. List of social and environmental GRI topics.

GRI 400: Social	GRI 300: Environmental
GRI 401: Employment	GRI 301: Materials
GRI 402: Labor/Management Relations	GRI 302: Energy
GRI 403: Occupational Health and Safety	GRI 303: Water and Effluents
GRI 404: Training and Education	GRI 304: Biodiversity
GRI 405: Diversity and Equal Opportunity	GRI 305: Emissions GRI
GRI 406: Non-discrimination	GRI 306: Effluents and Waste
GRI 407: Freedom of Association and Collective Bargaining	GRI 307: Environmental Compliance
GRI 408: Child Labor	GRI 308: Supplier Environmental Assessment
GRI 409: Forced or Compulsory Labor	
GRI 410: Security Practices	
GRI 411: Rights of Indigenous Peoples	
GRI 412: Human Rights Assessment	
GRI 413: Local Communities	
GRI 414: Supplier Social Assessment	
GRI 415: Public Policy	
GRI 416: Customer Health and Safety	
GRI 417: Marketing and Labeling	
GRI 418: Customer Privacy	
GRI 419: Socioeconomic Compliance (laws)	
Sustainability report Does a company have sustainability report/CSR policy?	

Furthermore, data on fleet size, affiliation to a larger shipping group, as well as location of company's headquarters, were collected as well. The data were presented using appropriate descriptive statistics (median and interquartile range, absolute and relative frequencies), and further analyzed using nonparametric Spearman's correlation, Pearsons Chi-Square Test and Mann Whitney sum-rang test. All analyses were performed in IBM SPSS Statistics 21.

3.2 Results and Discussion

One half of the 82 analysed companies are seated in the Free and Hanseatic City of Hamburg as the Europe's third-largest port and Gemany's maritime hub (Table 2). The other 41 companies are located in four German provinces on the North Sea and the Baltic coastline (Bremen, Lower Saxony, Mecklenburg-Western Pomerania and Schleswig-Holstein).

Table 2. List of German provinces where companies are located.

German province	Number of companies
Bremen	10
Hamburg	41
Lower Saxony	21
Mecklenburg-Western Pomerania	2
Schleswig-Holstein	8
Total	82

To understand the role of Hamburg as the main shipping province/city in the implementation of environmental and social stadnards, companies were disaggregated by location to Hamburg-based companies and those based in other German provinces.

The share of companies affiliated to a larger shipping group was comparable in Hamburg and other provinces (39.0% vs. 34.1% respectively; $\chi 2 = 0{,}210$, df = 1, p = 0,647).

Company Headquarters. The Table 3 presents the comparison of companies located in Hamburg vs. other German provinces regarding Total, Social and Environmental GRI covering points. Companies located in Hamburg have statistically greater number of the GRI Total and Social coverage points compared to the other German provinces. (Total GRI $Z = -2.508$, $p = 0.012$ and Social GRI $Z = -2.421$, $p = 0{,}015$). Difference in Environmental GRI covering points did not reach the treshold of statistical significance (Environmental GRI $Z = -1.928$, $p = 0.054$).

Number of Ships in Fleet. The fleet size correlated significantly with GRI covering points on the company level (Total GRI covering points: $\rho = 0.333$, $p = 0.003$; Environmental GRI covering points: $\rho = 0.274$, $p = 0.015$ and Social GRI covering points: $\rho = 0.340$, $p = 0.002$).

Affiliation to a Shipping Group. Table 4 presents the comparison of companies affiliated to a shipping groups vs. companies not being a part of a shipping group regarding Total, Social and Environmental GRI covering points.

There is statistically significant difference in Total, Environmental and Social covering points score between companies who are/not part of a shipping group (Total GRI

Table 3. Descriptive statistics location of the company's headquarters.

	Location of the company's headquarters					
	Hamburg			Other German provinces		
	Median	Percentile 25	Percentile 75	Median	Percentile 25	Percentile 75
GRI Total coverage points	6.0	2.0	14.0	2.0	1.0	7.0
GRI Environmental coverage points	1.00	0.00	7.00	0.00	0.00	3.00
GRI Social coverage points	4.00	2.00	9.00	2.00	1.00	5.00

Table 4. Descriptive statistics Company's affiliation.

	Company's affiliation					
	Part of a larger group			Not part of a larger group		
	Median	Percentile 25	Percentile 75	Median	Percentile 25	Percentile 75
GRI Total coverage points	7.0	2.0	16.0	2.0	1.0	7.5
GRI Environmental coverage points	1.00	0.00	7.00	0.00	0.00	2.00
GRI Social coverage points	5.00	2.00	10.00	2.00	1.00	5.00

score $Z = -2.745$, $p = 0.006$; Total Environmental score: $Z = -1.785$, $p = 0.074$ and Total Social score: $Z = -3.144$, $p = 0.002$).

The companies located in Hamburg, as the main hub of the German shipping industry, provide significantly better digital disclosure in comparison to the companies located in the other four German provinces. These results are not surprising whereas the companies based in Hamburg have better access to information pertaining to the industry's contemporary requirements.

The companies that were part of a larger group disclose more information related to the environmental and social issues than the single market players. These results are expected whereas the large shipping groups house the management and marketing teams that have better understanding on the importance of corporate sustainability, while the small shipping firms have limited human resources usually coping with the financial sustainability of their business operations.

The companies with larger number of ships in the fleet pay more attention to corporate and social responsibility topics thanks to their considerable organizational structure and financial resources.

Germany is a country with a strong public policy where environmental issues play a substantial social role. Their policies are in line with the widely accepted EU standards and prevailing practices when it comes to environmental issues and sustainable development, in particular. The shipping industry has begun to develop environmental and social consciousness, yet there are still many issues that need to be addressed.

4 Conclusion

CSR is a strategic business concern that indicates responsiveness towards economic and environmental issues striving for a favorable influence on the development of society, and consequently sustainable development. The concept of CSR requires more strategic approach, hence aimed at promoting sustainable development and the Agenda 2030, with a particular focus on data disclosure and transparency. Various research studies revealed that large companies have already started reviewing their social responsibility and legal compliance in the context of sustainable development and Agenda 2030. Furthermore, their contributions in achieving the SDGs are strategically planned to reinforce their core businesses. Along with the trend of integrating the CSR initiatives into daily business operations, the demand for more transparency on the companies' activities in these areas has been expanding (Taylor et al. 2018). Additionally, relevant institutions report the increase of disclosure of information related to their environmental and social impacts. Therefore, our study contributes to the growing body of research on voluntary disclosure of environmental and social initiatives. In the recent period, more and more companies have worked to monitor and manage environmental and social responsibility. Consequently, many different instruments to manage, measure, communicate and reward CSR performance have been developed. As the GRI Sustainability Reporting Standards are the first and most widely adopted global standards for sustainability reporting, we have used these standards to identify social and environmental topics covered by the research.

The study investigated the extent to which German freight shipping companies use websites as channels to publicly present their social and environmental achievements. The research revealed that companies based in Hamburg, companies with group affiliation and companies with large fleets outperform those located in the other German provinces, nonaffiliated shipping companies and medium-sized fleets. Nevertheless, there are still areas for improvements and further homogeneousness in digital disclosure of social and environmental information in the German shipping industry.

As previous studies have shown that CSR reporting is influenced by country context (Bonsón and Bednárová 2015), further studies may extend to other countries, as well as other industries. Additionally, future research may include the practice of reporting on the economic concerns as well. Finally, this study was done at a rather specific time period, thus a longitudinal study on the digital disclosure of CSR initiatives would capture more extensive and ample results.

Acknowledgement. This paper has been supported by the University of Rijeka under the project number Uniri-drustv-18-235-1399.

References

Aguilera, R.V., Rupp, D.E., Williams, C.A., Ganapathi, J.: Putting the S back in corporate social responsibility: a multilevel theory of social change in organizations. Acad. Manag. Rev. **32**(3), 836–863 (2007)

Aguinis, H., Glavas, A.: On corporate social responsibility, sensemaking, and the search for meaningfulness through work. J. Manag. **45**(3), 1057–1086 (2019)

Ait Sidhoum, A., Serra, T.: Corporate sustainable development. Revisiting the relationship between corporate social responsibility dimensions. Sustain. Dev. **26**(4), 365–378 (2018)

Albuquerque, R., Koskinen, Y., Zhang, C.: Corporate social responsibility and firm risk: theory and empirical evidence. Manag. Sci. (2018)

Bansal, P., Song, H.C.: Similar but not the same: differentiating corporate sustainability from corporate responsibility. Acad. Manag. Ann. **11**(1), 105–149 (2017)

Bonsón, E., Bednárová, M.: CSR reporting practices of Eurozone companies. Revista de Contabilidad **18**(2), 182–193 (2015)

Bonsón, E., Escobar, T.: Digital reporting in Eastern Europe: an empirical study. Int. J. Account. Inf. Syst. **7**(4), 299–318 (2006)

Bowen, H.R.: Social Responsibilities of the Businessman. Harper & Row, New York (1953)

Brundtland Report: World Commission on Environment and Development (1987). https://www.sustainabledevelopment2015.org/. Accessed 17 Apr 2019

Carroll, A.B., Shabana, K.M.: The business case for corporate social responsibility: a review of concepts, research and practice. Int. J. Manag. Rev. **12**(1), 85–105 (2010)

Carroll, A.B.: The pyramid of corporate social responsibility: toward the moral management of organizational stakeholders. Bus. Horiz. **34**(4), 39–48 (1991)

Coluccia, D., Fontana, S., Solimene, S.: Disclosure of corporate social responsibility: a comparison between traditional and digital reporting. An empirical analysis on Italian listed companies. Int. J. Manag. Financ. Account. **8**(3–4), 230–246 (2016)

Crowther, D., Seifi, S. (eds.): Redefining Corporate Social Responsibility. Emerald Group Publishing (2018)

Dahlsrud, A.: How corporate social responsibility is defined: an analysis of 37 definitions. Corp. Soc. Responsib. Environ. Manag. **15**(1), 1–13 (2008)

Deloitte: EU Parliament Adopts ESG Disclosure Directive for Large Companies and Groups. New York (2014)

Djokic, S.: The role of corporate social responsibility in the promotion of sustainable development of Southeastern European Countries, master dissertation, University of Belgrade and Middlesex University London (2019)

Elkington, J.: Cannibals with Forks: The Triple Bottom Line of 21st Century Business. Capstone (1997)

Ellen, P.S., Webb, D.J., Mohr, L.A.: Building corporate associations: consumer attributions for corporate socially responsible programs. Acad. Mark. Sci. J. **34**(2), 147–157 (2006)

Enders, J.C., Remig, M.: Theories of Sustainable Development. Routledge – Taylor and Francis Group (2015)

European Commission: Corporate Social Responsibility, Responsible Business Conduct, and Business & Human Rights: Overview of Progress. EC, Brussels (2019)

European Commission: ABC of the main instruments of Corporate Social Responsibility. EC, Directorate-General for Employment and Social Affairs (2004)

European Commission: White Paper on Corporate Social Responsibility. EC, Brussels (2003)
Freeman, R.E.: Strategic Management: A Stakeholder Approach. Pitman, Boston (1984)
Garriga, E., Melé, D.: Corporate social responsibility theories: mapping the territory. J. Bus. Ethics **53**(1–2), 51–71 (2004)
Georgeson, L., Maslin, M.: Putting the United Nations Sustainable Development Goals into practice: a review of implementation, monitoring, and finance. Geo Geogr. Environ. **5**(1) (2018)
Global Reporting Initiative (GRI) Sustainability Reporting Standards. https://www.globalreporting.org/Pages/default.aspx. Accessed 26 Apr 2019
https://www.globalreporting.org/. Accessed 5 May 2019
International standard ISO 26000: Guidance on Social Responsibility. ISO 26000:2010. https://www.iso.org/iso-26000-social-responsibility.html. Accessed 26 Apr 2019
ISO 26000: Guidance on social responsibility. ISO 26000:2010 (2010)
ISO Online Browsing Platform. https://www.iso.org/obp/. Accessed 08 Mar 2019
Kolk, A., Van Tulder, R.: International business, corporate social responsibility and sustainable development. Int. Bus. Rev. **19**(2), 119–125 (2010)
KPMG: KPMG International Survey of Corporate Responsibility Reporting. KPMG International Cooperative (2013)
Kumar, S., Kumar, N., Vivekadhish, S.: Millennium development goals (MDGS) to sustainable development goals (SDGS): addressing unfinished agenda and strengthening sustainable development and partnership. Indian J. Community Med. Off. Publ. Indian Assoc. Prev. Soc. Med. **41**(1), 1 (2016)
López, M.V., Garcia, A., Rodriguez, L.: Sustainable development and corporate performance: a study based on the dow jones sustainability index. J. Bus. Ethics (2007)
Mackey, A., Mackey, T., Barney, J.: Corporate social responsibility and firm performance: investor preferences and corporate strategy. Acad. Manag. Rev. **32**(3), 817–835 (2007)
Margolis, J.D., Walsh, J.P.: Misery loves companies: social initiatives by business. Adm. Sci. Q. **48**(2), 268–305 (2003)
Meadows, D.H., Meadows, D.L., Randers, J., Behrens, W.W.: The Limits to Growth, Potomac Associates. New American Library, Washington, DC (1972)
Mhlanga, R., Gneiting, U., Agarwal, N.: Walking the Talk: Assessing companies' progress from SDG rhetoric to action (2018)
Michelon, G., Pilonato, S., Ricceri, F.: CSR reporting practices and the quality of disclosure: An empirical analysis. Crit. Perspect. Account. **33**, 59–78 (2015)
Mihic, M., Petrovic, D., Vuckovic, A.: Comparative analysis of global trends in energy sustainability. Environ. Eng. Manag. J. **13**(4), 947–960 (2014)
Mihic, M., Shevchenko, S., Gligorijevic, E., Petrovic, D.: Towards strategic corporate social responsibility approach in international projects-review of South-South cooperation: a case study of Chinese projects in Angola. Sustainability **11**(10), 2784 (2019)
Millennium Development Goals. https://www.un.org/millenniumgoals/. Accessed 17 Apr 2019
Moon, J.: The contribution of corporate social responsibility to sustainable development. Sustain. Dev. **15**(5), 296–306 (2007)
Ortas, E., Álvarez, I., Garayar, A.: The environmental, social, governance, and financial performance effects on companies that adopt the United Nations Global Compact. Sustainability **7**, 1932–1956 (2015)
Orzes, G., Moretto, A.M., Ebrahimpour, M., Sartor, M., Moro, M., Rossi, M.: United Nations Global Compact: literature review and theory-based research agenda. J. Clean. Prod. **177**, 633–654 (2018)
Rasche, A., Waddock, S., McIntosh, M.: The United Nations Global Compact: retrospect and prospect. Bus. Soc. **52**(1), 6–30 (2013)
Rivera, J.J., Bigne, E., Curras-Perez, R.: Effects of corporate social responsibility on consumer brand loyalty. Revista Brasileira de Gestão de Negócios **21**(3), 395–415 (2019)

Sethi, S.P., Rovenpor, J.L., Demir, M.: Enhancing the quality of reporting in corporate social responsibility guidance documents: the roles of ISO 26000, global reporting initiative and CSR-sustainability monitor. Bus. Soc. Rev. **122**(2), 139–163 (2017)
Sustainable Development Goals – Knowledge Platform. https://sustainabledevelopment.un.org. Accessed 26 Apr 2019
Taylor, J., Vithayathil, J., Yim, D.: Are corporate social responsibility (CSR) initiatives such as sustainable development and environmental policies value enhancing or window dressing? Corp. Soc. Responsib. Environ. Manag., 1–10 (2018)
Theuws, M., van Huijstee, M.: A Comparison of the OECD Guidelines, ISO 26000 & the UN Global Compact. SOMO – Centre for Research on Multinational Corporations, December 2013
Thijssens, T., Bollen, L., Hassink, H.: Managing sustainability reporting: many ways to publish exemplary reports. J. Clean. Prod. **136**, 86–101 (2016)
United Nations Global Compact (UNGC). https://www.unglobalcompact.org. Accessed 17 Apr 2019
Vlastelica Bakic, T., Cicvaric Kostic, S., Neskovic, E.: Model for managing corporate social responsibility. Management **74**, 47–56 (2015)
Vlastelica, T., Cicvaric Kostic, S., Okanovic, M., Milosavljević, M.: How corporate social responsibility affects corporate reputation? The evidence from an emerging market. JEEMS J. East Eur. Manag. Stud. **23**(1), 10–29 (2018)
Vogt, M., Weber, C.: Current challenges to the concept of sustainability. Glob. Sustain. **2**(e4), 1–6 (2019)
Wang, H., Tong, L., Takeuchi, R., George, G.: Corporate social responsibility: an overview and new research directions: thematic issue on corporate social responsibility [from the editors]. Acad. Manag. J. **59**(2), 534–544 (2016). Research Collection Lee Kong Chian School of Business
Weber, M.: The business case for corporate social responsibility: a company level measurement approach for CSR. Eur. Manag. J. **26**(4), 247–261 (2008)
World Commission on Environment and Development (WCED): Our Common Future. Oxford University Press, New York (1987)
Zinenko, A., Rovira, M.R., Montiel, I.: The fit of the social responsibility standard ISO 26000 within other CSR instruments: Redundant or complementary? Sustain. Account. Manag. Policy J. **6**(4), 498–526 (2015)
WWF: Living Plant Report (2008)

Strength Calculation and Optimization of Boat Crane

Janko Morosavljević[1], Dražan Kozak[1](✉), Katarina Monkova[2], and Darko Damjanović[1]

[1] Mechanical Engineering Faculty in Slavonski Brod, J.J. Strossmayer University of Osijek, 35000 Slavonski Brod, Croatia
drazan.kozak@unios.hr

[2] Faculty of Manufacturing Technologies of the Technical University of Košice with a seat in Prešov, 080 01 Prešov, Slovak Republic

Abstract. By inspecting the existing constructional designs of boat cranes and the requirements which need to be fulfilled, the new constructional design of the boat crane is made. After determining the critical position of the crane in the form of a load, calculation of the inside diameters of cylinders is made and optimized depending on the maximum axial force. Numeric calculation of the stresses and total deformations in software ANSYS Workbench has proved that the design of crane has met all the requirements imposed on the strength. After designing the boat crane and preforming the strength calculations, it is necessary to analyze the motion of the designed crane construction. Boat crane is designed with hydraulic cylinders and rotation around the shaft using a hydraulic motor. Therefore, it is necessary to create a scheme of a hydraulic system of the crane. During the designing and strength analysis, no material exploitation was taken into account and reinforcements were added only to reduce the equivalent stresses and the maximum total deformation of the crane. Due to the possible oversizing of the reinforcements, the optimization was performed by the response surface optimization method. Optimization was performed to minimize the mass and it has been found that the construction was oversized and the mass was reduced by almost 14 kg.

Keywords: Strength · Optimization · Boat crane

1 Introduction

Boats with large drafts (measured from the lowest point of the hull to the structural water line at the point where the ship is most submerged) sometimes cannot reach close enough to the shore because of the small depth of the seaport. In that case, if there is no intended landing place near shore, boats are using anchors. To allow passengers to transport to the shore in such a case, there is a reserved place for smaller boats on the board. The number of the smaller boats depends on the size of boat and the number of passengers with the crew who are allowable to enter the boat at the same time. To make

N. Mitrovic et al. (Eds.): CNNTech 2020, LNNS 153, pp. 35–52, 2021.
https://doi.org/10.1007/978-3-030-58362-0_3

it easier for the crew to lift and lower smaller boat on the sea and enable passengers to transport to the shore, there are various devices and lifts mechanisms available for cargo manipulation.

Manual lowering and lifting can be very tedious and in some cases almost impossible. If there is already a device intended for that purpose, it can serve also for other cases such as lifting various cargoes with loading and unloading of the boat.

2 Designing and Dimensioning of the Basic Shape of the Boat Crane

The design and sizing of the basic shape of the crane will be carried out based on already existing models of cranes of this type from world-renowned companies such as Opacmare Co. According to their catalogs of existing cranes, most cranes are designed to carry smaller boats with a carrying capacity of 300 to 800 kg. Existing cranes of this capacity have an arm length from 2,3 to 3,6 m as seen in Table 1.

Table 1. Examples of existing cranes with their characteristics [1]

Model of the boat crane	Lenght of the arm, l/mm	Maximum allowable load, m/kg
Elbow crane, Gru serie 3076/34	3550	800
Elbow crane, Gru serie 3075/30	2830	500
Elbow crane, Gru serie 3092/26	2600	500
Elbow crane, Gru serie 3035/42	2305	550

Based on information from Table 1, the new crane design is designed to safely handle a load capacity of 500 kg. Therefore, the arm of the crane will be designed with a length of 2,6 m.

When lifting a boat or a water scooter, the crane must be able to carry the vessel over the fence on the edge of the boat and that will be achieved by rotating an arm about its axis of rotation.

For enabling the crane to change the height, a mechanical drive will be used with the help of a hydraulic pump and cylinder due to the large masses and lifting heights. Also, a hydraulic actuator will allow remote control of the crane so that the person can manipulate the load independently.

Since the crane should take as little space as possible on the outer surface of the boat, the arm would consist of three parts. It will consist of a movable mid-section of the crane arm to successfully overcome the height of the boat's fence. The remaining crane arm will cross over the fence to grab the cargo. To keep the remaining arm as small as possible while the crane is not used, it is desirable to reduce or to extend the arm as needed what can be achieved by using an additional hydraulic cylinder which will allow the arm to be telescopically extended. Telescopically extending the arm will be achieved by using an additional hydraulic cylinder which will be also remotely controlled. In conclusion, the arm will have three degrees of freedom of movement to successfully meet the stated functional requirements. Schematic of the crane with a load that fulfills stated requirements is shown in Fig. 1.

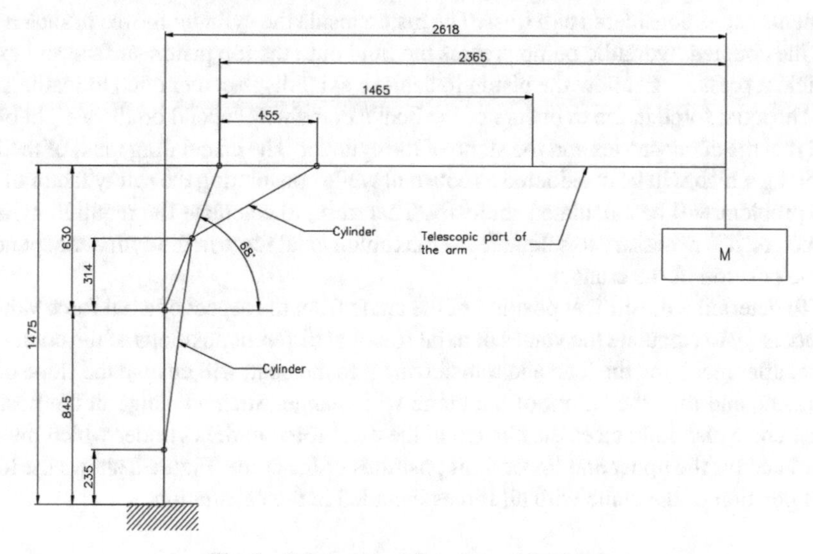

Fig. 1. Schematic of the crane with a load

According to the schematic of the crane with all dimensions needed, the basic model shape of the crane was created using the SolidWorks software package (Fig. 2).

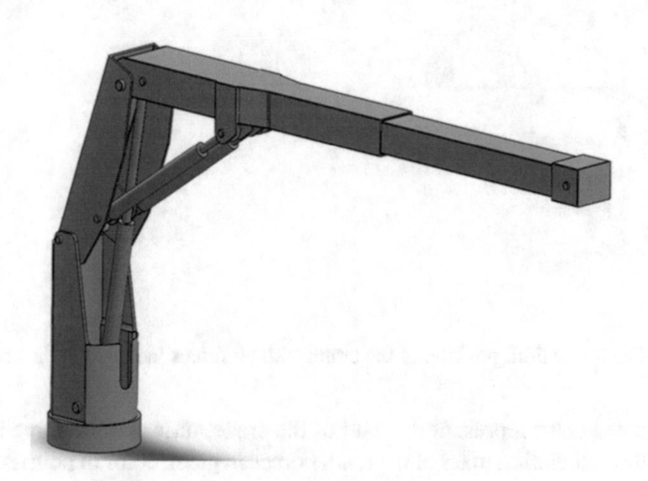

Fig. 2. The initial model of the boat crane

3 Optimization of Hydraulic Cylinder Angle Depending on Axial Force

The hydraulic cylinder can only transmit axial force due to the way it connects to other elements and is considered as a truss. The piston inside the cylinder moves in such a way that the coupled hydraulic pump presses the fluid onto the top piston surface and exerts sufficient pressure to allow the piston to achieve axial displacement due to loading.

The axial force in the cylinders of the boat's crane will depend on the weight of the load that the crane carries and the slope of the cylinder. The calculating mass of the load is 750 kg which will be considered as constant while considering the safety factor of 1,5. The problem will be calculated statically. Therefore, to calculate the required cylinder diameters, it is necessary to calculate the maximum axial force in the cylinder depending on the position of the crane.

To determine the critical position of the crane from the aspect of axial force value, it is necessary to calculate the values of axial forces at different positions of the crane. The crane, after receiving the load and transferring it to the boat, will change the slope of the cylinders, and thus the height of the crane will change. Such a change in the position of the crane also influences the change in the axial force in the cylinder which must be calculated for the upper and lower limit positions of the crane. Figure 3 shows the lower limit position of the crane with all forces included in the calculation.

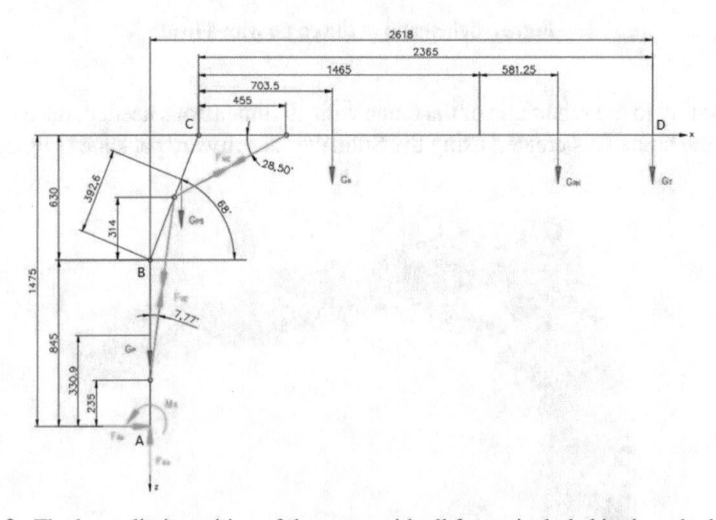

Fig. 3. The lower limit position of the crane with all forces included in the calculation

Forces in red color represent the load of the crane mass and the force in point D is the force of the calculating mass of the load. Forces in green color in point A are reaction forces of the fixed support. Other forces in green color are axial forces in the cylinders which values are needed to determine the critical position of the crane. Points B and C are cylinder joints. Using equations given in [2], calculation of the forces for this case has been made. The value of the axial force in the bigger cylinder is $F_{vc} = -258022$ N and the axial force in the smaller cylinder is $F_{mc} = -88053$ N.

The same calculation is made for the upper limit position of the crane which is shown in Fig. 4.

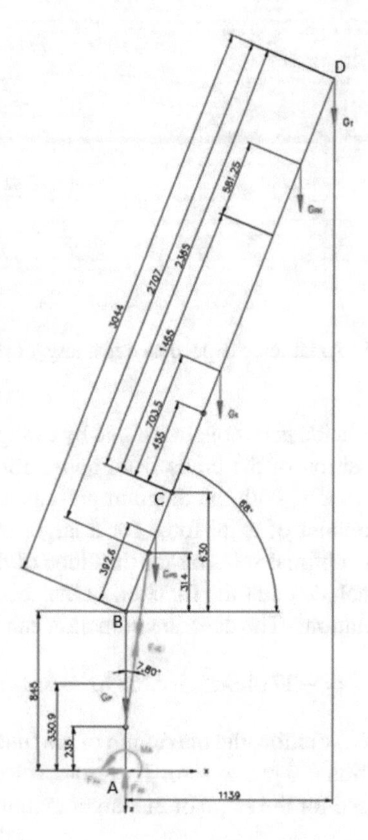

Fig. 4. The upper limit position of the crane with all forces included in the calculation

Using equations given in [2], calculation of the forces for this case has been made. The value of the axial force in the bigger cylinder is $F_{vc} = -109914\,\mathrm{N}$ and the axial force in the smaller cylinder is $F_{mc} = 0\,\mathrm{N}$.

It can be concluded that the axial force for the first load case is larger than the second case and that by changing the height of the crane from the first case to the second case, the axial force in the larger and smaller cylinder decreases. Therefore, the first load case is defined as a critical case on which will be determined the minimum required internal cylinders diameters.

After the analytical values of the axial forces in the cylinders and the reaction bonds in the support were calculated, to minimize the axial force in the cylinder for the critical case of the crane's position, it is necessary to check for which slope of the cylinder the force will be lesser. After analytical optimization, results can be shown with the diagram where the X-axis represents a slope of the cylinder and the Y-axis represents an axial force in the cylinder (Fig. 5).

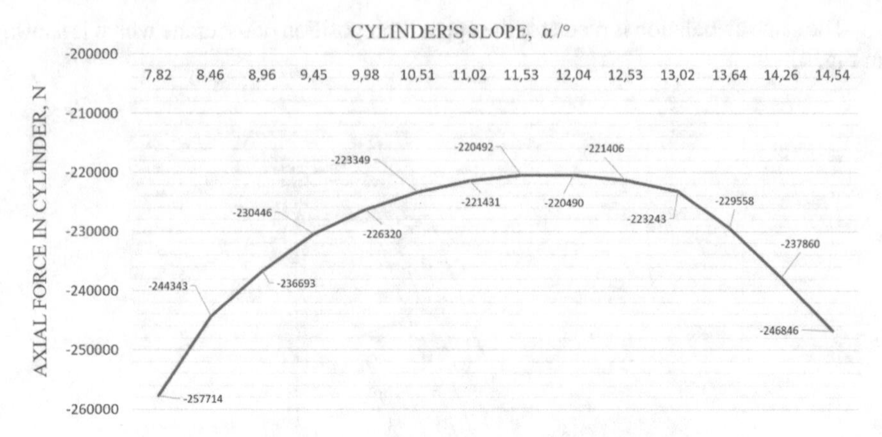

Fig. 5. Axial force dependence in a larger cylinder

From the shown diagram it can be concluded that by changing the slope of the larger cylinder in the critical position of the crane from approximately 8° to 12°, the axial force changes by almost 40 kN. With the diagram presented above, the approximated function of varying the amount of axial force for a larger cylinder depending on the slope is defined to find the optimal solution for the slope of the cylinder. According to the diagram, it can be concluded that the function of the second-order polynomial can describe the calculated solutions. The described function can be represented as:

$$F_{\mathrm{vc}} = -2714{,}9\alpha^2 + 62689\alpha - 581229 \tag{1}$$

To calculate the optimal solution, the maximum of the function is obtained when the first derivative of the function is equal to zero. Therefore, calculating with the described procedure, the optimal value for the slope of the larger cylinder is:

$$\alpha = 11{,}55^\circ$$

The same procedure is also made for the other (smaller) cylinder. It can be concluded that the function is also the second-order polynomial that can describe the calculated solution. The described function can be represented as:

$$F_{\mathrm{mc}} = -78{,}041\beta^2 + 4686{,}2\beta - 155435 \tag{2}$$

Calculating with the described procedure, the optimal value for the slope of the smaller cylinder is:

$$\beta = 30{,}02^\circ$$

After calculating the optimal slope of the larger and smaller cylinders, the slope value of the larger cylinder is 11,5° and the slope of the smaller cylinder is 30°. To calculate the inside diameter for the given case, it is necessary to take into account the axial force of the cylinder as a function of slope and pressure (Table 2).

Table 2. Calculated values of the required inside diameter of the cylinder depending on the number of cylinders

	Pressure in cylinder, *p*/MPa	Axial force, *F*/N	Inside diameter of cylinder in case for one cylinder, *d*/mm	Inside diameter of cylinder in case for two cylinders, *D*/mm
Larger cylinder	240	220492	108,15	76,48
Smaller cylinder	240	85898	67,51	47,73

According to the represented results, the inside diameter of a smaller cylinder is 70 mm and of larger cylinders is 80 mm for the case when two cylinders are installed. These values will be chosen for design reasons because it would be better and simpler to design two smaller cylinders with an inside diameter of 80 mm together with one cylinder which has an inside diameter of 70 mm.

After the cylinders for the boat crane were selected, a computer model of the crane was made with the required internal diameters of the cylinders and the dimensions of the crane components were adjusted to accommodate the cylinders inside the boat crane (Fig. 6).

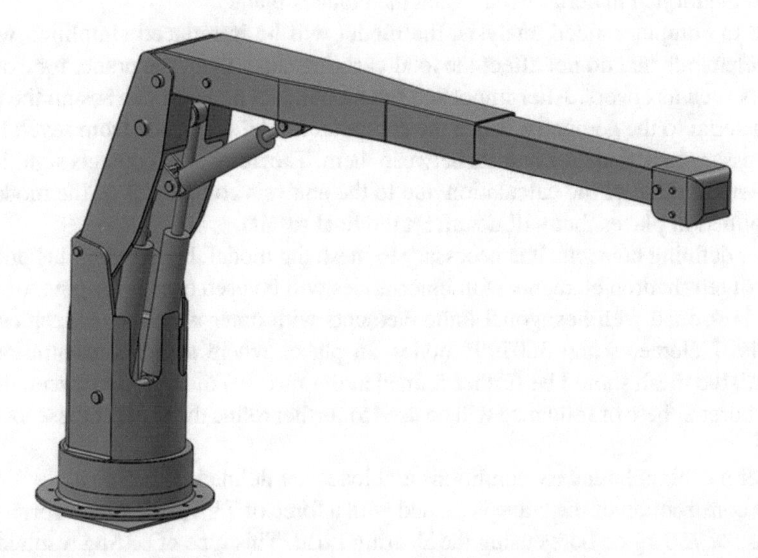

Fig. 6. Boat crane model after cylinder selection

4 Strength Calculation of Characteristic Parts of the Crane

After creating the computer model according to the technical requirements and sizing the cylinders, it is necessary to check the strength of the characteristic construction elements in the crane assembly. Before calculating the strength, it is necessary to determine the material from which the crane will be made. As it is located in a corrosive environment and must have good weldability properties, corrosion-resistant steel X5CrNi18-10 (AISI 304) is chosen.

Said steel is austenitic 18/10 Cr-Ni acid-resistant steel with very low carbon content leading to increased resistance to intercrystalline corrosion. Welding ability is good with all electrical procedures and gas welding should not be used. As this material, however, becomes more prone to intercrystalline corrosion after welding, the crane must be additionally protected against corrosion by a specific coating for additional protection [3].

The mechanical properties of this material are as follows:

- Young's modulus of elasticity: $E = 200\,\text{GPa}$
- Yield strength: $R_{\text{p0,2}} = 215\,\text{MPa}$
- Ultimate tensile strength: $R_{\text{m}} = 505\,\text{MPa}$
- Density: $\rho = 8000\frac{\text{kg}}{\text{m}^3}$
- Poisson factor: $\nu = 0{,}29$

The crane strength calculation will be made numerically using the Ansys 19.0 software package. It is necessary to first import the module for static analysis and to define the aforementioned material from which the crane is made.

Due to computer-aided analysis, the model will be introduced simplified without certain elements that do not affect the load-carrying capacity of the crane, for example, elements such as covers. After importing the model, it is necessary to assign the predefined material to the geometry. Since the computer model was made from several parts, it is necessary to define the contacts between them. The number of contacts significantly affects the duration of the calculation due to the analysis complexity so the model will be simplified in places that will not affect the final results.

After defining contacts, it is necessary to mesh the model. For solving this problem, a mesh of tetrahedron elements with inner-nodes will be used except for pins for which a mesh is defined with hexagonal finite elements with inner-nodes. The mesh consists of 2101967 elements and 3607991 nodes. In places where stress concentrations are expected, the mesh should be further refined at the surfaces of contact between the bolt and the bore. Sphere of Influence will be used to further refine the mesh at these locations (Fig. 7).

After meshing, boundary conditions and loads are defined.

The construction of the crane is loaded with a force of 7357,5 N, which corresponds to a mass of 750 kg on bores using the Bearing Load. This type of setting resulted from the crane carrying the load using a steel rope and the rope is supported by a pulley connected to the crane by bolts. In conclusion, the force is transmitted by the bolt to the crane, and since the bolt is cylindrical and is in the bore, the force distribution within the bore will correspond to the bearing load. The crane support is assumed as Fixed Support at the surfaces where the crane joins the boat with a screw connection. Label C in Fig. 8 represents the standard earth gravity acceleration by which the software includes the weight of the crane into the calculation.

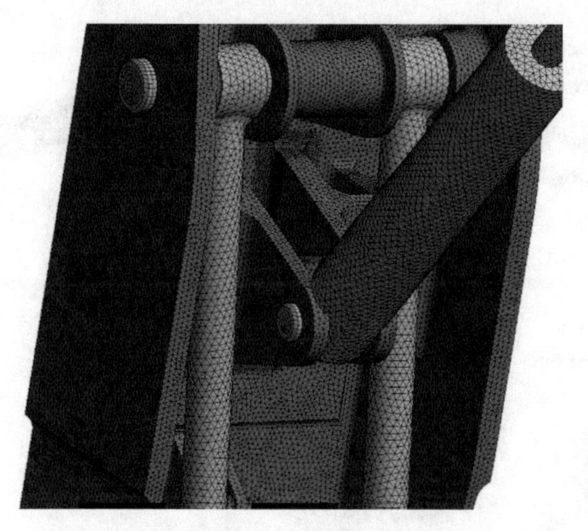

Fig. 7. Detail of meshed construction

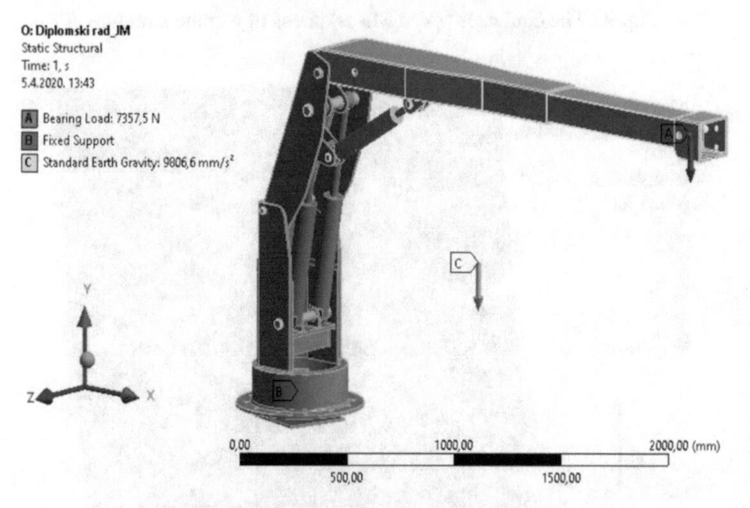

Fig. 8. Defined boundary conditions and loads

Figure 9 shows the results of equivalent stresses according to the Von Mises strength criterion, and Fig. 10 shows the detail of where the structure is reinforced.

Due to geometry, there are few stress singularities on the model, so the scale shows maximum stress of 430,57 MPa which is not representative result. On the area of the interest, maximum equivalent stress is around 150 MPa which is lower than the yield stress of 215 MPa. Therefore, the strength criterion is satisfied. Figure 10 shows a more visual representation of the stress field with the hidden cylinder for a better view.

In addition to equivalent stresses, the results of the total displacements of the structure must also be checked. Figure 11 shows the total displacements of the crane structure.

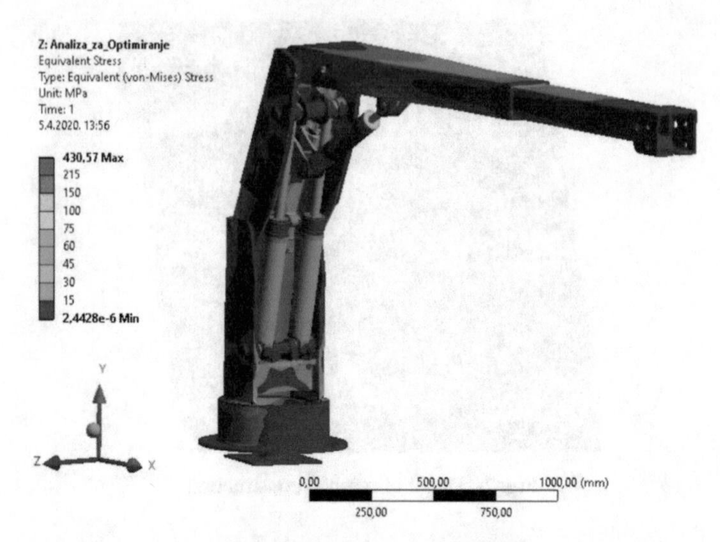

Fig. 9. The equivalent (Von Mises) stress of a crane structure

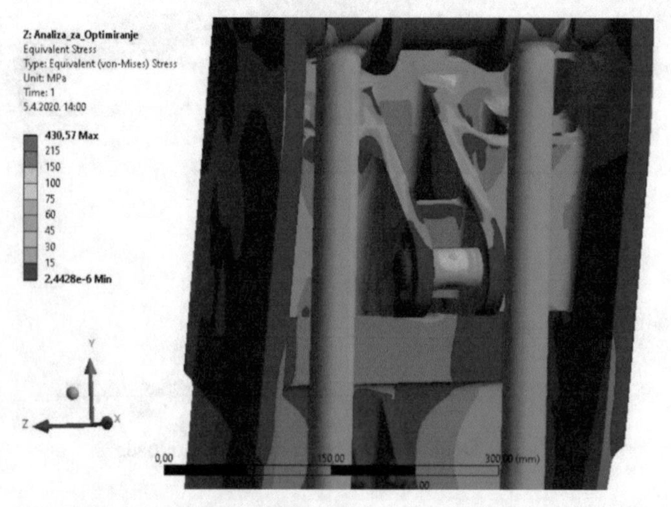

Fig. 10. Detail of the crane with equivalent stress distribution

From the results of the total displacements of the crane, the maximum displacement is 17,7 mm which for a total length of 2640 mm is a satisfactory displacement for the whole assembly.

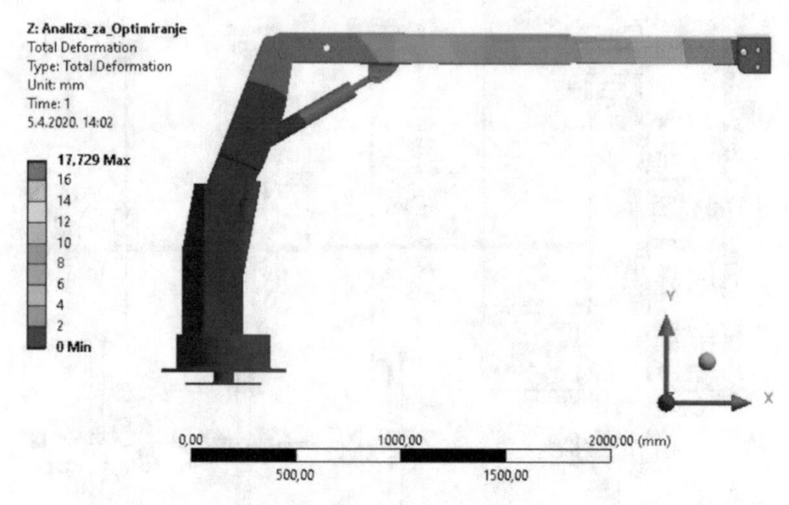

Fig. 11. Total deformations of boat crane

5 Hydraulic System of the Crane

Hydraulics are applied in mechanisms for high force, speed and acceleration requirements and small and uniform displacements. The basic elements of a hydraulic system are the pump, a working fluid, a piping, control and the hydraulic motor. In this case, the movement of the crane will be accomplished with the help of hydraulic cylinders and the rotation of the crane about an axis with a hydraulic motor where the said elements are driven by a hydraulic pump. Winch with an electric motor will be used to lift the load.

Hydraulic schemes are used for displaying hydraulic systems, where the manner of displaying hydraulic elements (symbols) and their connection is standardized. Therefore, it is necessary to first define the principle of operation of the hydraulic system within the crane, in example to make the hydraulic scheme of the system. Figure 12 shows the hydraulic scheme of the crane system [4].

The hydraulic scheme shows a pump unit with associated elements used to protect the pump against the risk of damage during loading. Therefore, the pump must be protected by a pressure relief valve (safety valve). Four main 4/3 manifolds are used, which are actuated by electromagnets (solenoids) and are centered by springs. In the central position, the pump flow is diverted back to the oil tank. The first from the left 4/3 manifold in Fig. 12 is connected on two larger cylinders which are previously calculated and described. The second 4/3 manifold is connected with the smaller cylinder and the third 4/3 manifold is connected to the cylinder which enables the arm elongation as previously described. A hydraulic motor which enables rotation of the crane about an axis is connected on the last 4/3 manifold.

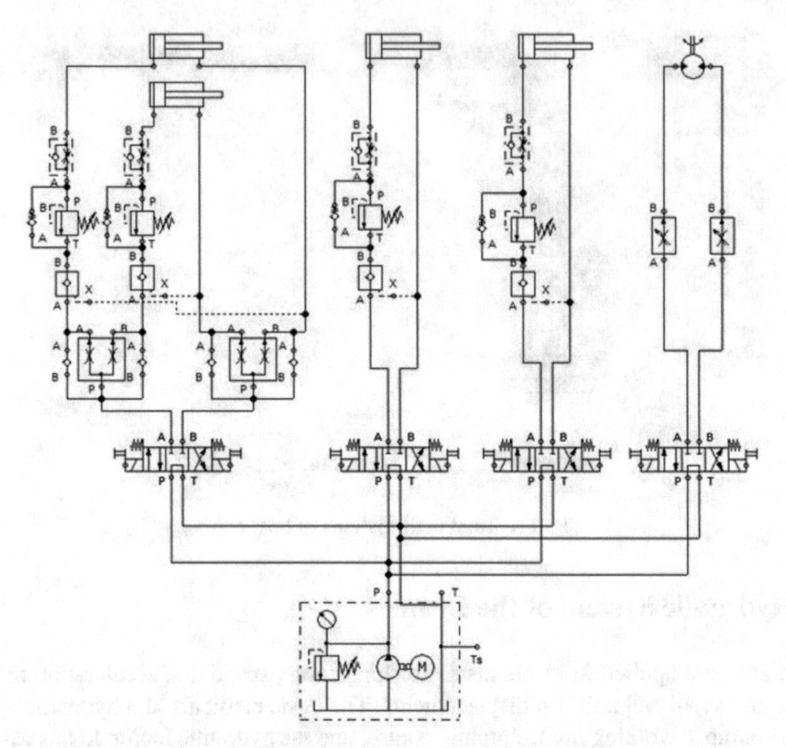

Fig. 12. Hydraulic scheme of the crane system

6 Optimization of Characteristic Elements of the Crane

While designing the crane, it was necessary to strengthen the connecting part of the crane to reduce the equivalent stresses and maximum total deformation at the end of the crane arm. No consideration was given to the usability of the material in the stiffening as much as possible what caused an increase of the mass of the connecting part of the arm. In order to reduce the mass of the structure, the optimizing procedure of the connecting part of the crane arm will be carried out.

Due to the complexity and size of the crane assembly, the optimization of the coupling part of the whole crane model will not be carried out, but the area of interest will be extracted by the submodeling technique to save computer resources and to reduce the analysis time. Submodeling is a modeling technique that enables the calculation of the desired solutions of a smaller part of a model with a finer mesh and more accurate results, which is an integral part of a larger model with a rougher mesh and less accurate results. It is necessary first to extract a submodel from the main model, taking into account that it must have the same coordinates as in the main model to be able to define boundary conditions. It is also necessary to define parameters on the submodel that will be optimized in further analysis. The parameters are defined as dimensions of some characteristic added stiffeners, which mostly contribute to the mass of the connecting part of the arm. Also, the parameter will be the thickness of the bracket to which the smaller cylinder attaches (Fig. 13).

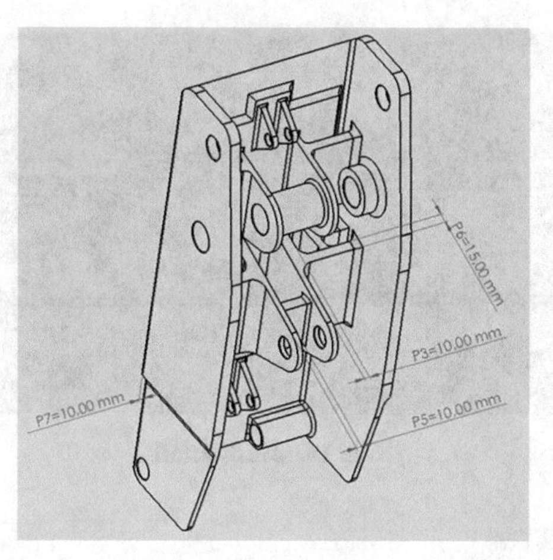

Fig. 13. Input parameters for optimizing the connecting part of the arm

The ranges of the displayed parameters are shown in Table 3.

Table 3. Ranges of input parameters

	Lower boundary,/mm	Upper boundary,/mm
Parameter P3,/mm	5	12
Parameter P5,/mm	10	15
Parameter P6,/mm	10	15
Parameter P7,/mm	5	10

After defining parameters, a system coupling is used in the ANSYS software package to automatically recognize that it is a submodel, to allow boundary conditions to be introduced to a submodel from the main model [5].

It is necessary to define all the settings in the static analysis of the connecting part of the arm starting from the mesh. A mesh of tetrahedron elements with inner-nodes will be used to solve this problem. The mesh consists of 1142285 elements and 1706491 nodes. It is necessary to further refine the mesh at places where stress concentrations are expected. Sphere of Influence will be used to further refine the mesh at these locations (Fig. 14).

After meshing, it is necessary to define boundary displacement conditions. When the system coupling is used, the software automatically recognizes that it is a submodel and offers the possibility to define boundary conditions with previously calculated results. The solutions of the displacement of the global model at the bolt joints at the connecting part of the arm will be defined.

Figure 15 shows a vector representation of the defined boundary displacement conditions at the points of connection of the bolt with the connecting part of the arm.

Fig. 14. Mesh detail

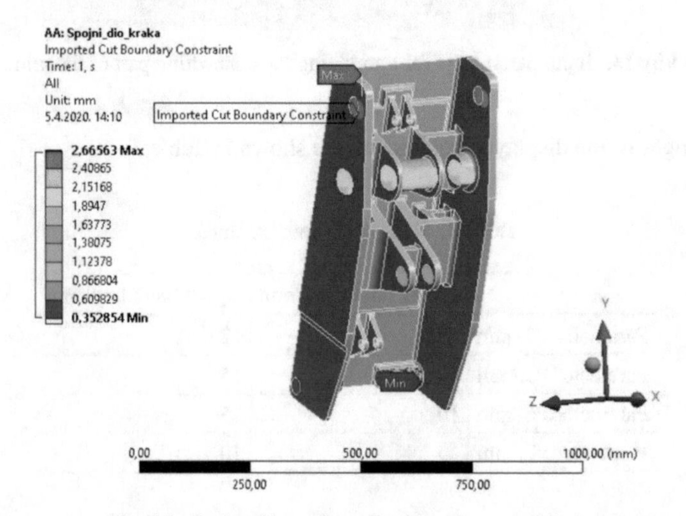

Fig. 15. Defined boundary displacement conditions

After defining boundary displacement conditions, it is needed to solve the analysis. The result of the equivalent (Von Mises) stress analysis is shown in Fig. 16.

The equivalent (Von Mises) stresses corresponding to the stresses previously calculated for the analysis of the global model. Also, stress singularity points are visible at the edges of the bore where the connecting part of the arm connects with other elements of the crane by bolts. Stress singularities will not affect the final optimization results and will be not further analyzed.

Figure 17 shows the solutions of the total displacements of the connecting part of the arm corresponding to the previously calculated total displacements for the whole crane.

In order to define the output parameters for the optimization procedure, it was also necessary to calculate the total displacement of the bracket's holes. Therefore, the maximum total deformation of the hole will be defined as the output parameter for the constraint in the optimization procedure because it significantly affects the maximum total deformation of the crane (Fig. 18).

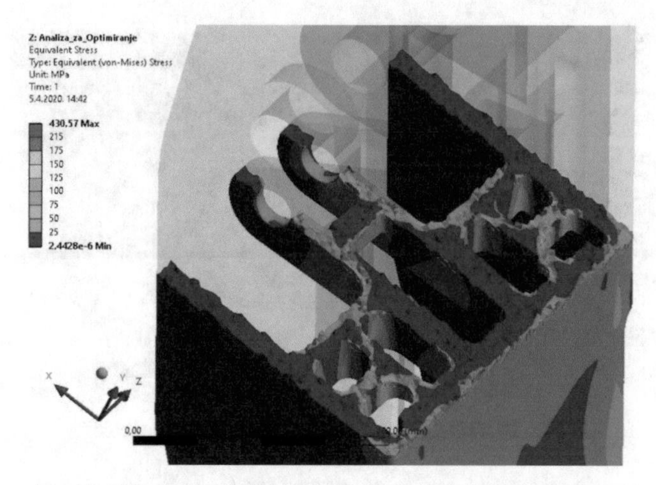

Fig. 16. Equivalent (Von Mises) stress of the submodel

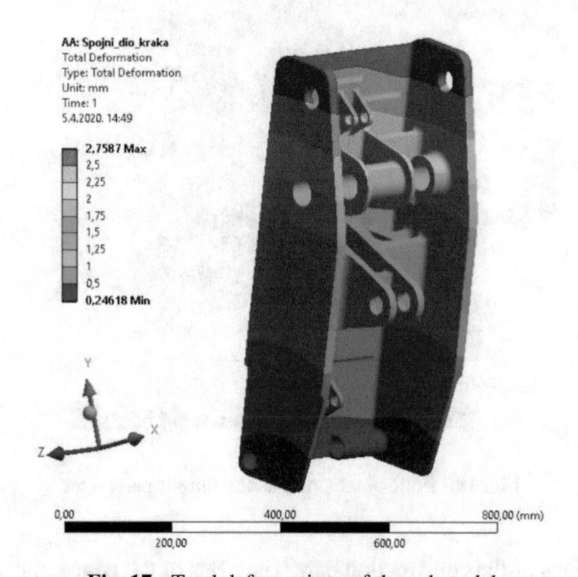

Fig. 17. Total deformations of the submodel

For the purpose of defining the output parameters of the optimization procedure, it was also necessary to calculate the equivalent stress on the surface where the bracket connects to the connecting part of the arm (Fig. 19).

The maximum surface equivalent stress will be defined as the output parameter for the limitation in the optimization procedure. This way of defining the parameter resulted from the fact that if the dimensions of the stiffeners reduce, in the calculation of the strength of the crane, equivalent stress on that surface will be the most affected. The optimization procedure will determine the goal to reduce the mass of the structure and

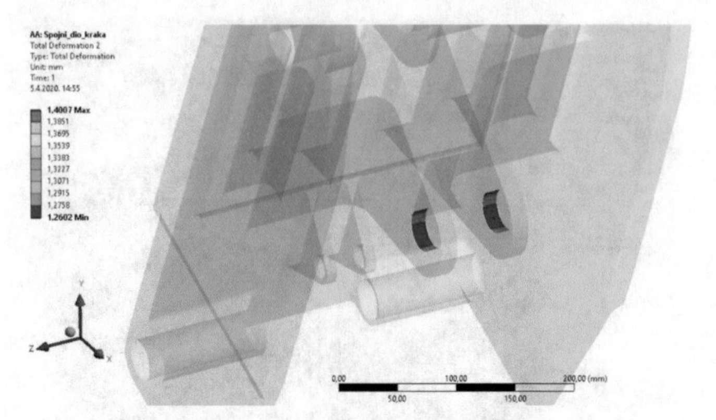

Fig. 18. Total deformation as the output parameter

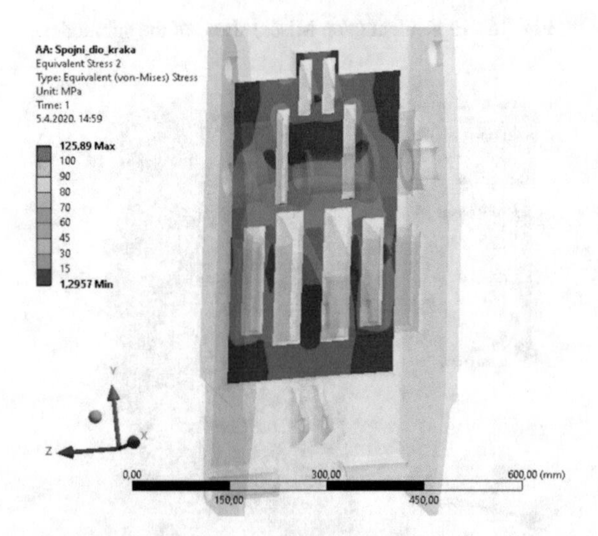

Fig. 19. Equivalent stress as the output parameter

overall dimensions of the construction part. The mass of the connecting part of the arm will be defined as the output parameter to allow the goal function to be specified.

The optimization process will be performed by the Response surface optimization method. The methodology of the method is a set of mathematical and statistical techniques for building a mathematical model that defines the relationships between input parameters and output parameters through a response function. Response surface optimization method has an important application in the design, development and formulation of new products, as well as in the improvement of existing product design. It defines the effect of the independent variables, alone or in combination, on the processes. In addition to analyzing the effects of the independent variables, this experimental methodology generates a mathematical model that describes the chemical or biochemical processes [6].

In order to perform the optimization procedure with the specified method, it is necessary to first determine the response surfaces. Response Surface module must be introduced to determine response surfaces. This module calculates all response areas depending on the input and output parameters and subsequently introduces the optimization module to define the objective and constraint function and to calculate the optimal solution [5].

The goal assigned to the optimization procedure is to minimize mass, while the limits are that the maximum equivalent stress on a predefined surface must not exceed the allowable stress of 143,3 MPa which is the value of the yield strength divided by safety factor of 1,5. The total maximum displacement of the holes of the bracket may not exceed 1,4007 mm. This value of maximum displacement is chosen because if this value changes it will result in a much bigger maximum displacement of the whole crane. By limiting displacement of the hole of the bracket with this value which is calculated previously in strength calculation, the maximum displacement of the crane should not change.

After solving the analysis, an optimal solution of the parameters is calculated. The mass of the model with initial parameter values was 103,59 kg and the solution with optimal parameter values is 89,73 kg. In conclusion, the optimization procedure reduced the mass of the connecting part of the arm by almost 14 kg, which is a significant mass reduction. Of course, if a larger number of input parameters were selected, the mass value could certainly be reduced by a slightly larger amount, but due to the limitation of the computational power used to perform this optimization procedure, this was not possible. Performing this optimization procedure with multiple input parameters for computer with average characteristics is a very time-consuming process that can take days, even weeks.

7 Conclusion

The structural change of the crane was shown gradually in the form of strength calculations, from the initial model of the crane which was designed and dimensioned based on the specified load and functional requirements. After the calculation of the required internal diameter and number of cylinders, a change in the design of the structure is visible and then it was analyzed in the Ansys Workbench 19.0 software package. A strength analysis was performed and it was concluded that the crane fulfilled all set requirements.

The movement of the crane is accomplished with the help of hydraulic cylinders and the rotation of the crane about an axis with a hydraulic motor where the said elements are driven by a hydraulic pump. Winch with an electric motor will be used to lift the load. Hydraulic systems are applied in situations requiring motion under the action of high loads and small and uniform displacements. In order to analyze the movements, it was necessary to schematically show and describe the hydraulic system of the crane.

The optimization procedure was performed with the Response surface optimization method on the connecting part of the crane arm. Four input parameters were determined for the optimization process, which included the dimensions of stiffeners in order to reduce the mass of connecting part of the arm while not increasing the maximum total deformation of the crane. The performed optimization reduced the mass of the connecting part of the arm by almost 14 kg.

References

1. Opacmare Co. Lifting systems catalogue page. https://www.opacmare.com/lifting-system/?lang=en. Accessed 05 Apr 2020
2. Matejiček, F., Semenski, D., Vnučec, Z.: Uvod u statiku sa zbirkom zadataka. Strojarski fakultet u Slavonskom Brodu, Slavonski Brod (2012)
3. Filetin, T., Kovačiček, F., Indof, J.: Svojstva i primjena materijala. Fakultet strojarstva i brodogradnje Sveučilišta u Zagrebu, Zagreb (2002)
4. Korbar, R.: Pneumatika i hidraulika. Veleučilište u Karlovcu, Karlovac (2007)
5. ANSYS Inc.: ANSYS Workbench User's Guide. Release 12.1. Copyright SAS IP Inc., U.S.A. (2009)
6. Baş, D., Boyacı, İ.H.: Modeling and optimization I: usability of response surface methodology. J. Food Eng. **78**(3), 836–845 (2007)

Flexible Manufacturing System Simulation and Optimization

Peter Pavol Monka[1,2(✉)], Katarina Monkova[1,2], Andrej Jahnátek[3], and Ján Vanca[1]

[1] Faculty of Manufacturing Technologies with the Seat in Presov, Technical University in Kosice, Presov, Slovak Republic
peter.pavol.monka@tuke.sk
[2] Faculty of Technology, Tomas Bata University in Zlin, Zlin, Czech Republic
[3] Jadrová energetická spoločnost' Slovenska, a.s., Tomášikova 22, 821 02 Bratislava, Slovak Republic

Abstract. The article deals with the simulation of real technological process using the computer technology to optimize the flexible manufacturing system (FMS). There was created a simulation model of the FMS for time-saving modelling of production aimed at the utilization of the system capacities for reaching the higher level of efficiency respecting delivery dates. The simulations provided a lot of other useful information describes the properties of the FMS in details. The most used information for the producer was time data (cycle, processing, waiting, lateness, etc.). A wide variation of clamping fixtures and missing buffers between CNC machines were the most problematic issues of the system, generally. Experimental verifications of the FMS confirmed the advantage of using an industrial simulation to verify different production strategies.

Keywords: Technological process · Manufacturing simulation · Production optimization

1 Introduction

The present situation in the industry is characterized as a period of intense progress of technologies at significant computer aid in all branches of the economy. The product has to be competitive, it has to be up to qualitative and functional standard, it must have a reasonable price, efficacious design; it must have regard for safety, ergonomic and another aspect, which decided about its marketability [1].

In connection with the technical advance, it increases the pressure on the manufacturers to develop and make the products as soon as possible at the minimal cost in the required quality. On the other hand, the evolution of the all-new product, or its innovation, is the process very difficult and time-consuming concerning requirements listed above. This process includes the inspection specialized to the verification of prototype functionality, which can be termed as simulation [2].

N. Mitrovic et al. (Eds.): CNNTech 2020, LNNS 153, pp. 53–64, 2021.
https://doi.org/10.1007/978-3-030-58362-0_4

Computer simulation is often used for definition of problems in the industry. Its fundamental principle is based on the fact that the real object is represented by the simulation model in software application. The results of the simulation experiments are backwards applied to a real object for improving its properties. The development of computer techniques in the last years allowed the simulation to become the tool of technological process efficiency increasing [3, 4].

According to Pantazopoulos [5], support and access to knowledge pools and use of modern technology, IT applications, and social media/special interest groups for knowledge diffusion are of vital importance. Institutional establishment, leadership drivers, and social vision can lead to the development of strong foundations of knowledge pools and shape their future perspectives.

2 Manufacturing Process Aspects

Generally, the production process is prepared on the base of process planes, the creation of which is subjected to the existence and interaction of factors that influence the manufacturing process design. The most important is [6]:

- product, technology, material, raw product,
- machine, production equipment,
- personnel (qualification and expertise),
- energy (type, method of transfer, amount),
- organization (time and space structure).

Although the classification of elements listed above is a greatly simplified, the decision making about used technological and production equipment is greatly difficult. Based on the impact of these factors and business possibilities, the suggestion of the suitable technology for part production is in progress, usually in this succession [7]:

1. Design- technological assessment of the product drawing – it is analyzed:

   a. starting and final state of a part material,
   b. the shapes of surfaces and dimensions,
   c. the prescribed tolerance,
   d. surface characteristics.

2. For the selection of a suitable variant of the production are on the base of the previous step determined:

   a. production areas,
   b. production technology,
   c. row product,
   d. technological methods of processing the various features of a component, the possibility of concentration of operations (minimizing of running production time) and technical-economic conditions.

3. Determination of sequence of operations and a detailed proposal:

   a. choice of production equipment,
   b. the scheme part set up,
   c. jigs and fixture preparations,
   d. sections and sequence of operations.

This sequence of steps eventuates into such structure of the process plan, which guarantees the best technical and economic conditions of the production. In this way, by analyzing of the input information (e.g. about the production object, technology, production equipment, …) the process plans have to be optimized to achieve the required output values in the fields of extremes functions optimization criteria [8].

Although the application of technological steps is a complex and difficult task, it cannot be done at once, but it can be carried out in several successive steps, in which some solutions are selected (technological methods of production and auxiliary equipment or process parameters). The choice or suggestion of the solution in a given stage depends on previous solutions. The sequence of decision steps may vary. The multi-stage decision gradually narrows the set of eligible solutions. E.g., the determining of the machine is given by the technology operations choice and the choice of instruments will be limited by the previous selection of machines. There are a large number of variants that are equivalent in terms of ensuring the production of all areas with the required properties. But they are not comparable in cost and labor productivity. According to the test function (minimum cost or maximum productivity), these variants will be optimized [1].

A method of production that creates the ability to easily adapt to changes in the type and quantity of the product is Flexible manufacturing system (FMS). This type of production system relatively simply customizes the products, too. Suitably designed FSM reduces material handling and transit times. By having the machinery to complete a certain process grouped in FMS, the product spends more time on the machinery and less time in transit between machines. Unlike batch processing, materials do not accumulate at a certain location to be worked or moved. This allows the operator the ability to move the unfinished product to the next station without the need of specialized equipment to move what would be, in a batch process, a larger load, and farther distances [9, 10].

FMS is a group of workstations, machines or equipment arranged so the product is processed progressively. The organization is based on the production process dividing into the modules which independently perform the defined manufacturing tasks and they are connected by information and material flow. The main advantages of FMS are continuous production time shortening; the increasing of production capacity; the achievement of production limpidity; the reduction of storage material and transport minimizing of the parts; the better utilization of the manufacturing equipment [3].

3 Manufacturing System Simulation

Manufacturing system simulation can be a very efficient tool for manufacturing system design via virtual beforehand evaluation of the system composition and testing the system behavior under various conditions. The manufacturing system simulation can serve for [11]:

- Evaluation of required resources (Number and types of machines/Material handling system/Necessary inventory location and size/Labor consumption/etc.)
- Performance analyses (Time in system/Number of produced parts/Bottlenecks in the system/Influencing of system performance by changes/etc.)
- Operations management (Supply chain management/Scheduling and planning/System reliability/Quality control, etc.)

The most used performance criterion for manufacturing simulation tools consist of usability criterion aspects [12, 13]:

- Effectiveness - the completeness and accuracy of task led to reach specific aims. Error rates and quality of solution are mostly used effectiveness indicators.
- Efficiency - the relation between (i) the completeness and accuracy of task with led to reach specific objectives and (ii) the resources spend to reach them. Efficiency indicators take in learning time and task completion time.
- User Satisfaction - user comfort and positive attitude towards using of the system. Attitude rating scales are usually used for users' satisfaction measuring.

3.1 Analyze of the Real FMS

For creating the FMS behavior model, it is needed [14, 15]:

- Creating of FMS and its environment composition model - determined by plant requirements
- material flow definition - determined by production base designed by plant
- defining FMS strategies - determined by process plans of production assortment
- defining of system variables and sets of their values – mainly operating times, auxiliary times, shifts, breakdowns, way of maintenance, etc.

Flexible manufacturing system model, shown in Fig. 1, was evaluated from system composition and the system behavior point of views under various conditions.

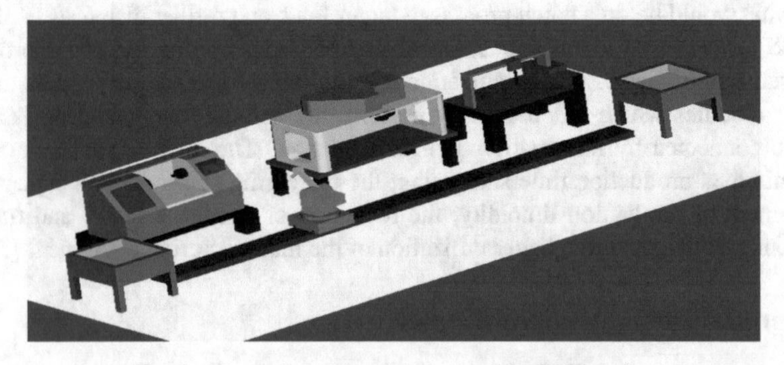

Fig. 1. Three-dimensional scheme of the FMS design

In modifications of the original technological procedures for conditions of the designed FMS, the following aspects were taken into account [16, 17]:

- possibility of using multiple tools in machining on production machines;
- the necessity of multiple entries of parts into FMS (for various type of processing).

Generally, the parts intended for processing in the real FMS can be classified into three groups according to Table 1.

Table 1. Classification of parts for the FMS.

	Classification		
	Group1	**Group2**	**Group3**
Description	Rotary objects manufactured on CNC lathe machine and measured on a CMM	Rotary objects manufactured on CNC lathe machine, CNC milling machine and measured on a CMM	Non-rotary objects manufactured on a CNC milling machine and measured on a CMM
Scheme	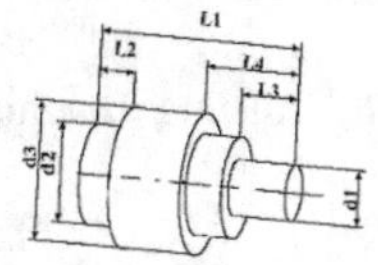	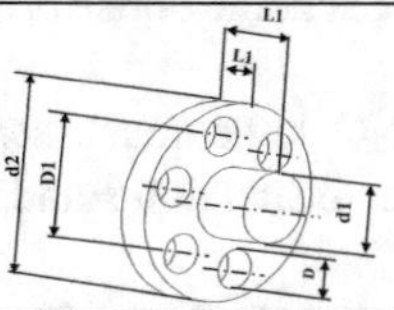	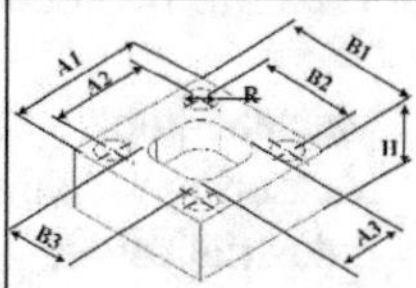
Number of parts in FMS	48	62	28
CNC Lathe	**YES**	**YES**	-
CNC Mill. m.	-	**YES**	**YES**
CMM	**YES**	**YES**	**YES**
	Sample parts produced in the FMS		
Part name	Ring A	Sleeve II	Plate I
Part No.	1001-19-10.10	1001-19-40.16	1001-19-70.06
Basic drawing details	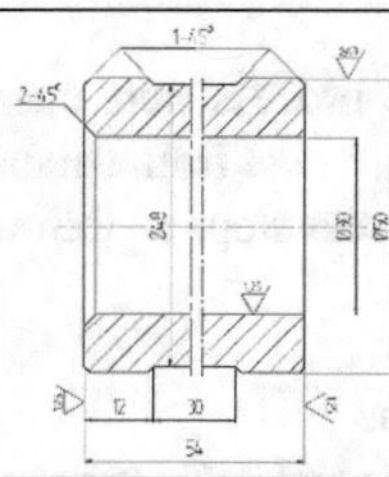	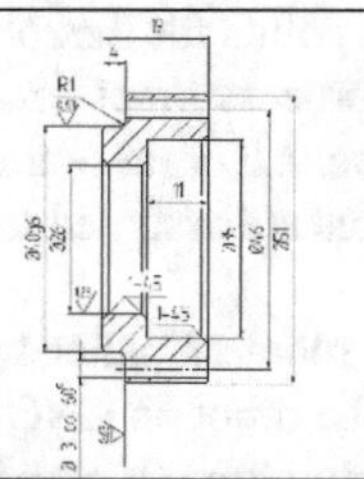	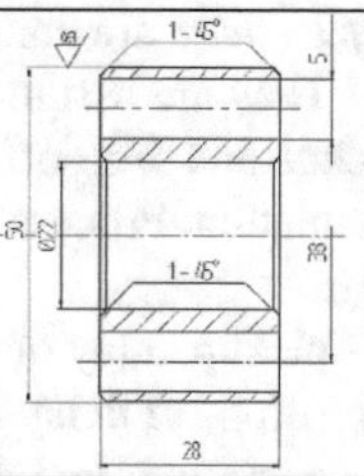

Set of parts for real production in the FMS is characterized by a very high degree of variation of required clamping fixtures. The clamping devices, used for parts produced in the FMS presented in Table 1, are shown in Table 2.

Following the classification of the real FMS processed parts into three groups is process plans scheme described in Table 3 and shown Fig. 2.

Table 2. Clamping devices for sample parts presented in Table 1.

Part name	Group	Clamping devices		
		CNC Lathe	CNC Mill. mach.	CMM inspection
Ring A	1	Mandrel ∅ 30	–	3-point clamping
Sleeve II	2	Chuck ∅ 50–100	Mandrel ∅ 26 mm	At a hole ∅ 26
Plate I	3	–	Hydraulic vice	Tripping dog to the corner

3.2 Simulation Model of the FMS

For a building of suitable simulation model were used information collected by the above described analyses. The properly prepared model allows a way to test the system from the level of simple revisions; through implementing of novel principles; to complete redesigns [18].

The FMS functional scheme consists of:

1. Machine tools (CNC lathe machine/CNC milling machine/CMM - Coordinate Measuring Machine -inspection stand);
2. Industrial robot;
3. Material transport units (Pallets – for material and finished parts store/Linear slide – for the robot moving);
4. Auxiliary devices (Manual measurement tools/Cutting tools/Clamping devices/Other devices necessary for the FMS).

Considering on preceding facts it was built a simulation model of the real FMS which graphical scheme is shown in Fig. 3.

Components of the FMS were described above. The robot is simulated as an automatic guided vehicle system - the linear slide is divided onto parts ***aseg1*** - ***aseg4***. Points ***acp1*** - ***acp5*** are control points for the robot.

They are assigned process times for each process step: operation time, setup time, pickup and drop-off time. All of these times are read from database tables in time of the simulation. Process times are assigned to process steps as follows:

t_0 - pickup delay of the robot on pallet 1
t_{11} - drop-off delay of the robot on CNC Lathe
t_1 - operation time + setup time (if necessary) on CNC Lathe
t_{12} - pickup delay of the robot on CNC Lathe
t_{21} - drop-off delay of the robot on CNC Mill. mach.
t_2 - operation time + setup time (if necessary) on CNC Mill. mach.
t_{22} - pickup delay of the robot on CNC Mill. mach.
t_{31} - drop-off delay of the robot on measurement CNC machine
t_3 - operation time + setup time (if necessary) on measurement CNC machine
t_{32} - pickup delay of the robot on measurement CNC machine
t_4 - drop-off delay of the robot on pallet 2

The shifts and maintenance time breakdowns of each component of the FMS are for simulation model necessary information, too. One strategy used in the FMS model simulation is presented in Table 4.

Maintenance was proposed for each machine as absolute - that means maintenance performing after every shift. If in simulation experiments is required to simulate the

Table 3. Process steps for process plans (Groups).

Process plan 1 (Group 1)	
Process step/Description	
1	New order which has to be realized
2	Move-between pallet 1 and CNC Lathe by the robot
3	Processing on CNC Lathe (setup, if necessary + operation)
4	Move-between CNC Lathe and CNC Mill. mach. by the robot
5	Processing on CNC Mill. mach. (setup, if necessary + operation)
6	Move-between CNC Mill. m. and CMM by the robot
7	Measuring on a CMM (setup, if necessary + operation)
8	Move-between CMM and pallet 2 by the robot
Process plan 2 (Group 2)	
Process step/Description	
1	new order which has to be realized
2	move-between pallet 1 and CNC Lathe by the robot
3	processing on CNC Lathe (setup, if necessary + operation)
4	move-between CNC Lathe and measurement CNC machine by the robot
5	measuring on a CMM (setup, if necessary + operation)
6	move-between CMM and pallet 2 by the robot
Process plan 3 (Group 3)	
Process step/Description	
1	New order which has to be realized
2	Move-between pallet 1 and CNC Milling machine by the robot
3	Processing on CNC Milling mach. (setup, if necessary + operation)

(*continued*)

Table 3. (*continued*)

Process plan 3 (Group 3)	
4	Move-between CNC Milling mach. and CMM by the robot
5	Measuring on a CMM (setup, if necessary. + operation)
6	Move-between CMM and pallet 2 by the robot

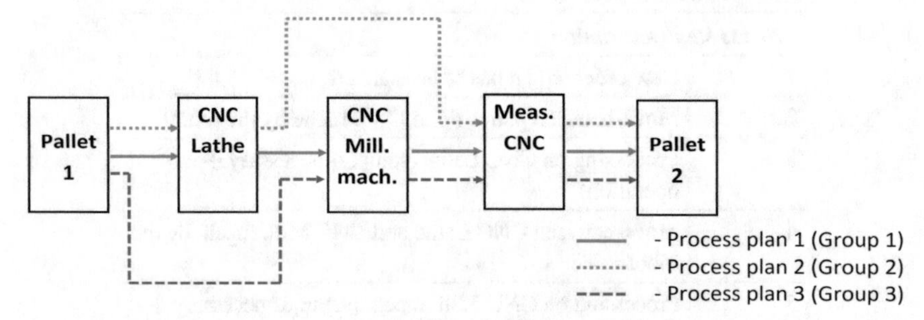

Fig. 2. The FMS process plans graphical scheme

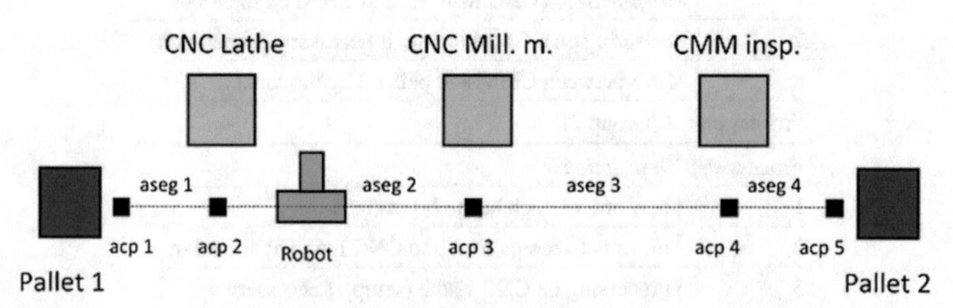

Fig. 3. Simulation model scheme

Table 4. Shift and maintenance definitions for selected parts.

Machine	Shift	Shift time	Maintenance	Maintenance time
CNC Lathe	Shift 1	8^{00}–14^{00}	Absolute	13^{30}–14^{00}
CNC Milling	Shift 2	8^{00}–14^{15}	Absolute	13^{45}–14^{15}
CMM Inspec.	Shift 3	8^{00}–14^{30}	Absolute	14^{00}–14^{30}

influence of random interruptions this could be realized through statistical functions describing it.

3.3 Simulation of the FMS

The simulation model preparation was done according to the above described analyses of the real FMS. As the most important information - from practice experiences and theoretical point of view, too - is a crucial function of the robot in the FMS. The robot is an only interconnecting element among all components of the FMS. That means the impossibility of production in the whole manufacturing system if the robot is down.

On-shift utilization of the CNC machines and ensuring of the delivery time were selected as the most relevant information for increasing of the FMS efficiency. All simulations and related analyses were focused on these two main optimization factors [19]. Samples of early-stage simulation results are shown in Fig. 4.

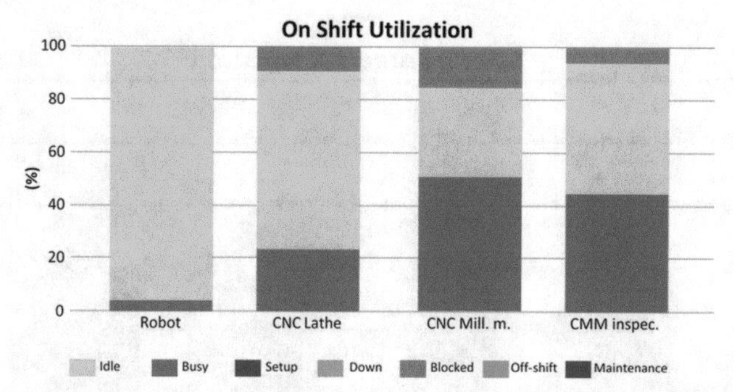

Fig. 4. Sample of an early-stage simulation On Shift Utilization results of the FMS components.

Detailed analyses of the obtained simulation results were used for manufacturing process optimization. The most important input information for the optimization were analyses: resources summary; on-shift utilization; lateness; scheduled cycle time; cycle time; processing time. The analyses utilisation led to understanding the interactions inside of the FMS – identification of utilization of equipment within the manufacturing operation; bottlenecks identification; and achievable range of production throughput given in stated constrains.

Manufacturing process optimization helped to increase on shift utilization of the FMS components (CNC machines). Graphical interpretation of optimized results from Fig. 4 are presented in Fig. 5.

From ensuring of delivery time point of view, the simulations for selected strategies were run. Some strategies the most problematic in order completions are shown in Fig. 6. The figure presented required order processing time and simulation results before and after optimization.

The simulations provided a lot of other useful information describes the properties of the FMS in details. The most used information for the producer was time data (cycle, processing, waiting, lateness, etc.). The wide variation of clamping fixtures and missing buffers between CNC machines were the most problematic issues of the system, generally.

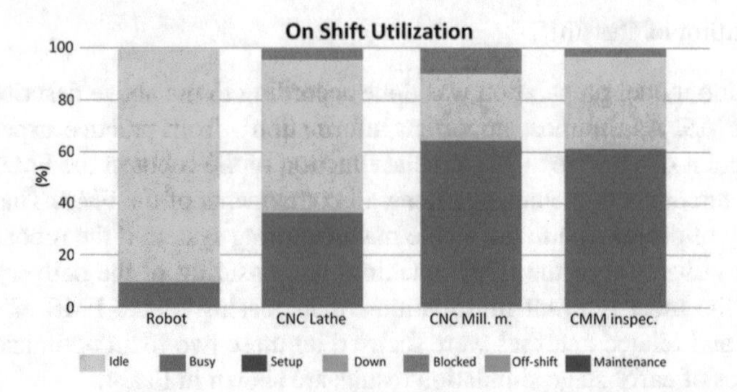

Fig. 5. Sample of a simulation result of the FMS components utilization after optimization

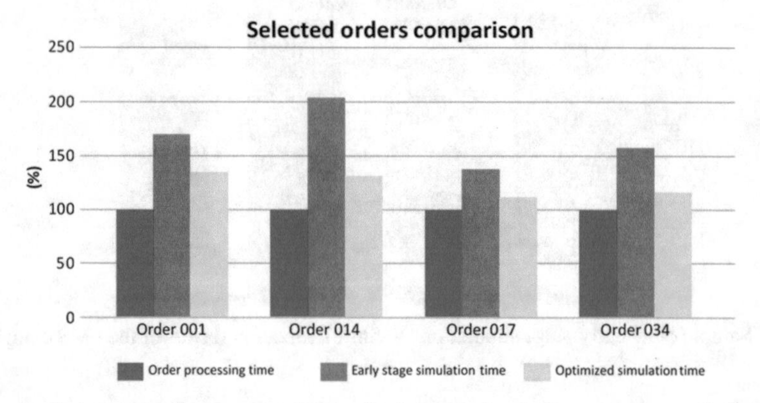

Fig. 6. Comparison of order completion of selected orders before and after optimization

4 Conclusions

According to [20], a Quality Management System (QMS) within a company or an organization as a series of interconnected processes is characterized by a holistic approach, where each process step aims to deliver the "best result" to the following one, in terms of efficiency and effectiveness, rendering the outcome of the entire system except for the interested parties involved, i.e., company, customers, employees, society. Minimizing waste and achieving maximum control to enhance process stability is key components of quality assurance.

The main aim of work described in the paper was to build a suitable simulation model of the FMS and demonstrate the influence of different manufacturing strategies on production and economic outputs. The most important part of the work was to analyze the existing system and create the corresponding model. The model was utilized for exploitation to optimize the workflow through the FMS cell. There were used a lot of simulations based on previous real orders. Optimizations of the manufacturing process were done through simulation results. Structure of the FMS without magazines between CNC machines was observed as a weakness of the system. The machines are often blocked because they are waiting to be operated by a robot in this constitution.

Another problem of the FMS is a processed product range. The range implies a wide range of fixtures used for their clamping for the processing. Time of conversion of clamping fixtures negatively influences of machines utilization efficiency. The flexibility of the FMS is at a low level for that reason. Experimental verifications of the FMS confirmed the advantage of using an industrial simulation to verify different production strategies.

Based on the above simulation's findings, it is possible to confirm that it is possible to optimize the structure and operation of FMS for obtaining higher production efficiency in compliance with the required delivery dates. It can be stated that the advantages of the FMS simulation were very useful for an understanding of relationships inside and near the system by constrains identification and substantial problems diagnosing. Visualization of the simulated processes is very beneficial in exploring possibilities, change preparing and investing consideration.

Acknowledgements. The paper is the result of the project implementation "Advanced planning and control of discrete production", code ITMS: 313012T109, supported by the Operational Program Research and Innovation funded by the ERDF; and it was prepared thanks to support by the Ministry of Education of the Slovak Republic through the grants APVV-19-0550 and KEGA 007TUKE-4/2018.

References

1. Milošević, M., et al.: Design and manufacture of reformer in polymer electrolyte membrane fuel cell. Struct. Integr. Life **17**(1), 21–24 (2017)
2. Mitrovic, N., et al.: Experimental and numerical study of globe valve housing. Chem. Ind. (2017). https://doi.org/10.2298/HEMIND160516035M(M23)
3. Panda, A., et al.: Progressive technology – diagnostic and factors affecting to machinability. Appl. Mech. Mater. **616**, 183–190 (2014)
4. Stoicovici, D.I., et al.: An experimental approach to optimize the screening in the real operating conditions. Manuf. Eng. **2**, 75–78 (2008)
5. Pantazopoulos, G.A.: Knowledge networks: a key driver for technological advancement and social progress. J. Fail. Anal. Prev. **17**(5), 823–824 (2017). https://doi.org/10.1007/s11668-017-0342-z
6. Monkova, K., et al.: Newly developed software application for multiple access process planning. Adv. Mech. Eng., 39071–39071 (2014)
7. Jerzy, J., Kuric, I., Grozav, S., Ceclan, V.: Diagnostics of CNC machine tool with R-test system. Acad. J. Manuf. Eng. **12**, 56–60 (2014)
8. Ceclan, V.A., et al.: Quality of the hydroformed tubular parts. Adv. Eng. Forum, 8–9 (2013)
9. Tanasic, I., et al.: An attempt to create a standardized (reference) model for experimental investigations on implant's sample. In: MEASUREMENT, 2015, vol. 72, pp. 37–42 (2015). ISSN 0263-2241
10. Zetek, M., Zetkova, I.: Increasing of the cutting tool efficiency from tool steel by using fluidization method. Procedia Eng. **100**, 912–917 (2015)
11. Kadnar, M., et al.: The design and verification of experimental machine for real journal bearing testing. Tech. Gaz. **18**(1), 95–98 (2011). ISSN 1330-3651
12. Monka, P., Monkova, K.: Basic mathematical principles for internal structure of new CAPP software application. In: Annals of DAAAM for 2009 & Proceedings of the 20th International Symposium, vol. 20, pp. 1281–1282 (2009). ISSN 1726-9679

13. Monka, P., et al.: Design and experimental study of turning tools with linear cutting edges and comparison to commercial tools. Int. J. Adv. Manuf. Technol. **85**, 2325–2343 (2016)
14. Baron, P., et al.: Proposal of the knowledge application environment of calculating operational parameters for conventional machining technology. Key Eng. Mater. **669**, 95–102 (2016)
15. Stojadinovic, S.M., Majstorovic, V.D.: Developing engineering ontology for domain coordinate metrology. FME Trans. **42**(3), 249–255 (2014)
16. Vuković, A., Perinić, M., Ikonić, M.: Conceptual framework for creating customized modular CAPP system. Eng. Rev. **31**(1), 35–43 (2011)
17. Markovic, J., Mihok, J.: Legal metrology and system for calibration and verification of the radar level sensors. Qual. Innov. Prosper. **20**(1), 95–103 (2016)
18. Turisova, R., et al.: Verification of the risk assessment model through an expert judgment. Qual. Innov. Prosper. **16**(1), 37–48 (2012)
19. Vychytil, J., Holecek, M.: The simple model of cell prestress maintained by cell incompressibility. Math. Comput. Simul. **80**(6), 1337–1344 (2010)
20. Pantazopoulos, G.A.: Failure analysis, quality assurance, and business excellence. J. Fail. Anal. Prev. **13**, 119–120 (2013)

Experimental and Numerical Methods for Concept Design and Flow Transition Prediction on the Example of the Bionic High-Speed Train

Suzana Linic[1](✉), Vojkan Lucanin[2], Srdjan Zivkovic[3], Marko Rakovic[4], and Mirjana Puharic[5]

[1] Innovation Center of the Faculty of Mechanical Engineering, University of Belgrade, 16 Kraljice Marije Street, Belgrade, Serbia
slinic@mas.bg.ac.rs

[2] Faculty of Mechanical Engineering, University of Belgrade, 16 Kraljice Marije Street, Belgrade, Serbia

[3] Military Technical Institute, 1 Ratka Resanovica Street, Belgrade, Serbia

[4] National History Museum, 51 Njegoseva Street, Belgrade, Serbia

[5] The Faculty of Civil Aviation, Megatrend University, 8 Bulevar marsala Tolbuhina, Belgrade, Serbia

Abstract. In the times of significant development of high-speed train transportation and taking the primacy over the others, one of the major designing tasks is to predict accurately and fast the vehicle's main performances, especially aerodynamics. Motivated by the bionic design of representative Japan's train Shinkansen, the presented bionic design was based on the observations of the kingfisher from Serbian national heritage collection. The specimen beak shape has been measured, by laser scanning, and converted to a mesh. Afterward, the longitudinal cross-section of the beak was implemented to a bionic high-speed train design. As the critical, the conditions of the train's forehead entering into the tunnel were selected, while the case scheme employed relative motion. The forehead contour distributions of the surface temperature, skin friction coefficient, and the pressure distributions, obtained with computational fluid dynamics, were used for the prediction of the transition zone extension. The anomalies in surface temperature behavior were additionally analyzed by pressure and density distribution inside the tunnel and over the forehead. Besides the gross time history of the pressure derivative of time is in correspondence with the referent, it was interpreted that the true biological form is not fully suitable for adoption for bionic design. The development of this design, employing contour simplification by the close parabolic function will continue in the future. This work suggests an economic and efficient approach to analyze the results of the Reynolds-Averaged Navier-Stokes equations adequate for the concept design stage.

Keywords: High-speed train · Aerodynamics · Bionic · CFD

N. Mitrovic et al. (Eds.): CNNTech 2020, LNNS 153, pp. 65–82, 2021.
https://doi.org/10.1007/978-3-030-58362-0_5

1 Introduction

Modern transportation faces numerous challenges where the requirements of speed, economy, and environmental protection are among the most demanding. High-speed transportation is taking on the dominant position powering the most dynamic development. Through many years the extensive knowledge base was developed by the theoretical, experimental, numerical, and operational researches in the field [1–8], based on classic transportation [9] and adopting the experiences from the other fields, mostly aeronautical. The very first high-speed train in operation, Shinkansen, Japan, have had incorporated many engineering and scientific solutions involving the bionics [10–12], fluid mechanic and heat transfer [13, 14], aircraft constructions and structure, to the highly advanced manual manufacturing, organization, and many others. Unfortunately, the knowledge base about the kingfisher does not cover fully the data needed for implementation in bionics thus multidisciplinary work and knowledge transfer in biology is a necessity [15]. The fluid dynamics and aerodynamics, significant for drag and power reduction [1, 2], are employed for profiling and shape optimization of the overall design as well as the all wetted details as the pantograph [10], aerodynamic brakes [16, 17] are, for instance, investigated numerically and by the wind tunnel testing [18].

From the noted and numerous other reference literature, it was learned that the case in which the HST is entering the tunnel is advisable to adopt as a critical case. Firstly because of the drag force increase over the values for open rail run, and secondly, because in the duration of this case the compression wave affecting the formation of the micro-pressure wave and the unpleasant noise after the wave exit. It is known that the overall pressure change inside the tunnel and the HST drag created during the HST enters into the tunnel are dependent mostly on the HST and tunnel geometry parameters, while on the contrary, the pressure change is not dependent on the HST fore-body shape. Therefore, the pressure and drag change in the tunnel are convenient parameters to follow in the researches of the HST open rail-tunnel configuration. Unfortunately, nowadays-required operational velocities are in order of several hundreds of km/h, when the consequent effect of the micro-pressure wave at the tunnel exit should not be neglected, but added to the list of criteria for designing. Herein, the strength of the micro-pressure wave, and thus the noise pollution, is dependent on the fore-body design [1–3], especially in the period of entering the fore-body into the tunnel. Furthermore, the time history behavior of the pressure derivative of time is a key parameter for decision-making in the selection process of HST design.

The goal of the presented work is to define the most economic and efficient method to design the HST fore-body concept, convenient for further optimization, and to predict the transition zone extension. The motivation was found in the biomimicry method applied to the Shinkansen train which aimed its performances. To gain knowledge, the same bird, the common kingfisher, from the heritage collection of the Natural History Museum was used as bio-inspiration. The laser scanning method was employed to create a digital model of the beak and the head while photographing and visual inspection was used to control and understand details for a reconstruction of the model. The obtained geometry data were incorporated in the bionic high-speed train, BHST, design. Following the experiences from the previous researches [1, 3, 5–7, 12] and intending to apply economic and efficient method, the analysis of the flow around the BHST fore-body entering the tunnel (V = 300 km/h) was selected. The short-length BHST (50 m) and the short tunnel (250 m) were selected because only partial train entrance was observed thus predict the flow behavior in the shortest time as possible. By this means, the bionic fore-body was tested to following flow parameters: (a) temperature, skin friction coefficient and pressure distributions along the contour outline of the fore-body in the cases outside and inside the tunnel, (b) pressure and density distribution in the field inside the tunnel in the vicinity of the BHST and over the BHST surfaces. The computational fluid dynamics employed the Reynolds-Averaged Navier-Stokes equations solved by the density-based solver, with a Realizable □□□ turbulence model. The relative motion of the BHST to the tunnel was supported by the sliding mesh. From the justified test of the infrared thermography application in the wind tunnel, the BHST designing method was shortened additionally because the static temperature and skin friction distribution, over the bionic forehead, aim prediction of the flow transition. These distributions are not commonly analyzed but they become powerful tools of transition detection under the RANS method, which, as known, is not capable to predict the boundary layer character in detail. The quality of design and the fulfillment of the expectations were checked by comparison of the pressure derivative of time in time-history with the reference data.

This work presents the efficient and economic method for definition the BHST concept design well suited for further optimization, which opened space for enrolment of different disciplines and contemporary techniques. At the same time, the further development of the BHST, implementation of the contemporary resources, and broader studies of the kingfisher are highlighted as needed for further research.

1.1 Background

3D Laser Measuring. The 3D laser measuring method is fast, usually mobile, non-contact, with high-quality, suitable for easy 3D digital modeling [19, 20], especially useful for biology and cultural heritage applications [21]. The method is based on triangulation, *i.e.* measuring of the constrained light beam vectors, elementary cells of the detector, and the spatial position of the laser device with the camera [21–24]. In common, with the complexity of the shape, the resulted data are affected by more influences. The main influencing factor on the measurement quality is the setup [22, 23]. The surface material, shape, and configuration, its surface quality and finish, roughness, and optical properties affect the scanning results [22, 24]. The reduced amount of light energy results in false measurement or the full absence of the input affection due to complete diffusion of the incidence beam, for very rough surfaces [25]. For the textile materials, [26] noted that captured data were influenced by the material structure, the direction of threads, and the base weaving. Furthermore, some investigations reported that darker surfaces caused the inability of data capturing [26].

Forced Convection. In the following, the short description of the forced convection and related phenomenon is given on the base of references [27–29], where further explanations may be found. The flow with forced convection is described with the energy equation involving internal energy, conduction, convection, and friction heat, neglecting the radiation from the surrounding bodies. Analogous to the flow characterization related to the free stream velocity, the extensive observations have shown that the flow is characterized as a laminar or a turbulent upon to the free stream temperature also. Therefore, the two boundary layers are recognized, the velocity and the thermal boundary layer. The mechanism of the heat transfer between the flow and the wetted surface is based on the friction rate of the fluid particles and the convection rate, strongly dependent on the thermal properties, both, of the model and the fluid. The boundary layers are growing and changing along the surface chord and orthogonal, after the fluid particles are acting to each other and the surface. Viewed in the downstream direction while near the wall, the uniform – laminar flow, after the critical Reynolds number is reached, begins to "re-organize" thus the wavy fronts arise (Tollmien-Schlichting waves) and further transform to the vorticity zone. Downstream, the small and sporadic vortices are growing and transforming the complex 3D shapes, which afterward experience the breakdown. In the next stage, the new turbulent cores are formed around which the turbulent spots are growing. At the end of flow transition, the groups of the turbulent spots are grown and tightly packed, covered the surface spanwise and the flow becomes fully turbulent. From the extensive analysis, it was learned that for the smooth flat plate immersed into the uniform flow, a component of the drag force coefficient originated from the surface friction, c_f, follows partially the two characteristic functions. The one described as $c_{fx} = 0.664/\mathrm{Re}_x^{1/2}$ ($\mathrm{Re_x} = \mathrm{Re_{critical}} \leq 5\ 10^5$), is corresponding to the laminar flow, and the other one $c_{fx} = 0.059/\mathrm{Re}_x^{1/5}$ ($5\ 10^5 \leq \mathrm{Re_{critical}} \leq 10^7$), to the fully developed turbulent flow, all the other values between describing the transitional flow. In general, the transition zone parameters depend on the immersed body geometry, wall roughness, free-stream velocity, and wall temperature. However, the local heat transfer coefficient, the local surface temperature, T, and c_{fx} follow similar behavior, their extreme values are pointing to the local position at which the fully turbulent flow is developed. This fact is of great help when analyzing the computational fluid dynamics, CFD, results obtained by Reynolds-Averaged Navier-Stokes, RANS, turbulent models, which resolve the problems of the large-eddies scale. Employing c_{fx} and T in RANS analyze add prediction of the transition zone confidently in the early stage of design.

Numerical Method. The CFD is resolving the fundamental equations of the fluid dynamics: continuity, momentum, and energy equations. Today, the numerical methods become the standard engineering and research tool after extensive development of the methods and their justifications [30–33]. The most used in engineering practice are the RANS equations. In this approach the flow variables, either the scalar or the vector, sourced by flow turbulence, are each represented by the sum of the two components. The first one is the mean value component and the other is the time-varying fluctuating component of the property. Therefore, when the flow variables of this form are incorporated into the continuity and momentum equations, written for the infinitesimal period, averaging over time leads to the unclosed system of equations The RANS equations are analogous to the instantaneous Navier-Stokes equations with a difference they contain the additional terms describing the presence of the turbulence, the Reynolds stresses. One way to solve the system of equations is by use of the density-based solver, which defines density field from the continuity equation, and afterward, applies it as an independent variable to calculate the pressure, with which it is coupled, from the momentum equations. Following this calculation procedure, the application of the density-based solver is mostly serving in cases with high compressibility and shocks. However, the selection of the solver depends on the flow conditions as a whole, and also local, so nowadays; the development of the solvers leads to high-quality results in the mid-cases also. The turbulence models were introduced to close the system of equations and they are grouped upon the number of transport equations where the time-quality compromise is the usual selection criterion. In the field of vehicle aerodynamics, the two-equation turbulence models, Standard $\kappa - \varepsilon$ or Realizable $\kappa - \varepsilon$ are common, following the recommendations.

To resolve the system of the governing equations iteratively, discretization of space and time is necessary. The CFD, based on the finite volume methods, uses the spatial discretization scheme with structured inflation layer, adjacent to the walls, and structured or non-structured 3D elements inside the domain, for simple steady and simple transient cases. On the contrary, the complex time accurate cases, which include the relative motion of the bodies of interest, the dynamic, and sliding meshes are applied. The latter is used for relative translation/rotation of two bodies or relative motion of the movable model to the steady model. The sliding mesh technique, appropriate in the case of the train passing the tunnel, applies a system of stable and movable meshes with the defined interface over the surface of sliding. The interface surface is divided between adjacent cells where their bounding cell zones overlay each other partially. The fluxes are calculated between every two temporary positions of adjacent cells represented on the interface surface.

2 Methods

Figure 1 shows the selected case setup intended for the study of the nose entry into the tunnel, motivated with works of [1–4]. The BHST train starts running from a distance equal to its length, $LC = 50$ m. Movement is uniform under standard atmospheric conditions. After a while, the train enters the tunnel. The selected tunnel hoods have clear cut, non-Shinkansen type, with blockage ratio $R = 0.22$ [1].

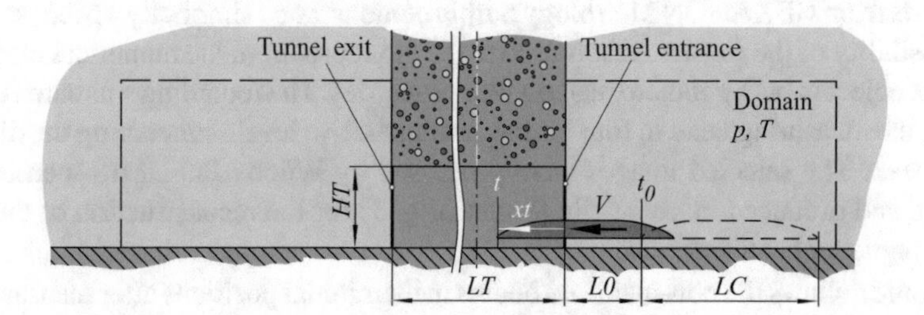

Fig. 1. The case scheme of the BHST passing the tunnel

2.1 Modeling and Designing of the BHST

For the BHST design, the inspiration has been drawn in the prepared kingfisher specimen (*Alcedo atthis*) from the Serbian heritage collection of the Natural History Museum, Belgrade, following the representative Shinkansen train, Japan [12].

Visual Inspection. The very first insight into the female kingfisher body has been made by visual inspection and analyses of the photographs. Visual inspection included measurements of the specimen and indirect measurements from the photographs, as shown in Fig. 2 [12]. Direct measurements had accuracy 0.1 mm.

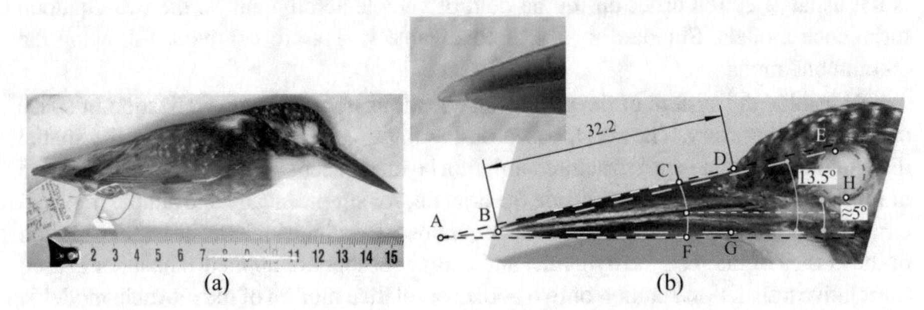

Fig. 2. Prepared kingfisher specimen: (a) Overall visage, (b) Detail of a kingfisher's beak

Photo shooting has been performed with Huawei Y530-U00 [12] and Lenovo K5 cameras for the series of combinations of light sources and specimen poses. Throughout this process, we face the main problems related to the setup of the clear orthogonally view to the beak sides and the capabilities of setting the correct lens depth. With intention to control the numerical model, later on, the specimen poses were slightly shifted between records, in the side and the front views. From the literature [34], it was learned that the kingfisher's feathers have specific optical properties, based on their structure and paint, which are expected to cause the laser scanning difficulties, besides the common oily water protection coating.

3D Laser Scanning of the Kingfisher. The scan of the kingfisher bird has been done by experts from HEXAGON Metrology Serbia demonstrated simplicity, speed, accuracy, and possibility of the Romer Absolute Arm [35] to reproduces the numerical model of a complex object in both geometry and surface properties. The recording was carried out by moving the measuring head in four lateral administrative levels concerning the direction of the beak. The selected images were processed by Wilcox PC-DMIS measurement software and produced in `.stl` file format (Fig. 3) for the reconstruction of the shape of the kingfisher head and beak.

Figure 3a shows the non-manifold bodies in their initial positions after measurements before gluing with the MeshLab software [36]. After gluing, the damages have been still present so the model has to be post-processed. Figure 3b shows a detail from the manual mesh reconstruction made by the Blender software [37] and the final mesh.

From the reconstructed scanned model, after control, the series of cross-sections were ordered and shown in Fig. 4. The complex shapes of the cross-sections and their distribution led to the comment that the beak has sword tip shape, with the furrow made by tomias, if viewed rotated from the front. The beak shape is in general, flattened from the sides. Along the first half-length, the cross-sections are semi-elliptic.

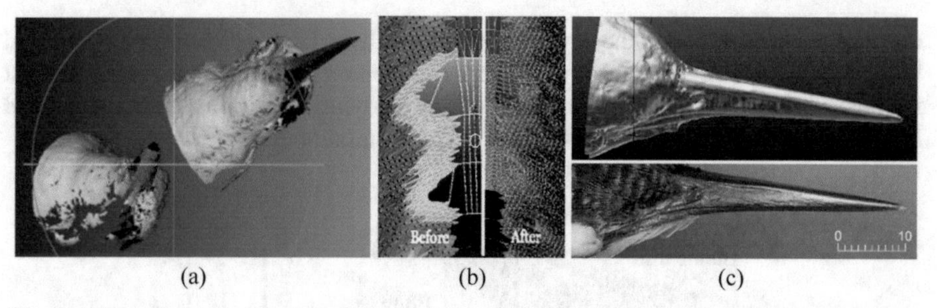

(a) (b) (c)

Fig. 3. Reconstruction of the kingfisher digital model [12]: (a) the two initial meshes, (b) post-processing. (c) model control.

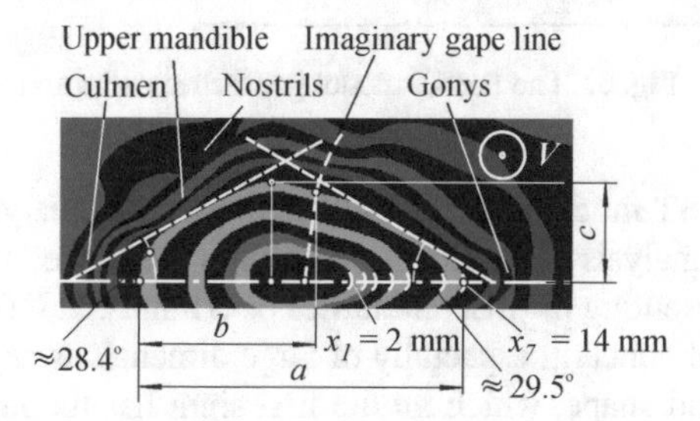

Fig. 4. The series of the kingfisher beak cross-sections (rotated for 90° counterclockwise)

From the middle to the beak's end, the inclined sides are almost flat. Viewing each mandible in a natural position, the cross-sections are semi-triangular with rounded contours at vertex places. The widths of the contours containing the outermost points along the culmen and gonys are in order of 1 mm, where the surface culmen is a little bit wider. From the scanned model and photos, the dimension relationships were described with the following fineness ratios: (a) the full profile from the top view up to nostrils' position $\varepsilon_{top/full} \approx 4.6$. (b) the half-profile $\varepsilon \approx 9.3$; (c) the full side-view profile $\varepsilon_{side/full} \approx 3.36$; (d) the upper mandible $\varepsilon_{up.mand} \approx 7.34$, and (e) the lower mandible $\varepsilon_{low.mand} \approx 6.44$. Following the marks in Fig. 4, for the first cross-section, at $x_1 = 2\,\text{mm}$ from the tip of the beak, $(a/2c)_1 \approx 2.2$ and for the seventh, at $x_7 = 14\,\text{mm}$, $(a/2c)_7 \approx 1.5$, while for the half-profiles are $(a/c)_1 \approx 4.3$ and $(a/c)_7 \approx 3.2$, respectively. From here, we commented that the main merits for kingfisher's non-rippling water entry are in the cross-section shape proportions, its flatness, and the fineness in both longitudinal planes.

The geometry of the beak was followed as much as possible in designing the 3D BHST, of course under the influence of subjective view, Fig. 5. The imaginary gape line was inclined thus the bionic part of the nose contour has been taken the advantage of the most of the beak's contour. The contour transition, over the upper contour line, has been done with simple rounding with respect to profiles' tangency. Over the sides, in the zone of a significant change of vertical profile contour (starting from the seventh up to the fifteenth meter), a slight narrowing of the horizontal profile with the largest width was performed. The BHST body narrowing was inspired by the Whitcomb area rule for supersonic flight by which the streamwise area distribution is adjusted to aim the drag reduction. Herein, the top-view contour with the largest dimensions is adjusted by narrowing to aim the control of the cross-section area distribution. As the interest is in developing the knowledge and methods by studying the flow around the forehead bionic contour part, the BHST design has a short central body.

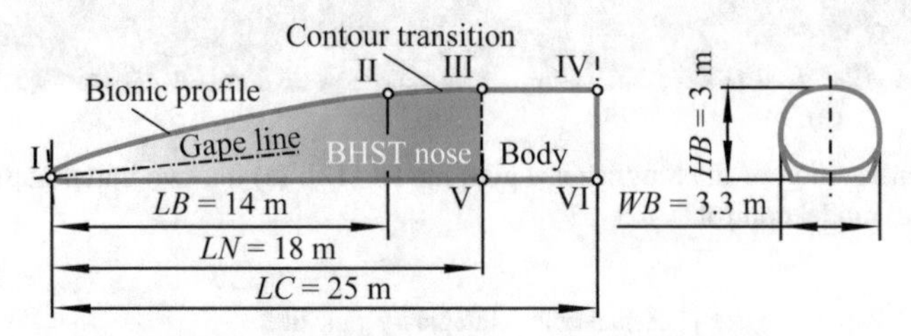

Fig. 5. The BHST model geometric definition

The resulting BHST model with traces of the reflection analysis is shown in Fig. 6 [38]. The reflection analysis aims reduction of time for modeling complex and high-quality bodies, by presenting the discontinuities of G1 and G2. Of course, manufacturability is the additional concern, especially of large dimensions as the BHST is. For the design proportions and shape, which on the first sight has the inclined flattened front and rear surfaces, we find the motivation in the reports [1–4] and many more.

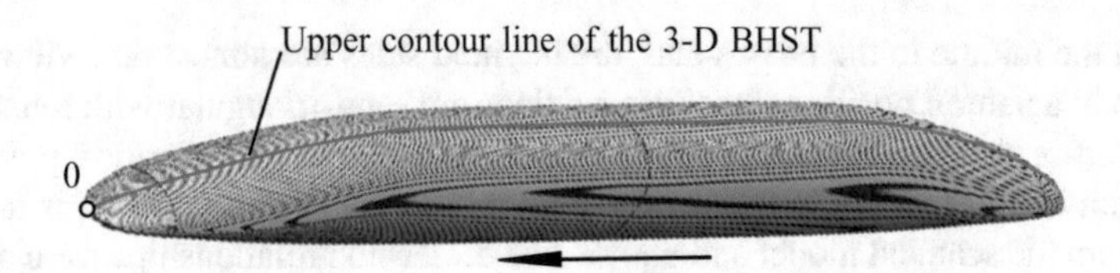

Fig. 6. The reflection analysis of the BHST surface [38]

2.2 Numerical Simulation

The numerical simulations were used for the boundary layer observations, more precisely the transition zone detection, by analyze of the surface temperature and the skin-friction coefficient distributions to contribute CFD cost reduction. The core of the used procedure, related to the density-based solver with employed heat transfer, is justified by [13]. The sliding mesh scheme is shown in Fig. 7 from the front and the side view, and a preview of the mesh details is shown in Fig. 8.

The mesh blocks have been made by composing of the smaller mesh parts, each related to the setup item. The BHST model was covered with a cylindric body representing the central section of the moving mesh, the only one with the unstructured mesh, and build over the inflation layer. At each end of this central section, the body of the moving mesh was extruded to the full length of the selected domain, filled with structured mesh elements, Fig. 8a.

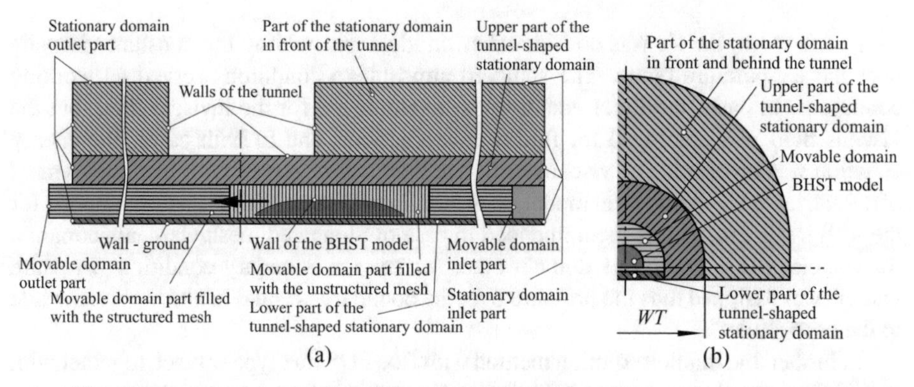

Fig. 7. The scheme of applied sliding mesh technique (a) side view, (b) front view.

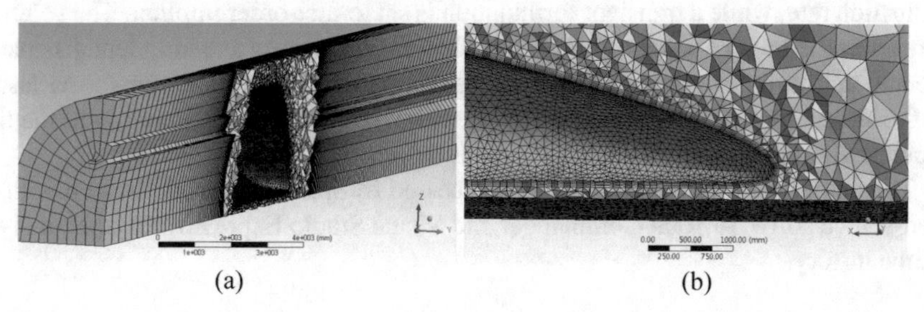

Fig. 8. The details of the mesh: (a) the movable mesh (b) the mesh around the nose tip [12, 31]

Further, all over the moving mesh, the base stationary mesh part was created, such that one part covers the ground beneath the moving mesh, and the other part covers an external rounded surface, Fig. 7a. This inner stationary mesh part took the shape of the tunnel, while stretched over the domain length. Finally, the surrounding was presented with front and back half-cylinder domains, of a total radius of about 100 m. The total stationary domain length is 500 m. Furthermore, the tunnel length is $LT = 250$ m, height 5.8 m, and width 8 m. The BHST model was distanced from the ground for 0.4 m. During the calculations, the moving mesh carried the BHST with a uniform velocity of $V =$ 300 km/h, from its beginning position equaled to the unit train length from the tunnel entrance, $LO = LC = 50$ m. in this case (Fig. 1).

In the end, the overall number of mesh elements was 2,000,000, of which 1,200,000 belonged to stationary mesh. The 97.7% of elements were elongated 9.81, next in order of 27.4 for 1.64%, while the maximal element elongation was 176 for just the several elements around the perimeters of the domain. The element skew was 0.041 for 83.1%, 0.125 for 8%, where the other values can be ignored.

The CFD calculations have been performed for the transient case involving the density-based solver, on one hand, because the significant flow compressibility was expected, and for the study of flow transition behavior based on forced convection, on the other hand [31]. The interface is set up over the sliding surface, placed between the stationary and the movable meshes. The flow changes were expected in areas adjacent to the tunnel entrance, at the tunnel, and areas behind the tunnel exit.

In this work, the air was considered as an ideal gas holding the constant viscosity over the temperature range. The standard atmosphere conditions served as working conditions. In basis, the Euler equations were employed for the inviscid fluid, but the viscous fluid was employed for further investigations, and in both cases, the energy equation was involved. The viscous turbulent flow was observed with the employment of a Realizable $\kappa - \varepsilon$ turbulent model, with Non-Equilibrium Wall Functions applied for the wall treatment, as it is recommended in case of elongated mesh elements common for vehicles [31]. To the one domain outer surface the boundary condition "Pressure Outlet" was assigned thus the pressure over the boundary is equal to atmospheric, while to the other "Inlet".

In further, the implicit solution method with Roe FDS flux type was set, together with spatial discretization parameters: the Green-Gauss Node Based for gradients, a second-order upwind scheme for flow, second-order for turbulent kinetic energy and turbulent dissipation rate, while a transient formulation is set to first-order implicit. The solutions were controlled with a value of 0.8 for turbulent kinetic energy and turbulent dissipation rate and unit turbulent viscosity. In the case of inviscid air, the time steps were lasting for 0.0025 s, employing 15 iterations each, while 0.0055 s was the duration of the time steps for viscous flow cases with 5 iterations employed.

The output data were collected and processed by specialized code written with the open-source software Total Commander and Visual Studio Express for further analyses in time history.

3 Results and Discussion

As common, the BHST geometry is here presented with contours (fitted) and the non-dimensional cross-section area (non-fitted), Fig. 9a, b [12], and more detailed analyses with gradients, Fig. 9c, d.

From here it may be commented that the combination of adjustments by bionic approach and area rule motivation was given the good correspondence with the results from [7]. The two bulges on the *z*-profile, which were following the natural contours of the kingfisher's culmen, have produced the recognizable changes of the first and second gradients both of the contour and area (Fig. 9c). In comparison to the reference [7], the gradient of the non-dimensional area distribution shows the change in front of the referent value (5–10% of the length), viewing in the streamwise direction. The fore-body to height ratio is in a range recommended in reference literature [1] as a measure for the reduction of pressure gradient, where the revolutionary parabolic body has shown the least reduction rates but also this change is prolonged in time. Herein, the fore-body contour may be assumed as a rounded bi-convex profile, which full potential is not expected to be shown at low velocity, and for that, the approximation with parabola would be a good simplification. Another possibility is present also, and it is related to the fact that the rhinotheca and the gnathotheca, the cover layers of the kingfisher's beak, assumed as common to birds, are growing all the time directionally from the base to the tip. As the kingfisher performs the extreme hunting habit by water entry in shallow water, it experiences sometimes difficulties, damages of the beak or even death. From Fig. 5b it may be interpreted that the beak of the observed specimen is slightly damaged on the

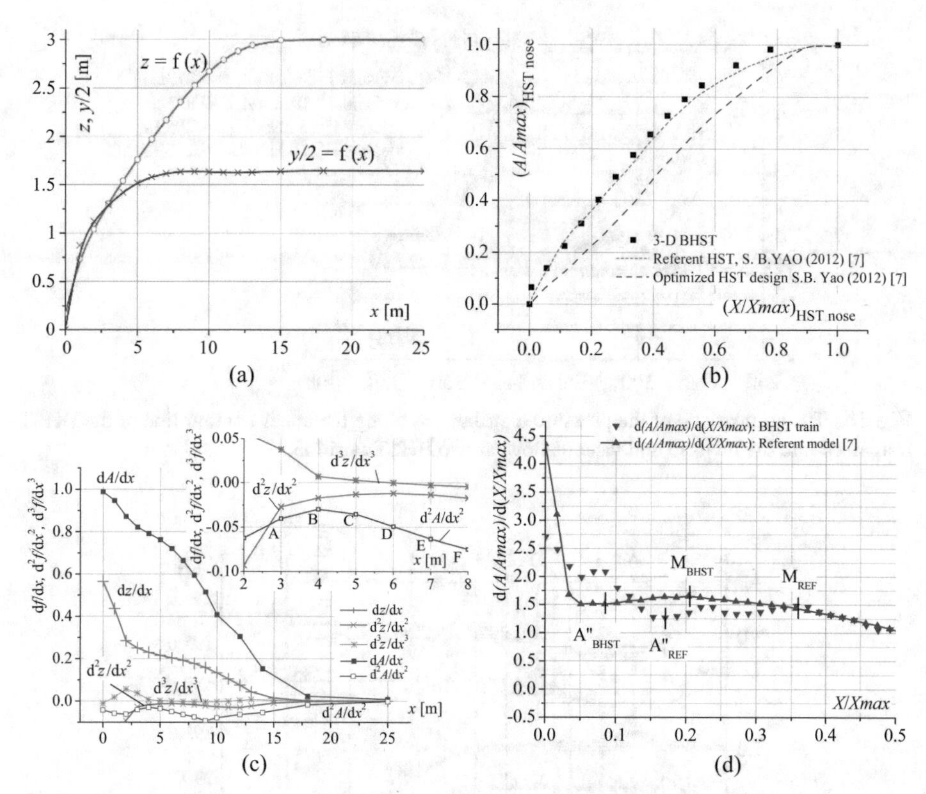

Fig. 9. The contour geometry and the area distribution of the BHST (a) the fitted contours in longitudinal directions [12] (b) the longitudinal area distribution compared with reference values from [7, 12] (c) the three gradients' orders of contours and area (d) the comparison of the normalized area gradients of BHST with reference from [7]

tip, rather than ideal. Thus, the approximation with the parabolic profile is then more justified by the actual specimen. Furthermore, the concept of design-for-drag-reduction, as studied in [5], in this example was adopted in the first place to remain the bionic characteristics of the fore- and after-body and creation of the concept design ready for further optimization.

Afterward, a decision about the air modeling stepped in. The comparison of the pressure values from Fig. 10 shows practically equal CFD results in both cases, viscous and inviscid flows if the BHST is running on the open air. The significant difference occur when the BHST entered the tunnel.

In Fig. 11 the distribution of c_{fx}, T, and p, were presented at the moment when the bionic segment of the fore-body entered into the tunnel [12].

From the previous studies of the forced convective flow over the flat plate [27–29] and the simple high-speed train [13], the analysis of the turbulent boundary layer around the BHST may be aided by the skin friction coefficient and temperature distribution analyze, Fig. 11.

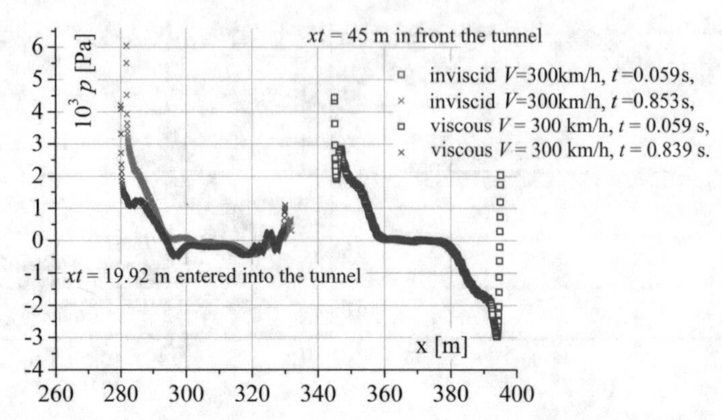

Fig. 10. The comparison of the pressure distributions along the upper contour line of the BHST immersed into the inviscid and viscous flow, at two BHST positions

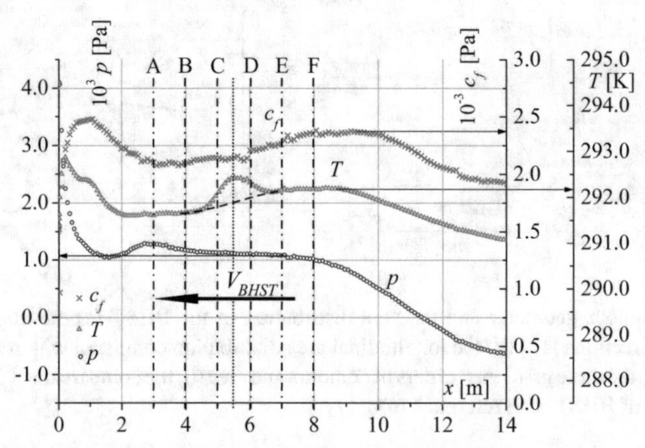

Fig. 11. The flow parameters' distribution along the fore-body bionic segment of the upper contour line that entered into the tunnel, $V = 300$ km/h, $xt = 14$ m, $t = 0.77$ s [12]

This indirect analysis of the boundary layer is only available when the compressible air is modeled with density-based solver since its advantage is in an ability to calculate the local surface temperatures. Therefore, we interpreted that the transition began at the third meter, while the fully turbulent flow occurred about the tenth meter from the nose tip according to the c_f values. The calculated values were slightly noisy, where the recognizable change occurred just before the sixth meter. Afterward, the analysis was broadened to static temperature when, unexpectedly was noticed that temperature bumped, for about 0.6 K–0.7 K, above the expected value and in the interval from 4 to 7.5 m from the nose tip. However, the T change began just in a place where c_f showed the signs of change. The temperature bump was interpreted as a consequence of non-adjusted contour to the temporal flow conditions, seems to be too even, and influenced by the surrounding flow. In further, the pressure was added to consideration, from which was noticed that the pressure had stagnation, in general, over the even contour length. However, a bump of pressure value shows a peak at approximately 2.7 m that is just in place of d^3z/dx^3 function peak from Fig. 9c. If we look backward, one may note that the peak of d^3z/dx^3 is directed by the natural shape, and in the example of the BHST it is not quite adequate. Of course, in this case, the subjectivity in modeling is of great influence also. In addition, it may be commented that set viscosity independence on temperature is justified since the compressibility effects did not make a significant influence. In some other cases, a dependency of viscosity to temperature, for instance after Sutherland function, is advisable For further work, it is recommended to simplify the BHST model by introducing the parabolic contour, in one hand easier for parameterization, and easier for modeling one the second. After the flow analyzes along the upper contour line, the pressure and density distributions around the BHST and by its side are analyzed, Fig. 12.

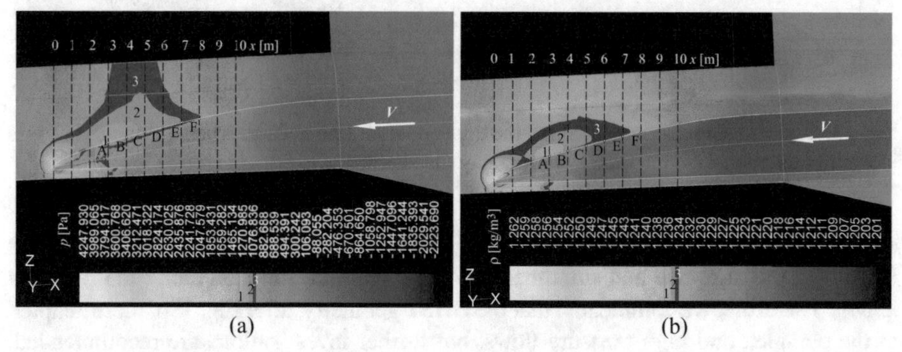

Fig. 12. The distribution of flow parameters over the train and the tunnel walls in case when the BHST bionic nose entered the tunnel, $V = 300$ km/h, $t = 0.77$ s (a) the pressure distribution (b) the density distribution

In Fig. 12a the pressure distribution colored zones (1, 2, 3), outlined with the adjacent iso-pressure lines, point to a sudden pressure change in relatively short length, beyond the nose curve bump. The pressure distribution, around the initial pressure bump in zone 1, looks unexpectedly deformed in length up to point F, where the zone 2 intruded into the field of higher magnitude zone. The pressure behavior looks mirrored onto the narrow zone following the upper contour zone one the BHST. Similarly, the adjacent iso-density lines show the magnitude discontinuity Fig. 12b. Again, we commented that the BHST profile shape and distribution of the cross-sections.

In Fig. 13 two groups of the results are given, one for the configuration when BHST is in front of the tunnel (at distances from the hood of 27 m and 10 m) and the other when it is running through the tunnel.

While approaching the hood, the BHST experienced pressure-loading slight drop over the fore-body, especially between the 5 m–15 m, that may be commented as a consequence of the flow over the contour of very small curvature at a middle part of the forehead, after what the curvature rose. Afterward, the calculations show that the BHST geometry lost influence when approached more to the hood, Fig. 13a. On the contrary, the BHST on the way through the tunnel, from 14 m to 45 m from the tunnel hood, experienced the higher pressure loading than outside the tunnel. The pressure loading over the forehead is similar, with shifted local extremes, until the whole forehead was

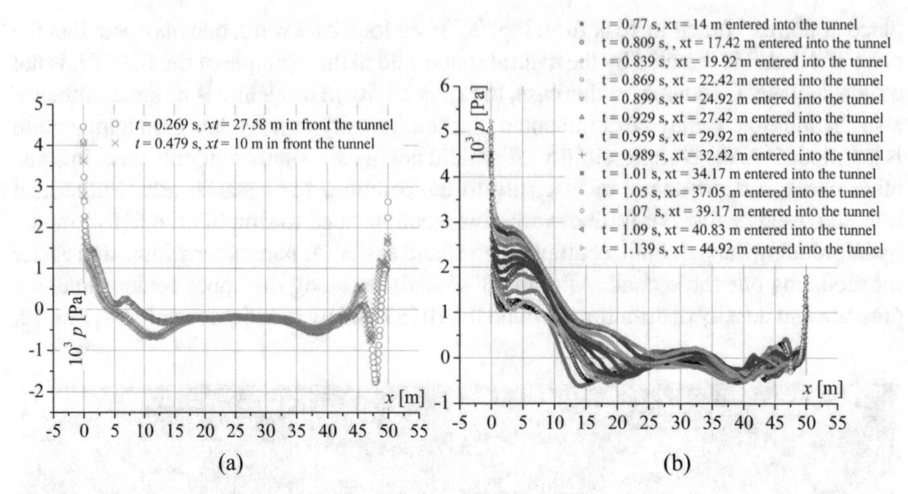

Fig. 13. The pressure distribution along the upper contour line in cases when: (a) the BHST was in front of the tunnel (b) the BHST was passing the tunnel

entered into the tunnel. Movement of the BHST deeper into the tunnel was resulted in gradual pressure rise and smoothening of the pressure function along the forehead length. Therefore, we commented that the BHST geometry, after Fig. 13b, more adapted to the complex and high pressure flows, but further investigations are recommended. Furthermore, the wavy behavior of the pressure along the length is recognized in Fig. 13b, where the shape symmetric of pressures over the front and rear curve transition segments (like in Fig. 13a) is last met when the fore-body bionic part entered into the tunnel. With each new step, the pressure distribution is less influenced by the BHST contour (forehead and central body). One may note that during this period the compression wave is still forming and propagating into the tunnel. In addition, it may be noted that the longer central body is needed to prevent influences of the flow over the train ends thus allowing pressure recovery over the central body, as well as employment of the longer tunnel, at least 1500 m according to [1]. However, the finer mesh may be added to the list of measures for the rising of the results' quality what in general leads to the necessity of the use of the advanced resources.

As presented in [3], the pressure difference for the selected gross dimensions of the high-speed train and the tunnel (type and dimensions) is an almost fixed value in general. The fore-body shape and proportions, and the tunnel entrance type and geometry are the major influencing factors to the pressure derivative behavior over the time of the fore-body tunnel entering [1, 7]. From here, to reduce the decision-making time and avoid the period of the complex flow formation in the tunnel, only the BHST entering is studied.

The summary of the BHST design quality is shown in Fig. 14 throughout the time history, t, of the pressure derivative on time, $\partial p/\partial t$. The measuring place of the pressure was 40 m deep inside the tunnel.

The comparison was made with the 74 m long TGV train, France, which generated $(\partial p/\partial t)_{\text{TGV experiment}} = 8400$ Pa and $(\partial p/\partial t)_{\text{max calculated}} = 9600$ Pa when entering with V

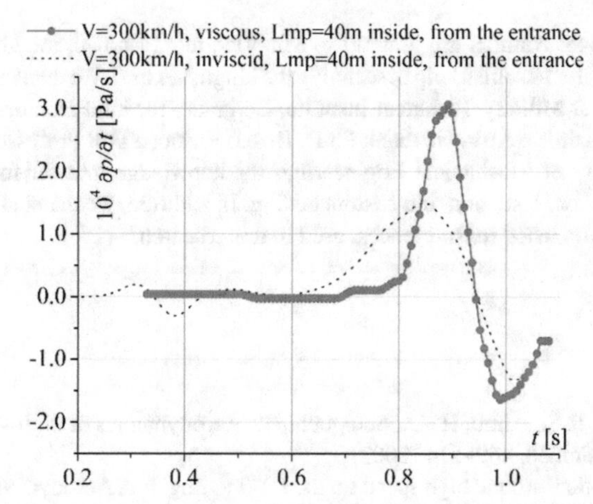

Fig. 14. The time history of the pressure derivative changes for BHST running at $V = 300$ km/h in the inviscid and the viscous flow. The pressure was measured at 40 m inside the tunnel from the entrance [12]

= 220 km/h through the tunnel with $R = 0.21$ (clear cut with rounded edges [39]) [7, 12]. The inviscid solution from this work shows 1.78 times larger $(\partial p/\partial t)_{max}$ value, about 0.13 s later than the $(\partial p/\partial t)_{TGV\ experiment}$, while the viscous solution is 3.6 times larger, and shifted for 0.18 s. Therefore, the $(\partial p/\partial t)_{max}$ values for the BHST are interpreted to be realistic in relation to the reference [7] as the BHST is running faster and blocking the tunnel in a larger degree.

4 Conclusion

This work presented the experimental and numerical methods that aimed designing of the bio-inspired high-speed train concept. The bio-inspiration was found in the kingfisher bird, well-known inspiration for Japan's Shinkansen train. The methods for natural shape recognition and reconstructions are described as well as the idea of train concept design. This work presented the idea of CFD study in a time sequence of high-speed train is running on the open rail and entering the tunnel. The BHST design shows that pure coping of the natural design creates some unexpected flow behaviors at a velocity of 300 km/h that might be avoided by the simplification of the longitudinal – base profile, as it is influenced by subjective view and selection of the bionic and base train parts. On the contrary, the actual design has to be studied further at higher velocities because it might be favorable for such operation ranges. This work opens new doors for employing new technologies as the actual additive manufacturing, in the process of studying the natural specimen. In addition, this work highlighted the possibilities of broader cooperation across science and engineering disciplines that is bringing quality through multidisciplinary knowledge.

Acknowledgments. The authors are thankful for the financial support of this work to the Ministry of Education, Science and Technological Development of the Republic of Serbia, Projects TR

34028 and TR 35045. Authors are grateful to Mr. Vladimir Ivanović, the General Manager of HEXAGON Serbia, for unselfish help in scanning the kingfisher bird. We thank Ph.D. Mirko Kozić, a Research Fellow at Military Technical Institute, Belgrade, for kind mentoring and knowledge transfer in fluid dynamics. Also, we thank Ph.D. Boško Rašuo, a Full Professor at the University of Belgrade, Faculty of Mechanical Engineering, for knowledge transfer in bionic. We thank Slaven Linić, MSc. for IT support and custom coding. In addition, we thank all the authors of the open-source software, listed in References, used in this research.

References

1. Raghunathana, R.S., Kimb, H.-D., Setoguchi, T.: Aerodynamics of high-speed railway train. Prog. Aerosp. Sci. **38**, 469–514 (2002)
2. Baker, C.: The flow around high speed trains. J. Wind Eng. Ind. Aerodyn. **98**, 277–298 (2010). https://doi.org/10.1016/j.jweia.2009.11.002
3. Martínez, A., Vega, E., Gaite J., Meseguer, J.: Pressure measurements on real high-speed trains traveling through tunnels. In: 6th International Colloquium on Bluff Bodies Aerodynamics and Applications, Milano, Italy, 20–24 July, pp. 1–11 (2008)
4. Kikuchi, K., Iida, M., Fukuda, T.: Optimization of train nose shape for reducing micro-pressure wave radiated from tunnel exit. J. Low Freq. Noise Vib. Act. Control **30**(1), 1–19 (2011)
5. Suzuki, M., Nakade, K.: Multi-objective design optimization of high-speed train nose. J. Mech. Syst. Transp. Logist. **6**(1), 54–64 (2013)
6. Yao, S.B., Guo, D.L., Yang, G.W.: Three-dimensional aerodynamic optimization design of high-speed train nose based on GA-GRNN. Sci. China Technol. Sci. **55**, 3118–3130 (2012)
7. Uystepruyst, D., William-Louis, M., Creusé, E., Nicaise, S., Monnoyer, F.: Efficient 3D numerical prediction of the pressure wave generated by high-speed trains entering tunnels. Comput. Fluids **47**, 165–177 (2011)
8. Puharić, M.: Aerodinamička istraživanja vozova (Aerodynamic research of the trains, in Serbian). Institut Goša, Beograd (2010)
9. Lucanin, V., Simic, G., Milkovic, D., Cupric, N., Golubovic, S.: Calculated and experimental analysis of cause of the appearance of cracks in the running bogie frame of diesel multiple units of Serbian railways. Eng. Fail. Anal. **17**(1), 236–248 (2010)
10. Sheppard, S.: Interview - Eiji Nakatsu: Lecture on Biomimicry as applied to a Japanese Train, AUTODESK. goo.gl/fBYHYu. Accessed 07 July 2016
11. Kobayashi, K.: JFS Biomimicry Interview Series: No. 6 "Shinkansen Technology Learned from an Owl?-the Story of Eiji Nakatsu" JFS Biomimicry Interview Series, Japan for Sustainability. https://www.japanfs.org/en/news/archives/news_id027795.html. Accessed 15 Mar 2020
12. Linić, S.: Biomimikrija kao metod aerodinamićkog dizajniranja voza velikih brzina (Biomimicry as a method of the high speed train aerodynamical designing, in Serbian). Dissertation, University of Belgrade, Faculty of Mechanical Engineering, Belgrade, Serbia (2018)
13. Linić, S., et al.: 2017 boundary layer transition detection by thermography and numerical method around bionic train model in wind tunnel test. Therm. Sci. **22**(2), 1137–1148 (2018)
14. Ristic, S., Linic, S., Samardzic, M.: Turbulence investigation in the VTI's experimental aerodynamics laboratory. Therm. Sci. **21**(3), S629–S647 (2017)
15. Raković, M., et al.: Geographic patterns of mtDNA and Z-linked sequence variation in the Common Chiffchaff and the 'chiffchaff complex'. PLoS ONE **14**(1), e0210268 (2019)

16. Puharić, M., Linić, S., Matić, D., Lučanin, V.: Determination of braking force of aerodynamic brakes for high speed trains. Trans. Famena **35**(3), 57–66 (2011)
17. Puharic, M., Matic, D., Linic, S., Ristic, S., Lucanin, V.: Determination of braking force on the aerodynamic brake by numerical simulations. FME Trans. **42**(2), 106–111 (2014)
18. Puharić, M., Lučanin, V., Linić, S., Matić, D.: Research of some aerodynamic phenomenon of high speed trains in low speed wind tunnel. In: The Proceedings of the 3rd International Scientific and Professional Conference "CORRIDOR 10 – A sustainable Way of Integrations", pp. 220–226. The R&D Institute "Kirilo Savic" a.d. and Association of Transport and Telecommunications of the Belgrade Chamber of Commerce, Belgrade, Serbia, 25 October 2012
19. Majstorovic, N., Zivkovic, S., Glisic, B.: The advanced model definition and analysis of orthodontic parameters on 3D digital models. Srpski arhiv za celokupno lekarstvo **145**(1–2), 49–57 (2017)
20. Majstorovic, N., Zivkovic, S., Glisic, B.: Dental arch monitoring by splines fitting error during orthodontic treatment using 3D digital models. Vojnosanitetski pregled **76**(3), 233–240 (2019)
21. Polić-Radovanović, S., Ristić, S., Jegdić, B., Nikolić, Z.: Metodološki i tehnički aspekti primene novih tehnika u zaštiti kulturne baštine (Methodological and technical aspects of the application of new techniques in the protection of cultural heritage, in Serbian). Institute Gosa, Belgrade (2010)
22. Gerbino, S., Del Giudice, D.M., Staiano, G., Lanzotti, A., Martorelli, M.: On the influence of scanning factors on the laser scanner-based 3D inspection process. Int. J. Adv. Manuf. Technol. **84**(9–12), 1787–1799 (2016)
23. Gerbino, S., Staiano, G., Lanzotti, A., Martorelli, M.: Testing the influence of scanning parameters on 3D inspection process with a laser scanner. In: Fischer, X., Daidié, A., Eynard, B., Paredes, M. (eds.) Conference: Joint Conference on Mechanical, Design Engineering and Advanced Manufacturing, Volume: Research in Interactive Design, vol. 4, pp. 314–320. Springer, Toulouse, France (2016)
24. Blanco, D., Fernández, P., Cuesta, E.: Influence of surface material on the quality of laser triangulation digitized point clouds for reverse engineering tasks. In: 2009 IEEE Conference on Emerging Technologies & Factory Automation, Mallorca, pp. 1–8 (2009)
25. Han, L.Q., Wang, Y.: Semiconductor laser multi-spectral sensing and imaging. Sensors **10**(1), 544–583 (2010)
26. Robinson, A., McCarthy, M., Brown, S., Evenden, A., Zou, L.: Improving the quality of measurements through the implementation of customised standards. In: Proceedings of 3rd International Conference on 3D Body Scanning Technologies, Lugano, Switzerland, pp. 235–246 (2012)
27. Cengel, Y.A., Cimbala, J.M.: Fluid Mechanics: Fundamentals and Applications, 2nd edn. McGraw-Hill, Boston (2010)
28. White, F.M.: Viscous Fluid Flow, 2nd edn. McGraw-Hill, New York (1991)
29. Schlichting, H.: Boundary-Layer Theory, 7th edn. McGraw-Hill, New York (1979)
30. Versteeg, H.K., Malalasekera, W.: An Introduction to Computational Fluid Dynamics. The Finite Volume Method. Longman, London (1995)
31. ANSYS Fluent Theory and User guide Documentation. ANSYS Ltd. (2009)
32. Blazek, J.: Computational Fluid Dynamics: Principles and Applications. Elsevier Alstom Power Ltd., Baden-Daettwil (2001)
33. Kozić, M.: Primena numeričke dinamike fluida u aeronautici (Application of numerical fluid dynamics in aeronautics, in Serbia). Naučnotehničke informacije L(3), Military Technilac Institute, Belgrade, Serbia (2013)
34. Stavenga, D.G., Tinbergen, J., Leertouwer, H.L., Wilts, B.D.: Kingfisher feathers – colouration by pigments, spongy nanostructures and thin films. J. Exp. Biol. **214**, 3960–3967 (2011)

35. Hexagon Metrology. ROMER Absolute Arm. Product Brochure. https://www.hexagonmi.com/-/media/Hexagon%20MI%20Legacy/hxrom/romer/general/brochures/ROMER%20Absolute%20Arm_overview_brochure_en.ashx. Accessed 16 Mar 2020
36. Cignoni, P., Callieri, M., Corsini, M., Dellepiane, M., Ganovelli, F., Ranzuglia, G.: MeshLab: an open-source mesh processing tool. In: Scarano, V., De Chiara, R., Erra U. (eds.) Sixth Eurographics Italian Chapter Conference, pp. 129–136 (2008)
37. Blender 2.82 Reference Manual. https://docs.blender.org/manual/en/latest/. Accessed 12 Mar 2020
38. Siemens PLM NX11. Documentation. https://docs.plm.automation.siemens.com/tdoc/nx/11/nx_help/#uid:index. Accessed 28 Feb 2018
39. Open Archives: Tunnel de Villejust (91) sur la ligne nouvelle TGV Atlatique (LN2). openarchives.sncf.com/archive/41116. Accessed 15 Mar 2020

Pre-fermentative Treatment of Grape Juice and Must from Vranec Variety with a Glucose Oxidase from *Aspergillus niger*

Verica Petkova[1](✉), Irina Mladenoska[2], Darko Dimitrovski[1], Trajce Stafilov[2], and Marina Stefova[2]

[1] Department of Food Technology and Biotechnology, Faculty of Technology and Metallurgy, University of Ss. Cyril and Methodius, Skopje, Macedonia
vericapetkova@gmail.com

[2] Institute of Chemistry, Faculty of Natural Sciences and Mathematics, University of Ss. Cyril and Methodius, Skopje, Macedonia

Abstract. High concentration of fermentable sugars in the grapes in the moment of harvest leads to obtaining wines with high concentration of alcohol. Certain grape varieties, such as the Vranec variety, are prone to reaching higher levels of sugar concentration, especially in warm and dry regions. This problem is accentuated in the recent years with the augmentation of the average temperature and shortening of the vegetative season due to global warming. One of the methods for reducing the alcohol level in these wines is by using an enzymatic pre-fermentative treatment with glucose oxidase from Aspergillus niger. This treatment was done on two modalities – crushed grapes (as grape must) and pressed juice. The efficacy of this treatment was demonstrated by measuring the level of gluconic acid in the juice and the must after the treatment using HPLC analysis. In the second modality higher efficiency was obtained, lowering the concentration of glucose for 28%. This treatment affects only part of the glucose, leaving the rest of it and the fructose available for fermentation by the yeasts. The final product is not a completely dealcoholized wine, but wine with lowered alcohol concentration. This type of product can be used for blending with traditionally fermented wines in order to make a correction of the final alcohol level.

Keywords: Glucose oxidase · Vranec · Alcohol reduction

1 Introduction

In Macedonia, the dominant variety, with approximately 50% of the production, is Vranec (*Vitis vinifer* L.). Vranec is an autochthonous grape variety from Montenegro [1], but it is very popular in other countries as well, such as Macedonia and Serbia. This variety is known for giving wines with intensive color and rich structure. However, Vranec is also a variety that is susceptible to reach high concentrations in sugar, which leads to producing wines with very high content of alcohol.

N. Mitrovic et al. (Eds.): CNNTech 2020, LNNS 153, pp. 83–90, 2021.
https://doi.org/10.1007/978-3-030-58362-0_6

Macedonia, as many other countries in the world, is under the influence of global warming. In recent years, on the whole territory of Europe, higher temperatures are shortening the vegetative season and create imbalance between the phenolic maturity and the accumulation of sugar in the grape berries. As a result, the produced wines are with significantly increased alcohol concentration, by about 2% v/v [2, 3]. Other factors that contribute to obtaining higher alcohol levels are healthier vines, leaving grapes on the vine for longer periods, and more efficient yeasts [4].

However, people are more conscious about the negative effects of alcohol, so the market of low-alcoholic and non-alcoholic wines is continually increasing. Low-alcoholic wines can be defined as dealcoholized (0.5% v/v), low alcohol (0.5%–1.2% v/v), and reduced alcohol (1.2% to 5.5%–6.5% v/v), although this can vary between countries [5]. It is proven that high alcohol content can alter the sensory profile having negative effects on fruit aromas in wines and increasing the perception of bitterness, astringency, hotness and roughness [6, 7]. These are some of the reasons for emerging of new technologies that are oriented towards lowering the alcohol levels in wines. This can be achieved by three approaches – removing the alcohol from the final wine (partial dealcoholisation) [8], use of novel yeast strains [9] or mixed strains of yeast [8] that will not use the whole quantity of available sugar for producing alcohol during fermentation, or by pre-fermentation strategies. Pre-fermentation strategies include early grape harvest [10], removal of a portion of fermentable sugar with membranes [11] and enzymatic methods such as utilization of glucose oxidase [12].

Glucose oxidases (GOX) are enzymes that have multiple scientific and industrial purposes [13]. They are FAD dependant enzymes that catalyze the oxidation of β-D-glucose to D-glucono-δ-lactone while producing H_2O_2 [14]. The D-glucono-δ-lactone is then converted into gluconic acid which is not metabolisable by wine yeasts [15].

This enzymatic procedure for removal part of the glucose by glucose oxidase was patented in 1986 by Villettaz [16]. However, most of the research for the use of this procedure in wine was done by Pickering [12, 17–19] on Riesling grape juice. Another type of preparation, in a form of bakery additive Gluzyme Mono® 10,000 BG, was used by Byiela et al. [20] in synthetic grape juice and must from Pinotage. The same enzymatic preparation was used for another trial in 2016 by Rocker et al. [21], but this time again on white grape varieties, Riesling and Pinot Blanc.

The efficiency of glucose oxidation depends on various conditions such as enzyme concentration, pH, oxygen concentration, temperature [12]. Glucose oxidase can be used as a crude enzyme or as an immobilized enzymatic preparation, depending on the purpose and the condition in which it's applied [14]. The immobilization of the enzyme should provide better stability in the medium and also give a possibility for easy removal and reutilization. Glucose oxidase can be immobilized in alginate beads [22] which are compatible for use in food industry.

2 Materials and Methods

2.1 Materials

Grapes from local red variety Vranec, from 2017 vintage, were used for this experiment. The vineyard is from the Gevgelija-Valandovo region in Macedonia. The grapes had

220 g/L fermentable sugars, which gives approximate potential alcohol level of 13%vol. The pH value of 4.00 and total acidity of 2.73 g/L are in accordance to the average values that have been obtained from this vineyard in previous vintages.

The used glucose oxidase (EC 1.1.3.4) which was lyophilized glucose oxidase from *Aspergillus niger.* This pure enzyme produced by Sigma Aldrich (GB) had activity of 252 100 U/g. The used liquid catalyse (Merck, Germany) was obtained from beef liver and had an activity of 1300000 U/mL.

All samples were inoculated with 10 g/hL dry yeast from the species *Saccharomyces cerevisiae* (Excellence XR; Lamothe Abiet, France).

The pre-fermentation treatment was done on two modalities – whole crushed grapes (as grape must) and pressed juice from the same grapes. Both modalities had their control (without enzyme), and two enzymatic treatments – one with glucose oxidase alone (GOX), and the other with a combination of glucose oxidase and catalase (GOX+CAT).

2.2 Enzymatic Reaction and Taking Samples

Reaction was carried out in aerated 300 mL Erlenmeyer flasks. Samples were taken just after crushing/pressing (point zero), every 12 h during 3-day pre-fermentation period, and at the end of the fermentation. All the samples were immediately frozen after sampling in liquid nitrogen.

2.3 Standard Wine Analysis

The standard parameters of the final wines, such as pH, titrable acidity and alcohol were estimated according to the Organisation Internationale de la Vigne et du Vin (OIV 2018). The concentration of iron and copper were also determined.

2.4 HPLC Analysis

The used HPLC system was Agilent 1200 (Hewlett-Packard) with UV VIS and RI detector. The used column was reverse phase RS 18 at room temperature. The composition of the mobile phase was acetonitrile/water in a ration of 80:20. The retention times were: gluconic acid 4.1 min, fructose 5.7 min and glucose 6.2 min (Fig. 1).

2.5 Spectrophotometric Analysis

A spectrophotometric method was used for determination of two indices. Total polyphenol Index (TPI or I280) and Folin-Ciocalteu Index (I_{FC}), both described in the Compendium of International methods of Analysis – OIV (IFC - OIV-MA-AS2-10) and the chromatic properties of final wines such as colour intensity, chromaticity (tint) and relative luminosity [23].

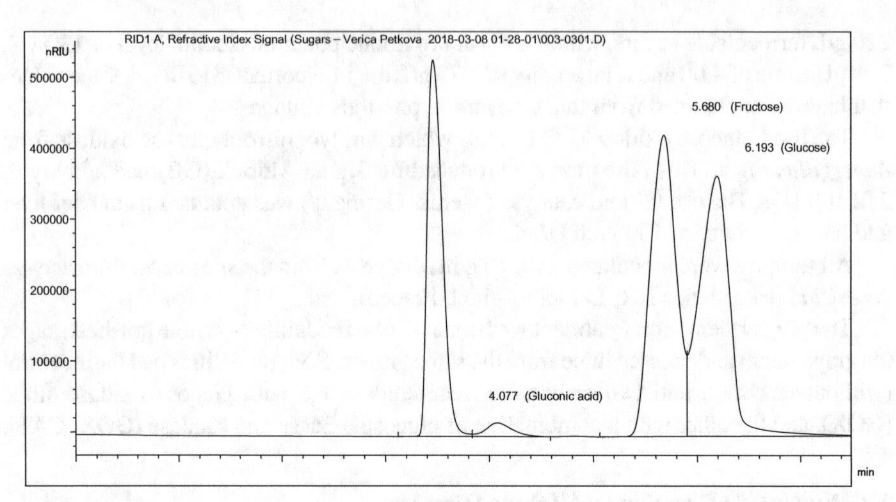

Fig. 1. Callibration chromatograph, showing the retention times of the three analyzed molecules - gluconic acid, fructose and glucose.

3 Results and Discussion

The basic analysis of the grapes (fermentable sugars, total acidity, pH) confirmed that this material is representative of the grapes that are usually harvested from this vineyard. The concentrations of iron and cooper were below the minimal legal authorized amounts, meaning that these metals will not interfere with the fermentation after the treatment.

The two modalities were chosen to represent two types of winemaking techniques. The first modality, the whole crushed grapes, is typical for wine production of red wine which is the one usually chosen for the Vranec variety. The second one, which is the juice without the solid parts of the grape berry (grape skin and seeds) is technology more often used for production of rose and white wine.

As expected, the results obtained for these two modalities were quite different. For the grape must, neither the enzymatic treatment with glucose oxidase alone, nor the one with the combination of glucose oxidase and catalase provide a measurable reduction of sugar concentration. This is probably due to the very complex matrix of the must, especially the presence of high concentration of polyphenols which can inhibit the activity of certain enzymes [24–26].

However, in the essay with the pressed juice, reduction of the sugar concentration was obtained. The juice lacks the solid parts of the grapes, hence the lesser extraction of phenolic compounds. The reduction of the sugar concentration is a result of the glucose concentration reduction, which is the substrate for the enzyme glucose oxidase. However, the level of fructose concentration remains unchanged. A significant result was obtained in the essay with GOX+CAT with 28% decrease in the glucose concentration. This decrease was more than two times higher than the decrease obtained in the essay using only GOX, which was 13% (Fig. 2). According these values for the sugar glucose, the final alcohol concentration should drop from the initially predicted 13% vol to around 10.9% vol and 12.2% vol, accordingly.

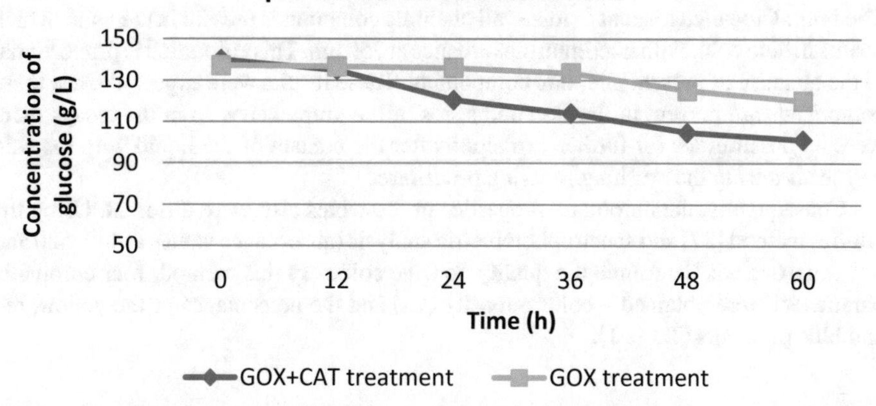

Fig. 2. Concentration of glucose in the grape juice essays after treatment with glucose oxidase (GOX) and combination of glucose oxidase and catalase (GOX+CAT).

The efficiency of the enzymatic treatment can also be demonstrated by the augmentation of the concentration of the gluconic acid. The gluconic acid is the final product of the glucose oxidases enzymatic reaction. In grape juice with GOX+CAT, the concentration of gluconic acid was 1.7 times the concentration in grape juice treated only with glucose oxidase (Fig. 3). Gluconic acid was not present in the control essay without enzymes.

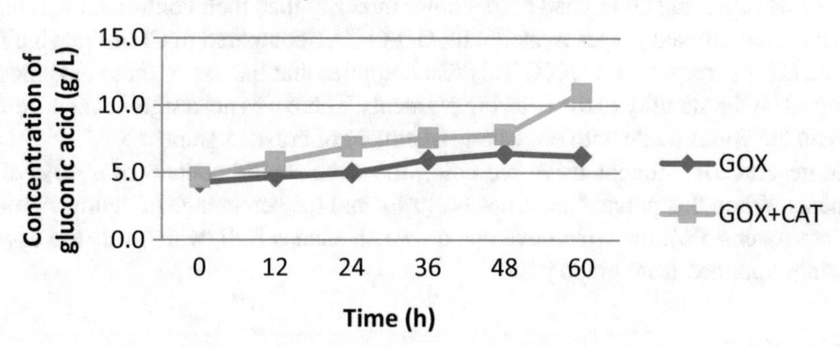

Fig. 3. Concentration of gluconic acid during enzymatic treatment of grape juice with glucose oxidase alone and a combination of glucose oxidase and catalase.

After inoculation with commercial dry yeast, all the wines have finished the fermentation and had less than 2 g/L fermentable sugars. Lower pH values were observed in the wines obtained from fermentation of the grape juice in comparison to those from the grape must.

Wines obtained from the crushed grapes had higher content of phenolic compounds, determined as the total polyphenol index (I_{280}) and the Folin-Ciocalteu Index (I_{FC}). The Folin-Ciocalteu reagent oxidizes all phenolic compounds present in the wine, which results in blue color with maximum absorbance at 750 nm. This coloration is proportional to the quantity of present phenolic compounds. These results were expected since these compounds are present in the skin and seeds of the grape berry, so in the essays there were no possibilities for further extraction after the contact of the liquid with the skin and seeds during the crushing/pressing procedure.

Consequently, the chromatic properties of the wines also were different. Using the Glories method [27] and spectrophotometric analysis (absorbance values at 420, 520 and 620 nm), one can determine the quality of wine color. By this method, four chromatic parameters were obtained – color intensity (CI) and the percentages of the yellow, red and blue pigments (Table 1).

Table 1. Color analysis of the wines.

	Crushed grapes			Pressed grapes		
	Control	GOX	GOX+CAT	Control	GOX	GOX+CAT
% of yellow pigments	35.75	35.58	36.89	53.31	53.71	51.30
% of red pigments	51.42	43.96	45.82	33.32	29.78	31.38
% of blue pigments	12.83	20.46	17.29	13.38	16.51	17.32

In the pressed juice there was a precipitation during the enzymatic treatment ant the fermentation that led to wines with very pale pink color compared to the clear red nuance of the other samples. The color intensity of the wines produced with enzymatic treatment (both GOX+CAT and GOX) had higher color intensity than their control. CI was higher for wine from pressed grapes treated with GOX+CAT compared to CI of the wine from pressed grapes treated with GOX. This result applies that the use of these enzymes has an impact on the stability of the coloring pigments. There was not a significant difference between the wines made with enzymatic treatment of crushed grapes.

In general case, for intensive red color, the wine should contain over 40% of red pigments. When this percentage drops (<40%), and the percentage of yellow pigments increase (over 45%), the wine develops brownish shades [28], which was the case for the wines obtained from grape juice.

4 Conclusion

The enzymatic treatment with glucose oxidase can have interesting applications in the process of winemaking. Although, the enzyme is not active enough when used in grape must, this problem may be overcome by using higher concentration of the pure enzyme. However, this pre-fermentative treatment shows good results for treating pressed grape juice. The wines obtained with this method have lower alcohol concentration and higher pH value, but the enzymatic activities influence the stability of the coloring pigments, leading to more pale and brownish color. The main purpose for treated wines would be blending them with traditionally made red wines in order to make a correction of the final alcohol concentration.

References

1. Maras, V., et al.: Research of origin and work on clonal selection of Montenigrin grapevine varieties cv. vranac and cv. kratosija. Agroznanje **13**(1), 103–112 (2012)
2. Jones, G.V., White, M.A., Cooper, O.R., Storchmann, K.: Climate change and global wine quality. Clim. Change **73**, 319–343 (2005).
3. Mira de Orduña, R.: Climate change associated effects on grape and wine quality and production. Food Res. Int. **43**, 1844–1855 (2010)
4. Conibear, H.: Rising alcohol levels in wine: is this a cause for concern? AIM Digest. **18**(4), 1–3 (2008)
5. Pickering, G.J.: Low and reduced-alcohol wine: a review. J. Wine Res. **11**(2), 129–144 (2000)
6. Goldner, M.C., Zamora, M.C., Di Leo Lira, P., Gianninoto, H., Bandoni, A.: Effect of ethanol level in the perception of aroma attributes and the detection of volatile compounds in red wine. J. Sens. Stud. **24**, 243–257 (2009)
7. King, E.S., Dunn, R.L., Heymann, H.: The influence of alcohol on the sensory perception of red wines. Food Qual. Prefer. **28**, 235–243 (2013)
8. Rolle, L., et al.: Alcohol reduction in red wines by technological and microbiological approaches: a comparative study. Aust. J. Grape Wine Res. **24**, 62–74 (2018)
9. Varela, C., et al.: Evaluation of gene modification strategies for the development of low-alcohol-wine yeasts. Appl. Environ. Microbiol. **78**(17), 6068–6077 (2012)
10. Kontoudakis, N., Esteruelas, M., Fort, F., Canals, J.M., Zamora, F.: Use of unripe grapes harvested during cluster thinning as a method for reducing alcohol content and pH of wine. Aust. J. Grape Wine Res. **17**(2), 230–238 (2011)
11. García-Martín, N., et al.: Sugar reduction in musts with nanofiltration membranes to obtain low alcohol-content wines. Sep. Purif. Technol. **76**(2), 158–170 (2010)
12. Pickering, G.J., Heatherbell, D.A., Barnes, M.F.: Optimising glucose conversion in the production of reduced alcohol wine using glucose oxidase. Food Res. Int. **31**(10), 685–692 (1998)
13. Moorthi, P.S.: Optimization, characterization and applications of glucose oxidase produced from Aspergillus awamori MTCC 9454 for food processing and preservation. Ph.D. thesis, Dr. M.G.R. Educational and Research Institute University (2009)
14. Bankar, S.B., Bule, M.V., Singhal, R.S., Ananthanarayan, L.: Glucose oxidase—an overview. Biotechnol. Adv. **27**(4), 489–501 (2009). https://doi.org/10.1016/j.biotechadv.2009.04.003
15. Van Rensburg, P., Pretorius, I.S.: Enzymes in winemaking: harnessing natural catalysts for efficient biotransformations-A review. S. Afr. J. Enol. Vitic. **21**, 52–73 (2000)
16. Villettaz, J.-C.: Patent EP 0194043 A1: method for production of a low alcoholic wine and agent for performance of the method. In: Volume EP 0194043 A1. Novo Nordisk A/S, Denmark (1986)
17. Pickering, G.J., Heatherbell, D.A., Barnes, M.F.: The production of reduced-alcohol wine using glucose oxidase-treated juice. Part I. Composition. Am. J. Enol. Vitic. **50**(3), 291–298 (1999)
18. Pickering, G.J., Heatherbell, D.A., Barnes, M.F.: The production of reduced-alcohol wine using glucose oxidase-treated juice. Part II. Stability and SO2-binding. Am. J. Enol. Vitic. **50**(3), 299–306 (1999)

19. Pickering, G.J., Heatherbell, D.A., Barnes, M.F.: The production of reduced-alcohol wine using glucose oxidase-treated juice. Part III. Sensory. Am. J. Enol. Vitic. **50**(3), 307–316 (1999)
20. Biyela, B.N., Du Toit, W.J., Divol, B., Malherbe, D.F., Van Rensburg, P.: The production of reduced-alcohol wines using Gluzyme Mono® 10.000 BG-treated grape juice. S. Afr. J. Enol. Vitic. **30**(2), 124–132 (2009)
21. Rocker, J., Schmitt, M., Pasch, L., Ebert, K., Grossmann, M.: The use of glucose oxidase and catalase for the enzymatic reduction of the potential ethanol content in wine. Food Chem. **210**, 660–670 (2016)
22. Ruiz, E., Busto, M.D., Ramos-Gómez, S., Palacios, D., Pilar-Izquierdo, M.C., Ortega, N.: Encapsulation of glucose oxidase in alginate hollow beads to reduce the fermentable sugars in simulated musts. Food Biosci. **24**, 67–72 (2018)
23. European Community.: Community methods for the analysis of wine. Commission Regulation (ECC) No. 26/76/90 of 17 September 1990. Off. J. Eur. Commun. **33**(L272), 1–191 (1990)
24. Yu, J., Ahmedna, M.: Functional components of grape pomace: their composition, biological properties and potential applications. Int. J. Food Sci. Technol. **48**(2), 221–237 (2013)
25. McDougall, G.J., Stewart, D.: The inhibitory effects of berry polyphenols on digestive enzymes. BioFactors **23**(4), 189–195 (2005)
26. Wittenauer, J., Mäckle, S., Sußmann, D., Schweiggert-Weisz, U., Carle, R.: Inhibitory effects of polyphenols from grape pomace extract on collagenase and elastase activity. Fitoterapia **101**, 179–187 (2015)
27. Glories, Y.: La Couleur Des Vins Rouges. 2éme Partie: Mesure, Origine Et Interpretation. Connaissance De La Vigne Et Du Vin. **18**, 253–271 (1984)
28. Poiana, M.-A., Moigradean, D., Gergen, I., Harmanescu, M.: The establishing of the quality of red wines on the bases of chromatic characteristics. J. Agroaliment. Process. Technol. **8**(1), 199–208 (2007)

Computing Aerodynamic Performances of Small-Scale Vertical-Axis Wind Turbines: Possibilities and Challenges

Jelena Svorcan(✉), Marija Baltić, and Ognjen Peković

Faculty of Mechanical Engineering, University of Belgrade, 11120 Belgrade, Serbia
jsvorcan@mas.bg.ac.rs

Abstract. The past few decades have been marked by an immense interest of the scientific community in making better use of renewable energy resources, particularly wind energy. One of the suggestions is to increase the number of small-scale vertical-axis wind turbines (VAWTs) in densely populated areas. Given that VAWTs primarily operate in adverse conditions (irregular wind speeds, Earth's boundary layer, vortex trail of surrounding objects), it is necessary to pay special attention to the numerical and experimental estimation of their aerodynamic performances. The conceptual design of small-scale wind turbines usually begins with detailed simulations of the encompassing flow field. There are three categories of most often employed computational methods for Darrieus-type VAWTs: *i*) quasi 1D momentum models frequently upgraded by blade element theory, *ii*) vortex models and *iii*) computational fluid dynamics (CFD) approach that enables modeling the complete flow field. Each of these methods is founded on a particular set of assumptions, has its advantages and disadvantages and can provide different numerical results, although engineers are usually mostly interested in accurately estimating power coefficient curves. This paper describes and references some of the realized and tested small-scale VAWTs. It also accentuates modeling possibilities as well as several dilemmas, challenges and errors that may appear while performing complex, unsteady aerodynamic analyses of fluid flows encompassing VAWTs. The main goal of the paper is to provide useful guidelines for successful design of small-scale VAWTs for urban environments.

Keywords: Vertical-axis wind turbines · Flow simulation · Power coefficient

1 Introduction

Driven by the ever-increasing inclination of the contemporary society towards the clean, renewable and cost-effective energy resources, numerous and diverse investigations of possible locations and wind turbines have been performed in the past few decades. As wind turbine's main purpose is to convert wind kinetic energy into usable electric energy, year 2018 was marked by an additional 49 GW being installed worldwide (of which approximately 11.5 GW is in Europe) as reported by International Renewable

N. Mitrovic et al. (Eds.): CNNTech 2020, LNNS 153, pp. 91–111, 2021.
https://doi.org/10.1007/978-3-030-58362-0_7

Energy Agency (IRENA) [1]. A comprehensive review of wind resource assessment, as a necessary prerequisite for wind turbines, can be found in [2]. Studies dealing with wind energy resources in Serbia identified eastern regions, especially around Vrsac, Danube basin east end and mountain tops, as the most favorable with average annual wind power density exceeding 300 W/m^2 [3] which is today reflected in several functioning wind farms across south-east Banat.

Wind turbines can be classified into horizontal- or vertical-axis structures depending on the direction of their rotational axis. Although horizontal-axis wind turbines (HAWTs) are much more employed for their higher efficiency, vertical-axis wind turbines (VAWTs) are also becoming more and more interesting for their omnidirectional operability, decreased noise, less complicated design since yaw mechanisms are not required, cheaper production of geometrically simpler blades and facilitated maintenance since almost all the components are located near the ground. Here also, we differentiate between drag- and lift-type structures. Brief descriptions of both categories can be found in [4, 5], while Bhutta et al. [6] provide a review of various configurations and design techniques applied to VAWTs. Figure 1 illustrates a typical straight-blade Darrieus-type VAWT that usually consists of 2–4 blades attached to a vertical rotating shaft with transmission, generator and all other necessary systems being located at the bottom of the turbine.

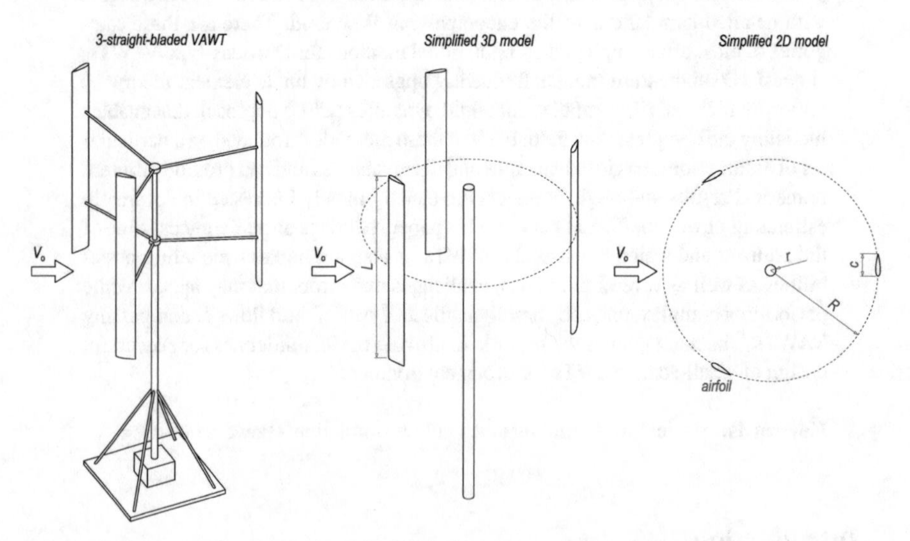

Fig. 1. Typical straight-bladed Darrieus-type VAWT.

Special attention should be paid to the estimation of aerodynamic performances of the VAWT rotor since it is the first element in the energy conversion chain. Therefore, its efficiency directly affects the operation of all the subsequent components of the complex wind turbine system. The aerodynamic analysis of VAWTs can be quite challenging since blades undergo a wide range of angles-of-attack during every revolution. This introduces numerous flow instabilities, separation and interaction between the oncoming blade and the vortex trail detached from the previous blade. Those are just some of the main reasons for the somewhat decreased efficiency of this type of wind turbines in comparison to HAWTs. However, their operability in low and changeable winds as well as reduced noise make them a very suitable choice for small and/or autonomous consumers in urban and densely populated areas. As observed by Tummala et al. [7], when designed properly and used at their optimal conditions, small-scale wind turbines become a reliable source of energy of particular socio-economic importance for the developing countries. Some views on the progress made recently in the utilization of small-scale wind turbines in urban habitat can also be found in [8].

The conceptual design of small-scale wind turbines usually begins with detailed simulations of the encompassing flow field. For that purpose, various mathematical models have been developed and used. They can roughly be divided into computationally less expensive potential flow models (including momentum and vortex models) and Navier-Stokes based computational fluid dynamics (CFD) models that take more physical details into account. These categories differ in starting assumptions, complexity, time required for computing as well as diversity and accuracy of final results. A very good review of simpler computational models for Darrieus-type straight-bladed VAWT together with accompanying general mathematical expressions can be found in [5, 9]. Momentum models serve to solve the decrease of velocity along the streamtube by equating the streamwise aerodynamic force on the blades with the rate of change of the air momentum. They are able to provide satisfactory averaged values of interference factors (that quantify the change of velocity), aerodynamic loads, torque and power in a short amount of time which still makes them quite popular, although they were first developed nearly 50 years ago. Different variants of momentum models can be extremely useful for preliminary analysis and optimization studies. Some examples can be found in [9–14]. Vertex models, on the other hand, regard the flow field through a distribution of vortices that represent both the moving blades (bound vortex) as well as the wakes behind them (free vortex). Each vortex is described by its intensity/strength, i.e. circulation, and induces velocity in every spatial point according to the Biot–Savart law. This category of models was also first applied to VAWTs in mid-seventies [5]. Generally, it enables considering the fluid flow as transient, capturing the instantaneous loads as well as performing wake analysis, but still incorporates significant simplifications and can require substantial computational resources since the wake develops in time. Some examples of its recent usage can be found in [12, 15–17].

Islam et al. [5] concluded that, ten years ago, the most widely used models were the aforementioned double-multiple streamtube model, free-vortex model and cascade model. However, the past decade has been marked by an increased number of CFD studies of VAWTs by finite volume method (FVM). Although this approach is more complex than momentum and vortex models, it enables considering the complete, transient, viscous (turbulent) fluid flow around the wind turbine. Diverse output results include full pressure, velocity, vorticity fields, etc. Both two- and three-dimensional analyses are almost equally employed as presented in [18–28]. However, although CFD approach is a powerful tool that enables simulating complex flow phenomena such as dynamic stall [29, 30], it is still not completely perfected for VAWTs, particularly for real i.e. variable and non-optimal working regimes, some examples are given in [31, 32]. There are still many aspects and features that should be carefully handled and this paper serves to point out to some of the possible challenges and errors that may appear while performing complex, unsteady aerodynamic analyses of flow fields encompassing VAWTs. Whenever possible, final validation of the utilized computational model should be obtained through comparison with available experimental data.

2 VAWT Geometry and Basic Parameters

Before continuing to detailed descriptions of computational models, the fundamental characteristics of VAWTs should be clearly defined. Rotor depicted in Fig. 1 contains $N_b = 3$ blades that are determined by their cross-section (usually a symmetric airfoil), chord c and length L. Rotor radius is R, shaft radius is r and its angular velocity is Ω. The oncoming, undisturbed wind speed is denoted by V_o. Rotor solidity is defined here as $\sigma = N_b c/R$. Another important dimensionless parameter is tip-speed-ratio $\lambda = R\Omega/V_o$ that corresponds to the operational regime.

Average mechanical power P generated by the turbine is estimated as the product of average torque Q per one rotation and angular velocity Ω. Power coefficient C_P, that indicates the efficiency of the rotor i.e. what part of available wind power was actually extracted, is computed as:

$$C_P = \frac{P}{0.5\rho AV_o^3} = \frac{Q\Omega}{0.5\rho AV_o^3} = \frac{Q}{0.5\rho AV_o^2 R}\frac{R\Omega}{V_o} = C_Q\lambda, \tag{1}$$

where A and ρ represent rotor reference area $A = DL$ and air density, respectively, while C_Q is the torque coefficient.

3 Simpler Computational Models Assuming Potential Flow

Although structurally simple, VAWTs can be quite tricky when it comes to their aerodynamic analysis due to the inherent flow unsteadiness, presence of turbulence, periodic changes of aerodynamic loads, boundary layer separation, dynamic stall, interactions between the blade and the wake, etc. During a single revolution, blade's angle-of-attack $\alpha(\psi)$ may change significantly with the angular coordinate ψ, thus causing diverse operating conditions, ranging from quite favorable to adverse. For that reason, we often divide the rotor into two halves, upstream and downstream, where the upstream zone can be considered as more beneficial (i.e. the contribution to the torque is greater). On the other hand, since velocities are small and air flow is periodical at operating conditions close to nominal, it is possible to assume potential flow that is incompressible, inviscid and irrotational, often coupled with small semi-empirical corrections, in the first approximation.

3.1 Single Streamtube (SST) Model

We start with the simplest approach [4]. Quasi-1D momentum models are based on the combination of conservation laws written for a streamtube and estimation of aerodynamic force acting on the blade from the known aerodynamic coefficients (i.e. 2D characteristics of the cross-sectional airfoil). We usually consider planar fluid flow where blades follow circular paths. Instantaneous blade position is determined by angular coordinate ψ. Depending on how many streamtubes are used to describe the flow (one or more) and whether each streamtube is considered as a whole or is split into upstream and downstream halves, we may differentiate between "single streamtube" and "double-multiple streamtube" models.

In its simplest form, momentum model assumes that the entire rotor belongs to a single streamtube as illustrated in Fig. 2a and that deceleration across the rotor is constant and marked by induction or interference factor a meaning that the induced axial velocity at the blade is $V_o(1-a)$. Relative velocity V squared at position ψ is then:

$$V^2 = [\Omega R + (1-a)V_o \sin\psi]^2 + [(1-a)V_o \cos\psi]^2, \tag{2}$$

where decelerated wind velocity $V_o(1-a)$ is projected into tangential ($\sin\psi$) and normal ($\cos\psi$) component with respect to the blade chord that is perpendicular to the rotor radius. In dimensionless form:

$$\frac{V}{V_o} = \sqrt{[\lambda + (1-a)\sin\psi]^2 + [(1-a)\cos\psi]^2}. \tag{3}$$

Local angle-of-attack α is then:

$$\alpha = \text{arctg}\frac{(1-a)\cos\psi}{\lambda + (1-a)\sin\psi}. \tag{4}$$

Since the total change in wind velocity infinitely fore and aft of the rotor is twice the reduction achieved at the rotor, it is possible to compute the change of air momentum per unit length as:

$$T = \dot{m}\Delta V = 4\rho Ra(1-a)V_o^2. \tag{5}$$

On the other hand, this force T also equals the averaged value of the axial aerodynamic force acting on all N_b blades during a single revolution (observe that only lift force is considered here, while drag force, as much smaller, is neglected):

$$T = \frac{N_b}{2\pi}\int_0^{2\pi} \frac{1}{2}\rho V^2 c_L \cos(\alpha+\psi) c\, d\psi. \tag{6}$$

By equating Eqs. (5) and (6), introducing a useful replacement $y = \lambda/(1-a)$ and performing adequate transformations, it is possible to reach an implicit equation:

$$\frac{1}{1-a} = 1 + \frac{1}{8}\frac{\sigma}{2\pi}\int_0^{2\pi}\left[(y+\sin\psi)^2 + \cos^2\psi\right] c_L \cos(\alpha+\psi)\, d\psi, \tag{7}$$

that can be solved iteratively. It should be borne in mind that local angle-of-attack α depends on tip-speed ratio λ, interference factor a and angular coordinate ψ, while lift coefficient c_L changes with the angle-of-attack α:

$$\alpha = \text{arctg}\frac{\cos\psi}{y+\sin\psi},\ c_L = c_L(\alpha). \tag{8}$$

Once a is determined, it is easy to estimate local tangential force F_T per unit length:

$$F_T = \frac{1}{2}\rho V^2 c(c_L \sin\alpha - c_D \cos\alpha). \tag{9}$$

Generated torque Q can be computed as:

$$Q = N_b H R F_T. \tag{10}$$

Finally, average power P is:

$$P = \Omega\frac{1}{2\pi}\int_0^{2\pi} Q d\psi. \tag{11}$$

3.2 Double-Multiple Streamtube (DMST) Model

Double-multiple streamtube (DMST) model, formulated by Paraschivoiu [10], is founded on the same assumptions regarding the flow, with the difference that more streamtubes are used to represent the rotor and that each streamtube is divided into two parts, upstream $_u$ and downstream $_d$. This requires the iterative solution of two interference factors, $a_u = V_u/V_o$ and $a_d = V_d/V_e$, per each streamtube as illustrated in Fig. 2b, since wind velocity decreases several times along the streamtube by interacting with the blade. The change of air momentum can be written separately for upstream and downstream parts:

$$\begin{aligned}\overline{\Delta F_{x,u}} &= (\rho V_u \Delta A)(V_o - V_e) = \frac{\rho}{2}\left(V_o^2 - V_e^2\right)\Delta A,\\ \overline{\Delta F_{x,d}} &= (\rho V_d \Delta A)(V_e - V_w) = \frac{\rho}{2}\left(V_e^2 - V_w^2\right)\Delta A.\end{aligned} \tag{12}$$

The axial aerodynamic force acting on the blade is now:

$$\overline{\Delta F_{x,u/d}} = \frac{N_b \Delta\psi}{2\pi}(\Delta F_N \cos\psi - \Delta F_T \sin\psi). \tag{13}$$

The relative velocities and angles-of-attack can be written as:

$$W_{u,d} = V_{u,d}\sqrt{(\lambda - \sin\psi)^2 + (\cos\psi)^2}, \alpha = \arcsin\left(\cos\psi\frac{V_{u,d}}{W_{u,d}}\right). \tag{14}$$

Again, by equating Eqs. (12) and (13) and after performing some mathematical manipulation, it is possible to obtain implicit equations for the computation of two interference factors as:

$$\begin{aligned} a_{i+1} &= f(a_i) = \frac{1}{1 + G(a_i)},\\ G(a_i) &= \frac{N_b c}{8\pi R|\cos\psi|}[c_N \cos\psi - c_T \sin\psi]\left(\frac{W_{u,d}}{V_{u,d}}\right)^2. \end{aligned} \tag{15}$$

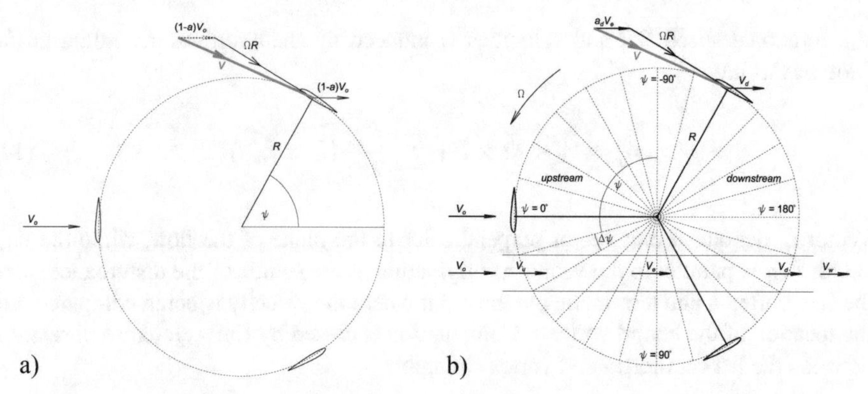

Fig. 2. Characteristic velocities and angles in: a) SST model, b) DMST model.

The model can be improved by considering the effects of the central shaft (as a potential flow around a rotating cylinder) or airfoil dynamic stall through semi-empirical corrections. It can also be made 3D by solving fluid flows in several cross-sectional planes along the length of the blade and incorporating losses from the tip vortices. However, some disadvantages of this class of computational models stem from the fact that they solve the flow along a single, axial direction. This implies that they do not provide reliable results in cases of higher rotor solidities and tip-speed ratios when interactions between blades and their wakes are strong and cross-flow is significant. That is why some more complex models are also used.

3.3 Vortex Models

This category of numerical approaches also assumes potential flow and serves to compute the velocity field around the rotor induced by the vortices that are meant to represent the blades and the wake. There are many variants, so here, free-vortex model that allows the wake to evolve during computation, as one of the most prevalent, is explained in more detail [15].

The effect of each blade j is replaced by a bound vortex of strength $\Gamma_{B,j}$ located at the first quarter of the airfoil chord. The wake trailing after each blade is presented by an array of free i.e. shed vortices $\Gamma_{S,1}^{j}$ as illustrated in Fig. 3. The intensities of free vortices are determined once, at the moment of their shedding, as the change in the circulation of the preceding bound vortex due to its revolution, in accordance with the Kelvin theorem:

$$\Gamma_{S,1}^{j} = \Gamma_{B,j}^{n+1} - \Gamma_{B,j}^{n} = \Gamma_{B,j}^{n+1} - \sum_{i=2}^{N} \Gamma_{S,i}^{j}. \tag{16}$$

The model assumes that intensities of the free vortices inside the wake do not change, but only travel downstream. However, each new computational step, i.e. angular increment $\Delta\psi$, induces the creation of new shed vortices, thus constantly increasing the wake. Relative velocity W_j at blade j is computed as the vector sum of undisturbed wind speed

V_o, rotational speed ΩR and velocities v_i induced by shed vortices according to the Biot–Savart law:

$$\vec{W}_j = \vec{V}_o + \vec{\Omega} \times \vec{r} + \sum_{i,j} \frac{\Gamma^j_{S,i}}{2\pi r} (\vec{z}_o \times \vec{r}_{o,ij}). \tag{17}$$

Vector $\vec{z}_o$ denotes a unit vector perpendicular to the plane of the flow, while the unit vector $\vec{r}_{o,ij}$ is parallel to the vector $\vec{r}_{ij}$ originating at the source of the disturbance, here the free vortex i, and terminating at the point where the velocity is being computed, i.e. the location of the bound vortex j. Computation is closed by Kutta–Joukowski relation between the lift coefficient and vortex strength:

$$\Gamma_{B,j} = \frac{1}{2} c_{L,j} W_j c. \tag{18}$$

At the end of each iteration, the wake should be allowed to move, which requires the computation of instantaneous velocities of each free vortex (as the sum of undisturbed wind speed and velocities induced by all the vortices in the fluid flow) according to the expression:

$$\vec{W}_k = \vec{V}_o + \sum_j \frac{\Gamma_{B,j}}{2\pi r} (\vec{z}_o \times \vec{r}_{o,jk}) + \sum_{i \neq k,j} \frac{\Gamma^j_{S,i}}{2\pi r} (\vec{z}_o \times \vec{r}_{o,ik}). \tag{19}$$

Now, the new position of each free vortex can be determined:

$$\vec{r}^{\,n+1}_k = \vec{r}^{\,n}_k + \vec{W}_k \Delta t, \tag{20}$$

and the computational cycle can be repeated.

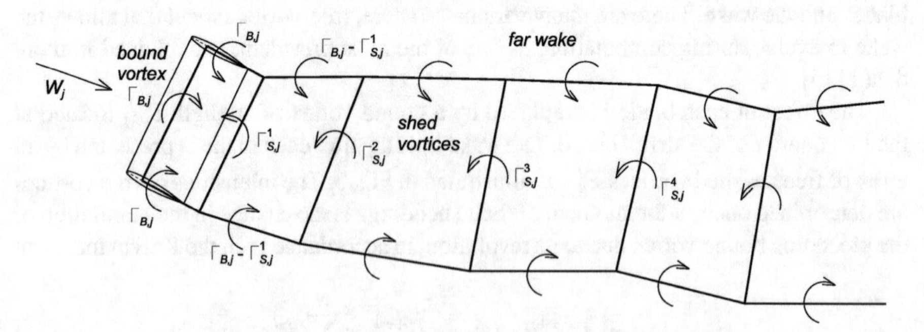

Fig. 3. Illustration of the blade segment and vortex wake.

Generally, this model enables obtaining transient and more accurate wake representation at medium values of tip-speed ratio. At higher λ, that corresponds to greater angular velocity Ω and slower wind speed V_o, the wake travels downwards too slowly so the interaction between the blades and the wake is too strong to be captured accurately by this potential flow approach. Another significant disadvantage is that required computational resources constantly increase thus prolonging the simulation. Also, if airfoil shape is to be modeled with satisfactory precision, a greater number of bound vortices for each blade should be employed. These detriments paved the path for the CFD simulations.

4 CFD Techniques Applied to VAWTs

Today, fluid flows around VAWTs are usually resolved by FVM that enables considering unsteady, turbulent, 3D flows. Its application range is somewhat wider than for momentum and vortex models since it can likewise be applied to rotors of greater solidity, shorter blades that induce greater losses, new airfoils, non-optimal working regimes, etc. Also, a greater amount of diversified output data can be obtained. The greatest problem that should be solved is how to adequately model the rotational motion of the blades.

4.1 Fundamentals of CFD Methodology

CFD approach, also based on the fundamental conservation laws, in particular mass, momentum and energy, implies solving flow quantities (pressure, three components of velocity, temperature, etc.) across the control volume. For Newtonian fluid, we can write:

$$\begin{aligned}
&\frac{\partial \rho}{\partial t} + \nabla \cdot (\rho \vec{v}) = 0, \\
&\rho \frac{\partial \vec{v}}{\partial t} + \rho(\vec{v} \cdot \nabla)\vec{v} = \rho \vec{F} - \nabla p + \mu \Delta \vec{v} + \frac{1}{3}\mu \nabla \mathrm{div}\, \vec{v}, \\
&\rho \frac{\partial e}{\partial t} + \nabla \cdot \left(\rho e \vec{v} + \bar{\bar{p}} \cdot \vec{v}\right) = -\nabla \cdot \left(\sum_j \vec{q}_j\right).
\end{aligned} \tag{21}$$

As these equations generally cannot be solved analytically, with some additional assumptions they can be solved numerically, and most often by FVM that solves the equations governing the fluid flow in integral form. The main reasons for its widespread usage are: applicability to complex geometries, conservativeness as well as the facts that it does not require transformations of coordinates and that all members appearing in the equations have physical meaning, hence it can be understood and implemented relatively easily [33]. The basic steps of a successful CFD simulation are: preparation of geometric model, generation of finite volume mesh across the computational domain, proper numerical set-up, computation and post-processing that requires analysis and validation of extensive results.

There are several possible ways to model the rotational motion of the wind turbine rotor differing in complexity and physicality. They range from quasi-stationary approaches when certain inertial terms are simply added to the equations, to the more advanced, transient ones, that involve the actual movement of the mesh nodes according to the specified rule (i.e. translation or rotation) [34]. Although the shape of the finite volumes does not change, this approach is computationally expensive since, in addition to solving the flow equations, it is necessary to compute the new coordinates of the mesh nodes in every time-step. However, it does enable the consideration of the transient interaction between the two (or more) zones – moving and stationary, and provides more accurate results.

A portion of work performed on simulating fluid flow around VAWTs by other authors can be summarized as follows. Howel et al. [18] performed a combined experimental and both 2D and 3D aerodynamic study of two-bladed and three-bladed small-scale VAWT rotor. They concluded that 3-bladed rotors outperform the 2-bladed ones, 2D analysis provides significantly higher power coefficients and that the effect of tip vortices should not be neglected. Castelli et al. [19, 20] proposed a new performance prediction model that combines BEM theory and CFD approach for both straight and helical bladed VAWTs. Nini et al. [21] performed 3D simulations of aerodynamic field around a 3-bladed VAWT at two tip-speed ratios. The system of flow governing equations was closed by Spalart–Allmaras turbulence model and dynamic stall was observed. Sun et al. [22] placed particular emphasis on the effects of interaction between vortices and blades on the aerodynamic performances of the simulated rotor and concluded that shed vortices can have considerable impact on the following blade. Balduzzi et al. [23] performed high-performance computations using 16000 processors in order to better understand the unsteady aerodynamics of Darrieus wind turbines and revealed that 3D flow effects are much more complex that assumed by the lower-fidelity models. Elsakka et al. [24] proposed a fast and accurate method for the calculation of the variable angle-of-attack from the computed velocity field, since angle-of-attack cannot be unambiguously computed in CFD simulations as it is possible when simpler models are used. Franchina et al. [25] performed both 2D and 3D computations of VAWT flow field while considering the effects of the domain dimensions, types of assigned boundary conditions, space and time resolutions and numerical accuracy. Guillaud et al. [26] performed large eddy simulations (LES) of vertical-axis hydrokinetic turbines of different solidities. Furthermore, the performances of real turbine vs. the idealized composed only of blades are compared. It is shown that blade tips and arms of low-solidity rotors greatly increase losses and that there are no particular gains in decreased solidities. Su et al. [27] used CFD approach to investigate the possible performance improvements of V-shaped blades in VAWTs. The results indicate a possible enhancement in power coefficient of approximately 20%. On the other hand, Karimian et al. [28] introduced a new configuration of Darrieus-type VAWT whose blades are segmented and its performances are examined by CFD techniques. Buchner et al. [29, 30] investigated the phenomenon of dynamic stall by performing planar analyses and comparing them to the experimental data obtained by particle image velocimetry (PIV).

4.2 Planar Flow Simulation

In the first step, although somewhat inferior, planar flow is considered as less computationally expensive than spatial flow. The rotor is simplified i.e. the effects of the central shaft on the flow are neglected, and solely blades are modeled and analyzed. The computational domain is shaped like a circle and its diameter is 50 times bigger than the rotor diameter D. Since fluid flow is highly changeable across the rotor, for better simulation of unsteady effects the computational domain is split into two zones: inner (moveable) – *rotor*, and outer (stationary) – *stator*.

Generated computational mesh is structured periodic and illustrated in Fig. 4. It contains 37500 finite volumes. Although this might seem course at first glance, this structured grid corresponds to an unstructured mesh of much greater number of elements. Its level of refinement was adopted after performing a grid convergence study. Prismatic cells adjacent to blade walls are sufficiently thin to produce dimensionless wall distance smaller than 1, $y^+ < 1$.

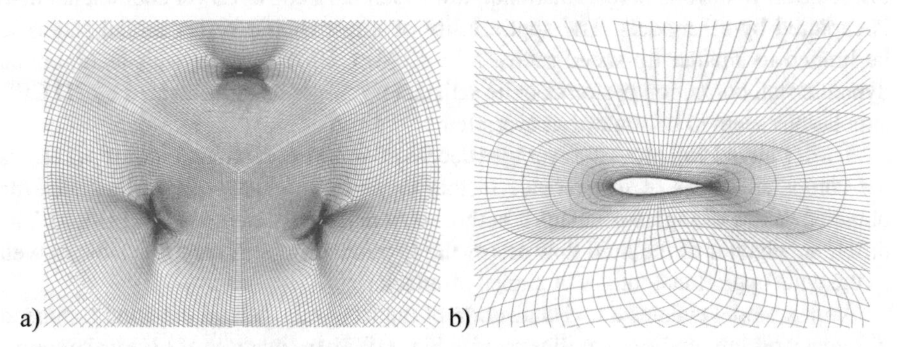

Fig. 4. An example of planar computational mesh: a) across the rotor, b) around the blades.

Fluid flow is transient, viscous and incompressible. It is solved by Reynolds-averaged Navier-Stokes equations closed by a two-equation $k-\omega$ SST turbulence model which performs well in flows including airfoils and adverse pressure gradients [35]. Undisturbed wind velocity V_o is defined at the inlet boundary, and zero gauge pressure Δp at the outlet. Angular velocity Ω is assigned to the inner rotational zone. Its rotational motion is modeled by the sliding mesh approach that requires the actual movement of the mesh by angular increment $\Delta\psi$ in every time-step Δt. Various operating conditions i.e. tip-speed ratios λ can be obtained by different combinations of V_o and Ω.

Pressure-based solver was employed with SIMPLEC pressure-velocity coupling scheme. All spatial derivatives are approximated by 2nd order schemes, and temporal derivatives by 1st order scheme. Number of iterations per time-step is 10–20. The computations were performed until reaching periodic convergence of torque coefficient C_Q, which usually required 8–10 revolutions.

5 Results and Discussion

In order to perform the validation and comparison of the described models a well-known, extensively tested geometry was used. The wind turbine in question is the so-called 3-bladed SANDIA 5 m whose experimental results can be found in [36]. Its main geometric characteristics are: diameter $D = 5$ m, number of blades $N_b = 3$, chord $c = 0.15$ m resulting in solidity $\sigma \approx 0.18$, blade length $L = 5.1$ m, constant airfoil NACA 0015 along the blade and angular velocity 125 rpm $\leq \Omega \leq$ 150 rpm. Used aerodynamic characteristics of the airfoil NACA 0015 were taken from [37].

Figure 5 compares the measured and computed values of power coefficient C_P with respect to tip-speed ratio λ. It should be mentioned that some discrepancies between the two sets of results are to be expected since several simplifications such as planar instead of spatial flow, neglected effects of central shaft and blade tips, etc. were introduced into computations. Furthermore, momentum and vortex models require the usage of airfoil characteristics (obtained on 2D stationary models) that certainly differ from the 3D behavior of airfoils in real rotational flows. General trend of power coefficient curve is captured by all models. The applicability of implemented vortex model seems to be limited to smaller tip-speed ratios, $\lambda < 4$. Also, DMST model with incorporated dynamic stall correction performs quite satisfactory, while results obtained by 2D CFD simulations seem to be somewhat overrated.

DMST model computes the averaged deceleration across the rotor. Figure 6a presents the computed values of interference factors a_u and a_d, while Fig. 6b illustrates the obtained values of angles-of-attack α at three different tip-speed ratios $\lambda = [3, 4, 5]$. The different behavior of the fluid flow across the two rotor halves is clearly visible as well as the wide operational ranges of angles-of-attack.

Vortex model enables the visualization of the free wake behind the wind turbine at different working conditions as illustrated in Fig. 7. With the increase of λ the wake begins to compress leading to significant interaction between the vortices and consequentially, computational inaccuracies.

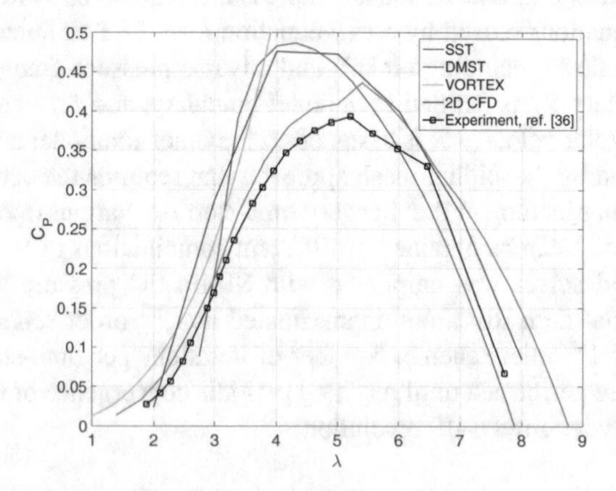

Fig. 5. Power coefficient $C_P(\lambda)$.

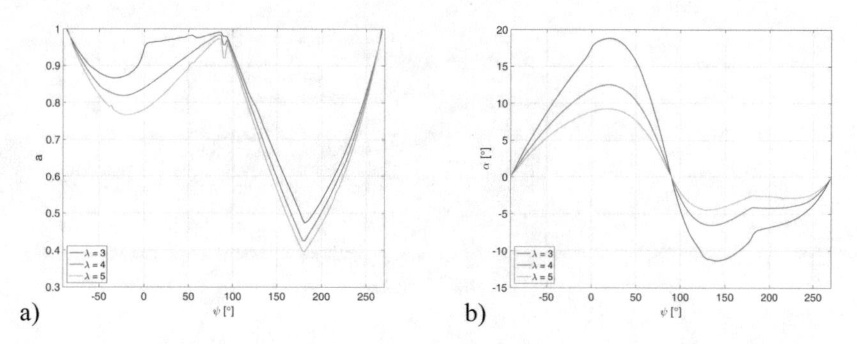

Fig. 6. a) Interference factors, b) angles-of-attack across the rotor at $\lambda = [3, 4, 5]$.

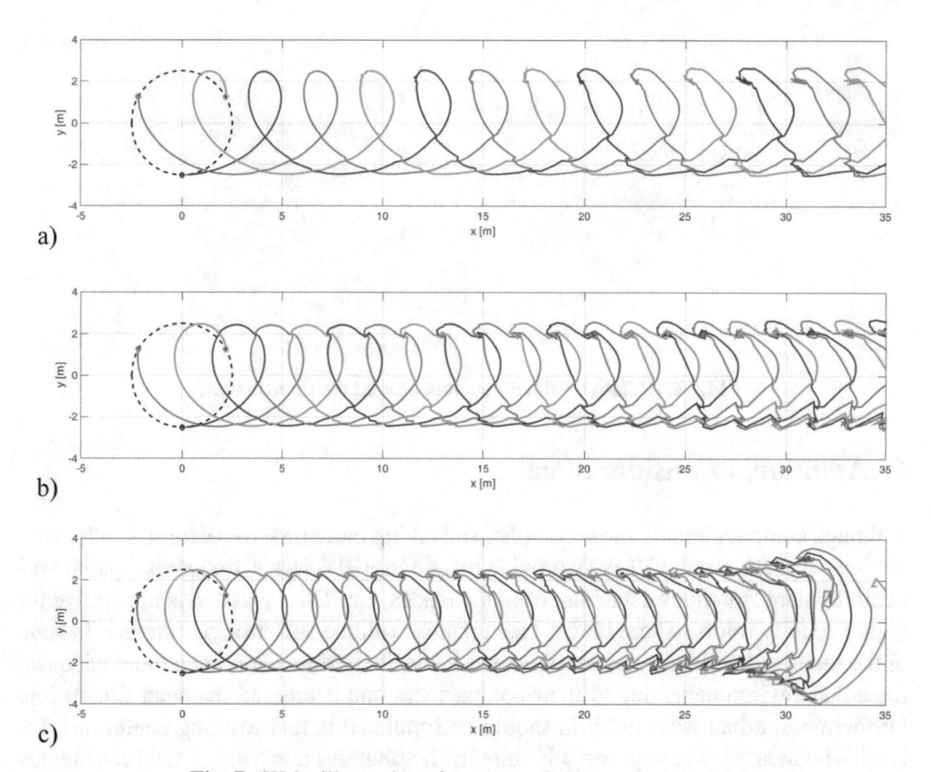

Fig. 7. Wake illustrations for: a) $\lambda = 2$, b) $\lambda = 3$, c) $\lambda = 4$.

Among other things, CFD simulations enable performing various qualitative analyses by velocity or pressure contours or instantaneous relative velocity vectors as depicted in Fig. 8. The blades are enlarged for clarity.

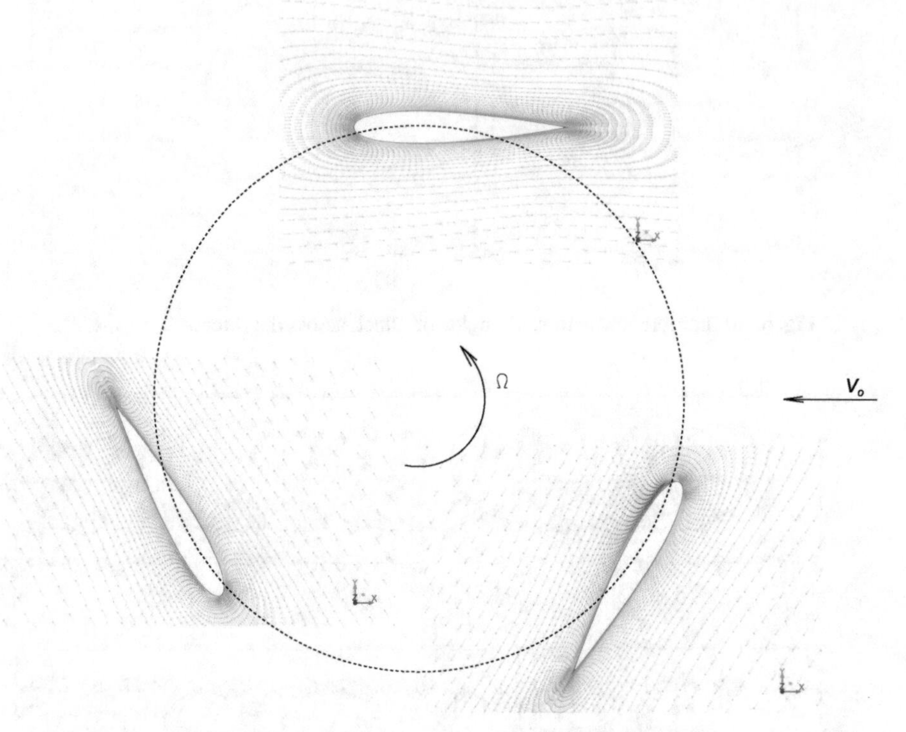

Fig. 8. Relative velocity vectors around the three blades.

6 Additional Considerations

Although computationally more complex and taking more time to perform, spatial simulations of flows around VAWTs provide insight into 3D effects, tip losses, spatial vortical/turbulent structures detaching from the blades, etc. This approach is imperative for higher solidity rotors, where blades take up more volume and strongly interact. In spite of the increased variability of aerodynamic loads, these rotors are much more adequate for urban environments due to their compactness and decreased diameter dimension. Furthermore, urban wind turbines should be simulated in real working conditions, i.e. Earth's boundary layer, with variable velocity distributions along the vertical coordinate.

6.1 Spatial Simulations

In this case, a small-scale VAWT designed for urban environments [38] (parks, rooftops, etc.) is considered. Its rotor diameter is $D = 3$ m (radius $R = 1.5$ m), blade length $L = 1.0$ m, number of blades $N_b = 3$, blade chord $c = 0.15$ m resulting in a somewhat higher solidity $\sigma = 0.3$ than previously investigated (which will also be noticeable through smaller power coefficient), constant airfoil NACA 0018 and nominal rotor angular velocity $\Omega = 200$ rpm.

In spatial simulations, it is possible to use the same assumptions as with planar case. The flow can be considered transient, incompressible and turbulent, and the rotor can be simplified to blades alone. Again, the computational domain should contain two adjacent zones with an interface surface between them. Inner, rotational part can be shaped like a cylinder, while the outer, stationary region is usually a cuboid extending sufficiently away from the rotor, Fig. 9. In this particular case, the geometric parameters of the cylindrical moving part of the computational domain are: diameter $D_R = 4$ m and height $H_R = 1.5$ m. Stator stretches $-3R$ fore and $+11R$ aft of the VAWT in the longitudinal direction, $\pm 4R$ in lateral direction and $\pm 2R$ below and above the rotor centroid resulting in the total stator height of $H_S = 6.0$ m.

Generated computational meshes will usually be hybrid unstructured for their better flexibility (allowing various cell shapes to be combined) and the fact they can be more easily automatically generated. They are also much bigger than the planar grids comprising at least several million cells. Since these meshes are not aligned with the flow, they should also incorporate thin layers of prismatic cells in the close proximity of the wall boundaries. Rotational zones can be additionally refined since the greatest gradients of the flow quantities appear precisely in these parts of the computational domain. A detail of the generated grid numbering approximately 3.2 million finite volumes is presented in Fig. 10.

As previously mentioned, the values of velocity (and turbulence quantities) and gauge pressure should be defined at the inlet and outlet boundaries, respectively. Angular velocity Ω should be assigned to the interior, rotational zone. Depending on the size of the mesh and required accuracy of the final results, time-step Δt (i.e. angular increment $\Delta\psi$) should be defined together with the adequate solver and discretization schemes.

Obtained output data may be presented in various forms. For instance, quantitative analysis can be performed by considering normal and axial forces acting on a single blade during a revolution at different wind speeds as illustrated in Fig. 11.

Here, again, it is possible to observe the differences between the upstream and downstream parts of the rotor, and how much more tangential force (and torque) is generated when $-90° \leq \psi \leq 90°$ in comparison to the region where $90° \leq \psi \leq 270°$.

Qualitative analysis, or the inspection of the 3D flow field, can be done by examining the turbulent structures' iso-surfaces as presented in Fig. 12, that appear as a consequence of the rotor rotation, flow unsteadiness, vorticity and separation, aerodynamic stalling, wake detachment etc. Greater structures colored in red point to the regions of increased losses and decreased efficiency. For better comprehension of the flow behavior, a non-optimal working regime is considered when $\lambda = 1.745$.

6.2 Effects of Ground and Velocity Profiles

Since the designed VAWT should operate in densely populated areas, it is also necessary to check its performance close to the ground, in the Earth's boundary layer, where velocity distribution can be approximated by power law:

$$V(h) = V_{ref}\left(\frac{h}{h_{ref}}\right)^{\alpha}. \tag{22}$$

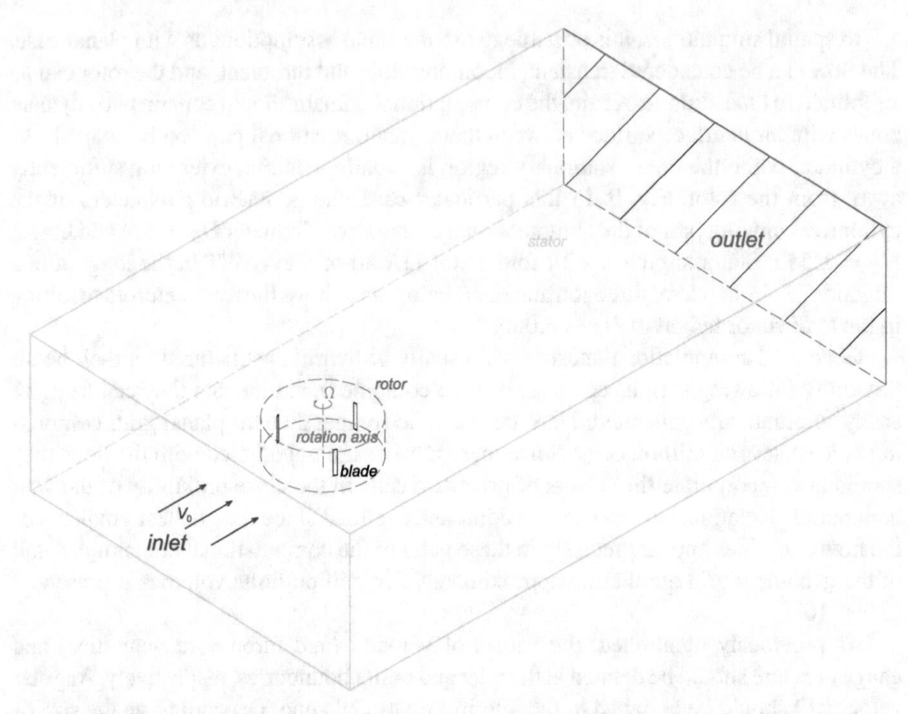

Fig. 9. Computational domain with boundaries and two sub-domains, taken from [38].

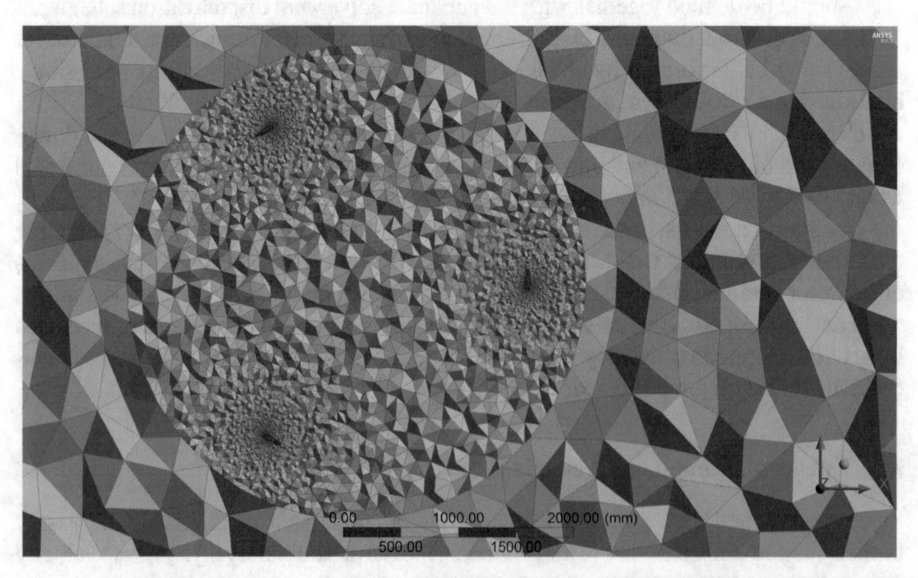

Fig. 10. A detail of 3D computational mesh, taken from [38].

It is assumed that rotor's centroid (mean, central point) is located at the height h_{ref} = 3.0 m, so this value is taken for the reference height necessary for the definition of

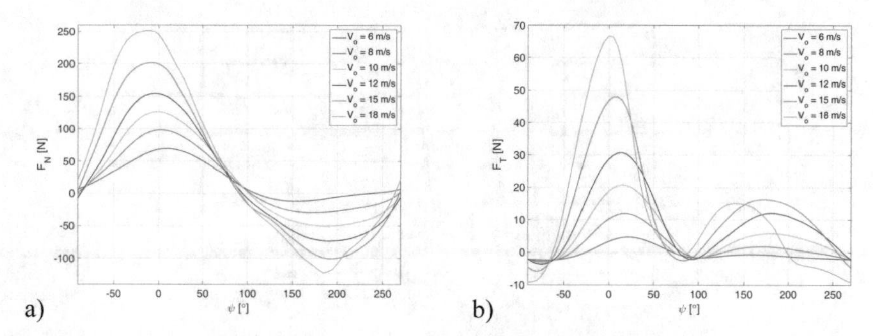

Fig. 11. Computed: a) normal and b) tangential forces acting on a blade during a revolution, taken from [38].

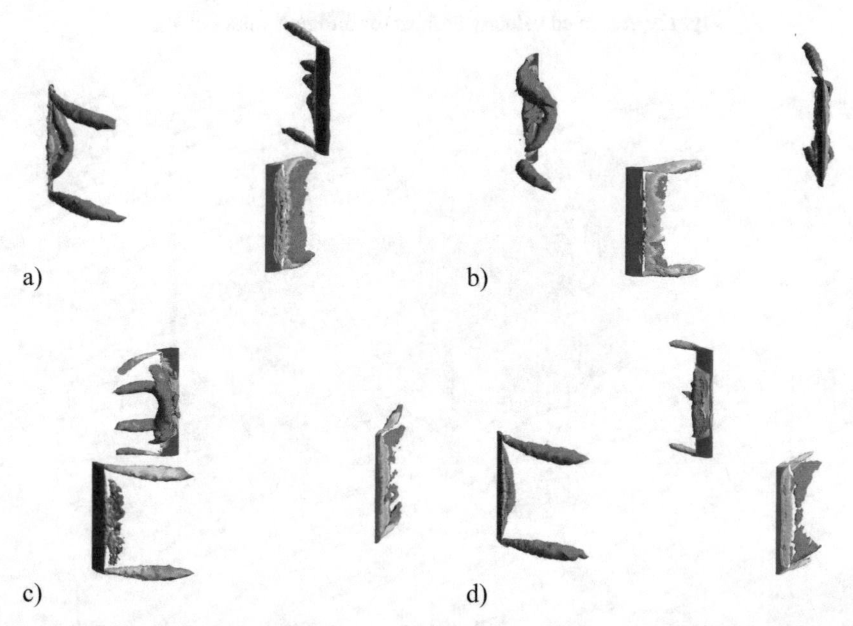

Fig. 12. Instantaneous vortical structures at four different moments in time detaching from the blades at $\lambda = 1.745$, taken from [38].

the velocity profile. The exponent $\alpha = 0.15$ corresponds to the terrain roughness class 2. Different working regimes were simulated by changing the values of reference velocity V_{ref}. Several of the used profiles are sketched in Fig. 13.

The shape of the computational domain was not changed. The only modification of the geometry is that now, lower surface presents the ground, i.e. wall boundary with no-slip boundary condition. Therefore, the mesh was also generated in a similar way to the previously described with the addition of boundary layer along the ground/lower surface which slightly increases the total number of elements.

Again, transient computations of incompressible, viscous flow around a small-scale VAWT have been performed by FVM for ten different working regimes, i.e. wind speeds.

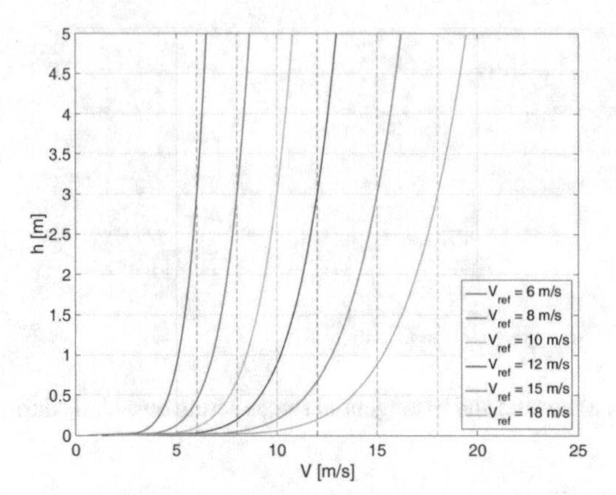

Fig. 13. Assumed velocity profiles for different values of V_{ref}.

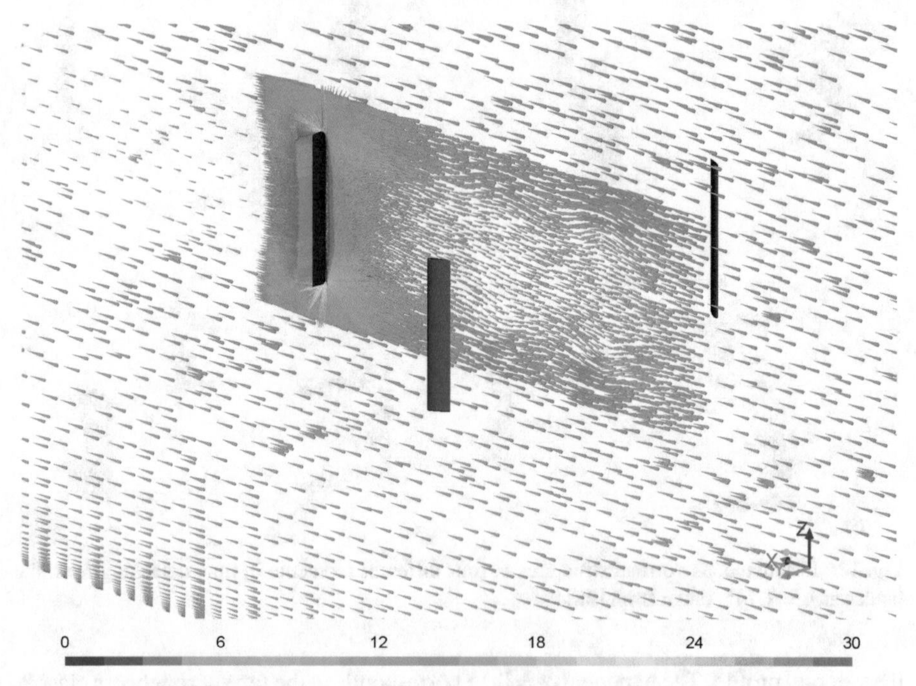

Fig. 14. Velocity vectors in the symmetry plane for $V_{ref} = 11$ m/s.

The goal of this part of research was to compare the aerodynamic performances of the investigated rotor in idealized, uniform velocity stream and a power-law profile corresponding to Earth's boundary layer. Computed values of power coefficient indicate that a performance reduction of 1–4% can be expected in real working conditions. Due to the small length of the blades, performance losses are not significant and can be neglected in conceptual design studies. Flow field in symmetry plane presented by velocity vectors is illustrated in Fig. 14. Inlet velocity profile and the effects of the rotor (velocity deceleration) are clearly visible.

7 Conclusions

Although flow fields around small-scale VAWTs are extremely unsymmetrical and unsteady, there are many computational models suitable for their aerodynamic investigation, ranging from simple momentum to more complex CFD based models. Each computational category is still quite employed today, for both wind and water/tidal turbines which is why this paper provides a review with the most notable advantages and disadvantages as well as basic equations of each model.

For preliminary analysis of low solidity rotors, momentum models provide quite useful and fast results regarding the average change in speed across the rotor, aerodynamic loads acting on the blades and generated power. However, if more complicated physical phenomena are of interest, probably CFD approach should be selected. Here also, some caution is needed. While some flow assumptions can be adopted without losing on the accuracy of the final results (e.g. that flow is incompressible), turbulence and blade/wake transient interference can hardly be neglected. For these reasons, CFD approach is still quite computationally expensive and the time required for performing the simulations is significant. As with other aerodynamic applications, it is often necessary to compromise between computational efficiency and accuracy. Therefore, the choice of the employed model will be dictated by the desired output, and in most cases, simpler and more complex models will be combined and/or compared.

Acknowledgement. The research work is funded by the Ministry of Education, Science, and Technological Development of Republic of Serbia through contract no. 451-03-68/2020-14/200105.

References

1. IRENA: Renewable capacity statistics 2019. International Renewable Energy Agency (IRENA), Abu Dhabi (2019)
2. Murthy, K.S.R., Rahi, O.P.: A comprehensive review of wind resource assessment. Renew. Sustain. Energy Rev. **72**, 1320–1342 (2017)
3. Komarov, D., Stupar, S., Simonovic, A., Stanojevic, M.: Prospects of wind energy sector development in Serbia with relevant regulatory framework overview. Renew. Sustain. Energy Rev. **16**(5), 2618–2630 (2012)
4. Manwell, J.F., McGowan, J.G., Rogers, A.L.: Wind Energy Explained, Theory, Design and Application, 2nd edn. Wiley, Chichester (2009)
5. Islam, M., Ting, D.S.-K., Fartaj, A.: Aerodynamic models for Darrieus-type straight-bladed vertical axis wind turbines. Renew. Sustain. Energy Rev. **12**(4), 1087–1109 (2008)
6. Bhutta, M.M.A., Hayat, N., Farooq, A.U., Ali, Z., Jamil, S., Hussain, Z.: Vertical axis wind turbine – a review of various configurations and design techniques. Renew. Sustain. Energy Rev. **16**(4), 1926–1939 (2012)
7. Tummala, A., Velamati, R.K., Sinha, D.K., Indraya, V., Krishna, H.: A review on small scale wind turbines. Renew. Sustain. Energy Rev. **56**, 1351–1371 (2016)

8. Stathopoulos, T., et al.: Urban wind energy: some views on potential and challenges. J. Wind Eng. Ind. Aerodyn. **179**, 146–157 (2018)
9. Brahimi, M.T., Allet, A., Paraschivoiu, I.: Aerodynamic analysis models for vertical-axis wind turbines. Int. J. Rotating Mach. **2**(1), 15–21 (1995)
10. Paraschivoiu, I.: Wind Turbine Design with Emphasis on Darrieus Concept. Presses Internationales Polytechnique, Montreal (2002)
11. Saeidi, D., Sedaghat, A., Alamdari, P., Alemrajabi, A.A.: Aerodynamic design and economical evaluation of site specific small vertical axis wind turbines. Appl. Energy **101**, 765–775 (2013)
12. Svorcan, J., Stupar, S., Komarov, D., Peković, O., Kostić, I.: Aerodynamic design and analysis of a small-scale vertical axis wind turbine. J. Mech. Sci. Technol. **27**(8), 2367–2373 (2013). https://doi.org/10.1007/s12206-013-0621-x
13. Posteljnik, Z., Stupar, S., Svorcan, J., Peković, O., Ivanov, T.: Multi-objective design optimization strategies for small-scale vertical-axis wind turbines. Struct. Multidiscip. Optim. **53**(2), 277–290 (2015). https://doi.org/10.1007/s00158-015-1329-6
14. Moghimi, M., Motawej, H.: Developed DMST model for performance analysis and parametric evaluation of Gorlov vertical axis wind turbines. Sustain. Energy Technol. Assess. **37**, 100616 (2020)
15. Ferrari, G.M.: Development of an aeroelastic simulation for the analysis of vertical-axis wind turbines. Ph. D. thesis, The University of Auckland, Auckland (2012)
16. Urbina, R., Peterson, M.L., Kimball, R.W., deBree, G.S., Cameron, M.P.: Modeling and validation of a cross flow turbine using free vortex model and a modified dynamic stall model. Renewable Energy **50**, 662–669 (2013)
17. Shi, L., Riziotis, V.A., Voutsinas, S.G., Wang, J.: A consistent vortex model for the aerodynamic analysis of vertical axis wind turbines. J. Wind Eng. Ind. Aerodyn. **135**, 57–69 (2014)
18. Howell, R., Qin, N., Edwards, J., Durrani, N.: Wind tunnel and numerical study of a small vertical axis wind turbine. Renewable Energy **35**, 412–422 (2010)
19. Castelli, M.R., Englaro, A., Benini, E.: The Darrieus wind turbine: proposal for a new performance prediction model based on CFD. Energy **36**, 4919–4934 (2011)
20. Castelli, M.R., Benini, E.: Effect of blade inclination angle on a Darrieus wind turbine. J. Turbomach. **134**, 031016 (2012)
21. Nini, M., Motta, V., Bindolino, G., Guardone, A.: Three-dimensional simulation of a complete vertical axis wind turbine using overlapping grids. J. Comput. Appl. Math. **270**, 78–87 (2014)
22. Sun, X., Wang, Y., An, Q., Cao, Y., Wu, G., Huang, D.: Aerodynamic performance and characteristic of vortex structures for Darrieus wind turbine. II. The relationship between vortex structure and aerodynamic performance. J. Renew. Sustain. Energy **6**, 043135 (2014)
23. Balduzzi, F., Drofelnik, J., Bianchini, A., Ferrara, G., Ferrari, L., Campobasso, M.S.: Darrieus wind turbine blade unsteady aerodynamics: a three-dimensional Navier-Stokes CFD assessment. Energy **128**, 550–563 (2017)
24. Elsakka, M.M., Ingham, D.B., Ma, L., Pourkashanian, M.: CFD analysis of the angle of attack for a vertical axis wind turbine blade. Energy Convers. Manag. **182**, 154–165 (2019)
25. Franchina, N., Persico, G., Savini, M.: 2D-3D computations of a vertical axis wind turbine flow field: modeling issues and physical interpretations. Renewable Energy **136**, 1170–1189 (2019)
26. Guillaud, N., Balarac, G., Goncalves, E., Zanette, J.: Large Eddy simulations on vertical axis hydrokinetic turbines - power coefficient analysis for various solidities. Renewable Energy **147**, 473–486 (2020)
27. Su, J., Chen, Y., Han, Z., Zhou, D., Bao, Y., Zhao, Y.: Investigation of V-shaped blade for the performance improvement of vertical axis wind turbines. Appl. Energy **260**, 114326 (2020)
28. Karimian, S.M.H., Abdolahifar, S.M.H.: Performance investigation of a new Darrieus vertical axis wind turbine. Energy **191**, 116551 (2020)

29. Buchner, A.-J., Lohry, M.W., Martinelli, L., Soria, J., Smits, A.J.: Dynamic stall in vertical axis wind turbines: comparing experiments and computations. J. Wind Eng. Ind. Aerodyn. **146**, 163–171 (2015)
30. Buchner, A.-J., Soria, J., Honnery, D., Smits, A.J.: Dynamic stall in vertical axis wind turbines: scaling and topological considerations. J. Fluid Mech. **841**, 746–766 (2018)
31. Kooiman, S.J., Tullis, S.W.: Response of a vertical axis wind turbine to time varying wind conditions found within the urban environment. Wind Eng. **34**(4), 389–401 (2010)
32. Rolin, V.F.-C., Porte-Agel, F.: Experimental investigation of vertical-axis wind-turbine wakes in boundary layer flow. Renewable Energy **118**, 1–13 (2018)
33. Ferziger, J.H., Perić, M., Street, R.L.: Computational Methods for Fluid Dynamics. Springer, Cham (2020). https://doi.org/10.1007/978-3-319-99693-6
34. ANSYS Fluent Theory Guide. Release 16.0. ANSYS, Inc., Canonsburg (2015)
35. Menter, F.R., Kuntz, M., Langtry, R.: Ten years of industrial experience with the SST turbulence model. In: Proceedings of the 4th International Symposium on Turbulence, Heat and Mass Transfer, pp. 625–632. Begell House Inc., West Redding (2003)
36. Sheldahl, R.E., Klimas, P.C., Feltz, L.V.: Aerodynamic performance of a 5-metre-diameter Darrieus turbine with extruded aluminum NACA-0015 blades. SAND80-0179. Sandia National Laboratories, Albuquerque (1980)
37. Sheldahl, R.E., Klimas, P.C.: Aerodynamic characteristics of seven symmetrical airfoil sections through 180-degree angle of attack for use in aerodynamic analysis of vertical axis wind turbines. SAND80-2114. Sandia National Laboratories, Albuquerque (1981)
38. Trivković, Z.: Multiobjective optimization of composite wind turbine blades. Ph.D. thesis, The University of Belgrade, Belgrade (2019)

Algorithm for Applying 3D Printing in Prototype Realization – Case: Enclosure for an Industrial Pressure Transmitter

Miloš Vorkapić[1](✉), Aleksandar Simonović[2], and Toni Ivanov[2]

[1] ICTM - CMT, University of Belgrade, Njegoševa 12, 11000 Belgrade, Republic of Serbia
worcky@nanosys.ihtm.bg.ac.rs

[2] Faculty of Mechanical Engineering, University of Belgrade, Kraljice Marije 16, 11000 Belgrade, Republic of Serbia

Abstract. Additive manufacturing technology helped many organizations to save money in the product design process by reducing prototype costs, and also by providing a means for early evaluation and decision making. The idea of this paper is to design an electronics enclosure for an intelligent industrial pressure transmitter, using the additive technology. All enclosure elements are made on a 3D printer WANHAO duplicator i3 plus, using PLA materials. The enclosure realization, from CAD drawings to the finished model, enables a designer to correct existing errors, or make certain modifications as required by end-users. A process is described that enables designers to review their decisions at any stage of product realization, thus providing much more freedom in rapid prototyping. In this example, the advantages and disadvantages of additive manufacturing over conventional manufacturing are outlined. Some deficiencies have also been observed, such as mechanical damage to surfaces, burning of surfaces, tearing of prints, and surface roughness. To mitigate such irregularities, both mechanical and chemical finishing methods were used. The example confirmed that the finishing methods can affect the final enclosure dimensions and shape. Further prototype development should focus more on print quality, which depends on the shape of surfaces, the accuracy of the geometry, the uniformity of structure and shape, material density, and the resolution of details.

Keywords: Prototype enclosure · Pressure transmitter · Additive manufacturing · 3D printing · PLA · Nozzle · Quality control

1 Introduction

The most common sources of innovative ideas for new product development are customers, competitors, fairs or exhibitions. According to previous research [1], ideas rarely come from universities or research institutes, which illustrates the low impact of the scientific community on industrial development. Also, the acquirement of new fabrication and prototyping techniques related to the development of new products is considered as

N. Mitrovic et al. (Eds.): CNNTech 2020, LNNS 153, pp. 112–129, 2021.
https://doi.org/10.1007/978-3-030-58362-0_8

the most important strategy for improving the business activities of the company. In this sense, additive technology becomes increasingly important.

According to ASTM International [2], additive manufacturing (AM) is also called "3D Printing". According to Gibson et al. [3], AM includes: 1) shape changes in existing products, 2) changes in design processes, 3) new model geometry, 4) realization of complex shapes, 5) complexity in material combination, and 6) variable scale model realization.

AM has helped many organizations to save time during the design process, by enabling faster decision-making, by reducing prototype implementation costs, and, in the end, by facilitating the manufacturing of modular and shortlife cycle products [4]. In this direction, the overall material cost is significantly reduced, while the waste and use of ancillary tools are minimized [5].

The use of AM enables fabrication of parts/models according to the digital data in the most direct manner, which, compared to conventional manufacturing (CM), represents a revolutionary advance, as it excludes machining, injection molding, and casting [6]. AM involves adding material in layers, while CM subtracts material from semi-finished products to get the desired shape. In support of this, Harris [7] examined the capabilities of AM processes, their applications in various industrial sectors, and the impact of AM in manufacturing systems, where they enable rapid and mass production.

In general, AM enables the realization of prototypes with complex geometries in relatively small quantities [8], and the favorable production of tools/molds to be used in mass production, especially when it comes to the aerospace industry and the development of complex mechanical-electrical products [9].

Zhang et al. [10] identified the advantages and disadvantages of AM. Among the advantages are: 1) realization of complex geometrical shapes; 2) the ability to easily change the model using a 3D program (shortening the time between iterative design changes and production launch); 3) low initial setup costs (easier manipulation of tools and equipment); 4) simple distribution of formats (implies the rapid realization of products using standardized 3D programs), 5) use of biodegradable materials (environmental protection is taken into account) and 6) space savings.

The main disadvantages of AM are: 1) low printing speed (which makes it unfavorable for mass production); 2) small desktop, i.e. parts are still being made by segment; 3) low uniformity of product quality due to the layer-by-layer production process; 4) quality, durability and tolerance of 3D printing devices; and 5) intellectual property and copyright issues.

Vayre et al. [11] showed that there are two major barriers in AM. First, the nozzle stays parallel to the vertical axis, and this causes a collision between the nozzle and the model. Secondly, when the nozzle speed is varied, there is a difference in the height of the material being applied.

The main parameters used for the comparison of CM and AM are energy optimization, waste reduction, conventional tools reduction, as well as design optimization. According to Davia-Arcil et al. [12], the comparison elements between CM and AM are given in Table 1.

In this paper, the example of the intelligent pressure transmitter enclosure will be used for the analysis of all AM elements. Also, this paper aims to show the importance of

Table 1. Comparison between CM and AM

Type	CM	AM
Geometry complexity	No	Yes
Materials range	Wide	Wide
Material price	Higher	Lower
Equipment price	Higher	Lower
Prototype price	High	Low
Manufacturing time	Shorter	Longer
Working time	Longer	Shorter
Waste	Yes	No

AM in prototype design, as well as the ability to modify a particular element/assembly on an existing product.

Prototypes made by 3D printing often have worse mechanical characteristics compared to final products manufactured by using traditional technologies. Plastic parts made on a 3D printer are cheaper and lighter compared to parts made of metal. When plastic parts are considered as a replacement for metal parts, there are four key properties of materials to be considered: heat resistance, chemical and corrosion resistance, wear resistance, and strength.

In this work, 3D printing was chosen because it takes much less time for the prototype to be realized. With the conventional technology, there are steps and procedures related to the materials procurement and storage, launching of the production, and the machines and tools amortization.

2 3D Printing Technology

A common feature of all 3D printers is that they do not process the material as conventional machining equipment; instead, their principle of operation is based on adding of the material in layers.

3D printing technology allows modifications of the model to be performed at any time until the very beginning of the manufacturing process. These changes are easy to implement, as they only require the use of a CAD software, thus simplifying the communication with the customer. The CAD model is subsequently realized as a 3D object on a printer [13].

However, according to Cupar et al. [14], 3D printing (with filament) requires advanced technical knowledge in preparation and parameter settings. 3D filament printing technology involves melting of a thermoplastic material through a heated nozzle.

The wound material (1) through the rollers (2) and the guide pipe (3) passes through the heater (4), where it melts at a defined temperature, and then, as a molten material, passes through the nozzle (5), after which it touches the panel (6) and adheres to it, where it is cooled and hardened as a finished part (7), see Fig. 1. The first layer is very important

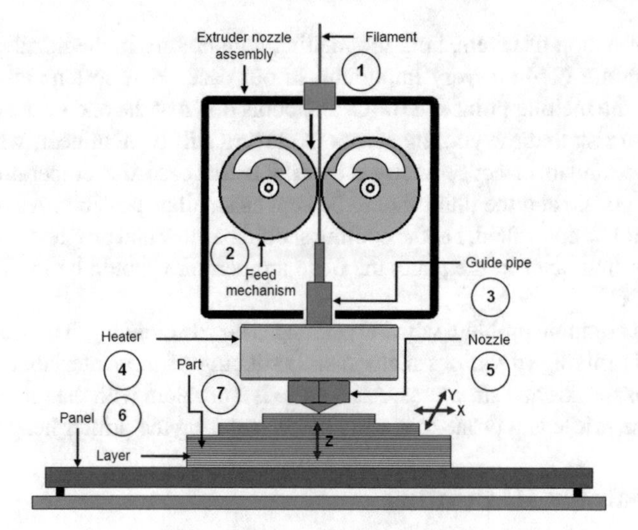

Fig. 1. 3D printer basic elements

because, during the cooling process, the molten material hardens, and the model can be deformed.

The defined distance between the nozzle and the plate is 0.1 mm. The molten material can not be properly bonded to the printer plate if the distance increases. Over time, such a part becomes loose and easily detaches from the plate.

When the first layer is formed in the XY plane, the nozzle rises along the z-axis, according to a given layer height (H), and begins the formation of the next layer, of the defined thickness (L), which merges with the previous layer, and then the same cycle is repeated [15–17] (see Fig. 2). However, in the case of a high-temperature setting, the previously formed layer may melt, which eventually leads to inaccuracies in the dimensions of the realized model.

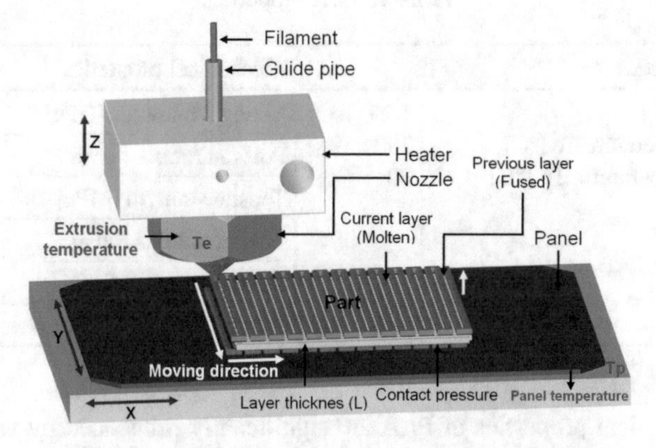

Fig. 2. Illustration of the layer formation process

In the formation of layers, both the melting temperature in the nozzle (Te) and the plate temperature (Tp) are very important. In our case, PLA softens at 150 °C, then warms up to the melting point at 210 °C, and cools down at the end of the process [18].

There is a risk that the cooling across the layers will be nonlinear, which gives an irregular shape and incorrect geometry at the end. In that case, the temperature difference between the nozzle and the plate should be kept as small as possible. Also, the cooling process should be controlled, i.e. the cooling should be slow and progressive [3]. In order to adhere the first layer to the plate, the plate temperature should be in the range from 50 °C to 60 °C.

The most common problem with 3D printing is nozzle clogging. The clogging occurs because of 1) mixing of various materials, 2) sticking of the materials to the channel walls and nozzle external structures. Also, there is a problem with the material aging, as it can become brittle and prone to tearing before entering the printer head [19].

3 Materials for 3D Printing

There is a wide variety of thermoplastic materials such as polylactic acid (PLA), acrylonitrile butadiene styrene (ABS), thermoplastic polyurethane (TPU), aliphatic polyamides (nylon), polyethylene terephthalate (PET) and others.

PLA is a material derived from renewable sources, such as corn starch or sugar cane [20]. It is an eco-friendly thermoplastic polyester, which naturally decomposes under external conditions [21].

It can be manufactured on pre-existing molding equipment, which makes it relatively cheap to manufacture [22], and is usually available in a large color palette. It can be machined.

PLA has applications in the packaging industry [23], biomedical sciences [24], automotive industry and electrical industry [25]. Physical and mechanical properties of PLA are given in Table 2 [26, 27].

Table 2. PLA properties

Physical properties		Mechanical properties	
Density, [g/cm^3]	1.25	Young's modulus [GPa]	3.5
Extrusion temperature Te [°C]	200	Poisson ratio	0,36
Transition temperature, Tg [°C]	60	Tensile strength [MPa]	73
		Shear modulus [GPa]	1,287
		Yield strenght [MPa]	49,6
		Elongation [%]	6

The mechanical properties of PLA are significantly influenced by various technological variables, such as the nozzle diameter, the formed layer thickness, the percentage infill value, the sample being filled, the filling rate and temperatures [21, 28]. Because of that, some deficiencies may occur: high brittleness, low toughness, and elongation [29].

4 Methodology of 3D Printing

Kim et al. [30] defined the activities that are essential for prototype realization, and they are:

Modeling - it includes material selection, input parameter control, design parameter optimization, analytical model development, algorithm, and database development based on quality prediction [31];
Printing - includes quality control and monitoring during the printing process;
Finishing - involves finishing control, finishing on pieces, defect corrections (by using thermal or chemical treatment, or by machining);
Analysis - involves interpreting of the obtained parameter values, and checking the operating parameters of the device;
Economics - includes validation of processes and parameters, defines the best orientation of a part, and provides an estimate of workability.

Generally, the methodology of 3D model development consists of the following steps [32, 33]:

The model is implemented by using one of the CAD software packages (Catia, Solid Edge, Solid Works, Pro/ENGINEER);
Then the 3D model is imported as a *.stl file into the specialized Ultimaker Cura program, which sets the operating parameters of the system. The STL file is a standard format for 3D printing, and it is readable by many 3D programs [34];
Transfer and manipulation of *.stl files in Ultimaker Cura: the advantage of the *.stl file format is in the fact that a majority of CAD application software supports it, while the disadvantage is the loss of the desired printer resolution, which is mainly related to the layer thickness [35];
After performing the proper printing settings and a simulation using Ultimaker Cura, G code (CNC machine code) is generated (with the gcod extension), which can be read by the 3D printer. The G code is a standard format that describes the tool trajectory: 1) the printer head movement along the x, y, and z directions, 2) the quantity of material displaced, and 3) the printer head speed. During this step, the G code is loaded into the printer, and the object is ready for production.

In this step, the realization of the 3D model is finished, and after the realization, the excess material is removed and the part is cleaned. Subsequently, additional processing can be performed if needed, after which the product can be used [9, 36].

The block diagram that describes the transformation of a CAD virtual 3D model into a finished/fabricated physical object is shown in Fig. 3. The virtual model allows both the designer and the customer to understand the design without the need for physical prototypes.

5 Enclosure Realisation

The intelligent industrial pressure transmitter (series TPas/rs-101) developed by ICTM-CMT, is an electronic instrument that consists of three modules: 1.) a measurement

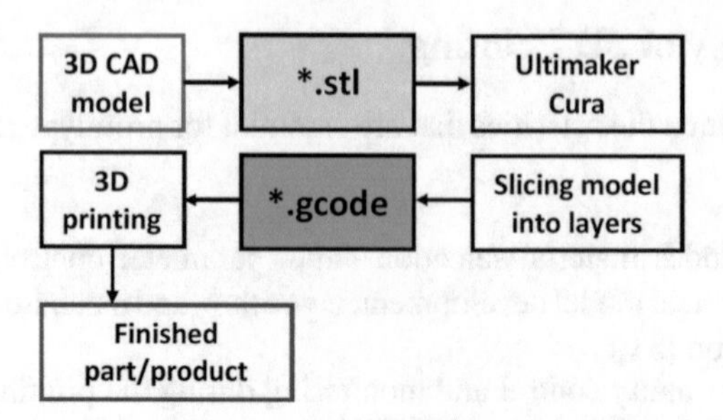

Fig. 3. Block diagram of the model realization

chamber (with sensor), 2.) a fixing element, and 3.) an enclosure with the electronics module.

A photograph of the enclosure for the intelligent pressure transmitter is shown in Fig. 4. The enclosure is made of aluminum alloy Al.Cu5.Mg1.55.

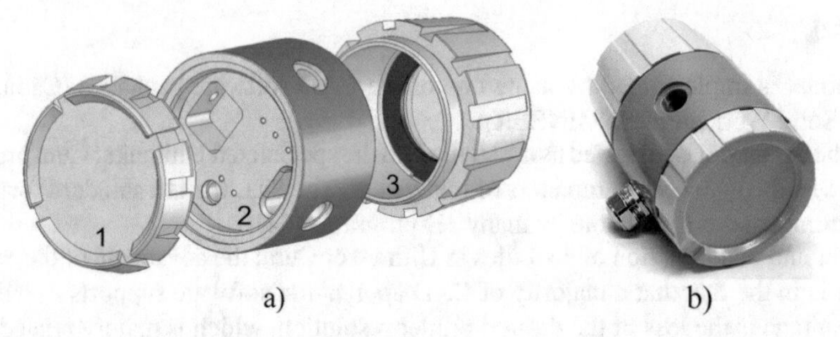

Fig. 4. CM: a) 3D view of the enclosure, b) Realized industrial transmitter enclosure

The idea is to fabricate the enclosure on a 3D printer using the existing technical documentation. The enclosure realization enables the designers to correct existing mistakes, and to modify the enclosure according to customer's requirements.

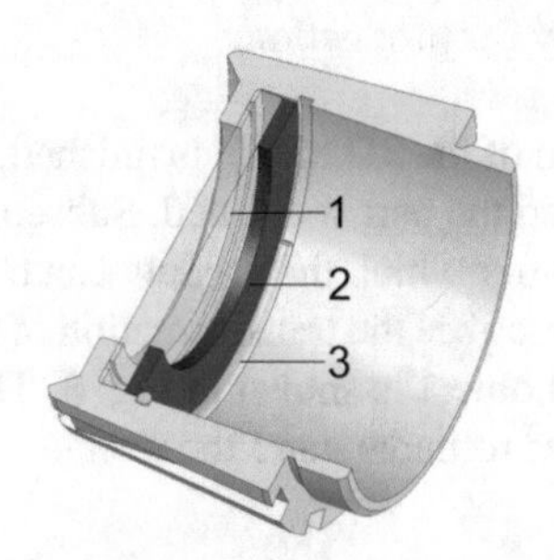

Fig. 5. Cross section of the front cover

The enclosure basic elements are 1) terminal block cover, 2) central part, and 3) front cover (see Fig. 4a). The front cover is deep enough to accommodate the electronics module, and it also enables readout of the content shown on the display. The front cover is fitted with 1) a transparent (glass) window, 2) a plastic spacer, and 3) a retaining ring (see Fig. 5).

The transmitter's electronics module is mounted on the front side of the central part of the enclosure (facing the front cover). The terminal block, mounted on the back side of the central part, and protected by the terminal block cover, provides for the instrument's connection with the control equipment. The measurement chamber is mechanically connected with the enclosure on the side of the central part, enabling at the same time the electrical connection of the sensor with the electronics module of the instrument.

The Ultimaker Cura software divides the mechanical model into thin layers, viewed vertically. Before the printing of the *.gcod file starts, the program offers the possibility of animation of the 3D model.

A graphical representation of the enclosure central part 3D model, with the nozzle theoretical trajectory during the fabrication, is given in Fig. 6. According to Bose et al. [37], in order to achieve the best possible precision and resolution of 3D printers, there should be as little distance as possible between the layers, solely dependent on the material characteristics and the design parameters.

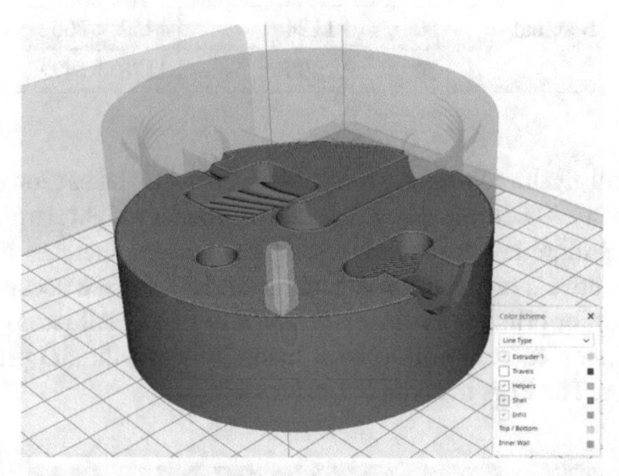

Fig. 6. Graphical representation of the enclosure central part 3D model, with the nozzle theoretical trajectory (Ultimaker Cura)

The Ultimaker Cura desktop space on the right has a list of parameters that are crucial in the implementation of the model, which are: 1) nozzle temperature, 2) plate temperature, 3) number of layers of the printed part, 4) infill type (various geometry shapes), 5) infill orientation, 6) infill build speed, 7) nozzle speed during printing, 8) idle head assembly speed.

In this regard, the following parameter values were chosen for the realization of the enclosure: 1) 215 °C for printing temperature, 2) 60 °C for plate temperature, 3) 50 mm/s, print speed, 4) 70 mm/s for the nozzle travel speed, and 5) 90/−45/0/+45 for infill orientation. In Table 3, printing durations are given as an example: 1) terminal block cover - A and 2) front cover - B. The printing process duration depends on two parameters: 1) the print speed, and 2) the percentage of infill. The required time to realize the covers, calculated by the Ultimaker Cura program, significantly deviates from the actual printing time, which is much longer.

Table 3. Priniting duration for the covers

Name	Print speed	Infill [%]/Calculated and actual time [min]		
	Travel speed [mm/s]	50	80	100
A calculated	50	540	686	778
	70	365	522	590
A actual	50	684	827	1024
	70	541	708	869
B calculated	50	947	1203	1360
	70	729	924	1043
B actual	50	1124	1452	1602
	70	995	1178	1422

The 100% infill gives a completely solid object, but it also takes a long time to finish. If the infill is less than 100%, the printing time and the model weight are reduced [38]. Low thickness requires more layers to complete the model. The increase in manufacturing time has negative effects on the quality (e.g. the model's surface becomes stepped).

The procedure for printing the covers is given in Fig. 7. All parts of the intelligent pressure transmitter's enclosure have been made by using the 3D printer WANHAO Duplicator i3 plus. The PLA material has been used.

Fig. 7. 3D printing of the enclosure cover

6 Quality Control of the Manufactured Parts

The manufactured parts were examined by using the Motic AE-2000 MET inverted microscope with 10x magnification. Some irregularities were observed, which are the consequence of the manufacturing process.

A photomicrograph (10x magnification) of the central part surface is shown in Fig. 8. The nozzle movement directions are indicated by numbers: 1 - [90°]; 2 - [−45°]; 3 - [+45°]; 4 - [0°].

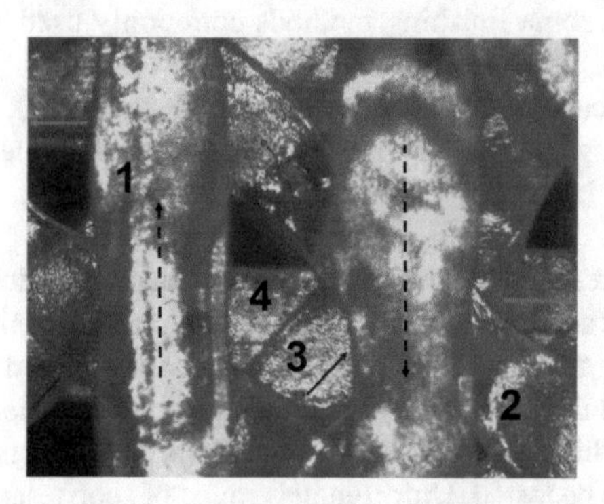

Fig. 8. Photomicrograph of the central part's surface (10x magnification)

Defects on the manufactured enclosure surfaces indicate that mechanical tearing of the printed part occurred because of the rapid nozzle movement. Furthermore, the high temperature of the nozzle burned the surface (it weakens the bonds in the polymer). A fragile and loose structure occurs due to the melting of the material and its layered arrangement. Because of this, interweaved threads of the dispensed material often occur in the XY plane, with layers forming a stepped structure, visible on the lateral surfaces (see Fig. 9).

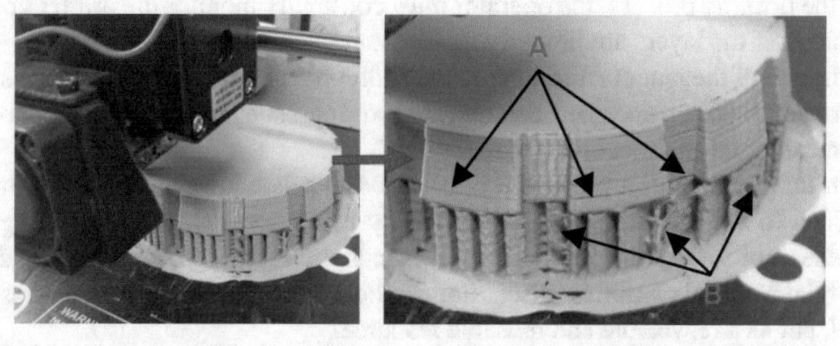

Fig. 9. Defects that occurred during manufacturing of the cover: A - improper layer arrangement, B - appearance of loose threads

Generally, the nozzle is in contact with the material, and is exposed to mechanical wear over time. The reason for the appearance of cracks can be irregular filament cooling after going through the nozzle.

According to Bourell et al. [39], the model quality depends on the realized surface shape, the geometry accuracy, the structure uniformity, and the shape, the density, and the resolution of the detail.

After model realization, the following parameters should be analyzed:

geometry - possible changes in dimensions (shape and size);
mechanical characteristics: strength, hardness, toughness, fatigue;
physical characteristics: residual stresses, surface roughness, porosity, defects.
According to [3, 40], some finishing methods commonly used to reduce to described irregularities are:
mechanical (infilling cavities with the same material, polishing);
chemical (protective layer and color application, use of perspirable solvents, epoxy resins application and metal layer application).

In this paper, the research primarily considers mechanical finishing. However, Sato et al. [41] showed that acetone, as an organic solvent, plays a significant role in chemical treatment of PLA. The main problem with finishing of 3D printed parts is, in most cases, degradation of mechanical properties, then changes in dimensions (tolerances), or the required time for finishing. The quality of manufactured products/prototypes is critical, and therefore needs to be tested. Unfortunately, most of such tests are destructive, which is a problem, since typically one prorotype (or a very small number of them) is made. In order to avoid destructive tests, it is suggested to control process operations, various methods, and tools [42–44].

7 The Algorithm of the Finishing Process

The algorithm of the finishing process is given in Fig. 10.

It encompasses the manufacturing process and the end of the production process. Before printing begins, the parameters must be defined, and the material must be chosen. During the printing process, the operator must constantly monitor the printer operation, e.g. to check if the layers are properly formed and glued along the z-axis.

At the end of the manufacturing process, additional geometry and surface inspection is performed. Finally, if accepted, the manufactured part needs to be additionally polished (mechanically and chemically).

With this treatment, a prototype is obtained that is equivalent in size, shape and dimensions to the one that would be made by using conventional machining. If the model does not satisfy the above criteria, it must be discarded. In the algorithm, the waste of the material caused by discarding a model is not seen as permanent loss of the material, but as a recyclable and reusable resource.

There is a significant difference between the visual representation and the realized model regarding the surface quality of the manufactured model, which depends on the printer parameters and characteristics, and the spatial orientation.

The prototype industrial transmitter enclosure realized by using a 3D printer is shown in Fig. 11. It can be said that there are many irregularities on its surfaces, so that additional processing is necessary.

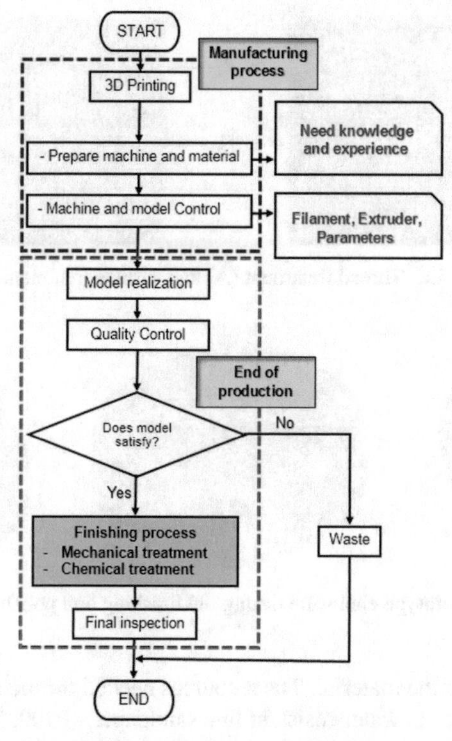

Fig. 10. The algorithm of the prototype manufacturing process

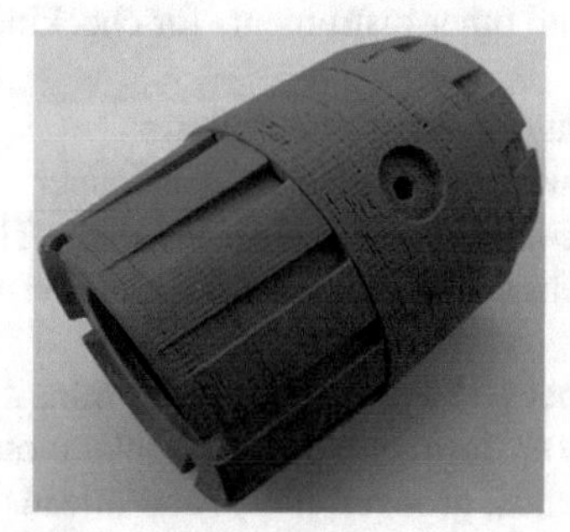

Fig. 11. Manufactured enclosure before processing

The procedure of rough processing involves the mechanical removal of excess material, as well as finishing at processing centers (scraping, milling, drilling, etc.), see Fig. 12.

One of the threads before and after further machining is shown in Fig. 13. Song et al. [45] proposed the screw connection realization using AM with control of dimensions and tolerances.

Further treatment of the manufactured parts was carried out by using three sandpapers of different granulations, with constant water cooling. The cooling is necessary to prevent

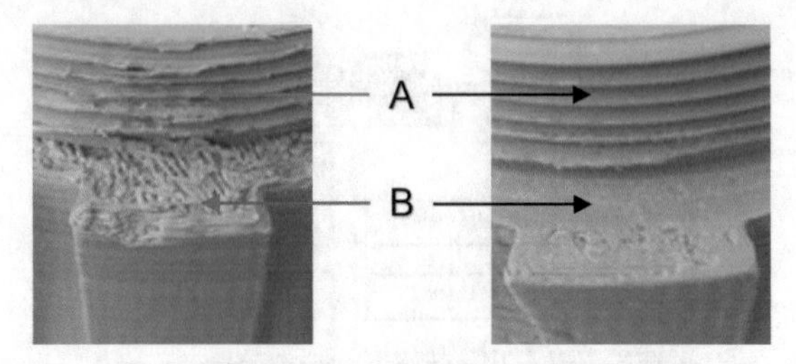

Fig. 12. Thread treatment (A) and surface treatment (B)

Fig. 13. Prototype enclosure during the finishing and polishing process

heating and melting of the material. The resources needed for mechanical processing of the enclosure parts are: 1) water-resistant fine sandpaper - P100, 2) water-resistant fine sandpaper - P180, 3) water-resistant fine sandpaper - P320, 4) continual cooling water source; 5) air source for cleaning and drying surfaces; 6) equipment for fine grinding and polishing; 7) calipers and other instruments for checking of dimensions before and after processing.

The sandpaper treatment process was as follows:

Surfaces were treated with water-resistant sandpaper - P100, by three operators, for 40 min. Each operator controlled all the treated parts. The grinder was also used to eliminate major imperfections and surplus material. After visual inspection, dimensional control was performed.

Water-resistant sandpaper P180 was used for 25 min. As in the previous case, the treatment was carried out by uniform pressure in circular motions, with continuous water cooling. Irregularities on bends, grooves, holes, and threads were removed in this way. The goal was to reduce the dimensions to the correct values.

Finally, water-resistant sandpaper - P320 was used to polish the surfaces as much as possible, and to prepare them for the gritting and painting process. The duration of this operation was 20 min.

The process of gritting in order to obtain a smooth surface for painting was carried out by using the water-resistant fine sandpaper P400. The finalization/polishing process lasted for a total of 20 min, after the putty was applied and dried (see Fig. 13).

The manual polishing process significantly increases the quality of the treated surfaces, but also increases the risk of exposing the employees to the atmosphere contaminated with fine particles [46].

In the end, the manufactured enclosure was painted, in order to give it the final appearance (see Fig. 14). Although the whole idea started from the existing industrial enclosure made of aluminum alloy, the goal of this work was to show the manufacturing workflow when AM is used. It is evident that there are issues to be solved from the beginning to the final stage. However, the savings in terms of the used material, time, and money are very important when the whole product development cycle is considered.

Fig. 14. Photograph of the realized prototype of the industrial transmitter enclosure

The required time for additional processing and inspection is given in Table 4.

Table 4. The total time in processing and control

Operation	Running time [min]	Number of operations
Mechanical treatment	105	1–3
Chemical treatment	20	1
Inspection	15	1

8 Discussion

In this example, we have tried to point out the advantages and disadvantages of AM over CM. A comparative analysis of CM and AM applications is given in the example of the realized prototype of the industrial pressure transmitter enclosure (see Table 5).

In the case of more complex surface shapes, where the distances between points are minimal and where frequent reorientation of the tool is required, AM is far superior to CM. The transmitter enclosure manufacturing is a good example of how designers can change their decisions at any stage of the product design. Also, this methodology gives a much greater freedom for modification of products, with minimal losses and without waste accumulation.

Table 5. CM and AM comparison: the example of the pressure transmitter enclosure

Production type	CM	AM
Usage	Manufacturing of high-quality prototypes in small series	Rapid prototyping for small series, suitable for the production of the casting mold
Material	A combination of different metallic materials: aluminum, steel, brass, copper	Thermoplastics, soluble metals, edible materials
Manufacturing	Closed system	Open system
Model quality	On a high level	At a much lower level
Tools	Several (main + auxiliary)	One (main)
Basic operations	Operations in processing centers (scraping, milling, drilling)	The material is applied in layers
Secondary operations	Threading and grooving, fine surface treatment, surface protection	Sanding, polishing and painting of surfaces
Processing time	Long - involves the use of more tools and accessories	Short - no wasted time on changing tools
Training	Long for each operation and eachtype of material	Short - just a few hours
Energy use	Throughout the process	Only during the material transformation process
Waste	Metal, various emulsions and oils	Minimal material waste, e.g. if the model does not fully meet the criteria after inspection
Environmental Protection	Partial - (care to recycle waste should be taken and hazardous material disposed of)	Full - 100% recyclable

9 Conclusion

The close connection between additive technologies and computing technologies through 3D software packages vastly facilitate the realization of products of complex geometry. The 3D CAD model can be realized by using a single device using and technology, without the need for auxiliary tools and additional manipulation of the model. This reduces the total cost through savings both at the design stage and in the processing.

The workspace size and AM machine speed allow for individual or small-scale production. Using more AM machines can be considered in production planning, as that can enable efficient realization of more products with different specifications. The communication between the manufacturer and the customer is more efficient due to the accelerated decision making process. This means that AM is a good strategic platform for technology-oriented organizations when competitors offer similar (or better) product at lower prices, and when the company has quality research and development.

With additive technology, it is possible to make a quick change in an existing product, perform experiments on an existing design, and permit the organization to adapt to the rapid prototyping concept. Market circumstances require new knowledge and skills from both designers and operators. The industrial transmitter enclosure, whose prototype was manufactured by using AM in this work, is an example of multifaceted engineering that includes mechanical, electronic and software aspects, which makes it a sophisticated design. In its realization, a care was taken to keep all the defined dimensions and geometry, but also to test the cost-effectiveness of the product realized by using AM.

The prototype of the pressure transmitter enclosure, made by using a 3D printer as described in this work, is equivalent to the conventionally fabricated enclosure in all aspects except the used material. Unfortunately, the realized enclosure did not satisfy all the criteria regarding the surface roughness quality and the material strength, so additional mechanical processing was done.

In this paper, the following assumptions were analyzed: 1) finishing has a key influence on the final dimensions and shape of the manufactured parts (the least effect is on elements with a round shape, and the largest on elements that have sharp edges), and 2) mechanical processing of surfaces has adverse effects in terms of deviation from designed dimensions.

In our future work, one of the main objectives in new product development will be to utilize all the benefits of rapid prototyping enabled by 3D printing technology.

Acknowledgement. This work was financially supported by the Ministry of Education, Science and Technological Development of the Republic of Serbia (Grants No. 451-03-68/2020-14/200026 and 451-03-68/2020-14/200105).

References

1. Vorkapić, M., Radovanović, F., Ćoćkalo, D., Đorđević, D.: NPD in small manufacturing enterprises in Serbia. Tehnički vjesnik **24**(1), 327–332 (2017)
2. ASTM F2792, Standard Terminology for Additive Manufacturing Technologies. ASTM (2014)
3. Gibson, I., Rosen, D.W., Stucker, B.: Additive Manufacturing Technologies. Springer, New York (2010). https://doi.org/10.1007/978-1-4939-2113-3
4. Van Wijk, A.J.M., van Wijk, I.: 3D Printing with Biomaterials: Towards a Sustainable and Circular Economy. IOS Press, Amsterdam (2015)
5. Gebler, M., Uiterkamp, A.J.S., Visser, C.: A global sustainability perspective on 3D printing technologies. Energy Policy **74**, 158–167 (2014)
6. Shah, R., Ward, P.T.: Lean manufacturing: context, practice bundles, and performance. J. Oper. Manage. **21**(2), 129–149 (2003)
7. Harris, I.D.: Additive manufacturing: a transformational advanced manufacturing technology-additive manufacturing represents a new paradigm and offers a range of opportunities for design, functionality, and cost. Adv. Mater. Process. **170**(5), 25 (2012)
8. Sachs, E., Cima, M., Williams, P., Brancazio, D., Cornie, J.: Three dimensional printing: rapid tooling and prototypes directly from a CAD model. J. Eng. Ind. **114**(4), 481–488 (1992)

9. Lu, B., Li, D., Tian, X.: Development trends in additive manufacturing and 3D printing. Engineering **1**(1), 85–89 (2015)
10. Zhang, L., Dong, H., Saddik, A.E.: From 3D sensing to printing: a survey. ACM Trans. Multimed. Comput. Commun. Appl. (TOMM) **12**(2), 1–23 (2015)
11. Vayre, B., Vignat, F., Villeneuve, F.: Designing for additive manufacturing. Procedia CIRP **3**, 632–637 (2012)
12. Davia-Aracil, M., Hinojo-Pérez, J.J., Jimeno-Morenilla, A., Mora-Mora, H.: 3D printing of functional anatomical insoles. Comput. Ind. **95**, 38–53 (2018)
13. Campbell, T., Williams, C., Ivanova, O., Garrett, B.: Could 3D printing change the world? Technologies, potential, and implications of additive manufacturing. Atlantic Council, Washington DC (2011)
14. Cupar, A., Pogačar, V., Stjepanovič, Z.: Shape verification of fused deposition modelling 3D prints. Int. J. Inf. Comput. Sci. **4**, 1–8 (2015)
15. Pham, D.T., Gault, R.S.: A comparison of rapid prototyping technologies. Int. J. Mach. Tools Manuf. **38**(10–11), 1257–1287 (1998)
16. Bassett, K., Carriveau, R., Ting, D.K.: 3D printed wind turbines part 1: design considerations and rapid manufacture potential. Sustain. Energy Technol. Assess. **11**, 186–193 (2015)
17. Tian, X., Liu, T., Yang, C., Wang, Q., Li, D.: Interface and performance of 3D printed continuous carbon fiber reinforced PLA composites. Compos. Part A Appl. Sci. Manuf. **88**, 198–205 (2016)
18. Shah, J., Snider, B., Clarke, T., Kozutsky, S., Lacki, M., Hosseini, A.: Large-scale 3D printers for additive manufacturing: design considerations and challenges. Int. J. Adv. Manuf. Technol. **104**(9), 3679–3693 (2019)
19. Petersen, E., Pearce, J.: Emergence of home manufacturing in the developed world: return on investment for open-source 3-D printers. Technologies **5**(1), 7 (2017)
20. Stephens, B., Azimi, P., El Orch, Z., Ramos, T.: Ultrafine particle emissions from desktop 3D printers. Atmos. Environ. **79**, 334–339 (2013)
21. de Ciurana, J., Serenóa, L., Vallès, È.: Selecting process parameters in RepRap additive manufacturing system for PLA scaffolds manufacture. Procedia CIRP **5**, 152–157 (2013)
22. Gupta, B., Revagade, N., Hilborn, J.: Poly (lactic acid) fiber: an overview. Prog. Polym. Sci. **32**(4), 455–482 (2007)
23. Sinclair, R.G.: The case for polylactic acid as a commodity packaging plastic. J. Macromol. Sci. Part A Pure Appl. Chem. **33**(5), 585–597 (1996)
24. Nampoothiri, K.M., Nair, N.R., John, R.P.: An overview of the recent developments in polylactide (PLA) research. Biores. Technol. **101**(22), 8493–8501 (2010)
25. Water, J.J., Bohr, A., Boetker, J., Aho, J., Sandler, N., Nielsen, H.M., Rantanen, J.: Three-dimensional printing of drug-eluting implants: preparation of an antimicrobial polylactide feedstock material. J. Pharm. Sci. **104**(3), 1099–1107 (2015)
26. Carneiro, P.M.C., Gamboa, P.: Structural analysis of wing ribs obtained by additive manufacturing. Rapid Prototyp. J. **25**(4), 708–720 (2019)
27. Almeida, T.C.D., Santos, O.D.S., Otubo, J.: Construction of a morphing wing rib actuated by a NiTi wire. J. Aerosp. Technol. Manage. **7**(4), 454–464 (2015)
28. Fernandez-Vicente, M., Calle, W., Ferrandiz, S., Conejero, A.: Effect of infill parameters on tensile mechanical behavior in desktop 3D printing. 3D Printing Addit. Manuf. **3**(3), 183–192 (2016)
29. Ljungberg, N., Wesslén, B.: Preparation and properties of plasticized poly (lactic acid) films. Biomacromol **6**(3), 1789–1796 (2005)
30. Kim, H., Lin, Y., Tseng, T.L.B.: A review on quality control in additive manufacturing. Rapid Prototyp. J. **24**(3), 645–669 (2018)

31. Peković, O., Stupar, S., Simonović, A., Svorcan, J., Komarov, D.: Isogeometric bending analysis of composite plates based on a higher-order shear deformation theory. J. Mech. Sci. Technol. **28**(8), 3153–3162 (2014)
32. Farin, G., Hoschek, J., Kim, M.S. (eds.): Handbook of Computer Aided Geometric Design. Elsevier, Amsterdam (2002)
33. Zivanovic, S.T., Popovic, M.D., Vorkapic, N.M., Pjevic, M.D., Slavkovic, N.R.: An overview of rapid prototyping technologies using subtractive. Addit. Formative Process. FME Trans. **48**(1), 246–253 (2020)
34. Wong, K.V., Hernandez, A.: A review of additive manufacturing. ISRN Mech. Eng. **1**, 1–10 (2012)
35. Nyman, H.J., Sarlin, P.: From bits to atoms: 3D printing in the context of supply chain strategies. In: 47th Hawaii International Conference on System Sciences, pp. 4190–4199. IEEE (2014)
36. Zeltmann, S.E., Gupta, N., Tsoutsos, N.G., Maniatakos, M., Rajendran, J., Karri, R.: Manufacturing and security challenges in 3D printing. JOM **68**(7), 1872–1881 (2016)
37. Bose, S., Vahabzadeh, S., Bandyopadhyay, A.: Bone tissue engineering using 3D printing. Mater. Today **16**(12), 496–504 (2013)
38. Griffiths, C.A., Howarth, J., De Almeida-Rowbotham, G., Rees, A., Kerton, R.: A design of experiments approach for the optimisation of energy and waste during the production of parts manufactured by 3D printing. J. Clean. Prod. **139**, 74–85 (2016)
39. Bourell, D., Kruth, J.P., Leu, M., Levy, G., Rosen, D., Beese, A.M., Clare, A.: Materials for additive manufacturing. CIRP Ann. **66**(2), 659–681 (2017)
40. Aguilar-Duque, J.I., Hernández-Arellano, J.L., Balderrama-Armendariz, C.O., Avelar-Sosa, L.: Reduction of the fused filament fabrication process time in the manufacturing of printed circuit board slots. Int. J. Comb. Optim. Probl. Inform. **11**(1), 59–75 (2020)
41. Sato, S., Gondo, D., Wada, T., Kanehashi, S., Nagai, K.: Effects of various liquid organic solvents on solvent-induced crystallization of amorphous poly (lactic acid) film. J. Appl. Polym. Sci. **129**(3), 1607–1617 (2013)
42. Straub, J.: Initial work on the characterization of additive manufacturing (3D printing) using software image analysis. Machines **3**(2), 55–71 (2015)
43. Mitrović, N.R., Petrovic, A.L., Milosevic, M.S., Momcilovic, N.V., Miskovic, Z.Z., Maneski, T.D., Popovic, P.S.: Experimental and numerical study of globe valve housing. HEMIJSKA INDUSTRIJA **71**(3), 251–257 (2017)
44. Baltić, M., Svorcan, J., Perić, B., Vorkapić, M., Ivanov, T., Peković, O.: Comparative numerical and experimental investigation of static and dynamic characteristics of composite plates. J. Mech. Sci. Technol. **33**(6), 2597–2603 (2019)
45. Song, P., Deng, B., Wang, Z., Dong, Z., Li, W., Fu, C.W., Liu, L.: CofiFab: coarse-to-fine fabrication of large 3D objects. ACM Trans. Graph. (TOG) **35**(4), 1–11 (2016)
46. Faludi, J., Bayley, C., Bhogal, S., Iribarne, M.: Comparing environmental impacts of additive manufacturing vs traditional machining via life-cycle assessment. Rapid Prototyp. J. **21**(1), 14–33 (2015)

Stability and Initial Failure Analysis of Layered Composite Structures

Ivana Vasovic Maksimovic[1(✉)], Mirko Maksimovic[2], and Katarina Maksimovic[3]

[1] Lola Institute, Kneza Viseslava Street 70a, Belgrade, Serbia
ivanavvasovic@gmail.com

[2] Belgrade Waterworks and Severage, Kneza Milosa Street 27, Belgrade, Serbia

[3] Secretariat for Utilities and Housing Services Water Management, Kraljice Marije Street 1, 11120 Belgrade, Serbia

Abstract. A nonlinear finite element method, based on the von Karman-High Order Shear Deformation Theory (HOST) and on the principle of minimal total potential energy, is used for the analysis of buckling and post buckling behavior and of the first-ply failure of an axially compressed laminated type composite structure. For this purpose and for the analysis of the instability responses an improved 4-node layered shell finite element is proposed. The finite element formulation is based on the third order shear deformation theory with four nodes shell elements having eight degrees of freedom per node. The first-ply failure of laminates and the delamination are some of the features incorporated in the geometric nonlinear formulation. The load-displacement relations for different types of graphite/ epoxy laminates are obtained. Stresses are computed in order to determine the first-ply failure of the mentioned axially compressed laminated composite structure. The buckling, post buckling and failure behavior of axially compressed curved composite panels are investigated numerically and compared with the experimentally obtained results. The buckling loads using linear and geometrically nonlinear analyzes are compared with experimental results and a good agreement between them is obtained. It is shown that the effect of stacking sequence on buckling load is evident.

Keywords: Finite elements · Composite structures · Instability · Initial failure load

1 Introduction

The multilayered fiber-reinforced composites due to their high elastic modulus, high strength, and low weight are very often used in the aerospace structures [1, 2]. The possibility of predicting the buckling behavior of anisotropic and laminated CFC shell structures, subjected to the compression loads is of great importance in the structural analysis [3, 4]. It is well-known from the experiments that the Poisson-Kirchhoff theory of plates underestimates the deflections and overestimates the natural frequencies and buckling loads. These results are the consequence of the neglected transverse shear

N. Mitrovic et al. (Eds.): CNNTech 2020, LNNS 153, pp. 130–146, 2021.
https://doi.org/10.1007/978-3-030-58362-0_9

strains in the classical plate theory (CPT). The improvements of the CPT, based on Kirchhoff's assumptions, were proposed by many authors (see e.g. [5, 6]). The first order, Reissner-Mindlin plate theories (FOST), included the effect of the transverse shear deformations on the buckling behavior of laminated composite plates, and were also used by a certain number of authors (see e.g. [6, 7]). The higher-order theories (HOST) were developed in [8, 9]. The classical laminate plate theory can be applied in many engineering problems. However, laminated shells made of advanced filament composite materials, with very high values of elastic to shear modulus ratios, are susceptible to thickness effects because their effective transverse shear module are significantly smaller than the effective elastic module along the fiber directions. These high ratios of elastic to shear modulus render the CPT inadequate for the buckling analysis of composite shells. The present work is focused on the development of the finite element models appropriate for the implementation into the instability analysis of thin and thick shell type composite structures. The higher-order shear deformation theory offers an accurate and global response just as the 3-D theory does but as it is computationally less demanding, it may be considered efficient for the buckling analysis and it is possible even to say that it provides an improved global response for such a kind of problems. Higher order shear deformation theory (HOST) with imposed condition on vanishing of the surfaced shear stresses is needed for laminated anisotropic shells [10, 11]. An improved 4-nodes shell element is derived combining HOST and the Discrete Kirchhoff's Theory (DKT) (see [15] and [16]) and membrane elements with drilling/rotation degrees of freedom (DOF) [13–19]. In this paper a special attention is paid to the buckling behavior of the laminated curved panels subjected to the axially compression load. A good agreement between the solutions obtained applying described finite elements and the experimental result was obtained.

2 Governing Equations

Composite laminates are modeled using the following theories: (i) equivalent single-layer 2-D theory, (ii) layer-wise 2-D theory and (iii) continuum-based 3-D and 2-D theories. The single-layer theory seems to be the simplest one and the most economical to be used between the three above-mentioned theories. The critical points are the choice of an appropriate theory and the way in which the transverse shear effects are included in it. Thus, in large-scale applications the choice of the theory depends on the relation between the accuracy of the results and the computational costs. From that point of view it seems that the displacements based theories have some advantages.

The formulation of the presented shell finite element is based on the single-layer 2-D theory because of its ability of an adequate representation of the global behavior (deflections, stresses, buckling loads) of thin composites. The HOST used here, assumed the parabolic distribution of the transverse shear stresses across the laminate thickness. The displacement field for the parabolic transverse shear deformation through the shell thickness is given by

$$u_1(x, y, z) = u + z\left[-a\frac{\partial w}{\partial x} + b\Psi_1 - c\frac{4}{3}\left(\frac{z}{h}\right)^2\left(\Psi_1 + \frac{\partial w}{\partial x}\right)\right]$$

$$u_2(x,y,z) = v + z\left[-a\frac{\partial w}{\partial y} + b\,\Psi_2 - c\,\frac{4}{3}\left(\frac{z}{h}\right)^2\left(\Psi_2 + \frac{\partial w}{\partial x}\right)\right]$$

$$u_3(x,y,z) = w \tag{1}$$

where u, v and w are translations of the points (x, y, z = 0) in the middle plane; w is the out-of-plane deflection; Ψ_1 and Ψ_2 are the rotations of a transverse normal about the y - and x - axes, respectively; h is the shell thickness.

The relations (1) are obtained assuming that the transverse shear stresses σ_4 and σ_5 are zero on the shell surfaces

$$\sigma_4(\mathrm{x},\ \mathrm{y},\ \pm\mathrm{h}/2) = 0;\ \sigma_5(\mathrm{x},\ \mathrm{y},\ \pm\mathrm{h}/2) = 0$$

The parameters a, b and c introduced in (1) can have the values zero and one. By combining their values, the displacement field given by (1) can very simply describe the third order shear deformation theory, the first order shear deformation theory and the classical Kirchhoff's plate theory. In that way it is possible to say that (1) represent the general expressions for the displacements of an arbitrary point of a multi-layered shell for the third order theory. This approach of displacements presentation is suitable for subsequent considerations of the formulation of a general shell finite element. This is particularly suitable for computer programme realization, since by combining parameters a, b and c, it is possible to obtain the desired type of the shell finite element able to describe the thin and thick multi-layered composite shells.

3 An Improved 4-Nodes Shell Finite Element

In the present analysis, a four-node quadrilateral from the 'serendipity' family of two-dimensional Co continuous isoparametric shell finite element with 8 degrees of freedom per node [4] is used. The formulation of a 4-nodes shell finite element that can be good enough also if applied to the thin multilayered plates/shells is by no means an easy matter. The authors' experience has shown that a good approach to the formulation of a 4-node shell finite element can be based on the application of the Discrete Kirchhoff's Theory (DKT) for bending behavior. DKT ensures C1 continuity at discrete points on inter-element boundaries. The improved 4-nodes layered shell element is derived combining HOST and DKT (Fig. 1). More details about that element can be found in [10] and [12].

In the Co finite element theory the continuum displacement vector within the element is defined by

$$a = \sum_{i=1}^{M} N_i(r,s)\ a_i \tag{2}$$

where Ni (r, s) is the interpolation function associated with the node i and expressed through the normalized coordinates (r, s); M is the number of nodes in the element and ai is the generalized displacement vector in the mid-surface.

In the case of the negligible mid-surface normal stress σz the stress-displacement relationships, stress resultants and the constitutive equations associated with the higher-order shear deformation theory are given in [2] and [17].

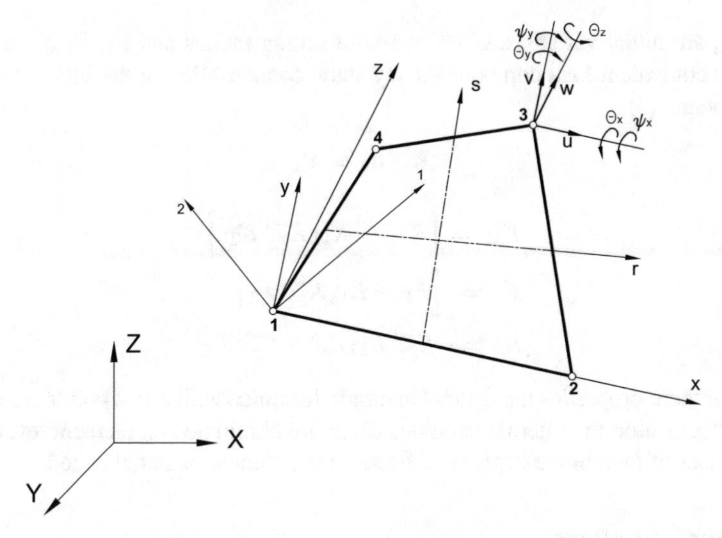

Fig. 1. The improved 4-nodes shell element

The total stiffness matrix of the element is obtained by the linear superposition of the following three independent parts:

Membrane stiffness matrix K_M
Bending stiffness matrix K_B, and
Rotational stiffness matrix $K_{\Theta z}$

In order to avoid irregular systems of equations in the case of completely plane systems, a very small rotational stiffness is adjoined to the variable Θz defining the rotation about the z-axis and it causes a larger stiffness of the system. The displacements (u, v) for the membrane element behavior are approximated by 6-term quadratic polynomials as shown in [4] and are defined by the expressions (3)

$$u = \sum_{i=1}^{4} N_i\,(r,s)\,U_i + (1 - r^2)\alpha_1 + (1 - s^2)\alpha_2$$

$$v = \sum_{i=1}^{4} N_i\,(r,s)\,V_i + (1 - r^2)\beta_1 + (1 - s^2)\beta_2 \tag{3}$$

where U_i, V_i are the nodal in-plane displacement; $(1 - r^2)$ and $(1 - s^2)$ are the incompatible shape functions. The displacements $b = [\alpha_i, \beta_i]^T$ can be taken as some internal displacements having a quadratic effect on actual displacement. The membrane element equilibrium relations are organized in a matrix form (4)

$$\begin{bmatrix} K_{11} & K_{12} \\ K_{21} & K_{22} \end{bmatrix} \left\{ \frac{d_n}{b} \right\} = \left\{ \frac{F_1}{F_2} \right\} \tag{4}$$

where d_n are nodal variables; b are non-conforming modes and F_1, F_2 are the corresponding equivalent load components. By static condensation of the internal variables b, one obtains (5)

$$K_M\ d_n = F_n \tag{5}$$

$$K_M = \left[K_{11} - K_{12} K_{22}^{-1} K_{21}\right]$$
$$F_n = \left[F_1 - K_{12} K_{22}^{-1} F_2\right]$$
$$b = -K_{22}^{-1} K_{21}\, d_n + K_{22}^{-1} F_2$$

For the element properties the Gauss Quadratic formulae with 2 × 2 points are used. By static condensation the internal variables αi, βi are eliminated on element level and the total number of membrane degrees of freedom per element is not changed.

4 Failure Analysis

Various first-ply failure theories are incorporated in the prebuckling, buckling and post-buckling failure analysis of the laminated fibrous composite structures. The Tsai-Wu criterion (see [6]) was considered here for the determination of the first-ply failure. Failure of an axially compressed panel was initiated near region with severe local bending gradients. Computational and experimental results indicate that local failures will occur in the regions of large radial displacements. These local failures are associated with the brittle failure characteristics of the graphite-epoxy material system. Lamina failure is said to have occurred when the state of stress at any point within the lamina not satisfies the tensor polynomial form of the Tsai-Wu criterion. The Maximum strain criterion is defined in terms of the limits in the strains in the principal material directions. These failure criterions are incorporated in geometric nonlinear postbuckling initial failure analyses.

Initial failure of layer within the laminate of a composite structure can be predicted by applying an appropriate failure criterion or first-ply failure theory. For the laminated composite structures is noticeable dependent of ply orientation and direction from load and geometry. For this structures four modes of failure: matrix cracking, fiber-matrix shear failure, fiber failure and delamination can be defined. Different failure theories are used in the pre-, buckling and post-buckling failure analysis of the laminated fibrous composite structures. Failure criteria in correlation with finite element results for the nonlinear postbuckling solutions to a significant extent predict the load value and the location of local failures in the laminates. These predictions are in very good correlation with experimental results. A finite element computational procedure is used for the first-ply failure analysis of laminated shells and the procedure is based on the higher-order shear deformation theory and tensor polynomial failure criterion

$$F_i\sigma_i + F_{ij}\sigma_i\sigma_j + F_{ijk}\sigma_i\sigma_j\sigma_k + \ldots \geq 1 \tag{6}$$

where σ_i are the stress tensor components in material coordinates and Fi, Fj and Fijk are the components of the strength tensors. Most failure criterion is based on the stress state in a lamina. The Tsai-Wu criterion [21] was considered in this work in order to determine first-ply failure. Failure of axial compressed panel initiated near region with severe local bending gradients. Computation and experimental results indicate that local failures occurred in regions of large radial displacements. These local failures are associated with the brittle failure characteristics of the graphite-epoxy material system. The procedure of calculating the first-ply failure load of laminated composite panels refers to calculating of stress and strains at all the nodes for each layer of laminate and then the maximum values of stress and strain are picked up. The failure loads for the weakest ply in the plate/shell are then calculated using various failure criteria using the iteration procedure. The increment in load level can be made suitable for predicting the failure load.

a) Tsai-Wu Criterion

The coefficients Fi and Fij in Eq. (6) are functions of the unidirectional lamina strengths and are presented below for Tsai-Wu criteria:

$$F_1\sigma_1 + F_{11}\sigma_1^2 + F_2\sigma_2 + F_{22}\sigma_2^2 + F_{12}\sigma_1\sigma_2 + F_{66}\sigma_6^2 \le 1 \quad (7)$$

where

$$\begin{aligned} F_1 &= X_t^{-1} - X_c^{-1}, & F_2 &= \left(Y_t^{-1} - Y_c^{-1}\right) \\ F_{11} &= (X_t X_c)^{-1}, & F_{22} &= (Y_t Y_c)^{-1} \\ F_{12} &= -(X_t X_c Y_t Y_c)^{\frac{1}{2}} & F_{66} &= S^{-2} \end{aligned}$$

Here are: X_t, X_c are the longitudinal tensile and compressive strengths, Y_t, Y_c are the transverse tensile and compressive strength, S- rail shear strength.

5 Numerical Examples

As an illustration of the proposed computational method for the buckling and postbuckling analysis and for the initial failure analysis of layered composite structure an axially compressed curved laminated panel is considered. The panel geometry and boundary conditions of the panel supports are shown in Fig. 2. The mechanical properties of CFC lamina and stacking sequence are given in Table 1.

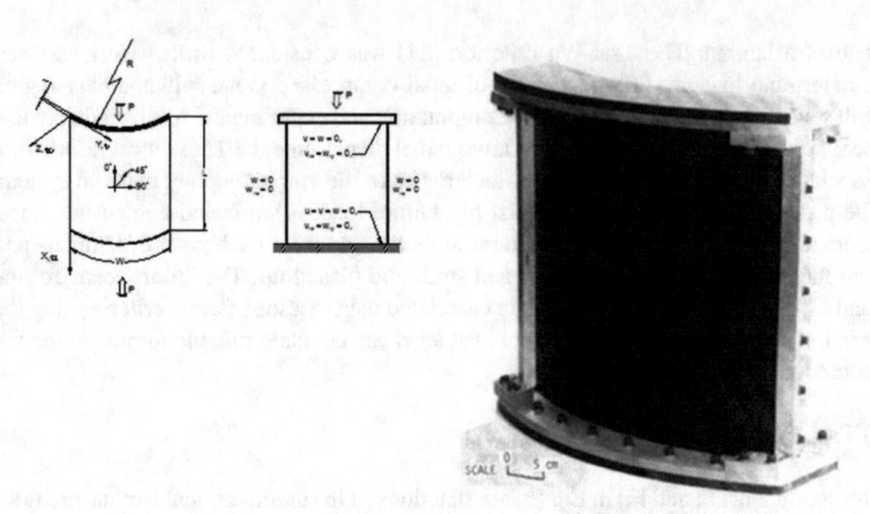

Fig. 2. Axially compressed curved composite panel

Table 1. Graphite-epoxy lamina properties

Material properties	Ultimate strains
$E_{11} = 124{,}500$ MPa	$\varepsilon_{11}^{t} = 0.0110$
$E_{22} = 12{,}450$ MPa	$\varepsilon_{11}^{c} = 0.0086$
$G_{12} = 6{,}100$ MPa	$\varepsilon_{22}^{t} = 0.0036$
$\nu_{12} = 0.38$	$\varepsilon_{22}^{c} = 0.0100$
$t_{layer} = 0.127$ mm	$\varepsilon_{12} = 0.0150$
Stacking sequences	
$A: [\pm 45/0/90/\pm 45/0/90]_s$	
$B: [\pm 45/90_4/\mp 45]_s$	
$C: [\pm 45/\mp 45/\pm 45/\mp 45]_s$	
$D: [\pm 45/\mp 45]_s$	

The nonlinear response (P-w) of axially compressed curved composite panel in the case of stacking sequences A, B, C, D as well as the difference between two initial failure criterions is presented in Figs. 3, 4 and 5 and the numerical and experimental results (see [20]) including buckling and postbuckling behavior are given in Table 2.

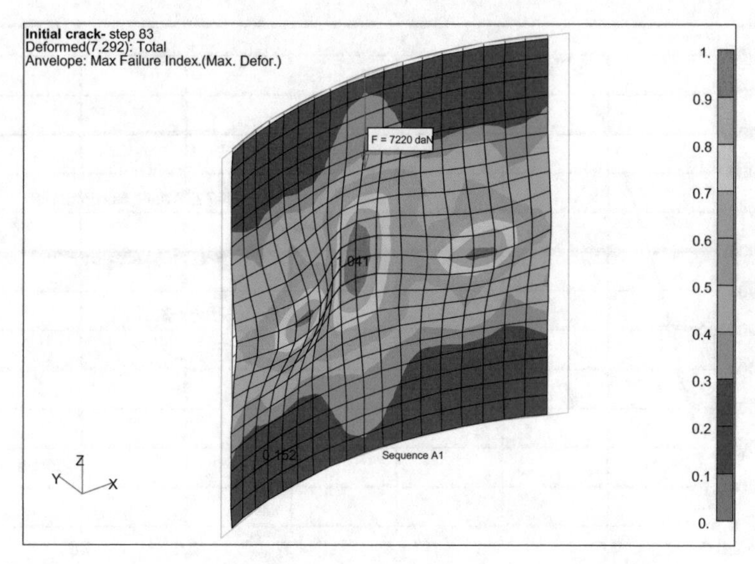

Fig. 3. Postbuckling response. Axial load – radial displacement (P-w)

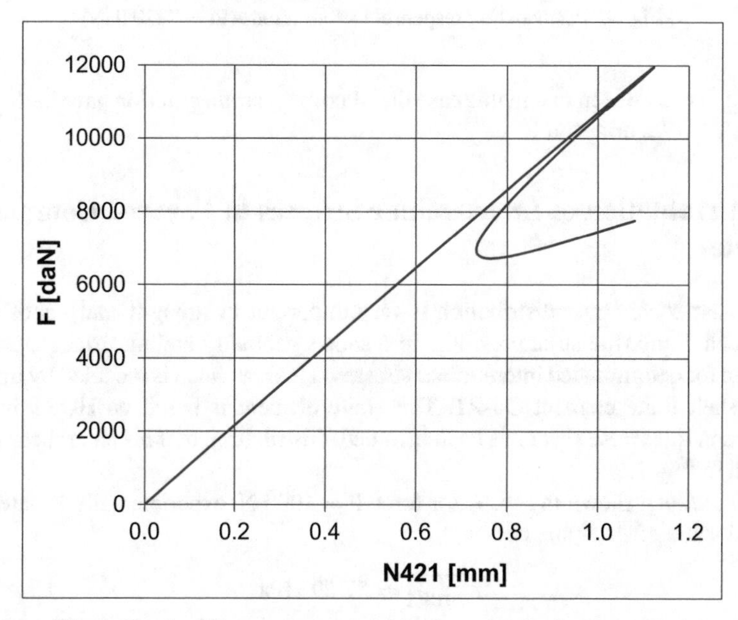

Fig. 4. Postbuckling response. Axial load – axial displacement (P-u)

Two initial failure criterions are used: (1) Maximum strain criterion and (2) Tsai-Wu criterion. Table 2 shows comparisons of experimental and numerical results (nonlinear approach) for initial failure load Pf and it can be seen that good agreement is obtained. From the practical point of view the numerically obtained results can be treated as

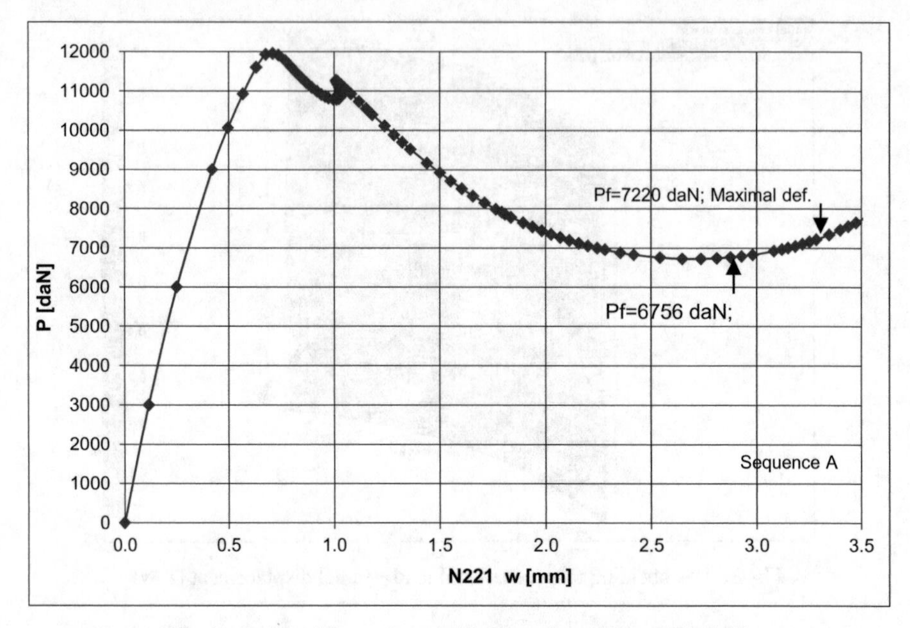

Fig. 5. Postbuckling response. Deformations (Pf = 7220 daN)

excellent. It can be seen that in this case the Maximum strain criterion gave better results than the Tsai-Wu criterion.

6 Determination of Interlaminar Stresses to Layered Composite Plates

Interlaminar shear stress distribution is very important in strength analysis of layered thin walled composite structures. Figure 6 shows geometry and material properties of specimen for determinated interlaminar stresses. This specimen is modeled by improved 4-node shell finite element Q4-RT. This finite element is based on HOST in which formulation transverse shear has been included. Distribution of transverse shear stresses is shown in Fig. 7.

For specimen shown in Fig. 7, for force F = 1000 N, experimentally is determined maximal value of shear stress:

$$\tau_{xz,\max}^{EXP} = 57.69\,\text{MPa}$$

$$\tau_{xz,\max}^{FEM} = 57.82\,\text{MPa}$$

On the basis of the comparison, of course a simple problem, a practically complete agreement of the FEM results with the experiment was obtained, which is very important for quality of finite element verification.

The second part of the analysis is concerned with determining the effect of stacking sequence on the distribution of interlaminar stresses, τ_{xz}, per the thickness of the layered

Table 2. Experimental and FE results of axial compressed curved panel

Speci-mens No.	Experiments [20]			Presented FE results			
	u_{cr} [mm]	P_{cr} [daN]	P_f [daN]	u_{cr} [mm]	$P_{cr,l}$ [daN]	$P_{cr,n}$ [daN]	P_f [daN]
A1	1.07	11332	8296				7220 (Max. def)
A2	1.12	11364	7932	1.124	12807	11956	
A3	1.22	11912	7226				6756 (Tsai-Wu)
B1	2.32	11101	5952				
B2	2.29	11352	5086	2.068	11460	10540	6050
B3	2.08	10210	5948				
C1	2.18	7706	6274	1.79	8030	7560	6017
C2	2.21	7562	5651				
D1	0.95	1791	1366	0.899	2005	1924	1315
D2	0.89	1661	1455				

ucr – Critical axial displacement (correspond to P_{cr})
$P_{cr,l}$ – Buckling load (linear "eigen" analysis)
$P_{cr,n}$ – Buckling load (geometrical nonlinear analysis)
P_f – Applied load at failure (Criterions: Maximum deformations and Tsai-Wu)

$$\tau_{xz,\max}^{EXP} = 57.69\,MPa$$

$$\tau_{xz,\max}^{FEM} = 57.82\,MPa$$

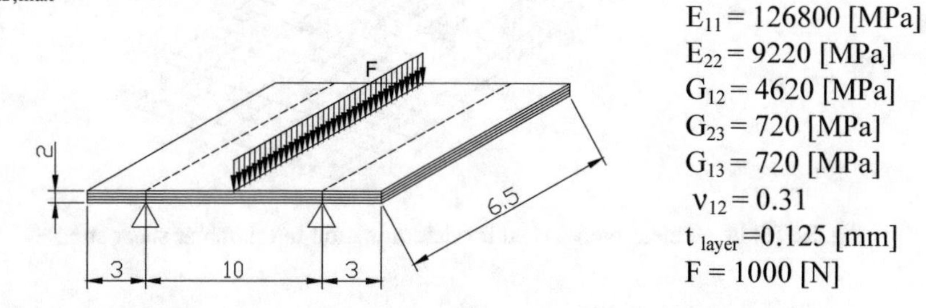

Fig. 6. Model of specimen for determination of interlaminar stresses

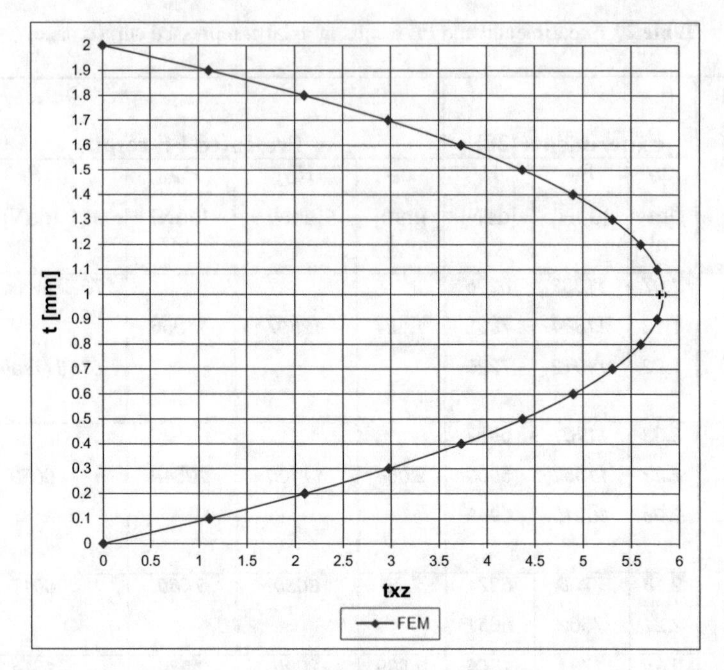

Fig. 7. Interlaminar shear stress distribution, τ_{xz}, in location of force action

composite plate. For this analysis the geometry of structural element and finite element model is used, as shown in Fig. 8. Three stacking sequences considered: sequences $[0_{16}]$, $[0_4/90_4]_s$ and $[90_4/0_4]_s$. and results are shown in Table 3 and Fig. 9.

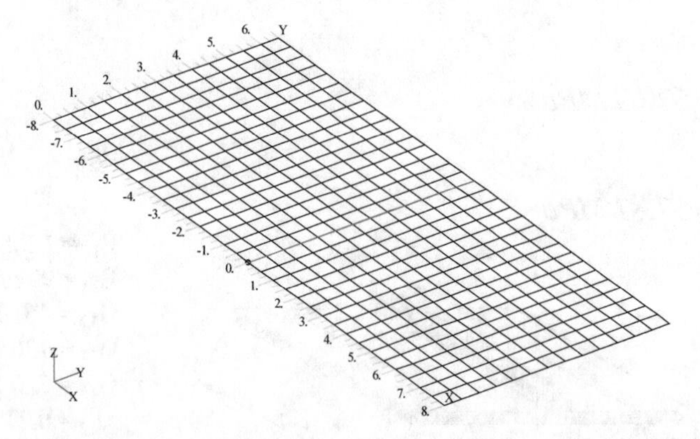

Fig. 8. Finite element model used for determination interlaminar shear stresses

The effects of stacking sequences of layered shell type composite laminate on interlaminar shear stress distribution are shown in Fig. 9. From diagrams shown in Fig. 9 is evident that influence of stacking sequence is very important especially on its maximal values. Thin walled layered shell type structures are very sensitive with respects of its

Table 3. Distribution transverse shear stresses, τ_{xz}, per thickness

t [mm]	$[0_{16}]$	$[0_4/90_4]_s$	$[90_4/0_4]_s$
0	τ_{xz} [MPa]	τ_{xz} [MPa]	τ_{xz} [MPa]
0.1	1.108	1.247	0.425
0.2	2.1	2.362	0.805
0.3	2.975	3.346	1.14
0.4	3.733	4.199	1.431
0.5	4.375	4.921	1.677
0.6	4.9	4.964	4.443
0.7	5.308	4.998	6.595
0.8	5.6	5.021	8.132
0.9	5.775	5.036	9.055
1	5.833	5.041	9.362
1.1	5.775	5.036	9.055
1.2	5.6	5.021	8.132
1.3	5.308	4.998	6.595
1.4	4.9	4.964	4.443
1.5	4.375	4.921	1.677
1.6	3.733	4.199	1.431
1.7	2.975	3.346	1.14
1.8	2.1	2.362	0.805
1.9	1.108	1.247	0.425
2	0	0	0

shear strength. Finite elements are able, based on HOST, to precise define shear stress distributions.

7 Numerical and Experimental Analysis of Axial Compressed Flat Laminated Panel

Attention in this section is focused to buckling and post-buckling analysis of an axially compressed the carbon epoxy composite (CFC) panel. A description of test specimen, testing equipment and instrumentation is given in Fig. 10. The problem of deriving the buckling load from the test data is discussed. A comparison between experimental and computation results is made for buckling load and post-buckling behavior of the layered composite panel. Experimental results are compared with computation results using Finite Element Method (FEM).

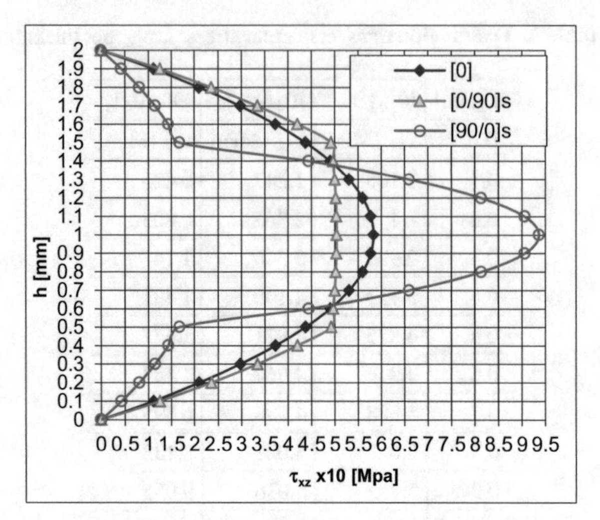

Fig. 9. The effect of stacking sequences to shear stress distributions, τ_{XZ}

8 Experimentally Determination Postbuckling Behavior of Axially Compressed Composite Panel

To determine buckling load of axially compressed composite panel here strain gauges are used. Strain gauges were applied on both sides of the layered composite plate in loading direction, Fig. 10.

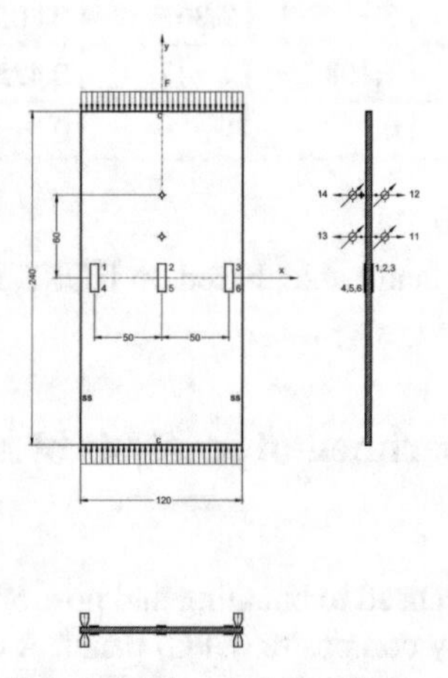

Fig. 10. Positions of strain gauges and transducers on testing specimen

Results of one buckling test are given in Fig. 11 and 12. The output of three strain gauge pairs as well as the membrane strains is shown.

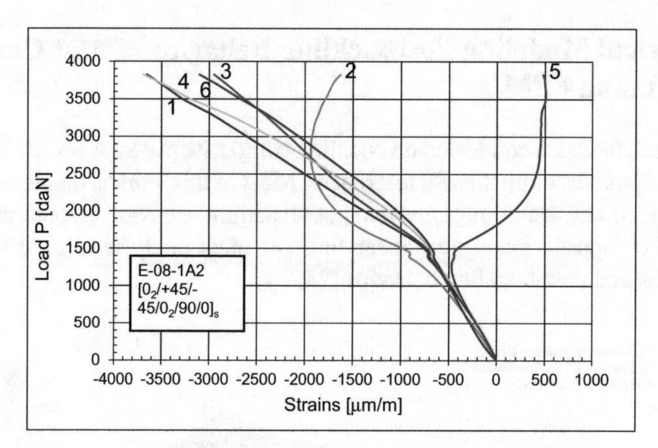

Fig. 11. Load (P) versus membrane strains (ε_m) (both sides)

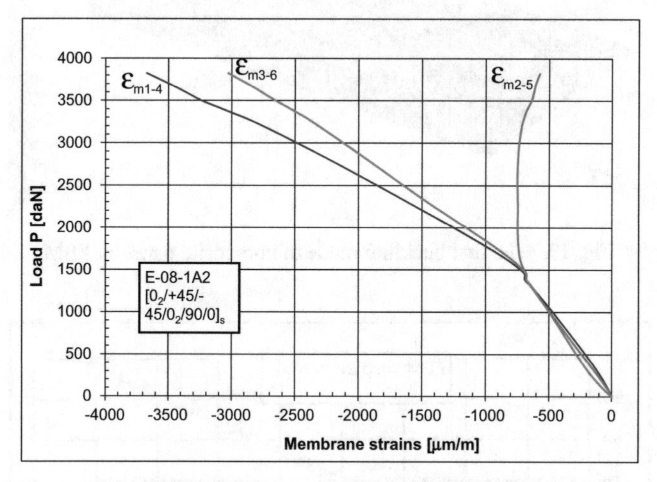

Fig. 12. Load (P) versus membrane strains (ε_m)

The load versus strain plots, Figs. 11 and 12, show the strain reversal behavior near the buckling load. The test data show a large transition area between the pre- and post-buckling situation so a well-defined buckling point cannot be given as one value. The membrane strains were plotted versus end load, Fig. 12. From Fig. 12 (Diagrams P-ε_m: loads versus membrane strains) buckling load of this composite panel is in range (1380–1450) daN. The composite plates were tested in a special tool that was attached to an SCHRENK test machine. To achieve clamped (c) and simply supported (ss) boundary conditions during loading of panel, as shown in Fig. 10, in this testing special tool is adopted.

9 Numerical Modeling Postbuckling Behavior of Flat Composite Panel Using FEM

Lot of research has been conducted on buckling composite plates. A review of the subject by Leissa [22] quotes more than 90 references. Most of this work is theoretical in nature, and very little, if any, attention is given to post-buckling behavior. To determine stability behavior of composite panel non-linear finite element analysis is used to determine buckling load and post-buckling behavior [23, 24].

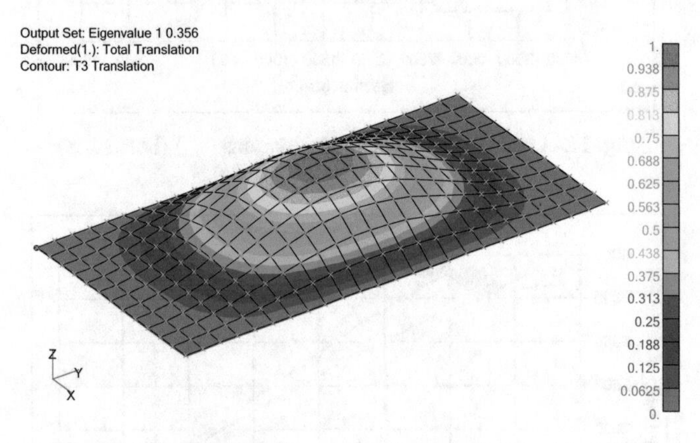

Fig. 13. The first buckling mode of composite panel by FEM

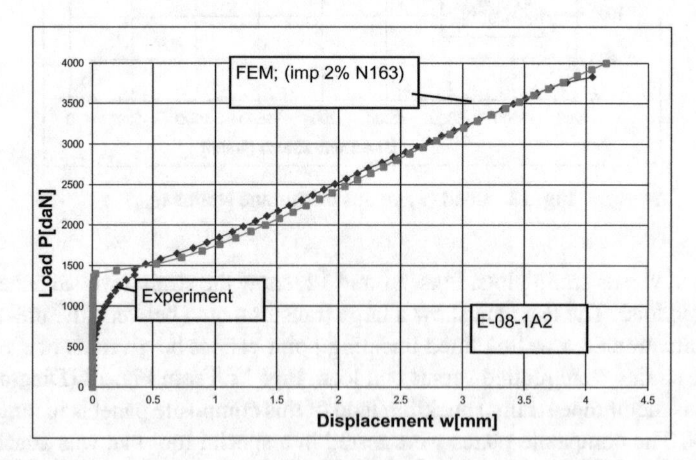

Fig. 14. Relations between load (P) and displacement - (w), with 2% imperfection

Finite element model and first buckling mode is shown in Fig. 13. In Fig. 14 experimental and computation results of buckling and post-buckling behavior using relation between load **P** and out of plane displacements (**w**) are given.

Maximum of plane displacements (w) measured for panels as defined in Fig. 10, is presented in Fig. 14. The load versus out of plane displacement plot, Fig. 14, shows a change of slope close to the buckling load. The pre-buckling part of this line is straight and if the post-buckling part of this line forms a straight line too, point of intersection can be taken as the buckling load. Buckling load using non-linear finite element analysis is $P_{cr} = 1405$ daN, Fig. 14. In Fig. 14 finite element predictions is shown for initial imperfection of 2% panel thickness.

10 Conclusion

This work present stability and initial failure analysis of layered composite structural components. The comparison of numerically obtained results with the experimental data shows that the proposed improved 4-nodes shell finite element, based on High Order Shear Deformation Theory (HOST), can be successfully applied for the prebuckling, buckling and postbuckling analysis as well as for the initial failure analysis and for the prediction of the location of local failure of compressed layered composite panels. In this research is shown that the FEM based on HOST is reliable method for determination interlaminar stresses in layered composite structures. It is concluded that the maximum strain criterion gives better results than the Tsai-Wu criterion. Computation methodology presented in this research can be efficient procedure in strength analysis of flight structures.

Acknowledgment. The authors would like to thank the Ministry of Education, Science and Technological Development of the Republic of Serbia for financial support.

References

1. Jankovic, D., Maksimovic, S., Kozic, M., Stupar, S., Maksimovic, K., Vasovic, I., Maksimovic, M.: CFD calculation of helicopter tail rotor airloads for fatigue strength experiments. J. Aerosp. Eng. **30**(5), (04017032-1)–(04017032-11) (2017). https://doi.org/10.1061/(ASCE)AS.1943-5525.0000734, ISSN (print): 0893-1321, ISSN (online): 1943-5525
2. Maksimovic, S., Kozic, M., Stetic-Kozic, S., Maksimovic, K., Vasovic, I., Maksimovic, M.: Determination of load distributions on main helicopter rotor blades and strength analysis of its structural components. J. Aerosp. Eng. **27**(6), (04014032-1)–(04014032-8) (2014). https://doi.org/10.1061/(ASCE)AS.1943-5525.0000301
3. Maksimović, S., Maksimović, K., Vasović, I., Maksimović, M., Stamenković, D.: Strength analysis of helicopter main rotor blade made from composite materials. In: 14th International Conference on Accomplishments in Mechanical and Industrial Engineering, Banja Luka, Republic of Srpska, Bosnia and Herzegovina, 24–25 March 2019, pp. 403–408 (2019). ISBN 978-99938-39-85-9, COBISS.RS-ID 8166456
4. Maksimovic, S.: Improved geometrically nonlinear finite element analysis and failure of fibre reinforced composite laminates. In: Brandt, A.M., Li, V.C., Marshall, I.H. (eds.) Proceedings of International Symposium on Brittle Matrix Composites 4, Warsaw, 13–15 September, Woodhead Publishing (1994)
5. Reddy, J.N.: Energy and Variation Methods in Applied Mechanics. John Wiley, New York (1984)
6. Reddy, J.N.: A generalization of two-dimensional theories of laminated composite plates. Commun. Appl. Numer. Methods **3**, 113–180 (1987)

7. Lo, K.H., Christensen, R.M., Wu, E.M.: A higher order theory of plate deformation, Part 1: homogenous plates, p. 44. J. Appl., Mechanics (1977)
8. Maksimovic, S., Komnenovic, M.: Improved nonlinear finite element analysis of layered composite structures using third-order theory. In: Topping, B.H.V., Papadrakakis, M. (eds.) The Second International Conference on Computational Structures Technology, Athens, Greece. Civil Comp Press (1994)
9. Maksimovic, S.: Some computational and experimental aspects of optimal design process of composite structures. Int. J. Compos. Struct. **17**, 237–258 (1990)
10. Maksimović, M., Vasović, I., Maksimović,, K., Maksimović, S., Stamenković, D.: Crack growth analysis and residual life estimation of structural elements under mixed modes. In: 22nd European Conference on Fracture - ECF 22, Belgrade, Serbia, 26–31 August 2018
11. Ružić, D., Maksimović, S., Milosavljević, D.: Postbuckling response and failure analysis of layered composite structural components. In: Mechanics and Material Conference, USA, 17–20 June 2003
12. Maksimović, S., Maksimović, K., Vasović, I., Đurić, M., Maksimović, M.: Residual life estimation of aircraft structural components under load spectrum. In: 8th International Scientific Conference on Defensive Technologies, OTEH 2018, Belgrade, Serbia, 11–12 October 2018
13. Allman, D.J.: A compatible triangular element including vertex rotations for plane elasticity analysis. Comput. Struct. **19**, 1–8 (1984)
14. Mac Neal, R.H., Harder, R.L.: A refined four-noded membrane element with rotational degrees of freedom. Comput. Struct. **28**, 75–84 (1989)
15. Vasovic, I., Maksimovic, S., Maksimovic, K., Stupar, S., Bakic, G., Maksimovic, M.: Determination of stress intensity factors in low pressure turbine rotor discs. Math. Probl. Eng. **2014**, Article ID 304638, 9 p. http://dx.doi.org/10.1155/2014/304638
16. Vasovic, I., Maksimovic, S., Maksimovic, K., Stupar, S., Maksimović, M., Bakic, G.: Fracture mechanics analysis of damaged turbine discs using finite element method. Therm. Sci. **18**(Suppl. 1), S107–S112 (2014)
17. Maksimović, M., Nikolić-Stanojević, V., Maksimović, K., Slobodan, S.S.: Damage tolerance analysis of structural components under general load spectrum. Tehnicki Vjesnik-Tech. Gazette **19**(4), 931–938 (2012)
18. Vidanović, N.N., Rašuo, B.B., Kastratović, G., Grbović, A., Puharić, M., Maksimović, K.: Multidisciplinary shape optimization of missile fin configuration subject to aerodynamic heating. J. Spacecraft Rockets (2019). https://doi.org/10.2514/1.A34575
19. Vasovic, I., Maksimovic, M., Maksimovic, K.: Residual fatigue life estimation of structural components under mode-I and mixed mode crack problems. In: Mitrovic, N., Milosevic, M., Mladenovic, G. (eds.) CNNTech 2018. LNNS, vol. 90, pp. 3–21. Springer, Cham (2020). https://doi.org/10.1007/978-3-030-30853-7_1
20. Knight Jr., N.F., James, H., Starnes Jr., J., Allen Waters Jr., W.: Postbuckling behavior of selected graphite-epoxy cylindrical panels in axial compression. In: 27th Structures, Structural Dynamics and Material Conference, San Antoni (1986)
21. Tsai, S.W., Wu, E.M.: A general theory of strength for anisotropic materials. J. Compos. Mater. **5**, 58–80 (1980)
22. Leissa, A.W.: Advances in vibration, buckling and postbuckling studies on composite plates. In: Composite Structures, p. 312. Applied Science Publishers, London (1981)
23. Maksimović, S.: Instability analysis of layered composite structures using shell finite elements based on the third order theory. J. Appl. Compos. Mater. **3**, 301–309 (1996)
24. Buskell, N., Davies, G.A.O., Stevens, K.A.: Postbuckling failure of composite panels. In: Proceedings of 3rd International Conference on Composite Structures, Paisley College of Technology, 9–11 September 1985

Design and Analysis of an Optimal Electric Propulsion System Propeller

Toni D. Ivanov(✉) and Ognjen M. Peković

Faculty of Mechanical Engineering, Department of Aeronautical Engineering, University of Belgrade, 11000 Belgrade, Serbia
tivanov@mas.bg.ac.rs

Abstract. There has been an increased interest in the field of electric power aircraft propulsion systems in recent years mostly due to the technological improvements in power electronics, energy storage and permanent magnet electric motors. This is not unusual since these systems can provide a large increase in efficiency and reliability while decreasing the carbon footprint. In most small and medium size UAVs the fully electric propulsion system is implemented. To obtain a propulsion system with an overall high efficiency one must take into consideration the efficiencies of all individual components as well as their matching meaning that the individual characteristics of all components have to be considered when designing the propulsion system. In this paper a propeller design methodology for a fully electric propulsion system is presented. The electric components were represented via simplified mathematical models while the propeller design was performed by the blade momentum theory with vortex wake deflection including corrections for the tip losses. The airfoil aerodynamic characteristics for different Reynolds numbers were determined by XFOIL and corrected for Mach number and post stall effects due to rotation. An off-design point analysis of the designed propeller was then performed and discussed. It was shown that an efficient fully electric propulsion system for initial design purposes can be created with the methodology. Some recommendations regarding the limitations and the possible improvements are also given in the paper.

Keywords: Propeller design · Electric propulsion · BLDC motor

1 Introduction

A lot of attention has been oriented towards electric propulsion powered aircraft in recent years. The advantages achieved in the fields of power electronics, most significantly in electric motors, motor drivers and batteries have provided engineers with the ability to implement these elements into relatively simple yet reliable and efficient aircraft propulsion systems. Considering the advantages and disadvantages of the current state of technology, different variants for realization of propulsion systems have been derived, some of which are: the full electric, the hybrid serial and parallel, the partial

N. Mitrovic et al. (Eds.): CNNTech 2020, LNNS 153, pp. 147–166, 2021.
https://doi.org/10.1007/978-3-030-58362-0_10

serial/parallel hybrid, the turboelectric and partial turboelectric system [1]. The possibility of decoupling the motor from the power source which can be a battery, battery cell, gas generator etc. means that electric propulsion systems can provide much easier implementation of power distribution and are therefore considered by many research and commercial organizations as a solution to the implementation of distributed propulsion systems. Obvious examples for this notion are: Boeing SUGAR, Nasa N3-X, Nasa LEAPtech, EADS E-thrust, HES Element one, DARPA X-plane, Lilium, etc.

Since their introduction to remote controlled aircraft models in the 1970s electric motors have become somewhat a standard in the RC community. Large breakthrough was done with the introduction of neodymium magnets in the late 1980s which, along with the advances made in scaling production and size of MOSFETs, made the high power to weight BLDC motor possible. With this and with the introduction of lithium-ion batteries in the early 1990s the use of electric propulsion for a wide range of different applications was made possible. The first to introduce this kind of propulsion in the aeronautical sector on a larger scale were manufacturers of small unmanned aerial vehicles or UAVs. First introduced primarily for military applications their use was extremely well received in the civil sector as well. Nowadays UAVs are used in a large range of applications in different industries in the civil sector such as: agriculture, transportation, utility, energy and infrastructure, real estate and construction, insurance, mining, film, media, public safety, delivery, scientific research etc.

This increase of interest has stimulated existing aircraft manufacturers to enter the UAV market as well as the appearance of a large number of new startups and companies specifically specialized in UAV design and manufacturing. The positive market share growth that has been going on in the recent period is expected to continue and even grow in the future with some projections being that the aircraft electrification market is to increase by more than twofold by 2030 while the UAV market is estimated to more than double by 2025. While still in its infancy, especially large growth is expected in the fixed wing VTOL market which is estimated to increase more than seven times from 2019 to 2030 (Fig. 1)[1].

[1] https://www.marketsandmarkets.com/Market-Reports/fixed-wing-vtol-uav-market-173456250.html (Accessed: May 2020)

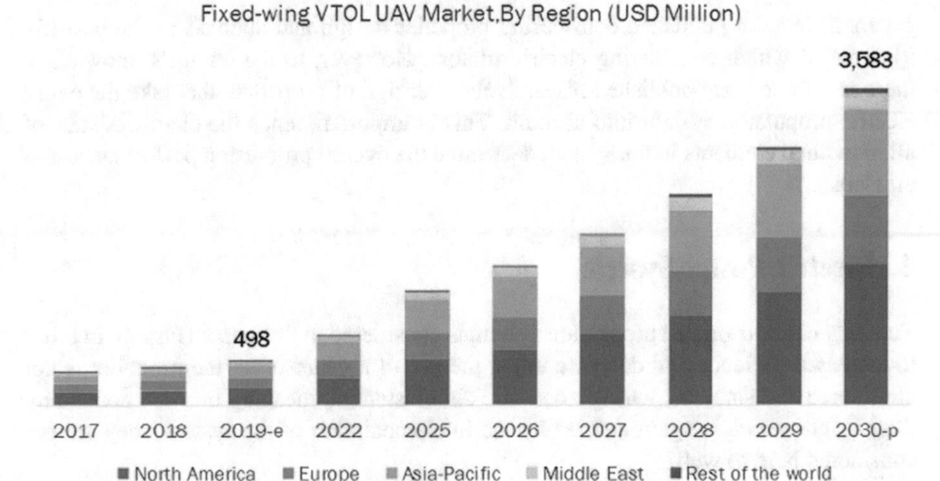

Fig. 1. Fixed-wing VTOL UAV Market estimation, Source: MarketsAndMarkets (see Footnote 1)

The idea behind fixed-wing vertical take-off lift-off or VTOL aircraft is to achieve the cruise efficiency of fixed wing aircraft while at the same time providing the advantages of rotorcrafts that is, eliminating the need for a runway and accomplishing the ability to hover. This idea is not new. In fact, some of the first aircraft of this kind were presented as early as the 1920s. During the 1950s and 1960s large efforts were made in the development of fixed wing VTOL resulting in the creation of the Harrier jet aircraft. In the following years many experimental aircraft have been developed mostly for testing purposes and military applications. Due to the high complexity and price, this kind of aircraft hasn't caught on in the civil sector with limited use for military applications as well. The advances made in electronics coupled with the advances in electric propulsion systems allowed for a decrease in complexity and price which made researchers reconsider fixed-wing VTOLs. Electric propulsion enabled the possibility to consider small fixed-wing VTOL UAVs as well which are of interest to the author.

As previously mentioned there are a lot of manufacturers of BLDC motor and propellers however in most cases there specialized for one or the other and is up to the user to select the optimal configuration. In the literature one can find useful guidelines and algorithms for propeller/BLDC motor matching for electric propulsion systems [2–4]. Although some manufacturers offer combined propulsion systems or recommendations for choosing different components this is almost exclusively for static thrust conditions i.e. for take-off and hover. Even more there is very little information about propellers behavior in axial flight conditions.

The advances made in 3D printing technology and the increase in availability of such printers have made possible for hobbyist and professionals to design and manufacture their propellers. For this reason, the author's intention is to provide a simple methodology for the design and analysis of propellers for a certain electric power system. A lot of papers have been published considering propeller design and analysis in the past [5–9] some of which considering electric motors. However, to the author's knowledge, there is little to none published research about design of propellers that take the entire electric propulsion system into account. This is important since the characteristics of all individual elements in the system determine the overall propulsion performance and efficiency.

2 Electric Power System

The fully electric aircraft propulsion system is considered in this paper (Fig. 2). In order to increase efficiency and decrease mass, the use of a gearbox for the propeller is not considered and since the wiring losses are circumstantial, meaning they are defined by their length which is again defined by the implementation of the system, they are not considered here as well.

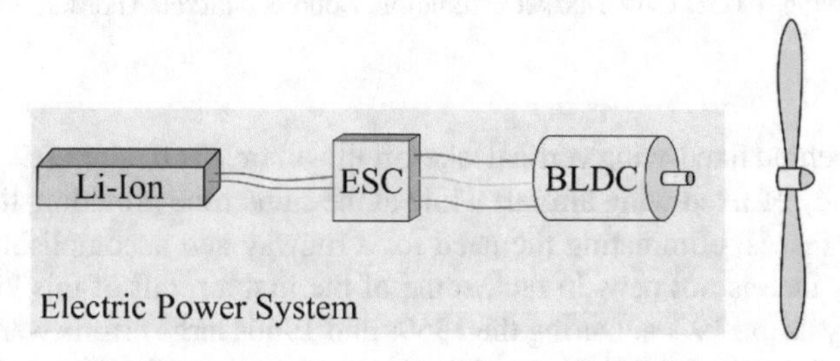

Fig. 2. Electric propulsion system schematics

2.1 Brushless DC Motor

As mentioned previously Brushless direct current or BLDC motors are the most widely used motors today for the propulsion of micro, small and medium aircraft. These motors are essentially permanent magnet synchronous motors with trapezoidal back electromagnetic force or back EMF. This allows them to have 15.4% higher power density than sinusoidal PMSM [10] although with more expressed torque ripples. With the use of an electronic speed controller BLDC motors can be powered by a pulse modulated width - PWM DC power hence the name. It is usual to represent the BLDC motor as a DC motor with the electromechanical circuit shown in Fig. 3 (left). Generally, the schematic in the figure would be equivalent to one coil but if the motor is isotropic it will be valid for the entire motor [11]. Analyzing the circuit, we obtain the following equations which prescribe the behavior of the BLDC motor:

$$\mathrm{U} = Ri + L\frac{\mathrm{d}i}{\mathrm{d}t} + E_{EMF} \tag{1}$$

$$T = J\frac{\mathrm{d}\omega}{\mathrm{d}t} + K_f\omega \tag{2}$$

where the first equation is more commonly called the electrical equation in which U is the terminal phase voltage in [V], R is the armature resistance in [Ω], i is the phase current in [A], L is the armature self-inductance in [H] and EEMF is the back-EMF in [V]. Equation (2) is the mechanical equation and in it T is the motor torque in [Nm], J is the rotor inertia in [kgm^2], ω is the rotor speed in [rad/s] and Kf is the friction coefficient in [Nm/rad/s].

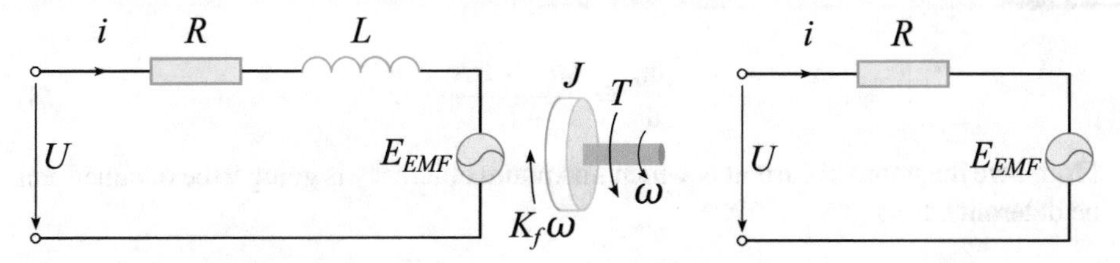

Fig. 3. BLDC motor coil schematic (left) and simplified motor schematic (right)

Although there are many procedures for motor parameter estimation [12–14] most manufacturers do not give information for the rotor inertia, armature inductance and friction coefficient hence for practical applications, it is more usual to represent the BLDC motor via a simplified schematic (Fig. 3, right). This representation, although not completely accurate can be considered satisfactory for initial analysis of stationary regimes. Now introducing a motor velocity or speed constant KV [rpm/V]:

$$K_V = \frac{2\pi}{60K_e} = \frac{n_{no-load}}{U_{peak}} = \frac{2\pi}{60K_T} \tag{3}$$

where Ke is the back-EMF constant, KT = T/i is the torque constant [Nm/A] and $n_{no-load}$ is the rotor velocity at U_{peak} [V] without load in [rpm], the BLDC motor equation can now be given as:

$$U = Ri + n/K_V \tag{4}$$

It should be noted that the motor velocity $n = (U - Ri)K_V$ in Eq. 4 is in [rpm]. Also, while the notion that KT = Ke is correct for trapezoidal BLDC motors, for sinusoidal PMSM motors $K_T = K_e\sqrt{3}/2$ and care must be given for the RMS values in the equation which will depend on the phase connection. The definition for the torque constant would imply that no current will flow when there is no load on the motor shaft. This is however not true since current will flow through the armature even when the motor is in idle i.e. without load due to the friction and other losses in the motor that need to be overtaken for it to rotate. It is therefore standard practice for manufacturers to give the idle current I0 [A] for a motor at certain voltage. Given this, the motor torque can be calculated as:

$$T = K_T(i - I_0) = \frac{30}{\pi K_V}(i - I_0) \tag{5}$$

Having in mind that the motor shaft power $P = T\omega = Tn\pi/30$ in [W] it can be written:

$$P = (U - Ri)(i - I_0) \tag{6}$$

and since the motor electrical power $P_E = Ui$, the efficiency of the electric motor can be determined:

$$\eta = \frac{P}{P_E} = \frac{(U - Ri)(i - I_0)}{Ui} \tag{7}$$

where i^2R are the thermal power losses. If we differentiate Eq. 7 according to the current, we get:

$$\frac{d\eta}{di} = \frac{I_0 U - i^2 R}{i^2 U} \tag{8}$$

From here the nominal current at which maximum efficiency is going to be obtained can be determined:

$$\frac{d\eta}{di} = 0 \rightarrow i_{nom} = \sqrt{\frac{UI_0}{R}} \tag{9}$$

Similarly, from Eq. 6, the peak current i_{peak} at which the motor will have maximum power can be determined:

$$i_{peak} = \frac{U + RI_0}{2R} \tag{10}$$

This current is not realistic however since the motor max current and power limitations are much lower mostly due to overheating.

Let us consider some of the parameters in the previous equations. The electrical resistance R is a function of the material electrical resistivity ρ [Ωm], wire length l [m] and cross-sectional area A [m^2]:

$$R = \rho \frac{l}{A}$$

For metal conductors the material electrical resistivity is often represented as a linear function of temperature (excluding self-heating effects):

$$\rho(T) = \rho(T_0)[1 + \alpha(T - T_0)]$$

where T is the material temperature, T_0 is the base temperature and α is the electrical resistance temperature coefficient. For example copper has electrical resistivity of $\rho(T_0) = 1.68\text{E}-8$ [Ωm] at $T_0 = 20$ °C and $\alpha = 0.00393$ at 20 °C. From here for a fixed l/A, the electrical resistance of the motor as a function of temperature can be calculated:

$$R(T) = R(T_0)[1 + \alpha(T - T_0)] \tag{11}$$

Assuming that the temperature change is proportional to the Ohmic thermal power losses i^2R and that there is a thermal equilibrium [12] the resistance can be expressed as a function of the current:

$$R(T) = R(T_0)\Big[1 + \Big[\alpha(T_{max} - T_0)/i_{max}^2\Big]i^2\Big] \tag{12}$$

where T_{max} is the maximum temperature at i_{max}. Some manufacturers give the maximum temperature at a given maximum current while for other an approximation can be made considering the temperature class of the winding insulation and the maximal allowed current.

As mentioned, the idle or no-load current needs to account for the zero load losses. There are multiple factors that influence this parameter such as bearing friction, air friction etc. so its value will change for different speeds. This change can be easily measured and the I_0 can be represented as a linear or second order function of ω. It should be noted that the back EMF constant and the torque constant will change for different loads and speeds as well [4]. Their behavior can be approximated by some fitting but it would be more practical to use a more complicated mathematical representation of the motor if one wishes to obtain a better model [11, 15, 16]. In this paper K_V and K_T are considered constant with Eq. 3 being valid.

2.2 Electronic Speed Controller

For the BLDC electric motor to be able to run on a DC supply a motor driver or ESC is needed. One of the benefits of BLDC motors is that the armature current is discrete and at a certain angle only two of the phases are supplied with current which allows for simpler motor drivers in comparison to PMSM. However, there are controllers that utilize all three phases in order to obtain higher power values. In order to be able to achieve the electronic motor commutation information about the rotor position is needed which can be obtained via a position sensor (most often Hall effect sensors are used) or by the induced back EMF current in the unpowered phase when a two-phase controller is used. Since the use of an additional sensor increases the price, the back-EMF or sensorless controller is more often used in cheaper motors.

Electronic commutation is performed by the use of switching transistors, from which most popular are the bipolar junction – BJT, insulated gate bipolar – IGBT and metal-oxide semiconductor field effect transistor – MOSFET. The MOSFET transistor has the highest switching frequency and a low resistance for small voltages which makes it suitable for motors with a smaller power. Its internal resistance is a function of the voltage R~U2.5 so for higher voltages IGBT transistors are more often used. There are few switching topologies for achieving the necessary commutation most popular of which is the full bridge topology (Fig. 4).

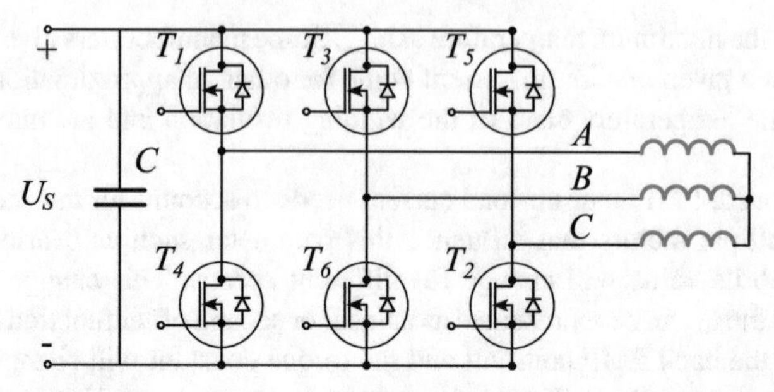

Fig. 4. Full bridge topology for a star or Y connected 3 phase BLDC motor

The losses in ESCs can be mostly accounted by the conduction losses and the switching losses in the transistors. There are few methodologies for estimation of these losses as well as techniques for their measurement [17, 18]. Manufacturers most often give the ESC internal resistance. Even though the switching frequency is often given the on and off time are dependent on the throttle, rpm and number of motor poles so these losses can be hard to determine. For this reasons and because the switching losses are almost an order of magnitude smaller than the conduction losses at higher throttles [17], in this paper the ESC losses are presented via the following equation:

$$P_{loss} = dU_{BAT} R_{ESC} i^2 \tag{13}$$

where R_{ESC} is the ESC internal resistance, i is the current and d is the PWM duty cycle.

2.3 Battery

Batteries are electrochemical devices which are used to store electric energy. Due to their power density limitations they are the weakest point in the electric propulsion system, yet, having their reliability and simplicity for use as well as the advances achieved in Li Ion batteries in mind, they represent a viable solution for small UAVs. The most effective way to represent the batteries behavior is through charge/discharge curves obtained by experimental investigation. However, this is an expensive and time consuming process so it is not always practical. Therefore, large number of mathematical models has been developed for the estimation of their performance [19, 20]. Probably the most often used in initial design for design point analysis is the simple equivalent electric circuit or EEC model. In this model the battery is simply defined as an ideal power source with an internal resistance R_{BAT}:

$$U_{BAT} = U - R_{BAT} i$$

This is an oversimplification of the battery behavior and for more detailed analysis a more complex model should be used, especially if one is interested in flight endurance or similar. For the purpose of an optimal propeller on-point design however it can be considered satisfactory.

3 Propeller Design Methodology

The propeller design methodology used in this paper is based on the work of Adkins and Liebeck [5, 6] who developed an inverse design methodology using blade element theory in which the momentum equations are set equivalent to the circulation equations. Their methodology has proven to be highly efficient and has been used by researchers for propeller design in the past [21].

According to the momentum equations for a disk annulus the incremental thrust can be expressed as:

$$dT = 2\pi r dr \rho V(1 + a)2VaF, \tag{14}$$

and the incremental torque as:

$$dQ/r = 2\pi r dr \rho V(1 + a)2\Omega r a' F. \tag{15}$$

where: r is the radial position of the annulus in [m], ρ is the fluid density [kg/m^3], V is the velocity [m/s], Ω is the angular velocity in [rad/s], a is the axial interference factor, a' is the tangential interference factor and F is the momentum loss factor used to account for the radial flow of the fluid.

From (Fig. 5, left) the incremental trust and torque can be expressed as:

$$dT = dL \cos\phi - dD \sin\phi = dL \cos\phi(1 - \varepsilon \tan\phi) \tag{16}$$

$$dQ/r = dL \sin\phi + dD \cos\phi = dL \sin\phi(1 + \varepsilon/\tan\phi) \tag{17}$$

where dL and dD are the incremental lift and drag forces acting on a given blade station respectively while $\varepsilon = dD/dL$ is the drag-to-lift ratio and ϕ is the flow angle.

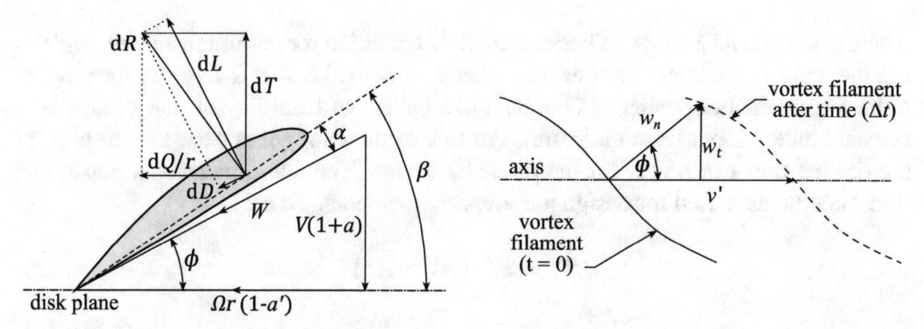

Fig. 5. Force and velocity diagram of blade element at radial station *r* (left) and propeller wake displacement velocity v' definition (right)

From circulation theory the total incremental lift at a given blade station, r, can be expressed as a function of the circulation Γ, number of blades B and the total velocity W:

$$dL = B\rho W\Gamma dr \tag{18}$$

while the circulation in the wake of the corresponding annulus would be:

$$B\Gamma = 2\pi r \mathrm{F} w_t \tag{19}$$

where $\mathrm{w_t}$ is the tangential (swirl) velocity which can be represented as:

$$w_t = V\zeta \sin\phi \cos\phi \tag{20}$$

where $\zeta = v'/V$ is the corresponding displacement velocity ratio. From Eqs. 18, 19 and 20 the circulation can be expressed as:

$$\Gamma = 2\pi V^2 \zeta \mathrm{G}/(B\Omega) \tag{21}$$

$$\mathrm{G} = \mathrm{F}x \sin\phi \cos\phi$$

where $x = \Omega \mathrm{r}/\mathrm{V}$ is the speed ratio. Replacing the circulation in Eq. 21 with Eq. 18 the lift in Eqs. 16 and 17 can be represented as a function of the displacement velocity ratio ζ and the flow angle ϕ. Now, combining Eqs. 16 and 17 with Eqs. 14 and 15 relation between the axial and tangential interference factors with the displacement velocity ratio ζ can be obtained:

$$a = (\zeta/2)\cos^2\phi(1 - \varepsilon \tan\phi) \tag{22}$$

$$a' = (\zeta/2x)\cos\phi \sin\phi(1 + \varepsilon/\tan\phi) \tag{23}$$

From Eqs. 22 and 23 and from Fig. 5 (left) it can be derived that:

$$\tan\phi = (1 + \zeta/2)/x = (1 + \zeta/2)\lambda/\xi \tag{24}$$

where $\xi = r/R$ and $\lambda = RV/\Omega$. Since the Betz condition for minimum energy, neglecting the wake contraction requires that $r \tan\phi = const$ i.e. that the vortex sheet needs to be a constant independent of radius it can be shown from Eq. 24 that ζ must be a constant independent from the radius. For this methodology it is necessary to provide the desired thrust or power for the propeller design. The non-dimensional power and thrust coefficients used for design purposes are then obtained as:

$$T_c = 2T/\left(\rho V^2 \pi R^2\right) \tag{25}$$

$$P_c = 2P/\left(\rho V^3 \pi R^2\right) = 2Q\Omega/\left(\rho V^3 \pi R^2\right) \tag{26}$$

Now Eqs. 16 and 17 can be written as:

$$T_c' = I_1'\zeta - I_2'\zeta^2 \tag{27}$$

$$P_c' = J_1'\zeta + J_2'\zeta^2 \tag{28}$$

Where the primes are derivatives with respect to ξ and:

$$I_1' = 4\xi G(1 - \varepsilon \tan\phi)$$

$$I_2' = \lambda\left(I_1'/2\xi\right)(1 + \varepsilon/\tan\phi)\sin\phi\cos\phi$$

$$J_1' = 4\xi G(1 + \varepsilon/\tan\phi)$$

$$J_2' = \left(J_1'/2\right)(1 - \varepsilon\tan\phi)\cos^2\phi$$

As mentioned the displacement velocity ratio ζ needs to be constant for optimal design. Therefore, two different scenarios can be considered. One for a specified thrust:

$$\zeta = (I_1/2I_2) - \sqrt{(I_1/2I_2)^2 - T_c/I_2} \tag{29}$$

$$P_c = J_1\zeta + J_2\zeta^2 \tag{30}$$

and for a specified power:

$$\zeta = -(J_1/2J_2) + \sqrt{(J_1/2J_2)^2 + P_c/J_2} \tag{31}$$

$$T_c = I_1\zeta - I_2\zeta^2. \tag{32}$$

Where the integration of I_1, I_2, J_1 and J_2 is carried out over the region $\xi = r_{hub}/R$ to $\xi = 1$.

When the local lift force for a single blade is represented through the local lift coefficient and equaled to Eq. 18:

$$\mathrm{d}L/\mathrm{d}r = \rho W\Gamma = \frac{1}{2}C_L\rho W^2 c$$

and having in mind Eq. 21 it follows that:

$$Wc = 4\pi\lambda GVR\zeta/(C_L B) \tag{33}$$

Since the local Reynolds number $Re = Wc/\nu$, where ν is the kinematic viscosity, for a given design local C_L the local Reynolds number can be determined and then the lift-to-drag ratio ε can be obtained. For an optimal rotor ε should have the smallest value so in order to achieve this the whole procedure should be done iteratively. The total velocity can now be determined from the following equation:

$$W = V(1 + a)/\sin\phi \tag{34}$$

Once the total velocity is determined the airfoil chord can be obtained from Eq. 33. The local blade twist angle can be obtained from Fig. 5 as $\beta = \phi + \alpha$ where α is the local angle of attack corresponding to the obtained C_L.

In the design procedure Adkins and Liebeck [5] propose that either Goldstein's relations which are more accurate or the Prandtl's approximate solution can be used for calculating the momentum tip loss. In this paper Prandtl's formulation for tip losses is going to be used since it is computationally less demanding [22]:

$$F = (2/\pi)\arccos\left(e^{-f}\right) \tag{35}$$

where:

$$f_{tip} = \frac{B}{2}\frac{R - r}{r\,\sin\phi}$$

It should be noted that the Prandtl formulation for hub losses cannot be used in the design procedure since it will yield a zero chord in the hub section (Fig. 6).

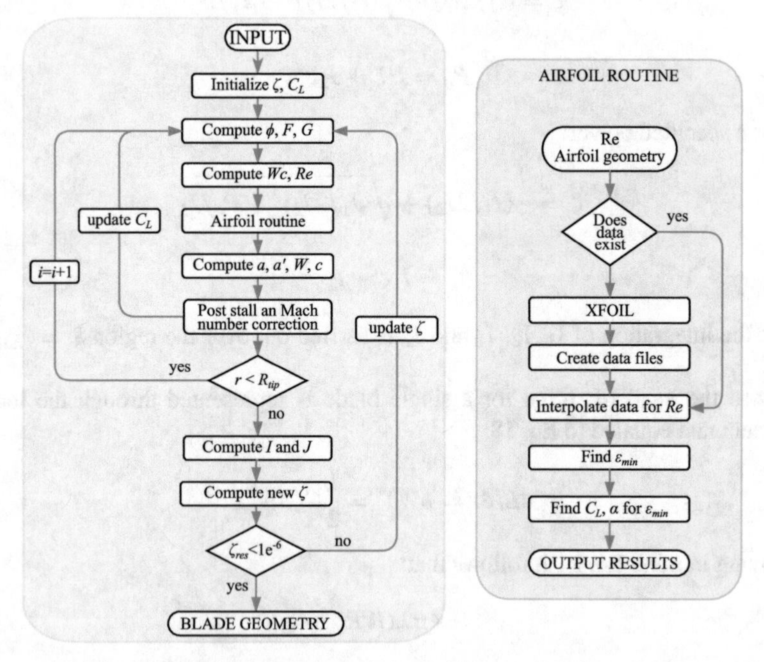

Fig. 6. Propeller design procedure flowchart

3.1 Airfoil Section

As can be seen from the previous section the airfoil characteristics are extremely influential to the efficiency of the propeller hence the overall efficiency of the propulsion system. It is therefore essential to have as close approximation to their realistic behavior as possible. In order to determine the aerodynamic coefficients, the XFOIL program was used [23]. Due to its simplicity and accuracy for wide range of Reynolds numbers, since its creation this program was used in numerous researches. Although CFD methods based on finite volume analysis are more accurate and have been used for airfoil optimization purposes [24] their use was considered inefficient here due to the high computational cost.

In order to save computational time, having in mind the variation of Reynolds number along the blade. it was considered impractical to run the XFOIL program in every iteration of the design procedure so a multidimensional interpolation routine was created. Files containing the polar data for different Re numbers were created and then data from it was extracted for the interpolation. It should be noted that even though XFOIL is often used for small Re analysis in the case of extremely small Re numbers (for ex. smaller than 10^4) the obtained results can have noticeable disagreements with experimental results hence use of experimental or CFD values would be recommended [25]. However, since the author's idea was to create a tool for fast initial design that can be later used for optimization purposes and having in mind that there are disagreements between experimental results from different sources as well and that XFOIL tended to underestimate the airfoil efficiency this approach was considered conservative and reasonable.

The data obtained with XFOIL is for planar two-dimensional airfoil sections. As a consequence of the blade rotation the boundary layer is going to be affected in such manner that the stall is delayed and shifted to higher angles of attack. There are numerous correction models in the literature that account for this three-dimensional effect [21]. Here the Du – Selig [26] method was used:

$$C_{L,3D} = C_{L,2D} + \Delta C_L \tag{36}$$

$$C_{D,3D} = C_{D,2D} - \Delta C_D \tag{37}$$

where ΔC_L and ΔC_D are the incremental increase of lift and decrease of drag respectively and can be obtained by the following correction formula:

$$\Delta C_L = f_L \left(C_{L,p} - C_{L,2D} \right)$$

$$\Delta C_D = f_D \left(C_{D,2D} - C_{D,0} \right)$$

where $C_{L,p} = 2\pi(\alpha - \alpha_0)$, $C_{D,0} = C_{D,2D}$ for $\alpha = 0$ while the functions f_L and f_D can be obtained from the following equations:

$$f_L = \frac{1}{2\pi}\left[\frac{1.6(c/r)a - (c/r)^{\frac{d}{\Lambda}\frac{R}{r}}}{0.1267b + (c/r)^{\frac{d}{\Lambda}\frac{R}{r}}} - 1\right]$$

$$f_D = \frac{1}{2\pi}\left[\frac{1.6(c/r)a - (c/r)^{\frac{d}{2\Lambda}\frac{R}{r}}}{0.1267b + (c/r)^{\frac{d}{2\Lambda}\frac{R}{r}}} - 1\right]$$

where a, b and d are empirical correction factors which were found by Du and Selig to be equal to unity with comparison to experimental data. $\Lambda = \Omega R / \sqrt{W^2 + (\Omega R)^2}$ represents a modified tip speed ratio.

Along the blade the Mach number changes as well. The XFOIL software already has the Karman-Tsien correction implemented into its calculations however this would mean that another set of calculations for different Mach numbers has to be done for the multidimensional interpolation and in order to avoid this and having in mind that the maximum Mach number should not exceed 0.7 in the design point, the Glauert correction for compressible flow for the Lift coefficient was used:

$$C_{Lcomp} = C_L/\sqrt{1 - M^2} \tag{38}$$

4 Results and Discussion

In order to test the procedure a propeller was designed for an electric system consisted of the KDE4014XF-380 BLDC motor, the Blackline 2600-35C battery and the Castle Phoenix 45 ESC. The specifications of the individual electric components are given in Table 1. The propeller blade was designed with the NACA 4415 airfoil for the first 35% and the NACA 4412 for the rest of the blade. Two blade propeller with an outer diameter of 0.3 [m] and a hub diameter of 0.0465 [m] which is the diameter of the electric motor for the design point conditions: flight velocity of 30 [m/s] at height of 1000 [m] was considered. The changes of atmospheric density and temperature were calculated according to the ISA model. The duty cycle was set to 75% meaning 0.75 full throttle. Since all components are in serial connection the resistance for the determination of optimal current of the entire power system (Eq. 9) was set to be the sum of all components resistances ($R = R_M + R_{BAT} + R_{ESC}$).

Table 1. Specifications of the electric system components

KDE4014XF-380			Castle Phoenix 45			BlackLine 2600-35C		
K_V	380	[rpm/V]	R_{ESC}	2.6	[mOhm]	*Cells*	5S	
R	7.5	[mOhm]	$I_{cont.}$	45	[A]	R_{BAT}	5.8	[mOhm]
I_0	0.5	[A]	I_{max}	60	[A]	*Capp.*	2600	[mAh]
I_{max}	36	[A]	$V_{inp.\,max}$	19.2	[V]	I_{max}	78	[A]
P_{max}	1065	[W]				V_{cell}	3.7	[V]

The airfoil characteristics for different Re numbers are given in Fig. 7 where the red line corresponds to the interpolated values for $Re = 8 \times 10^5$. Note that this Re number was selected for presentation purposes only while in the design procedure the corresponding Re is going to be interpolated.

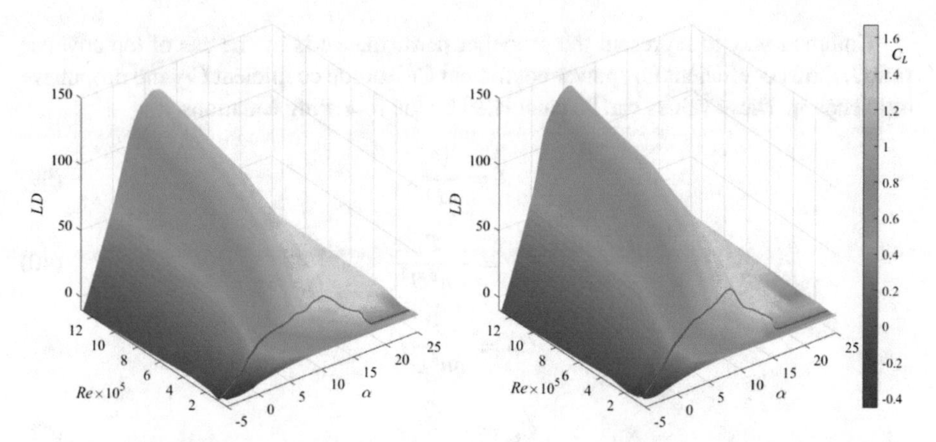

Fig. 7. Aerodynamic characteristics of NACA 4415 (left) and NACA4412 (right) airfoils

The optimal blade geometry obtained by the design procedure is given in Fig. 8. It should be noted that the airfoil chord and data closest to the 35% were interpolated between the two used airfoils in order to avoid sharp and unrealistic transitions in the blade geometry.

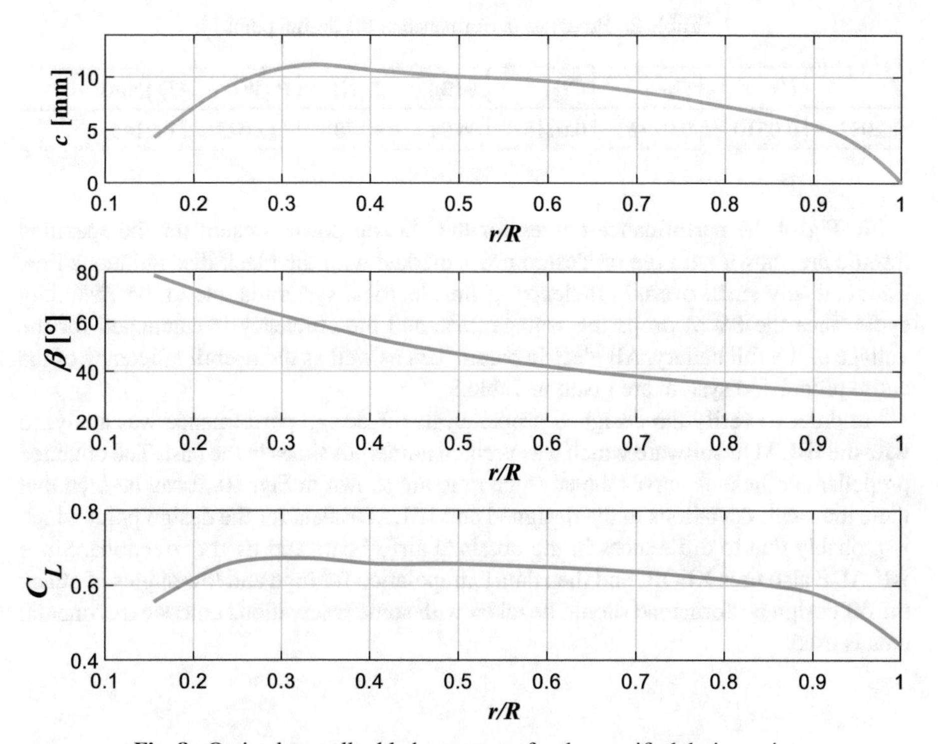

Fig. 8. Optimal propeller blade geometry for the specified design point

Common way to represent the propeller performance is by the use of the advance ratio J, thrust coefficient C_T, power coefficient C_P, torque coefficient C_Q and propulsive efficiency η. These values can be calculated by the following equations:

$$J = \frac{V}{nD} \tag{39}$$

$$C_T = \frac{T}{\rho n^2 D^4} \tag{40}$$

$$C_P = \frac{P}{\rho n^3 D^5} \tag{41}$$

$$C_Q = \frac{Q}{\rho n^2 D^5} \tag{42}$$

$$\eta = \frac{P}{TV} = \frac{C_T J}{C_P} \tag{43}$$

It should be noted that here n is in [rpm/s]. The designed propeller coefficients for the set design point are given in Table 2.

Table 2. Propeller performance at the design point

J	C_T	C_P	C_Q	η [%]	T [N]	P [W]	Q [Nm]
1.2032	0.0510	0.0722	0.0115	84.94	3.178	112.023	0.2145

In Fig. 9 the performance curves for the electric power system for the specified throttle are shown with the on design point marked with the black discontinuous line. The relatively small overall efficiency of the electrical system is due to the 75% duty cycle since the PWM drops the voltage 25% and the efficiency is calculated for the voltage of the full battery. All electric parameters as well as the overall efficiency of the entire propulsion system are given in Table 3.

In order to verify the designed propeller, its off-design performance was analyzed with the JBLADE software which was verified numerous times in the past. The obtained propeller coefficients versus the advance ratio are shown in Fig. 10. It can be seen that there are small deviations in the designed and JBLADE data for the design point which is probably due to differences in the obtained airfoil data and used corrections. Since JBLADE also uses XFOIL and then data extrapolation for high and low angles of attack the off design performance should be taken with some reservations unless experimental data is used.

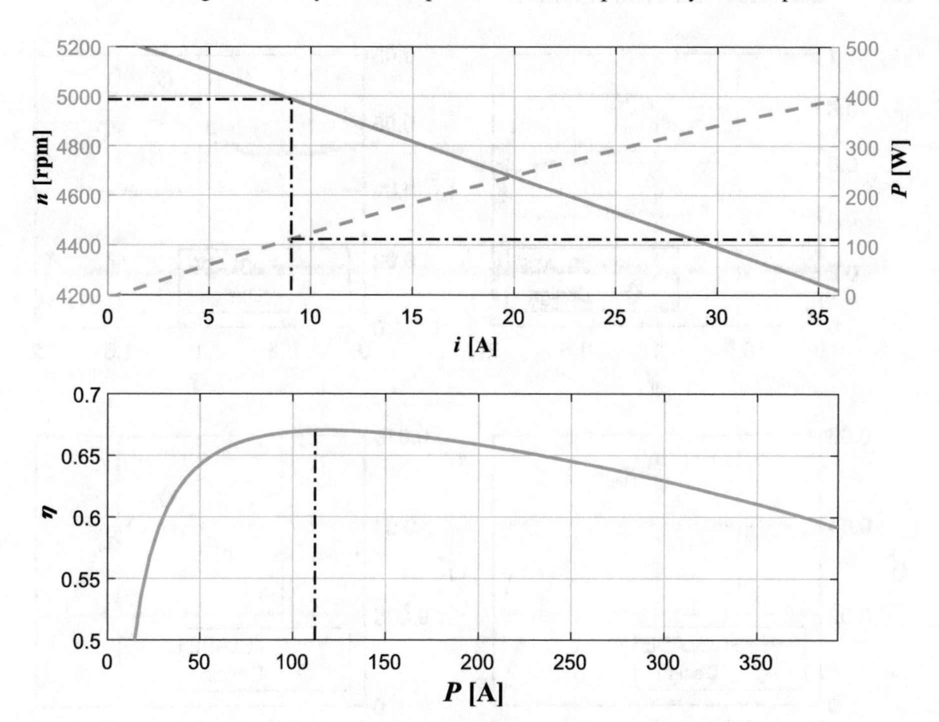

Fig. 9. Electric power system performance for the specified design point

Table 3. Electric propulsion system parameters and efficiency

U [V]	U_{ESC} [V]	I_{OPT} [A]	P_E [W]	n [rpm]	η_E[%]	η_{TOT}[%]
18.5	13.8	9.037	167.18	4986.6	67.01	56.92

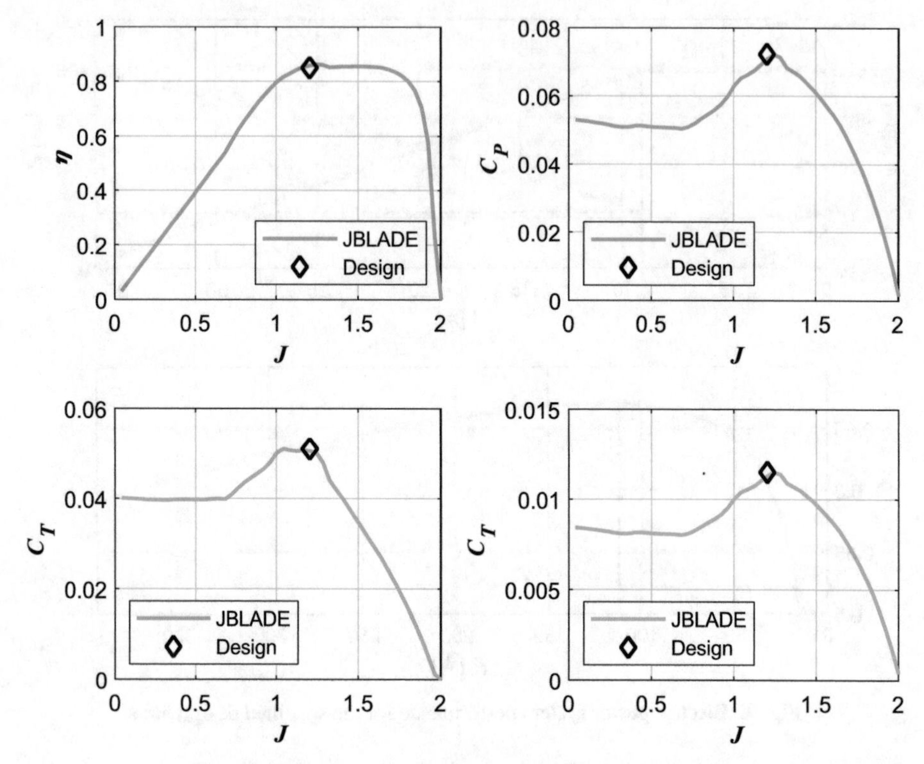

Fig. 10. Off-design performance of the designed propeller

5 Conclusion

In this paper a simple and fast propeller design methodology for fully electric propulsion systems was presented. The electrical components were presented by simplified mathematical models while the propeller design was performed according to a modified blade momentum theory with vortex wake deflection with tip loss correction first introduced by Adkins and Liebeck [5, 6]. The design methodology was then tested through an example. The propeller coefficients were then verified by the JBLADE software with good agreement in the obtained results.

This approach gives a good starting point in the initial design of more efficient electric propulsion systems with certain limitations that were presented. Recommendations for more accurate analysis are stated and references for better and more complex formulations for the electrical system are given.

References

1. National Academies of Sciences, Engineering and Medicine, Commercial Aircraft Propulsion and Energy Systems Research, Reducing Global Carbon Emissions. The National Academies Press, Washington (2016)

2. Ivanov, T.D., Fotev, V., Petrović, N., Svorcan, J., Peković, O.: Optimization of BLDC motor/propeller matching in the design of small UAV's. In: 8th International Scientific Conference on Defensive Technologies OTEH 2018, Belgrade, Serbia (2018)
3. Anggraeni, D., Sumaryanto, A. R., Sumarna, E., Rahmadi, A.: Engine and propeller selection for propulsion system LAPAN surveillance UAV – 05 (LSU-05) using analytic and experimental test. In: International Seminar of Aerospace Science and Technology II, Jakarta (2014)
4. Chaput, A.J.: Small UAV motor and propeller methods - a parametric system engineering model based approach. In: AIAA SciTech Forum 2018 AIAA Aerospace Sciences Meeting, Kissimmee, Florida (2018)
5. Adkins, C.N., Liebeck, R.H.: Design of optimum propellers. J. Propul. Power **10**(5), 676–682 (1994)
6. Adkins, C.N., Liebeck, R.H.: Design of optimum propellers. In: AIAA 21st Aerospace Sciences Meeting, Reno, Nevada (1983)
7. Leesang, C., Jaemin, Y., Cheolheui, H., Jinsoo, C.: Aerodynamic design and analysis of a propeller for a micro air vehicle. J. Mech. Sci. Technol. **20**(10), 1753–1764 (2006)
8. Rutkay, B., Laliberté, J.: Design and manufacture of propellers for small unmanned aerial vehicles. J. Unmanned Veh. Syst. **4**, 228–245 (2016)
9. Prathapanayaka, R., Vinod, K.N., Krishnamurthy, S.: Design, analysis, fabrication and testing of mini propeller for MAVs. In Symposium on Applied Aerodynamics and Design of Aerospace Vehicle (SAROD 2011), Bangalore (2011)
10. Krishnan, R.: Permanent Magnet Synchronous and Brushless DC Motor Drives. CRC Press, Taylor and Francis Group, Boca Raton (2010)
11. Vukosavić, S.: Electrical Machines. Springer, New York (2013)
12. Drela, M.: Second Order DC Electric Motor Model (2006). http://web.mit.edu/drela/Public/web/qprop/motor2_theory.pdf. Accessed Apr 2020
13. Kumpanya, D., Thaiparnat, S., Puangdownreong, D.: Parameter identification of BLDC motor model via metaheuristic optimization techniques. Ind. Eng. Serv. Sci. **4**, 322–327 (2015)
14. Campa, R., Torres, E., Salas, F., Santibanez, V.: On modeling and parameter estimation of brushless DC servoactuators for position control tasks. In: Proceedings of the 17th World Congress, The International Federation of Automatic Control, Seoul, Korea (2008)
15. Krishnan, R.: Electric Motor Drives, Modeling, Analysis and Control. Prentice Hall Inc., Upper Saddle River (2001)
16. Pillay, P., Krishnan, R.: Modeling, simulation, and analysis of permanent-magnet motor drives, Part II: the brushless DC motor drive. IEEE Trans. Ind. Appl. **25**(2), 274–279 (1989)
17. Gong, A., Verstraete, D. and Macneill, R.: Performance testing and modeling of a brushless DC motor, electronic speed controller and propeller for a small UAV. In: 2018 Joint Propulsion Conference, Cincinnati, Ohio (2018)
18. Gong, A., Verstraete, D.: Experimental testing of electronic speed controllers for UAVs. In: 53rd AIAA/SAE/ASEE Joint Propulsion Conference, Atlanta, Ga (2017)
19. Piller, S., Perrin, M., Jossen, A.: Methods for state-of-charge determination and their applications. J. Power Sources **9**, 113–120 (2001)
20. Einhorn, M., Conte, F.V., Kral, C., Fleig, J.: Comparison, selection, and parameterization of electrical battery models for automotive applications. IEEE Trans. Power Electron. **28**(3), 1429–1437 (2013)
21. João, M. P.: Development of an open source software tool for propeller design in the MAAT project. Ph.D. thesis, University of Beira Interior, March 2016
22. Wald, Q.R.: The aerodynamics of propellers. Prog. Aerosp. Sci. **42**, 85–128 (2006)
23. Drela, M.: XFOIL: an analysis and design system for low reynolds number airfoils. In: Low Reynolds Number Aerodynamics, Lecture Notes, vol. 54, pp. 1–12. Springer (1989)

24. Ivanov, T.D., Simonović, A.M., Petrović, N.B., Fotev, V.G., Kostić, I.A.: Influence of selected turbulence model on the optimization of a class-shape transformation parameterized airfoil. Therm. Sci. **21**(3), S737–S744 (2017)
25. Posteljnik, Z., Stupar, S., Svorcan, J., Peković, O., Ivanov, T.: Multi-objective design optimization strategies for small-scale vertical-axis wind turbines. Struct. Multi. Optim. **53**(2), 277–290 (2015)
26. Du, Z., Selig, M.S.: A 3-D stall delay model for horizontal axis wind turbine performance prediction. In: ASME Wind Energy Symposium, American Institute of Aeronautics and Astronautics, Reston, Virginia (1998)

Detection of Jet Engine Viper Mk 22-8 Failure in Vibration Spectra

Miroslav M. Jovanović(✉)

Serbian Armed Forces, Technical Test Center, Belgrade, Serbia
jovanovic.miroslav.77@gmail.com

Abstract. This paper presents a method for detecting abnormal content in vibration spectra obtained from jet engine. Jet engines are machines with high speed turbine-compressor assembly (spool). In order to ensure the functionality of the engine core, a large number of bearings, gears, intermediate shafts are required that are interconnected and depend on the basic engine speed. Many aircraft today have engines that do not have vibration measurement and monitoring devices integrated. Most often, these are "old" engines, in which, in addition to failures of individual bearings, gear assemblies and shafts, they can be caused by material characteristics and structural integrity. Most often, the cause of vibration problems of old engines can be detected and eliminated only by using test bench equipped with vibration measuring and analysis devices. The paper is described on an analysis of vibration frequency spectrum at Rolls&Royce Viper 22-8 engine with aim to identify the engine malfunction.

Keywords: Jet engine · Test cell · Vibration · Measurement · Frequency spectrum

1 Introduction

Jet engine have been developed from 1930s and integrated on aircraft platform. Nowadays, the jet engines are the most frequently used power unit of an aircraft. Jet turbines are the most important pivotal rotating machines belonging to the family of internal combustion engines that are used intensively in aerospace industrial. Indeed, it is used to achieve the main function of generating mechanical energy in form of shaft rotation from kinetic energy of gases produced in the combustion chamber [1] and generates the thrust. These motors are machines that are highly susceptible to vibration due to the high speed of shaft rotation and all associated components.

Today, the engines are equipped with digital computer and digital-analog sensors with aim to monitor and maintain in allowable operating modes. This system is called FADEC (full authority digital engine control) and it consists of a digital computer, called an "electronic engine controller" (EEC) or "engine control unit" (ECU), and its related accessories that control all aspects of aircraft engine performance [2]. FADEC works on the principle of multiple input (air density, throttle lever position, engine temperatures,

N. Mitrovic et al. (Eds.): CNNTech 2020, LNNS 153, pp. 167–183, 2021.
https://doi.org/10.1007/978-3-030-58362-0_11

engine pressures and many other parameters), with its own algorithm determines the most optimal parameters and distributes them to multiple management actuators to regulate the fuel flow, stator vane position, air bleed valve position and others. Its basic purpose is to provide optimum engine efficiency for a given flight condition. FADEC not only provides for efficient engine operation, it also allows the manufacturer to program engine limitations and receive engine health and maintenance reports [2].

Each of the jet engine rotating shafts or spools can include a compressor-turbine assembly which in turn includes fan and turbine blades. This is reason to involve different techniques to monitor the vibrations of rotating machinery and to signal dangerously high vibration amplitudes or, at least, signal large incremental changes over the otherwise normal patterns of vibration [3]. The system that performs engine vibration monitoring and is commonly referred to as Airborne Vibration Monitoring (AVM) System. The AVM measure rotor out of balance with high indication reliability, and they also provide further functionality that brings large operational and commercial advantages. The AVM system is based to monitor the shaft rotation and its harmonics, which most often coincide with the frequency of blades passing frequency (BPF) of compressor and turbine stages and indicate engine shaft (spool) unbalance.

The FADEC and AVM represent the systems for jet engine conditions and data monitoring in flight, but the failures that can be caused by damage of other mechanical elements (gearboxes, bearings) with indications can only be guessed as the above systems do not perform full vibrodiagnostics with maps of permitted vibration levels per frequency spectrum [4, 5].

Despite modern maintenance concepts for technical systems, a large number of jet engine installed on aircrafts are maintained with hard time maintenance philosophy. This means that after a certain period of time, the components that have been predicted by operating time must be replaced. In this case, the overhaul of "old" engine require serious checks with sophisticated equipment as the integrity of the structure becomes a problem. In this paper is described the approach of checking the tuning of the jet engine after overhaul and the tuning of its elements based on the recording and analysis of the vibration spectrum.

2 Overhauled Jet Engine and Its Condition

2.1 Viper Mk 22-8 Jet Engine

The Viper series of jet engine is developed and produced by Armstrong Siddeley in late 1950's. The Rolls-Royce has taken the rights and continued the development of this engine series. The Viper Mk 22-8 jet engine is a straight-flow turbojet, embodying a seven stage axial compressor, coupled directly to a single stage turbine (Fig. 1). This rotating assembly is supported in three main bearings [6].

Combustion is effected in a straight flow, annular chamber which is equipped with twelve fuel feed pipe units (burners), supplied by a Lucas variable flow pump via the fuel control system.

Starting is by a combined starter/generator mounted on the front of the air intake casing. This drives the compressor through a reduction gearbox and a bevel gearbox is an aircraft installation item. This unit, together with the hydraulic pump and tachometer generator, are the only engine-driven auxiliaries.

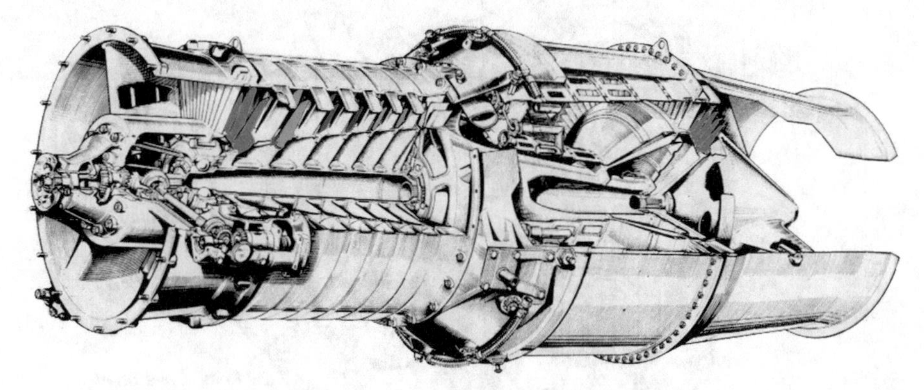

Fig. 1. Cross section of the engine Viper Mk 22-8

The front main bearing housing retains the labyrinth seal stationary sleeve and the outer race, cage and balls of the thrust bearing. The split inner race of the bearing, together with a rotating labyrinth seal unit are mounted on the front end of the compressor rotor. Starter/generator gearbox mounted at the front of the air intake casing. Starter/generator is coupled with the internal gear box. It transmits the drive from the starter/generator through a quill coupling shaft to a step-up internal gear.

The internal gearbox is located on air intake casing and is driven by compressor. The internal gearbox serves the following purposes:

- provides a mounting for the starter/generator gearbox on its front face;
- transmits the drive from the starter/generator gearbox to the compressor rotor during starting;
- transmits the drive from the compressor rotor to the accessory drives bevel box, oil pump and starter/generator during normal start;
- provides a mounting for the engine breather pipe support bracket; and
- embodies an oil jet for the lubrication of the starter/generator gearbox gears.

Details of the internal gearbox are shown in Fig. 2. The mainshaft engages the compressor rotor extension shaft at its rear end and the internal gear of the starter/generator at its front end. Bevel gears transmit the drive to the accessory drives bevel box and the oil pump drive shafts which are contained the hollow vanes of the air intake casing [6].

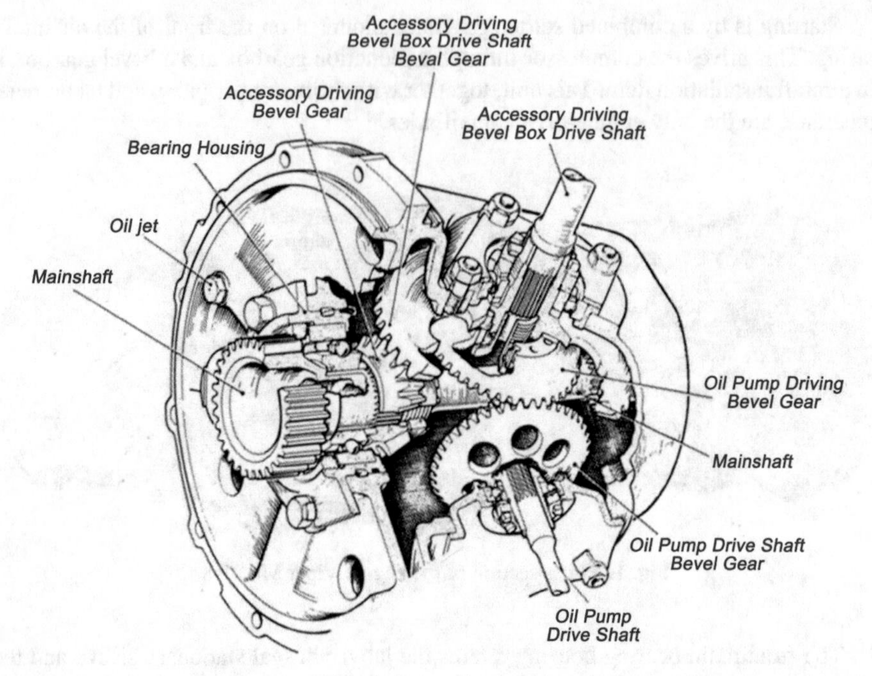

Fig. 2. Internal gearbox

Bolted to the side of the air intake casing and driven by the shaft in the left-hand hollow vane of the casing, this gearbox contains the drives to the fuel pump, tachometer generator and hydraulic pump.

The starter/generator box and internal gearbox are the sources of vibration because a large number of bearings and gears are placed in this location. Of course, the most influential source of vibration is still the compressor and its stages at engine speed, but the vibrations of bearings and gears can interference the vibration spectrum on ground [7, 8] and in flight [9, 10].

2.2 Test Cell for Jet Engine

Test cell is an enclosed facility that should provide a necessary conditions of environment to evaluate turbine engine performance after overhaul or other maintenance of jet engines.

The test table with all associated elements should neutralizes the presence of all atmospheric phenomena (the presence of vortices, turbulence, and non-uniform temperatures and pressures variation) that can affect the engine performance and test repeatability. Therefore, all test cell configurations should be designed to provide stable testing conditions and minimize turbulent flow by minimizing pressure loss and temperature and pressure variations. Under most environmental conditions, a test cell configuration should not allow recirculation of engine exhaust gases from the cell exhaust stack into the cell inlet. It should also prevent reingestion of engine exhaust gases at the rear of the engine back into the engine inlet [11].

The following system elements are common to most cell configurations described in Advisory Circular (AC) No: 43-207 Correlation, Operation, Design and Modification of turbofan/jet engine test cell. The difference between the rehearsal tables is usually shown in the intake section, "L", "U" or "folded" type of test cell. The configuration of jet engine test cell, which is used in this work, is shown in Fig. 3.

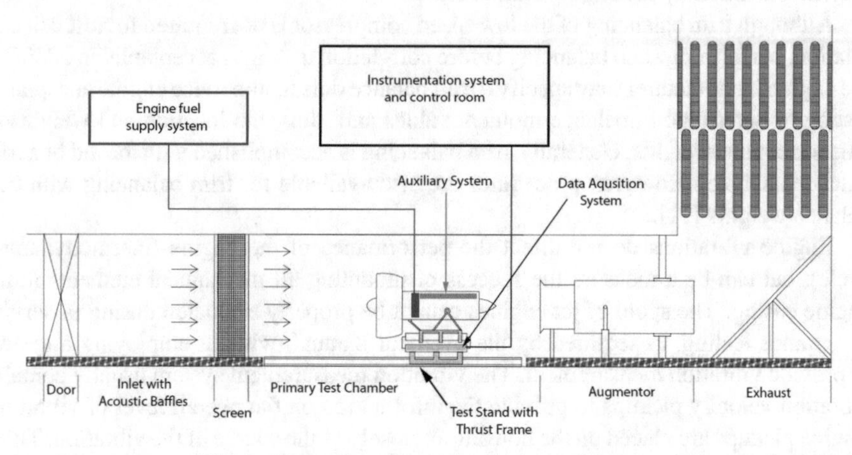

Fig. 3. Test cell configuration for jet engine tests

The instrumentation quantity, position, range and accuracy of test cell are usually specified in an engine manufacturer document, such as a facility planning manual or overhaul manual. The data acquisition and instrumentation system of jet engine test cell usually contain the systems for measure the pressure, temperature, engine speed, fuel flow, trim balance, thrust and vibration [11].

A few pressures must be measured on the test table. It has to be measured atmospheric pressure, pressure before and pressure after compressor. Other pressures are recommended but are not mandatory. Pressures are measured by simple pressure gauges, individual dedicated pressure transducers, or a scanning-valve system.

Temperature measurement systems must be able to measure temperatures in range of low to high temperatures. The temperatures of engine cold section are measured with E- or J-type thermocouples, while the temperatures of engine warm section are used K-type thermocouples. The number of thermocouples for measuring the temperature depends on the engine manufacturer requirements for testing. The junction of thermocouple should be calibrated to reduce system noise and error.

The engine speed of all spool are measure by the engine speed measurement system. The engine speed measurement system may consist analog or digital devices for measuring engine speeds in revolutions per minute or percentages.

A typical fuel flow measurement system may consist of one or more turbine flow meters (or comparable flow measuring device(s)), each connected to a digital frequency counter and display gauge. This total measurement system, from the flow meter through the frequency counter to the display gauge, must be checked end-to-end for the entire operating range of the system. For example, a defective frequency counter could provide an acceptable error in the low flow range, while generating a divergently increasing unacceptable error in the high range [11].

Thrust measurement or force measurement can be measured in two ways. The first method is by using hydraulic measurement systems, and the other is the use of electronic strain gauge (Wheatstone bridge) measuring tapes. The type of thrust measurement system is defined by the engine manufacturer.

Although trim balancing of the low speed compressor is not required for test cell correlation, performing such balancing before correlation testing is acceptable. In addition, the engine manufacturer may specify a trim balance during in-service engine acceptance testing to determine imbalance moment values and, thus, the location and weights of balance counterweights. Generally, trim balancing is accomplished with the aid of a trim balance analyzer. However, procedures are also available for trim balancing with trial balance weights [11].

Engine vibrations do not affect the performance of the engine (thermodynamics cycle), but can be a measure the success of mounting all mechanical elements in the engine casing. The spool of jet engines cannot be properly evaluated during in-service acceptance testing, as required by the overhaul manual, without employing a system to provide vibration measurement. The vibration measurement system usually contains vibration velocity pickups to provide the information on the overall level of vibration. Engine pickups are placed on the housing as closely as the source of the vibration. These systems usually don't provide a spectral analysis of vibrations

2.3 The Test Cell and Addition Test Equipment for Viper 22-8

The engine Viper 22-8 mounted at test cell is given in Fig. 4. The parameters which are monitoring in working cycle at instrumentation system and control room is given in Table 1.

Fig. 4. Engine Viper 22-8 at test cell

Table 1. Measuring parameters of jet engine test cell.

No	Parameter	Label	Range
1	Inlet temperature	T_I	−40 °C–70 °C
2	Exhaust temperature	T_E	−20 °C–800 °C
4	Oil temperature	T_O	−40 °C–120 °C
5	Atmosphere pressure	p_a	750 mbar–1050 mbar
6	Compressor outlet pressure	p_2	0.1 bar–10 bar
7	Inlet fuel pressure	p_F	0.1 bar–10 bar
8	Start fuel pressure	p_{SF}	0.1 bar–10 bar
9	Basic fuel pressure	p_{BF}	0.1 bar–10 bar
10	Inlet oil pressure	p_{OP}	0.1 bar–5 bar
11	Inlet hydraulic pressure	p_{IH}	0.1 bar–5 bar
12	Outlet hydraulic pressure	P_{OH}	0.1 bar–200 bar
13	Engine speed	n	0%–100%
14	Engine thrust	T	0 dN–3000 dN
15	Fuel flow	f_F	200 l/h–3000 l/h
16	Vibration	A_V	0 ips–7 ips

In addition to the basic instrumentation of the test cell, for vibration testing of the overhauled jet engine, equipment for vibration measurements and analysis of the frequency spectrum up to 5000 Hz is mounted.

These additional equipment have the ability to monitor vibrations in the frequency spectrum. This is an important fact because in addition to the amplitudes at the basic engine speed, it is necessary for us to see the vibration levels and their interrelationships at frequencies corresponding to other rotating elements that are an integral part of the engine.

The addition equipment are:

- HBM manufacturer acquisition system, QuantumX model, type MX840A,
- Uniaxial piezoelectric accelerometer manufactured by PCB Piezotronics, model 352C03,
- Piezoelectric accelerometer connection adapter with SOMAT acquisition system, and
- Software Analysis and Software Catman V4.1.1.

The piezoelectric accelerometer was mounted in front of velocity meter data acquisition system on air intake casing, above the front main bearing housing and internal gearbox. The position of sensors is shown in Fig. 5.

Fig. 5. Position of vibration sensors

Data were captured in the frequency range 0–5000 Hz, using 12800 point Fourier transforms with a Hanning window and a 50% overlap. This resulted in a frequency resolution of 0.39 Hz.

Test was performed during one working cycle from minimum engine running speed (40% or 5504 rpm) to maximum engine running speed (100% or 13760 rpm). The engine running speed was changed in steps by approximately 2–5% of rpm, with aim to catch all disturbance in vibration spectrum. After each change of engine running speed, the engine was allowed to stabilize and reach steady state conditions before the sets of acceleration and velocity readings were taken. Including the time taken to change the engine settings, the engine was allowed to settle to steady-state conditions and the data were captured. The measurement time for each regime was in the range of 30–45 s. The data for each regime were taken for about 20 s and captured as such.

3 Results of Engine Viper 22-8 Test

Testing of the overhauled engine was performed according to the procedure of the engine manufacturer and it was determined that there was a problem with the vibration levels. The vibration level at this check was higher than allowed over 68.8% of engine speed and were from 0.98 ips (inch per second) till 1.05 ips. The maximum permissible vibration level on the test bench for this engine is 0.95 ips.

A test conducted with an engine equipped with vibration measuring equipment had the results shown in Fig. 6. Engine parameters, as well as the vibrations acting on air intake casing, were measured throughout the whole starting and working cycle with increase from minimum to maximum engine running speed and inversely.

Based on Fig. 6, the frequency spectra for four engine speed, 88%, 93.5% 97.5% and 100%, shown in Fig. 7, 8, 9 and 10, respectively.

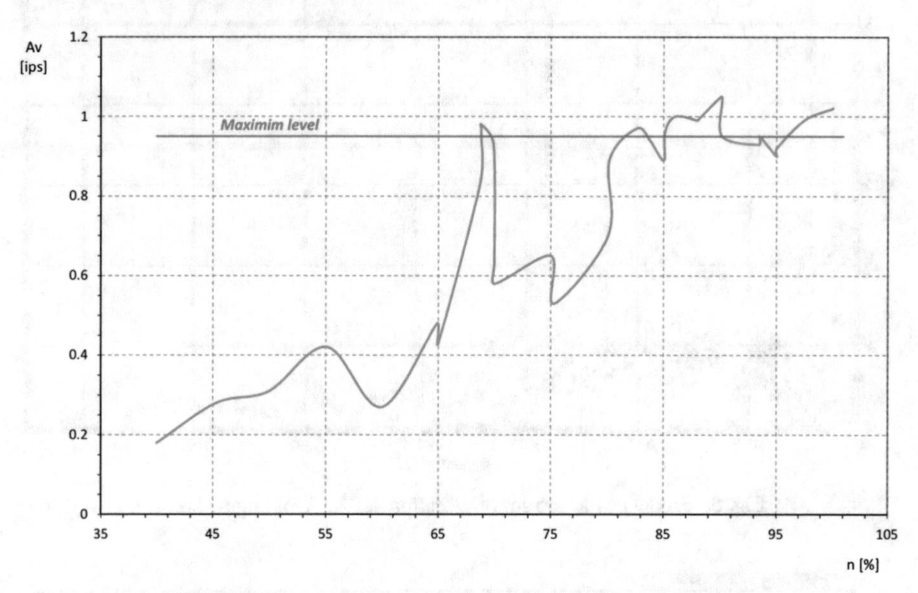

Fig. 6. The vibration level recorded from the test cell sensor

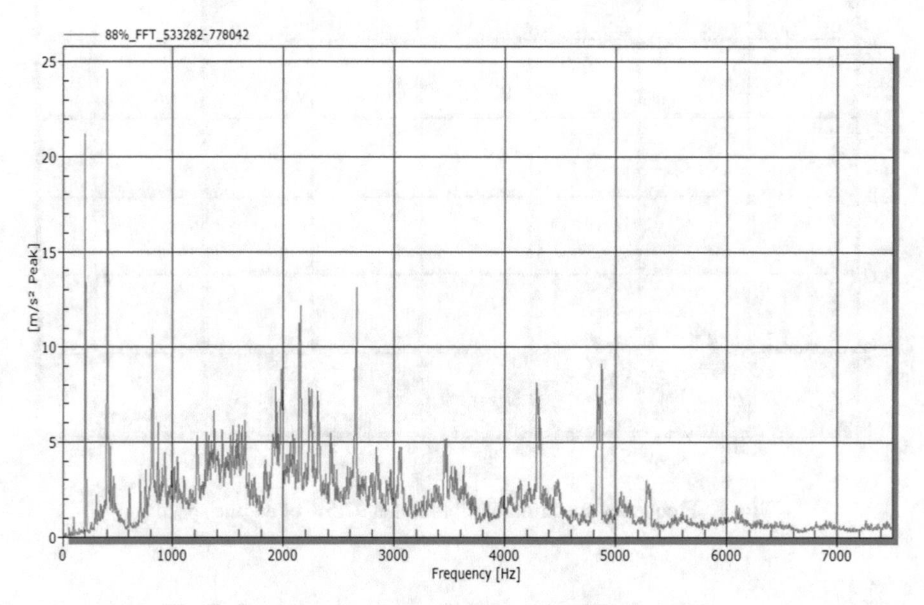

Fig. 7. Frequency spectrum of vibration at 88% of engine speed

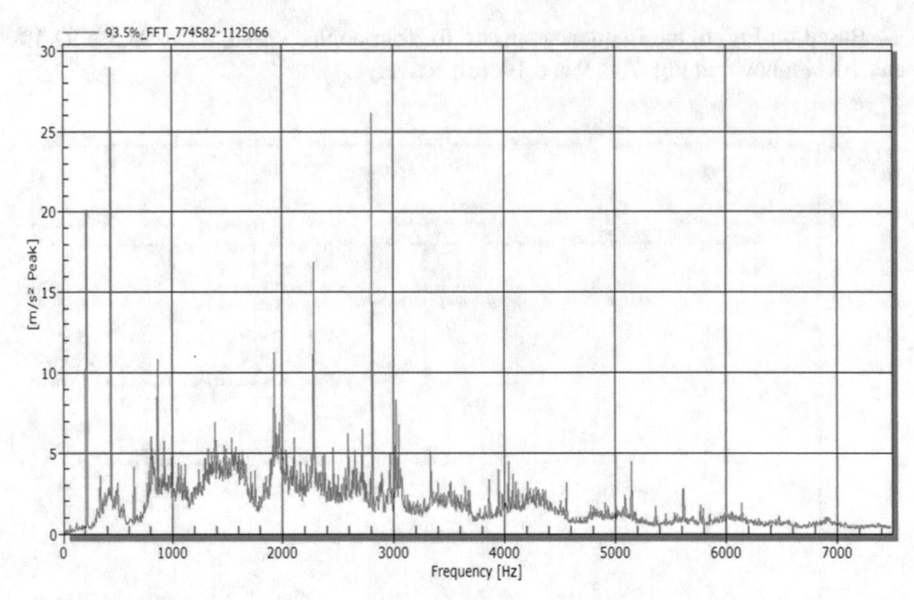

Fig. 8. Frequency spectrum of vibration at 93.5% of engine speed

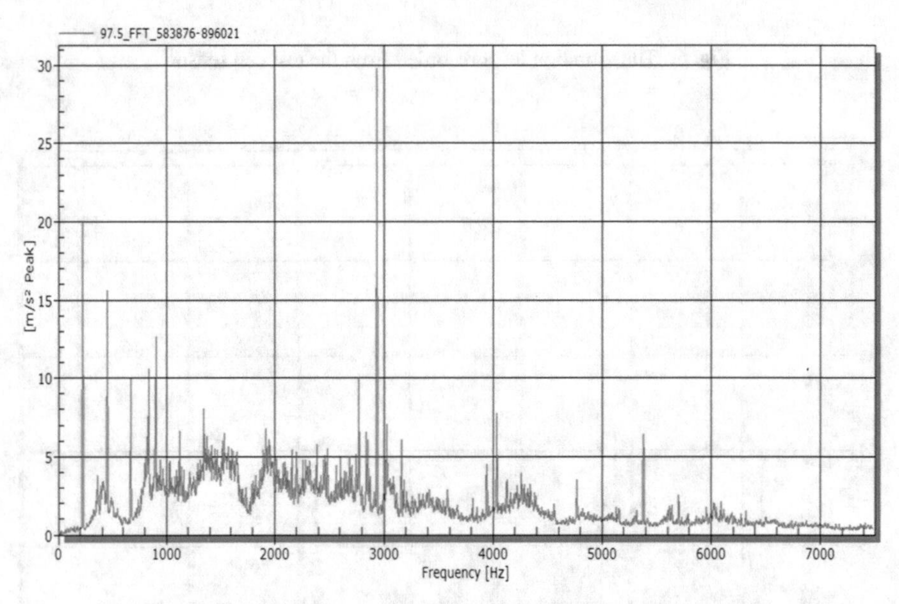

Fig. 9. Frequency spectrum of vibration at 97.5% of engine speed

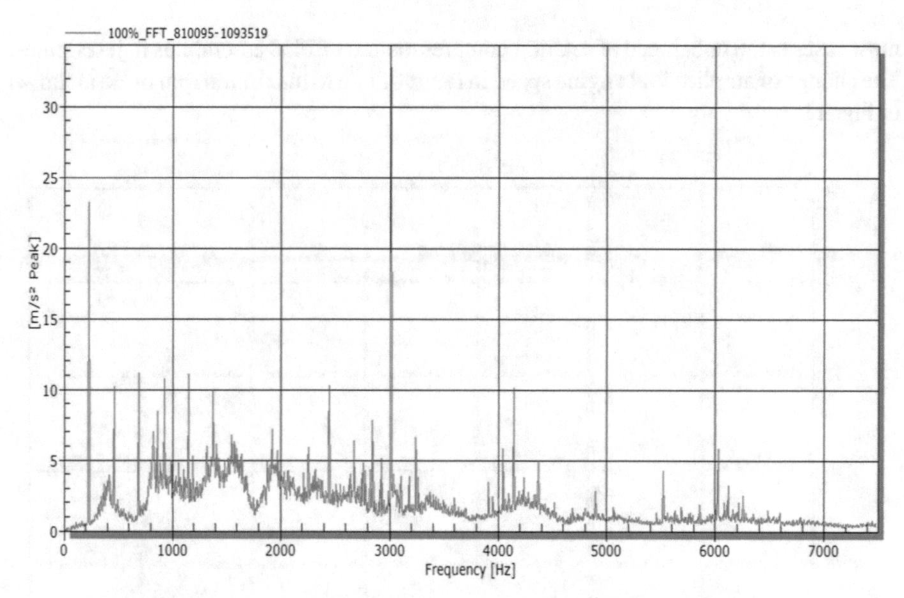

Fig. 10. Frequency spectrum of vibration at 100% of engine speed

In the Table 2 shows the values of the dominant amplitudes in the frequency spectrum for the observed 4 modes of engine speed and are given in relation to the harmonics (rows) of working frequency (Hz).

Table 2. Dominant amplitudes in the frequency spectrum of engine speed.

88% 12145 RPM, 202.5 Hz		93.5% 12900 RPM, 215 Hz		97.5% 13450 RPM, 224 Hz		100% 13800 RPM, 230 Hz	
Order	[m/s^2]	Order	[m/s^2]	Order	[m/s^2]	Order	[m/s^2]
2	24.61	2	28.99	13	29.8	13	32.76
1	21.16	13	26.15	1	20.03	1	23.2
13	13.12	1	20.08	2	15.61	3	11.18
10.63	12.14	10.63	16.9	4	12.66	5	11.08
4	10.62	9	11.22	3.73	10.57	4	10.78
21.17	8.12	4	10.81	12.37	10.05	10.63	10.28
24	9.106	14	8.27	3	9.92	2	10.22
				18	7.73	18	10.14

Presentation of the results of vibration measurement in jet engine is usually very difficult due to the large number of interdependent elements. When processing the results of vibration measurements, the most common problems are first eliminated, which in most cases is the unbalance of rotating compressor and turbine assemblies in jet engines. The change of amplitude at engine speed in range of idle to maximum rpm in Hz is shown in Fig. 11.

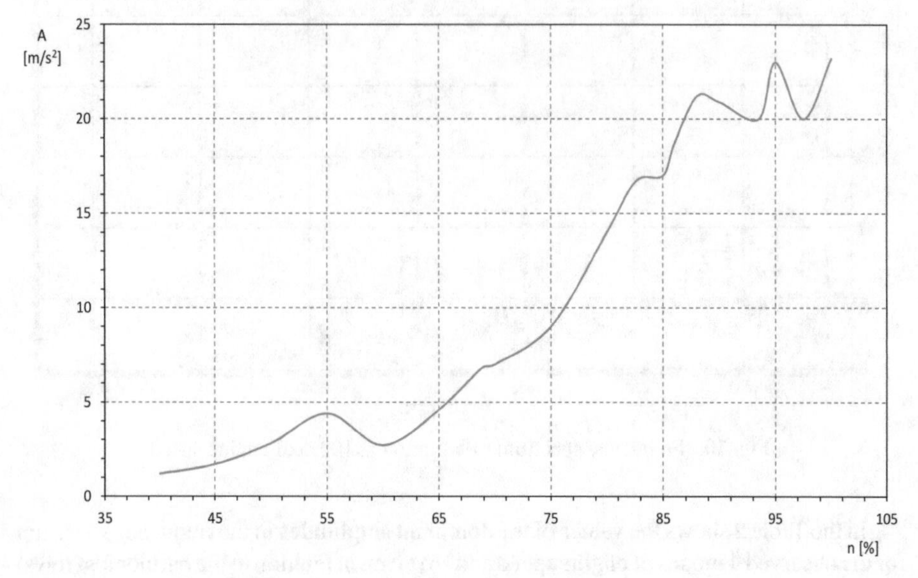

Fig. 11. The vibration amplitude at frequency of engine speed

Figure 11 shows that as the rpm increases, there is no exponential increase in amplitude indicating that the overhauled motor has no imbalance problems.

4 Analysis of Vibration Results

Based on the results given in the previous section, it can be concluded that the vibration level is above the permitted limits in the range of 65% to 69% and above 85%. This section will analyze vibrations over 85%, because in this range a potential problem can be adequately identified.

At a speed range of 88% to 93.5%, the amplitude at a frequency of 2xRPM is dominant, followed by an amplitude corresponding to the RPM frequency. Significant amplitude is also at a frequency corresponding to 13xRPM (Fig. 7 and Fig. 8).

At a speed of 95% the amplitude at a frequency of 2xRPM decreases, the amplitude at a frequency of 13xRPM increases intensively. Higher order harmonics (2x, 3x, 4x, 5x) appeared.

At a speed 100%, the frequency corresponding to 13xRPM is amplitude modulated with frequencies 0.360xRPM (corresponding to the speed of the fuel pump) and 1xRPM (corresponding to the speed of the motor shaft).

The amplitude modulation is shown in the Fig. 12.

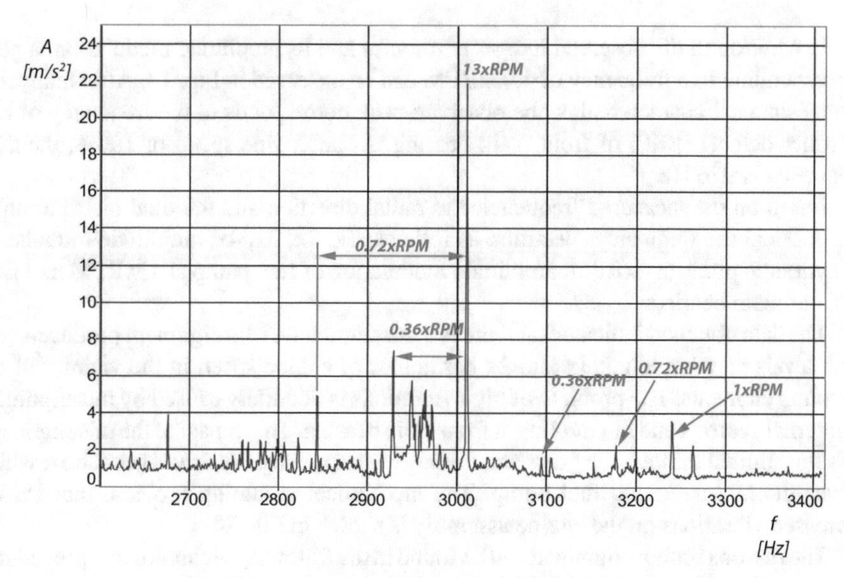

Fig. 12. Amplitude modulation at frequency 13xRPM

In addition to the mentioned amplitude modulation, another phenomenon was observed at a frequency corresponding to 4xRPM. The amplitude modulation corresponding to 0.286xRPM on both sides of the 4xRPM harmonics corresponds to the operation of a hydraulic pump located on the same shaft as the fuel pump. The frequency spectrum at speed of 100% to 1200 Hz is shown in the Fig. 13.

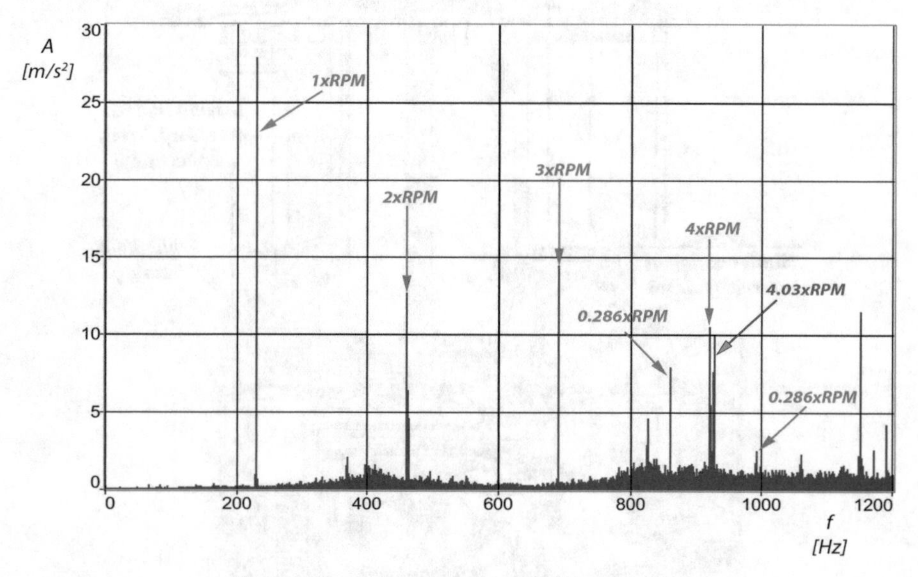

Fig. 13. Frequency spectrum at 100% to 1200 Hz

In addition to the observed integer harmonics and its amplitude modulation, a peak corresponding to a frequency of 4.03xRPM can be observed in Fig. 13. After analyzing the mechanical characteristics, the observed peak corresponds to the frequency of ball or roller defects (BSF) of front main bearing. At an engine speed of 100%, the BSF frequency is 926 Hz.

Based on the measured frequencies in radial direction and the final distribution of the peaks in the frequency spectrum, as follows: 1x, 2x, 3x, 4x, amplitude modulation of hydraulic pump at 4xRPM, amplitude modulation of fuel pump at 13xRPM and BSF of front main bearing.

The data obtained indicate that there is a misalignment. Misalignment produced very high levels of vibration at 1x and 2x harmonics of engine speed in the vicinity of the coupling elements. The problem of high vibrations is definitely caused by misalignment in internal gearbox and its coupling at front main bearing. The impact of the misalignment was transmitted through the gears and driving shaft to accessory drives bevel box, which drives the hydraulic and fuel pump. The mechanical mounting problem that caused increased vibrations on the engine assembly is shown in Fig. 14.

The reasons for misalignment can be found in the following technological procedures in the process of engine overhaul:

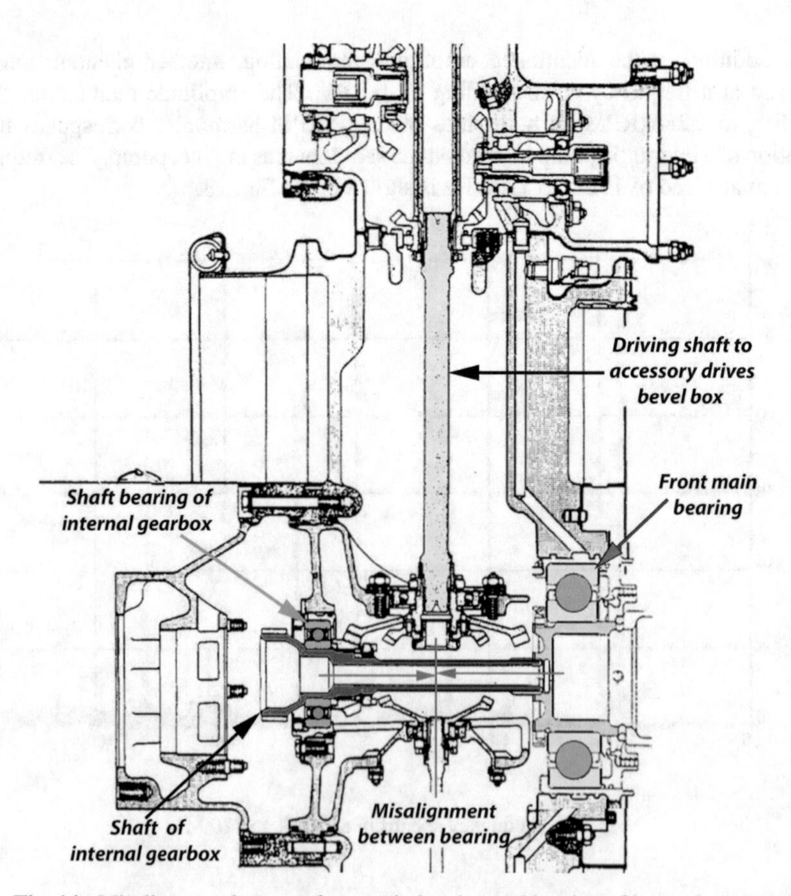

Fig. 14. Misalignment between front main bearing and bearing of internal gearbox

- inadequate gaps between main shaft of internal gearbox and compressor drum after assembly,
- improperly positioned front bearing of the internal gearbox main shaft (incorrectly seated), which caused misalignment in relation to front main bearing, and
- inadequate clearance when mounting the internal gearbox main shaft to the compressor drum.

It should also be mentioned that the existence of a large number of harmonics in the frequency spectrum of the basic RPM, 2xRPM, 3xRPM, 4xRPM, indicates that there is also a looseness in the rotating bearing of the internal gearbox.

Based on the performed analysis and the given conclusions, the Viper 22-8 engine was checked and adjusted the elements in the internal gear unit. After mechanical corrections, a test was performed on the engine on the test cell, without the use of additional equipment for measuring vibrations. The vibrations of the motor were lower than the vibration shown in this paper and the motor was sent to the user. The results of vibration measurements after correction are shown in Fig. 15.

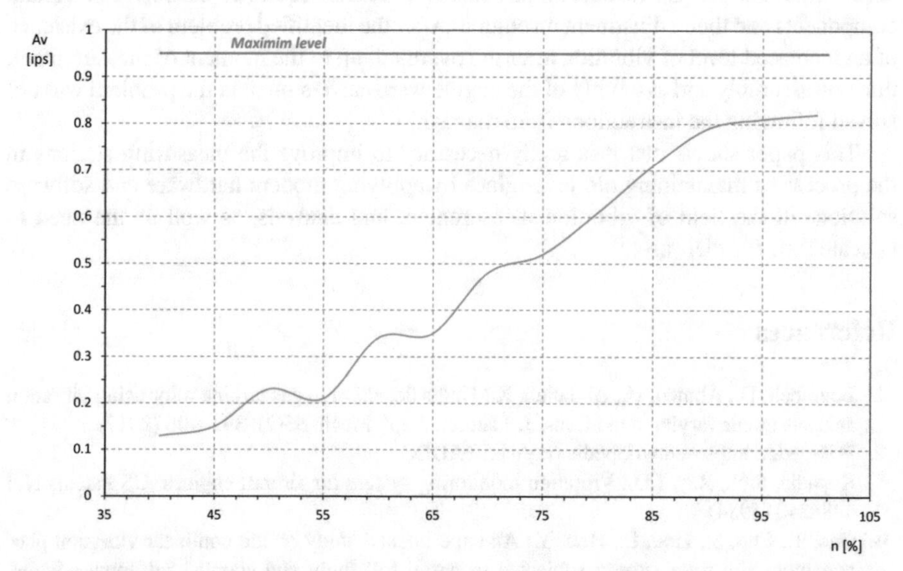

Fig. 15. The vibration level recorded from the test cell sensor after mechanical correction in internal gearbox

5 Conclusion

In this study, the importance of measuring the vibration of a jet engine on a test cell with equipment capable of performing frequency analysis in real time is shown. The understanding of the frequency spectrum in jet engine allows identification and localization of the source of the vibration anomaly as well as quantification of the defect.

The existence of this equipment is extremely important in situations where overhauled are made to old jet engine models for which manufacturers do not provide full maintenance support (spare parts). In addition to standard problems with engine components that may be damaged and require repair or replacement, there may also be a problem with the integrity of the structure due to aging. The integrity of the jet engine structure, due to life cycle and age, requires that a frequency vibration spectrum be recorded prior to each overhaul, which would be compared to the frequency spectrum after the overhaul.

The possibility of splitting the spectrum overall value into multiple frequency bands allows us to know in advance the areas in which the most typical problems are usually manifested and help us to identify them even before we can visualize the spectrum in frequencies and wave in the time.

This paper describes how to troubleshoot the increased vibration level of a Viper 22-8 engine after repair. The problem was successfully solved by applying the theory of vibration analysis and the recorded parameters in the time and frequency domains. This approach to vibrodiagnostic on overhauled jet engine is important in terms of saving time and human resources, as existing manuals require replacement of certain components and their adjustment on engine. After the identified problem of the existence of an increased level of vibration after the overhaul, up to the moment of measurement, three disassembly and assembly of the engine were carried out and the problem was not solved following the instructions from manual.

This paper shows that it is really necessary to improve the measuring stations in the process of maintaining old jet engines by applying modern hardware and software solutions in the field of vibration measurement and analysis, as well as the need to educate staff for their use.

References

1. Benrabeh, D., Ahmed, H., Abdallah, K.: Faults detection in gas turbine rotor using vibration analysis under varying conditions. J. Theoret. Appl. Mech. **55**(2), 393–406 (2017)
2. Wikipedia. https://en.wikipedia.org/wiki/FADEC
3. Kapadia, S.N., Ray, J.D.: Vibration monitoring system for aircraft engines. US Patents No: 4488240 (1984)
4. Gao, T., Cao, S., Hou, L., Hou, Y.: An experimental study on the nonlinear vibration phenomenon of a rotor system subjected to barrel roll flight and coupled rub-impact faults. Measurement **153**, 107406 (2020)
5. Khan, M.A., Shahid, M.A., Ahmed, S.A., Khan, S.Z., Khan, K.A., Ali, S.A., Tariq, M.: Gear misalignment diagnosis using statistical features of vibration and airborne sound spectrums. Measurement **145**, 419–435 (2019)
6. Maintenance manual, Viper Mk 22-8, Rolls-Royce, Bristol (1987)
7. Stupar, S., Simonović, A., Jovanović, M.: Measurement and analysis of vibrations on the helicopter structure in order to detect defects of operating elements. Sci. Tech. Rev. **62**(1), 58–63 (2012)
8. Ilić, Z., Rašuo, B., Jovanović, M., Janković, D.: Impact of changing quality of air/fuel mixture during a flying plane equipped with piston propeller group with respect to vibration low frequency spectrum. FME Trans. **41**(1), 25–32 (2013)

9. Ilić, Z., Rašuo, B., Jovanović, M., Jovičić, S., Tomić, L., Jankovic, M., Petrašinović, D.: The efficiency of passive vibration damping on the pilot seat of piston propeller aircraft. Measurement **95**, 21–32 (2017)
10. Žakić-Nedeljković, D., Simonović, A., Stupar, S., Lukić, N.: Measurement and analysis of vibrations of the axial - flow compressor caused by inlet flow instability during the flight of aircraft. Sci. Tech. Rev. **65**(2), 34–39 (2015)
11. Federal Aviation Administration (FAA): Correlation, Operation, Design and Modification of turbofan/jet engine test cells, Advisory Circular (AC) No: 43-207 (2002)

LDA Experimental Research of Turbulent Swirling Flow Behind the Axial Fans in Pipe, Jet and Diffuser

Đorđe S. Čantrak(✉), Novica Z. Janković, and Dejan B. Ilić

University of Belgrade, Faculty of Mechanical Engineering, Kraljice Marije 16, Belgrade 35, 11120 Belgrade, Serbia
djcantrak@mas.bg.ac.rs

Abstract. In this paper is analyzed turbulent swirling flow generated by axial fans. Three cases will be presented. The first one is the axial fan impeller in-built in the pipe 27.74D long, where D = 0.4 m is pipe inner diameter. The second case is study of the turbulent swirling flow in the free jet generated by the axial fan impeller with casing. The third case is conical diffuser with the inlet diameter D = 0.4 m and total divergence angle of 8.6°, studied in the test chamber followed by the flow meter. Axial fan, even in the case of the same impeller geometry, had various duty points in all these installations. All regimes were adjusted by the continuous fan rotation speed regulation, as well as with the impeller blade angles. Only in the case of the installation with diffuser, additional regulating valve and booster fan were sometimes applied. In all these cases was used laser Doppler anemometry (LDA), one- and three-component systems. All generated flows are extremely turbulent and in the case of the pipe and diffuser occur reverse flows in the central flow region. In the paper will be discussed calculated integral values, such as average circulation or swirl number. Time-averaged circumferential velocity profiles revealed Rankine vortex type in pipe, while in the diffuser is generated, with the different type of impeller, solid body profile. Turbulence levels and their distributions are analyzed. In addition, statistical moments of the higher order, like skewness and flatness factors are discussed.

Keywords: Turbulent swirling flow · Axial fan · Laser Doppler anemometry

1 Introduction

Turbulence is still unsolved problem of the classical physics and attracts researchers worldwide. "A completely formal theory of turbulence still doesn't exist." [1], but few definitions are listed here. "So we might picture turbulence as a seething tangle of vortex tubes, evolving under the influence of their self-induced velocity field." [2]. "Turbulence is a three-dimensional time-dependent motion in which vortex stretching causes velocity fluctuations to spread to all wavelengths between a minimum determined by viscous forces and a maximum determined by the boundary conditions of the flow. It is the

N. Mitrovic et al. (Eds.): CNNTech 2020, LNNS 153, pp. 184–202, 2021.
https://doi.org/10.1007/978-3-030-58362-0_12

usual state of fluid motion except at low Reynolds numbers." [3]. "Turbulence is a non-stationary spatio-temporal evolution of the random motion of complex eddy structures." [4].

Turbulent swirling flow is one of the most complex forms of turbulence. It exists in various forms in nature and technical systems. However, it has great scientific and technical importance. Research of turbulent swirling flow in gas cyclones is presented in [5]. Anyhow, focus of this paper is study of the turbulent swirling flow behind the axial turbomachine, namely axial fan impeller without guide vanes, due to the necessity in technical practice to estimate flow losses with and without guide vanes, as well as to study turbomachinery energy efficiency.

In [6, 7] theoretical and experimental research of the turbulent swirling flow in pipe and diffuser, generated by guide vanes is presented. In [6] experimental investigations of statistical properties and structure of turbulent swirling flows in pipes and diffusers, as well as non-gradient turbulent transfer are presented. Experimental research of turbulent swirling in a straight pipe, with an introduced free vortex type is presented in [8], classifying turbulent swirling flow after distribution of the circumferential velocity in four groups: forced vortex, free vortex, Rankine vortex, which is combination of the previous two vortices, and the wall jet.

Experimental investigation of the turbulent swirling flow is a challenging task for classical probes, so, new geometries are developed and extensively used in [9, 10].

Theoretical considerations of the turbulent swirling flow in pipe, determination of the turbulent shear stresses and turbulent viscosity on the basis of the experimentally determined pressure and velocity fields are presented in [10–30].

Numerous papers on the topics concerning turbulent swirling flow research in pipes and diffusers are listed in paper [31]. In these papers are presented various experimental techniques. Original classical probes used in combination of angle probe with two symmetrically positioned holes which determine flow direction in quasy-two-dimensional flow and one combined probe which measure total pressure and with attached sleeve static pressure in the same point [9, 10, 32]. These two, i.e. three probes, determine dynamic pressure and consequently velocity. A subminiature DISA 55A53 hot-wire anemometry (HWA) probe was successfully used in research of the structure of turbulence of flows in pipes and diffusers [6, 7, 26, 27]. Original V-geometry HWA probes were used in research of the turbulent swirling flow in pipe behind the axial fan [11, 33].

Optical measurement techniques are introduced in the research of turbulent swirling flow behind the axial fan in various geometries (pipe, diffusers, jet), due to reverse flow regions and vortex core dynamics [34–62]. It is possible to resolve velocity and vorticity fields and correlate with fan duty points. It is possible, after criterion of total average velocity minima, to observe vortex core dynamics [34, 35, 39]. Invariant maps were determined on the basis of the time resolved high stereo particle image velocimetry data and flow anisotropy was studied on the basis of them [34, 35, 39].

Extensive numerical simulations in turbulent swirling flow research have been developed [8], as well as Large eddy simulations (LES) [63]. Namely, investigation of the turbulent swirling flow in circular pipe in the wake behind an axial fan by URANS and LES using two different CFD codes SPARC and OpenFOAM [63, 64]. Results of simulations are compared with experimental results obtained with LDA in [34]. Turbulent swirling flow in circular pipe, generated by guide vanes, is computed as 2D axisymmetric, with various turbulent models and focus on computations with Reynolds stress transport models Launder-Gibson (LG) and Speziale-Sarkar-Gatski (SSG) [65]. Obtained CFD results were validated by Čantrak S. experimental results [6]. It was shown that Reynolds stress models provide good prediction of mean velocities [65].

In [66] is performed experimental and numerical research of the turbulent swirling flow in pipe with focus on the turbulent transport processes and problems in its modeling. Numerical modeling of non-local transport in turbulent swirling flow [67, 68] is conducted, also, on the basis of the experimental results [6]. CFD simulations of the turbulent swirling and non-swirling incompressible flow in straight conical diffusers, on the basis of the same experimental results [6], are presented in [69].

Turbulent swirling flow investigation in three straight conical diffusers for various axial fans' rotation speeds and blade angles, with various measuring techniques, is reported in [40, 46, 56, 70]. Angle probe and the original construction of the Pitot-static probe with the detachable front sleeve were used for determination of the pressure fields [10, 32, 46, 48, 53, 56, 70]. In addition, one- and two-component LDA systems were used for velocity components and Reynolds stresses measurements [40, 46, 56, 70]. In [56] are presented distributions of the inlet Boussinesq number and outlet Coriolis coefficient and discussed flow energy loss coefficients.

Determination of the flow losses in pipes with inbuilt axial fans without guide vanes, as well as their duty points, is discussed in papers [60, 71–73]. These flow conditions could result with unstable fan duty points.

Influence of the axial fan blade angle, as well as other aspects of the axial fan impeller geometry on the turbulent swirl flow characteristics in pipe was discussed in, besides other papers in [34, 74]. It was shown that, for the specially designed axial fan impeller, which generates turbulent swirling flow with Rankine swirl distribution, inbuilt in pipe, that Rankine combined profile consisted of two vortices, i.e. solid body and free vortex, survives downstream and in one measuring section forms similar velocity profiles [9, 10, 34, 35, 39, 41, 47, 48, 51, 58, 59, 61, 62, 71–74].

Investigation of the free turbulent swirling jet behind the axial fan by use of the three-component LDA system is presented in [42, 75]. The majority of the obtained experimental results are positioned in the region of the three-component isotropic turbulence and axisymmetric expansion in Lumley invariant maps [42].

In this paper test rigs and measuring LDA equipment for turbulent swirling flow investigations are presented in three geometries pipe, diffusers and jets. Obtained and processed data, like average velocity fields, flow integral parameters are presented and discussed afterwards, as well as statistical moments of the second and higher orders.

2 Experimental Test Rigs and LDA Systems

Two various axial fan impellers with adjustable outlet blade angle (β_R) were used. Both impellers had rotation speed $n = 1000$ rpm. All installations had profiled mouth inlet and inbuilt diameter of $D = 0.4$ m.

Experimental results in pipe [34] and diffuser [70], obtained with one-component LDA system, which measured subsequently all three velocity components, are presented.

In these both cases axial fan impeller is in-built like "Category B: free inlet and ducted outlet" [76]. In the case of the research of turbulent swirling jet, i.e. wake behind the axial fan impeller was applied three-component LDA system [75]. In this case installation was following construction "category A: free inlet and free outlet" [75]. In this chapter are described test rigs used in these installations.

2.1 Test Rig for Investigation of the Turbulent Swirling Flow in Pipe

Test rig design is presented in Fig. 1. It is 27.74·D long, where inner pipe diameter is approximately D = 0.4 m. It has two transparent sections. Position of the measuring section (Fig. 1, position 3) is x/D = 3.35, where x is coordinate along pipe axis, with origin at the tangential surface of the profiled bell mouth test rig inlet (Fig. 1, position 2). Measurements in other measuring sections are not discussed here, due to the paper scope.

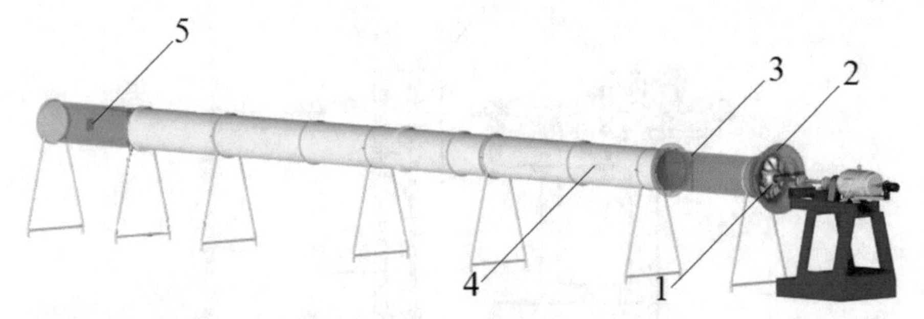

Fig. 1. Test rig for investigation of the turbulent swirling flow in pipe: 1 - axial fan impeller, 2 - profiled bell mouth inlet, 3 - measuring section (x/D = 3.35), 4 - non-transparent pipe with measurement positions for classical techniques and 5 - downstream measuring section.

Axial fan impeller (Fig. 1, position 1) is designed, by Protić†, to generate Rankine vortex [34, 35, 51, 54]. It has nine adjustable blades. Ratio of the hub and outer diameter is 0.5 and it belongs to the group of the medium pressure axial fans. In this paper are presented results for the case when the blade angle at impeller diameter Da = 0.399 m was adjusted to $\beta_{Ra} = 30°$. So, this fan is briefly denoted ZP30. The axial fan rotation speed was regulated by a fully automated thyristor bridge with error up to ±0.5 rpm, designed and made by Stojiljković [34, 35, 75].

One-component LDA system Flow Explorer Mini LDA, by Dantec, with BSA Flow Software was used in these measurements, as well as for measurements in diffuser. Laser has 35 mW power, red light wave length 660 nm, focus 300 mm, measuring volume size 0.1013 × 0.1008 × 1.013 [mm^3] with specified measuring uncertainty of 0.1%. Laser was positioned on the precise linear guide [34]. Axial and radial velocities were measured from pipe side along vertical diameter of the measuring section (Fig. 1, position 3), while circumferential from above and below with overlapping of the measuring points in the central region [34]. This was done due to the fact that laser focus length was only 300 mm. Measurements were conducted successively for each velocity component in backscatter mode.

Transparent pipe had approximately 5 mm wall thickness. Measurement dislocations, especially in the region r/R > 0.6, where r is pipe radial coordinate measured from the pipe axis and R = D/2, are discussed and presented in [34, 49, 52]. Flow was seeded by thermal fog generated by Antari Z3000II, and liquid "Heavy Fog". It was positioned at the test rig inlet and naturally distributed and mixed by the axial fan impeller. Measurements lasted 10 s for each point. Measuring points were distributed on 10 mm along the vertical diameter. Sampling rate and validation varied along the vertical diameter, but they both reached high values.

2.2 Test Rig for Investigation of the Turbulent Swirling Flow in Jet

Experimental test rig for investigation of the turbulent swirling flow in the free jet generated by the axial fan impeller, i.e. in its wake, is presented in Fig. 2.

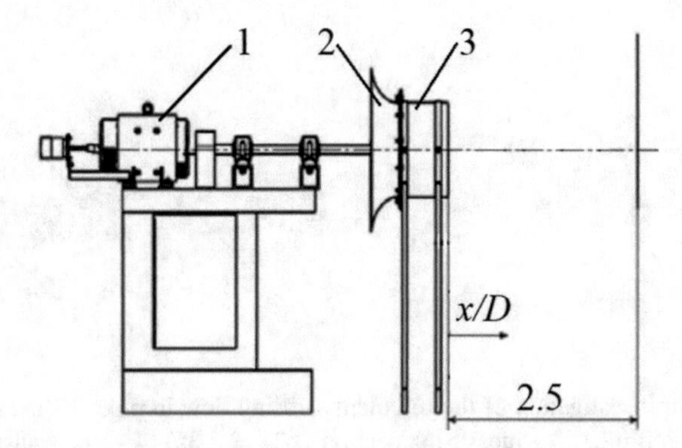

Fig. 2. Test rig for investigation of the turbulent swirling flow in jet: 1 - DC motor, 2 - profiled bell mouth inlet and 3 - axial fan casing with inbuilt impeller.

In this case the same fan impeller with the adjusted blade angle $\beta_{Ra} = 30°$ is used, as well as DC motor with fully automated thyristor bridge, like in the investigation of the turbulent swirling flow in pipe. Profiled bell mouth inlet (Fig. 2, position 2) was also the same. Seeding was, also, applied in the same manner on the axial fan suction side, as in the previous case. Only liquid EFOG, Density Fluid, Invision was used [35, 75]. In this paper results of the three velocity components measurements in the measuring section x/D = 2.5 (Fig. 2), i.e. x = 1000 mm, are presented. However, more complex three-component LDA system, by TSI, was applied here. Laser Continuum Ar-Ion of 5 W, by Coherent was used. During measurements laser power was 916 mW. Laser wavelengths are 514.5 nm, 488 nm and 476.5 nm. Two TSI laser probes TR60 with beam expanders XPD60-750 were used to form measurement volume. The measurement volume diameter was approximately 70 μm, while measurement volume length was approximately 280 μm. The measurement focus with attached optics was on 757.7 mm, so probes didn't generate any disturbance in the measurements. One of the probes was measuring two velocity components, while the second only one. Both of them were operating in backscatter mode. Probes position and measurement uncertainty quantification is presented in [43]. The TSI Flow Sizer software was used for acquisition and preliminary data analysis. Recording time was 20 s for all measurements. Data frequency and validation varied along the measuring sections. Measurements have been performed in the software coincidence mode and the highest sampling rate was 3.5 kHz. Most of the measurements had approximately 1.5 kHz [75].

2.3 Test Rig for Investigation of the Turbulent Swirling Flow in Diffuser

Test rig is designed and presented in Fig. 3 [40]. The incompressible swirl flow field is induced by the axial fan impeller (Fig. 3, position 2) with outlet blade angle of 29°, which is set in the initial part of the straight pipe section followed by a conical diffuser (4). Axial fan geometry is presented in [34, 70]. It has seven adjustable blades. Fan was powered and controlled by the DC motor and adequate controller (1). Test rig, like in Fig. 1, had profiled bell mouth inlet (2). The conical diffuser is with the inlet diameter D = 0.4 m, length L = 1.8 m and total divergence angle of 8.6°, studied in the test chamber (5) followed by the flow meter – nozzle (7). Measurements were performed in one measuring section in conical diffuser (z_2 = 0.2 m, z/R_0 = 1) with one-component laser Doppler anemometry (LDA) system, presented in Sect. 2.1, for one flow regime C and for impeller rotation speed 1000 rpm.

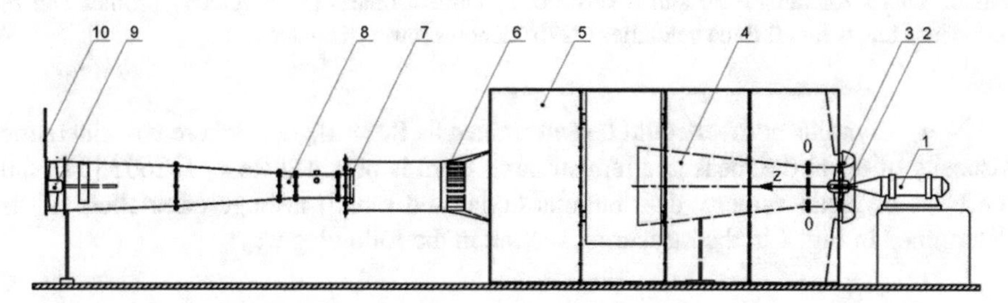

Fig. 3. Test rig for investigation of the turbulent swirling flow in diffuser.

LDA measurements of axial and circumferential velocities have been performed successively along one measuring radius with 21 measuring points. The same seeding was applied, as in the case of the turbulent swirling flow investigation in the pipe. In some cases was used also booster fan (Fig. 3, position 9) and outlet valve (Fig. 3, position 10). This installation could also follow ISO 5801 [76] for testing axial fans.

3 Time Averaged Velocities, Integral Flow Characteristics and Turbulence Levels

In this chapter experimentally obtained profiles of the time-averaged velocities will be presented and discussed for all three cases of investigations of the turbulent swirling flow in pipe, jet and diffuser. In addition calculated integral flow characteristics and turbulence levels will be presented.

3.1 Turbulent Swirling Flow in Pipe

In Fig. 4 non-dimensional time-averaged (mean) axial (U/U_m), radial (V/U_m) and circumferential (W/U_m) velocity profiles are presented for measuring section x/D = 3.35 (Fig. 1), n = 1000 rpm and fan ZP30, where U_m is an averaged velocity by area obtained as $U_m = 4Q/D^2\pi$, where Q is volume flow rate numerically determined on the basis of the mean axial velocity profile in the measuring section. In this case is obtained: Q = 0.86 m^3/s and U_m = 6.68 m/s [34].

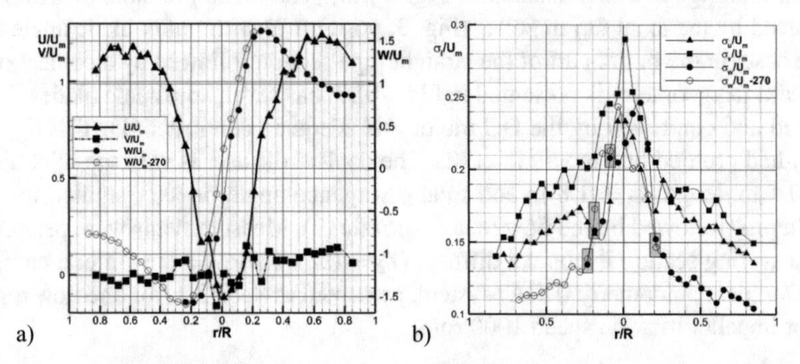

Fig. 4. Charts for fan ZP 30 and n = 1000: a) Dimensionless mean velocity profiles and b) turbulence levels for all three velocities ("270" denotes lower diameter part).

Now, Reynolds number could be determined as $Re = U_m D/\nu$, where ν is kinematic viscosity of air at the measured temperature. Here is obtained Re = 182602 [34]. On the basis of mean velocity distributions (axial and radial) average circulation (Γ) is determined in Fig. 4 in the measuring section in the following way:

$$\Gamma = 4\pi^2 R^3 \int_0^1 UW\left(\frac{r}{R}\right)^2 d\left(\frac{r}{R}\right)/Q = 5.41\frac{m^2}{s}. \qquad (1)$$

Swirl number is defined by Strscheletzky for turbomachinery [77] in the following way $\Omega = Q/(R\Gamma) = 0.79$ and it was already used in this form in [10, 34, 53, 56]. This value is obtained on the basis of the experimental results in this case.

Axial and circumferential velocity profiles are similar, but non-homogeneous. Reverse flow is obvious in central region (Fig. 1a). Mean axial velocity has negative values here. All four flow regions are obvious in circumferential velocity profile. Namely, central solid body region is characterized with linear circumferential velocity distribution. This is surrounded with shear stress region characterized with W_{max}. Sound flow region follows with increase of radius r. Here is rW = const., and mean axial velocity is almost constant. The last, boundary layer region, is not revealed due to extensive optical curvature. Radial velocity exists, but it is not high, although it has great gradients in the vortex core region. In these diagrams some points are omitted. Overlapping of the twice measured circumferential mean velocity values in the vortex core region is more-less obvious in Fig. 4, even for turbulence levels. Gray rectangular mark good overlapping in Fig. 4b. This chart reveals high anisotropy in all flow regions. The highest turbulence levels for all three velocities are reached in the vortex core region. The lowest turbulence levels are reached in the sound flow region for circumferential velocity, while the highest in almost all measuring points is for radial velocity.

3.2 Turbulent Swirling Flow in the Axial Fan Jet

Distributions of the mean velocities and turbulence levels for turbulent swirling flow in jet, generated by the axial fan impeller, are presented in Fig. 5.

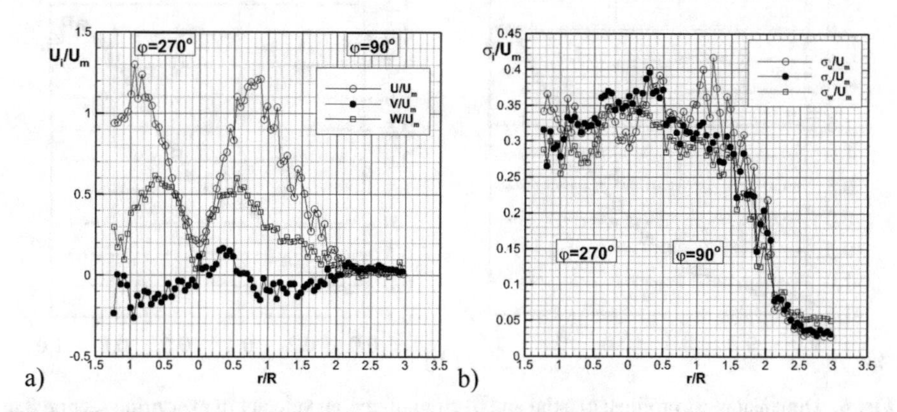

Fig. 5. Distributions of the a) mean velocities and b) turbulence levels for fan ZP 30 and n = 1000 rpm in the turbulent swirling jet ("90" denotes upper, while "270" lower part of the measuring section; R = 0.2 m).

Experimentally obtained axial and circumferential velocity distributions are axisymmetric, but non-uniform (Fig. 5a). The effect of circumferential velocity is very strong. Reverse flow doesn't exist and U_{max} is reached in the region $r/R > 0.7$, while W_{max} closer to the pipe axis. Axial velocity profile is concave and it means that swirl development process is not accomplished in this measuring section. In the jet axial velocity is dominant, and radial velocity is higher than in the case of research in pipe. Following circumferential velocity distribution, it could be concluded that vortex structure doesn't have the same structure in jet and in pipe, despite the fact that the same axial fan impeller is inbuilt. Anyhow, it is inbuilt in two different ways, so it generates two various velocity distributions. Average velocity $U_m = 6.62$ m/s is here calculated on the basis of the axial velocity profile and volume flow rate $Q = 0.853$ m^3/s in the region up to $r = R = 0.2$ m. These values are calculated on the basis of the pipe diameter, and they are very close to the ones obtained in pipe. Obtained average circulation is lower $\Gamma = 2.55$ m^2/s, while swirl number is higher $\Omega = 1.65$. Reynolds number, calculated in the same way, is Re = 189254.

Turbulence levels are significantly higher in the whole region (Fig. 5b) than in the case of the turbulent swirling flow in pipe. Maxima are reached for the axial fluctuating velocity. Anisotropy is obvious also in this case. Values in both cases are significantly lower than in the case of the turbulent swirling flow in diffuser in the whole measuring section for axial and circumferential velocities (Fig. 7).

3.3 Turbulent Swirling Flow in Diffuser

Experimentally obtained time averaged dimensionless axial (U) and time averaged dimensionless circumferential (W) velocities, measured with LDA systems, in one measuring section 2 ($z_2 = 0.2$ m, $z/R_0 = 1$) in diffuser are presented in Fig. 6. Regime C is the case described in Table 1.

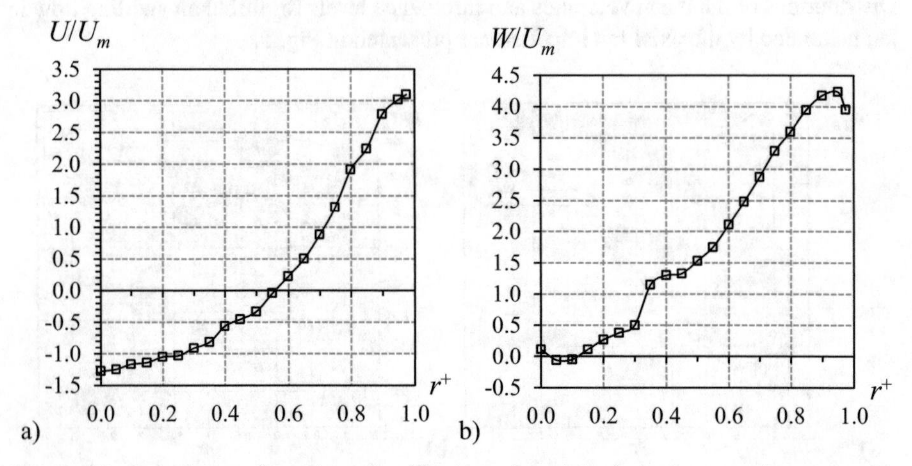

Fig. 6. Dimensionless profiles: a) axial and b) circumferential velocity in measuring section 2 in diffuser, for regime C.

In the vortex core, the axial velocity components have negative values. Namely, non-dimensional axial velocity profile shows existence of the reverse flow in the vortex core region, like it was shown also in the case of the turbulent swirling flow in pipe. It is confirmed that in the cases of high vortex strength, recirculation currents can be formed. Value of the axial velocity component increases with increasing of r^+ ($r^+ = r/R_2$), i.e. in the direction of the boundary layer region.

Based on the circumferential velocity profile in measuring section 2 for regime C, it is observed that the vortex is asymmetrically positioned with respect to the diffuser axis, i.e. the center of the vortex core region is displaced at $r^+ \approx 0.05$. The highest value of circumferential velocity is achieved for high value of radius, namely $r^+ = 0.95$. The circumferential velocity distribution is directly proportional to the distance from the axis, i.e. with the radius, by the law of the forced vortex distribution. It is shown in [34] that this fan impeller forms solid body circumferential velocity profile also in the pipe. Volume flow rate in diffuser is calculated in the following way:

$$Q = 2\pi \int_0^{R_2} rU dr, \tag{2}$$

where R_2 is diffuser radius in the measuring section 2. Average circulation is calculated as it is given in (1), but the outer radius in this section is R_2. Swirl number is calculated in the similar way, but $R = R_2$. Average velocity is now calculated for the case $D/2 = R_2$, while Reynolds number $R_e = U_m D_2/\nu$. All these calculated integral flow parameters are presented in Table 1.

Table 1. Calculated integral flow parameters in measuring section 2.

Regime	Q [m^3/s]	Γ [m^2/s]	Ω [-]	U_m [m/s]	Re [-]	Booster fan
C	0.241	8.69	0.13	1.62	49740	No

Distributions of the turbulence levels for axial and circumferential fluctuating velocities are presented in Fig. 7.

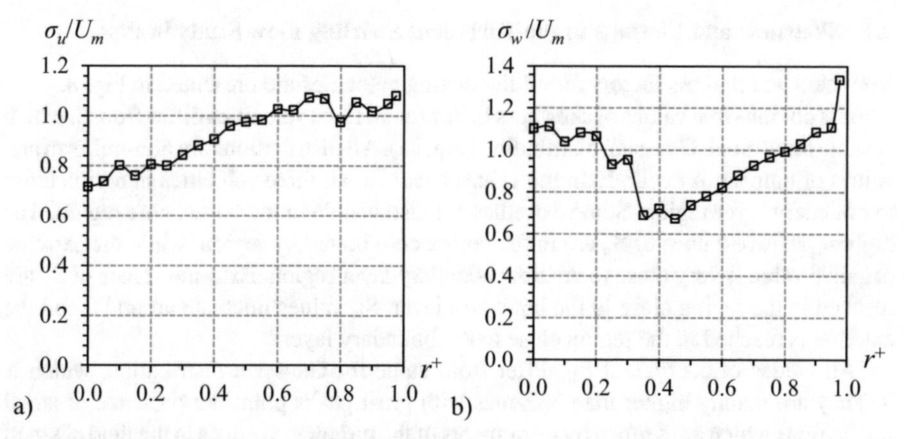

Fig. 7. Turbulence levels in diffuser (measuring section 2) for regime C for: a) axial and b) circumferential fluctuating velocities.

The highest levels of turbulence for axial velocity are reached in the region $0.6 < r^+ < 0.95$, in the vicinity of the diffuser wall, i.e. boundary layer region. The highest levels of turbulence for circumferential velocity are reached in the vortex core region $0 < r^+ < 0.2$ and near in the boundary layer vicinity region $0.85 < r^+ < 0.95$ (Fig. 7).

4 Statistical Moments of the Third and Fourth Order

Statistical moments of the third (skewness) and fourth (flatness or kurtosis) order for all three fluctuating velocities behind the axial fan impellers in pipe, jet and diffuser are discussed in this chapter.

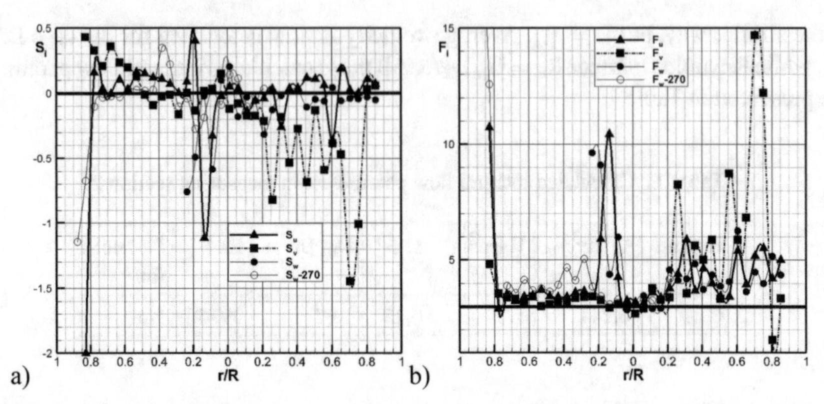

Fig. 8. a) Skewness and b) flatness factors for all three fluctuating velocities in the pipe measuring section x/D = 3.35 for ZP30 and n = 1000 rpm.

4.1 Skewness and Flatness in the Turbulent Swirling Flow Study in Pipe

Skewness and flatness factors for all fluctuating velocities are presented in Fig. 8.

It is obvious that values of skewness factor for all three velocities differ from 0, which is value for normal, Gaussian distribution (Fig. 8a). All distributions are non-uniform and values of both signs occur. Extreme values occur for all three velocities in region close to boundary layer region. Some extremes are also visible in the vortex core region. The highest positive values of S_u are in the vortex core boundary region, while the extreme negative value is very close to the flow boundary layer region. Extreme values of S_v are reached in the region close to the boundary layer. S_w values fluctuate around 0 and the extreme is reached in the region close to the boundary layer.

All values of coefficient F_w differ from value for Gaussian distribution, which is 3. They are mainly higher than 3. Zones with great peaks point out presence of small fluctuations which are formed by movements of the turbulent vortices in the field of small velocity gradients. All three velocities have their extremes. So, various size turbulence structures occur in flow.

4.2 Statistical Moments of the Third and Fourth Order in the Study of the Turbulent Swirling Flow in Jet

In Fig. 9 distributions of the skewness and flatness factors for turbulent swirling flow in the wake of the axial fan impeller, i.e. in the free jet, are presented.

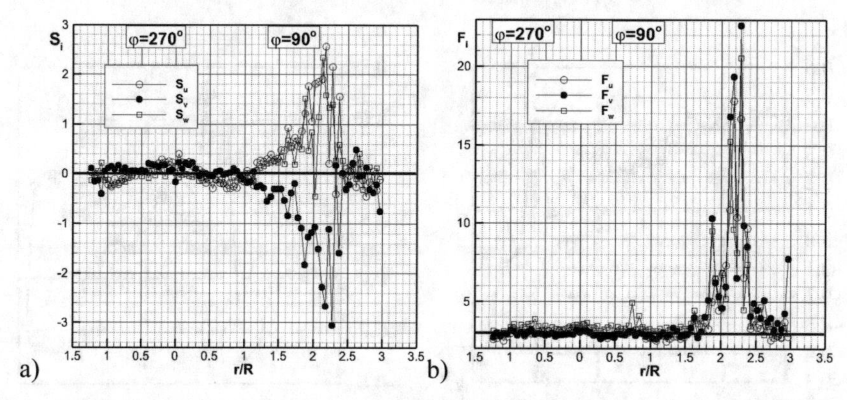

Fig. 9. a) Skewness and b) flatness factors for all three fluctuating velocities in the jet section x/D = 2.5 for ZP30 and n = 1000 rpm.

In this case skewness and flatness also differ for all three fluctuating velocities from values for Gaussian distribution. All three factors change sign and have lower values in the zone up to r/R = 1, where maximum for all fluctuating velocities is $S_{u,max} = 0.3945$ in measuring point r/R = 0.0494, while minimum is, again, reached for the axial velocity $S_{u,min} = -0.3018$ in r/R = 0.7407. In the complex flow zone r/R ∈ (1, 2.5) distributions of the S_u and S_w on one side and S_v on the other side are almost axisymmetric. Value $S_{u,max} = 2.5557$ is reached in r/R = 2.1728, while $S_{v,min} = -3.0534$ in r/R = 2.2716.

Flatness factors for all fluctuating velocities, again, reach higher values in the region r/R ∈ (1, 2.5). Value $F_{u,max} = 17.8129$ occurs, again, in the same point r/R = 2.1728, where $S_{u,max}$ was reached. Anyhow, the highest flatness factor $F_{v,max} = 22.6483$ is reached in the vicinity, in point r/R = 2.2716, while $F_{u,min} = 2.4351$ in r/R = 1.2346. It is interesting to notice that $F_{u,max} = 3.4677$ for region r/R ∈ (0, 1) is reached in r/R = 0.0494, where $S_{u,max} = 0.3945$ for the same flow region.

4.3 How Do Statistical Moments of the Higher Order Behave in Turbulent Swirling Flow in Diffuser?

Distributions of skewness and flatness for both velocities in measuring section 2 are presented in Figs. 10 and 11.

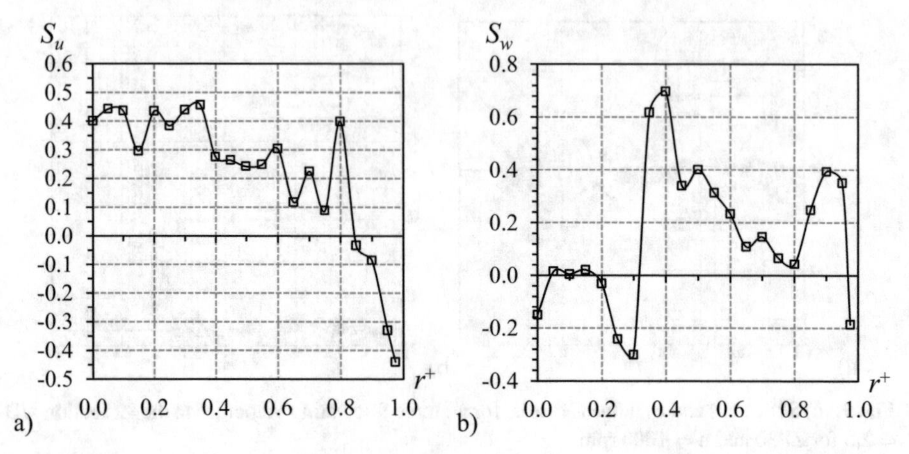

Fig. 10. Skewness factors in diffuser (measuring section 2) for regime C for: a) axial and b) circumferential fluctuating velocities.

Distribution of the skewness factor (Fig. 10) is, again, not equal to Gaussian probability distribution value ($S_w = 0$). It is obvious that negative and positive values occur. Skewness factor for axial fluctuating velocities has maxima in the center of the diffuser (central flow region - vortex core region) and in $r^+ = 0.8$. It has minimum in region near to the wall of diffuser. Skewness factor for circumferential fluctuating velocities has values close to 0 in the center of the diffuser (Fig. 10b).

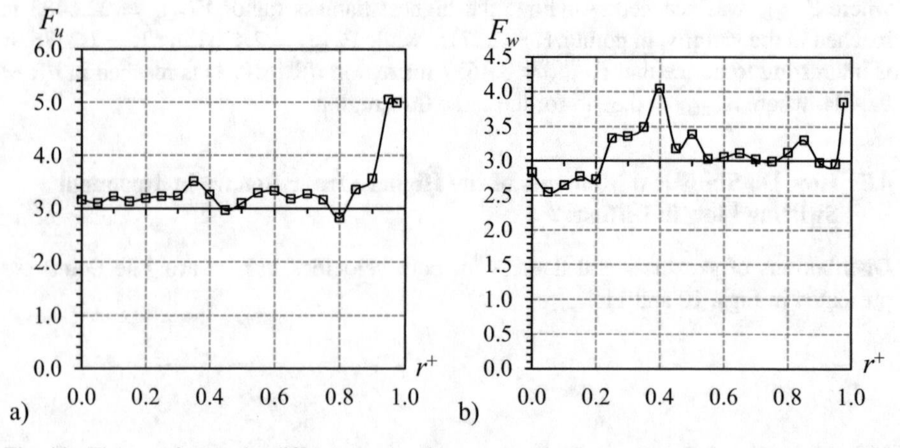

Fig. 11. Flatness factors in diffuser (measuring section 2) for regime C for: a) axial and b) circumferential fluctuating velocities.

The flatness factor for axial fluctuating velocity is slightly greater than 3, except in region near to the wall of diffuser where has maxima (Fig. 11a). Value of flatness factor for circumferential fluctuating velocity oscillates around the value for normal, i.e., Gaussian probability distribution ($F_w = 3$). In vortex core region value of flatness factor is slightly less than 3 and has maxima for $r^+ = 0.4$ and near to the wall of diffuser (Fig. 11b).

5 Conclusions

In this paper turbulent swirling flow research behind the axial fan impeller is presented in three various geometries: pipe, jet and diffuser. The first and the third cases could be classified as category B "free inlet and ducted outlet" according to standard ISO 5801 [76], while the second case as category A "free inlet and free outlet". In the first and the second case were used the same impellers, while in all cases impeller rotation number was identical. In the first two cases blade angle at the impeller outer diameter was the same 30°, while in the research in diffuser, it was 29°. In all three cases high Reynolds numbers are reached.

Various vortex types are generated. Anyhow, all obtained, time-averaged velocity profiles are non-homogeneous. Rankine vortex structure is revealed in pipe, while in the jet complex vortex structure with lower values of circumferential velocity is unveiled. In diffuser is generated circumferential "solid body" velocity profile. It is noticed that almost identical average U_m velocities are reached in the first two cases, and, consequently, volume flow rates. Significantly lower flow rate is measured in the case of the turbulent swirling flow in diffuser. Average circulations are $\Gamma_{diff} > \Gamma_{pipe} > \Gamma_{jet}$, while calculated swirl numbers are in order $\Omega_{jet} > \Omega_{pipe} > \Omega_{diff}$. So, the average circulation is reached in diffuser, while swirl number in jet. The highest turbulence levels are reached in the case of diffuser, while the lowest in the pipe. In the first two cases turbulence anisotropy is revealed, what is partially proved in the case of diffuser, because the third, radial, component was not measured, but it is also expected. In all cases skewness and flatness factors for three fluctuating velocities differ from the values for normal Gaussian distributions. Extremes are reached in the case of the pipe and diffuser.

In addition, one can conclude that LDA technique with adequate flow seeding, due to high sampling rate and validation, was successfully applied in all these complex geometries and investigations of the turbulent swirling flow generated by the axial turbomachine.

Acknowledgement. Axial fan with nine blades, applied in this research, is designed and constructed by Prof. Dr.-Ing. Zoran D. Protić† (1922–2010). Prof. Dr Zoran Stojiljković, designed and built a very precise original fan rotation speed regulator used in experiments in this paper. This works is financially supported by the Ministry of Education, Science and Technological Development Republic of Serbia (MESTD RS), contract number 451-03-68/2020-14/200105 (subproject: TR35046) and by the Bilateral project "Joint Research on the Development Technology of Low-head Run-of-the-river Hydropower", between MESTD RS and Ministry of Water Resources in China and Renewable Energy and Rural Electrification Zhejiang International Science Center which authors hereby gratefully acknowledge.

References

1. Tennekes, H., Lumley, J.L.: A First Course in Turbulence. The MIT Press, Cambridge (1972)

2. Davidson, P.A.: Turbulence, An Introduction for Scientists and Engineers. Oxford University Press, New York (2004)
3. Bradshaw, P.: An Introduction to Turbulence and Its Measurement. Pergamon Press, Oxford (1971)
4. Čantrak, S., Lečić, M., Ćoćić, A.: Fluid mechanics B, University of Belgrade, Faculty of Mechanical Engineering, Belgrade, Serbia (2009). http://fluidi.mas.bg.ac.rs/mfb/handout/mfB-handout.pdf. Accessed 15 Mar 2020
5. Hoekstra, A.J., Derksen, J.J., Van Der Akker, H.E.A.: An experimental and numerical study of turbulent swirling flow in gas cyclones. Chem. Eng. Sci. **54**(13–14), 2055–2065 (1999)
6. Čantrak, S.: Experimental investigations of statistical properties of turbulent swirling flows in pipes and diffusers. Ph.D. thesis, University Karlsruhe, Karlsruhe, Germany (1981). (in German)
7. Čantrak, S., et. al.: Influence of swirl on structural parameters of turbulent pipe flow. ZAMM **67**(5), T271–T272 (1987)
8. Kitoh, O.: Experimental study of turbulent swirling flow in a straight pipe. J. Fluid Mech. **225**, 445–479 (1991)
9. Benišek, M.: Investigation on the existence of stable profile vortex flow through cylindrical long lined circular pipes. ZAMM **57**, T173–T175 (1977)
10. Benišek, M.: Investigation of the swirling flow in long lined circular pipes. Ph.D. thesis, University of Belgrade, Faculty of Mechanical Engineering, Belgrade, Serbia (1979). (in Serbian)
11. Lečić, M.R.: Theoretical and experimental investigation of turbulent swirling flows. Ph.D. thesis, University of Belgrade, Faculty of Mechanical Engineering, Belgrade, Serbia (2003). (in Serbian)
12. Ćoćić, A.S.: Investigation of the structure of inhomogeneous turbulence using invariant theory. Mag. thesis, University of Belgrade, Faculty of Mechanical Engineering, Belgrade, Serbia (2007)
13. Algifri, A.H., Bhardwaj, R.K., Rao, Y.V.N.: Eddy viscosity in decaying swirl flow in a pipe. Appl. Sci. Res. **45**, 287–302 (1988)
14. Algifri, A.H., Bhardwaj, R.K.: Prediction of the heat transfer for decaying turbulent swirl flow in a tube. Int. J. Heat Mass Transf. **28**(9), 1637–1643 (1985)
15. Kreith, F., Sonju, O.K.: The decay of a turbulent swirl in a pipe. J. Fluid Mech. **22**(2), 257–271 (1965)
16. Scott, C.J., Bartelt, K.W.: Decaying annular swirl flow with inlet solid body rotation. J. Fluids Eng. **98**(1), 33–40 (1976)
17. Einstein, H.S., Li, H.: Steady vortex flow in a real fluid. In: Proceedings HT & Fluid Mech. Institute, pp. 33–43 (1951). Stanford University, California
18. Murakami, M., Kito, O., Katayama, Y., Iida, Y.: An experimental study of swirling flow in pipes. Bull. JSME **19**(128), 118–126 (1976)
19. Akiyma, T., Ikeda, M.: Fundamental study of the fluid mechanics of swirling pipe flow with air suction. Ind. Eng. Chem. Process Des. Dev. **25**, 907–913 (1986)
20. Rochino, A., Lavan, Z.: Analytical investigations of incompressible turbulent swirling flow in stationary ducts. J. Appl. Mech. **36**(2), 151–158 (1969)
21. Scott, C.J., Rask, D.R.: Turbulent viscosities for swirling flow in a stationary annulus. J. Fluids Eng. **95**(4), 557–566 (1973)
22. Yoshizawa, A., Yokoi, N., Nisizima, S., Itoh, S.-I., Itoh, K.: Variational approach to turbulent swirling pipe flow with the aid of helicity. Phys. Fluids **13**(8), 2309–2319 (2001)
23. Reader-Harris, M.J.: The decay of swirl in the pipe. Int. Heat Fluid Flow **15**(3), 212–217 (1994)

24. Čantrak, S.: Structure of turbulent swirling flow (introductory lecture). In: Proceedings Symposium “Contemporary Problems of Fluid Mechanics, pp. 41–59 (1992). University of Belgrade, Faculty of Mechanical Engineering, Belgrade, Serbia
25. Hosseini, A., Analysis of the influence of Rankin vortex on turbulent transfer processes. Mag. thesis, University of Belgrade, Faculty of Mechanical Engineering, Belgrade, Serbia (2011)
26. Čantrak, S., Benišek, M., Nedeljković, M.: Turbulent viscosity of the swirling flow in light of Boussinesq’s assumption of turbulent stresses. In: Proceedings, 16th Yugoslav Congress of Theoretical and Applied Mechanics, B 6–3, pp. 177-184, Bečići, Yugoslavia (1984)
27. Čantrak, S., Vukašinović, B., Lečić, M.: Investigation of the mechanism of turbulent transfer in hydromechanical processes in the field of centrifugal force. Procesna tehnika **3**, 21–26 (1995)
28. Benišek, M., Čantrak, S., Nedeljković, M.: Theoretical and experimental investigation of the turbulent swirling flow characteristics in a conical diffuser. ZAMM **71**(5), T453–T456 (1991)
29. Lilley, D.G., Chigier, N.A.: Nonisotropic turbulent stress distribution in swirling flows from mean value distributions. Int. J. Heat Mass Transf. **14**(4), 573–585 (1971)
30. Parchen, R.R., Steenbergen, W.: An experimental and numerical study of turbulent swirling pipe flows. J. Fluids Eng. **120**(1), 54–61 (1998)
31. Benišek, M.H., Lečić, M.R., Čantrak, Đ.S., Ilić, D.B.: The school of the turbulent swirling flow at the Faculty of Mechanical Engineering University of Belgrade. Therm. Sci. **21**(S3), S899–S911 (2017)
32. Benišek, M.H., Lečić, M.R., Ilić, D.B., Čantrak, Đ.S.: Application of new classical probes in swirl fluid flow measurements. Exp. Tech. **34**(3), 74–81 (2010)
33. Lečić, M., Radojević, S., Čantrak, Đ., Ćoćić, A.: V-type hot wire probe calibration. FME Trans. **35**(2), 55–62 (2007)
34. Čantrak, Đ.S.: Analysis of the vortex core and turbulence structure behind axial fans in a straight pipe using PIV, LDA and HWA methods. Ph.D. thesis, University of Belgrade, Faculty of Mechanical Engineering, Belgrade, Serbia (2012). (in Serbian)
35. Čantrak, Đ., Nedeljković, M., Janković, N.: Turbulent swirl flow characteristics and vortex core dynamics behind axial fan in a circular pipe. In: Proceedings CMFF 2012, pp. 749–756 (2012). Department of Fluid Mechanics, Faculty of Mechanical Engineering, Budapest University of Technology and Economics, Budapest, Hungary
36. Čantrak, Đ., Stamatios, P., Janković, N.: Stereoscopic PIV measurements and visualization of a turbulent swirl flow behind an axial fan in a pipe. In: Proceedings, 3th International Symposium Contemporary Problems of Fluid Mechanics, pp. 289–300 (2011). University of Belgrade, Faculty of Mechanical Engineering, Belgrade, Serbia
37. Čantrak, Đ., Nedeljković, M., Janković, N.: Turbulent swirl flow dynamics. In: Proceedings IConSSM 2011, The 3rd International Congress of Serbian Society of Mechanics, pp. 251–261 (2011). Serbian Society of Mechanics, Vlasina Lake, Serbia
38. Ristić, S., Ilić, J., Ristić, O., Čantrak, Dj., Tašin, S.: Overview of uncertainty sources in flow velocity vector measurement by LDA. In: Proceedings, 5th International Scientific Conference on Defensive Technologies, OTEH 2012, pp. 43–48 (2012)
39. Čantrak, Đ.S., Janković, N., Lečić, M.R.: Laser insight into the turbulent swirl flow behind the axial flow fan. In: Proceedings of ASME Turbo Expo 2014: Turbine Technical Conference and Exposition, GT 2014, Technical track: Fans and Blowers, ASME TURBO EXPO 2014, Paper No. GT2014-26563, Düsseldorf, Germany, pp. V01AT10A024, 10 p. (2014)
40. Ilić, D.B., Čantrak, Đ.S., Janković, N.Z.: Reynolds number influence on integral and statistical characteristics of the turbulent swirl flow in straight conical diffuser. In: Proceedings of the 6th International Congress of Serbian Society of Mechanics, Serbian Society of Mechanics, Mountain Tara, Serbia, paper No. M2e, 6 p. (2017)

41. Čantrak, Đ.: LDV and PIV in turbomachinery. In: Proceedings of the Sixth Regional Conference, Industrial Energy and Environmental Protection in South Eastern Europe Countries, Session: Pump Units and Systems - Good Practice and Solutions for Increasing Energy Efficiency, paper No. 068SS, Introductory Lecture (2017). Serbian Society of Thermal Engineers, Zlatibor, Serbia
42. Janković, N.Z., Čantrak, Đ.S., Nedeljković, M.S.: Three-components LDA investigation of the turbulent swirl jet behind the axial fan. In: Proceedings of the Conference on Modelling Fluid Flow (CMFF 2018), The 17th International Conference on Fluid Flow Technologies, Paper No. CMFF18-101, 8 p. (2018). Department of Fluid Mechanics, Faculty of Mechanical Engineering, Budapest University of Technology and Economics, Budapest, Hungary
43. Ilić, J.T., Janković, N.Z., Ristić, S.S., Čantrak, Đ.S.: Uncertainty analysis of 3D LDA system. In: Proceedings of the 7th International Congress of Serbian Society of Mechanics, paper No. M3j, 8 p. (2019). Serbian Society of Mechanics, Sremski Karlovci, Serbia
44. Čantrak, Đ., Janković, N., Tašin, S.: Laser anemometry in axial fans testing. Energy Econ. Ecol. **15**(3–4), 89–96 (2013). (in Serbian)
45. Čantrak, Đ., Janković, N., Nedeljković, M.: Coherent vortex structure investigation behind the axial fan in pipe. In: PAMM, 90th Annual Meeting, 2 p. (2019). GAMM, TU Wien, Wien, Austria
46. Ilić, D., Čantrak, Đ., Janković, N.: Integral and statistical characteristics of the turbulent swirl flow in a straight conical diffuser. Theoret. Appl. Mech. **45**(2), 127–137 (2018)
47. Čantrak, Đ.S., Janković, N.Z., Ilić, D.B.: Statistical characteristics and time autocorrelation coefficients of the turbulent swirl flow in pipe. In: PAMM, pp. 579–580 (2016). GAMM, TU Braunschweig, Braunschweig, Germany
48. Čantrak, Đ., Janković, N., Ilić, D.: Investigation of the turbulent swirl flow in pipe generated by axial fans using PIV and LDA methods. Theoret. Appl. Mech. **42**(3), 211–222 (2015)
49. Ilić, J., Ristić, S., Čantrak, Đ., Janković, N., Srećković, M.: The comparison of air flow LDA measurement in simple cylindrical and cylindrical tube with flat external wall. FME Trans. **41**(4), 333–341 (2013)
50. Čantrak, Đ., Gabi, M., Janković, N., Čantrak, S.: Investigation of structure and non-gradient turbulent transfer in swirl flows. In: PAMM, pp. 497–498 (2012). GAMM, TU Darmstadt, Darmstadt, Germany
51. Čantrak, Đ.S., Čolić Damjanović, V.M.Z., Janković, N.Z.: Study of the turbulent swirl flow in the pipe behind the axial fan impeller. Mech. Ind. **17**(4), 412-1–412-13 (2016)
52. Ristić, S.S., Ilić, J.T., Čantrak, D.S., Ristić, O.R., Janković, N.Z.: Estimation of laser-Doppler anemometry measuring volume displacement in cylindrical pipe flow. Therm. Sci. **16**(4), 1027–1042 (2012)
53. Benišek, M.H., Ilić, D.B., Čantrak, Đ.S., Božić, I.O.: Investigation of the turbulent swirl flows in a conical diffuser. Therm. Sci. **14**(S), S141–S154 (2010)
54. Protić, Z.D., Nedeljković, M.S., Čantrak, Đ.S., Janković, N.Z.: Novel methods for axial fan impeller geometry analysis and experimental investigations of the generated swirl turbulent flow. Therm. Sci. **14**(S), S125–139 (2010)
55. Ilić, J., Čantrak, Dj., Srećković, M.: Laser sheet scattering and the cameras' positions in particle image velocimetry. Acta Physica Polonica A **112**(5), 1113–1118 (2007)
56. Ilić, D.B., Benišek, M.H., Čantrak, Dj.S.: Experimental investigations of the turbulent swirl flow in straight conical diffusers with various angles. Therm. Sci. **21**(S3), S725–S736 (2017)
57. Čantrak, Đ., Ilić, J., Hyde, M., Čantrak, S., Ćoćić, A., Lečić, M.: PIV measurements and statistical analysis of the turbulent swirl flow field. In: Proceedings ISFV 13 – 13th International Symposium on Flow Visualization, FLUVISU 12 – 12th French Congress on Visualization in Fluid Mechanics, ID 183-080420 (2008). ISFV, Nice, France

58. Čantrak, Đ., Ristić, S., Janković, N.: LDA, classical probes and flow visualization in experimental investigation of turbulent swirl flow. In: DEMI 2011, 10th International Conference on Accomplishments in Electrical and Mechanical Engineering and Information Technology, pp. 489–494 (2011). University of Banja Luka, Faculty of Mechanical Engineering, Banja Luka, Republika Srpska
59. Čantrak, Đ., Janković, N.: Use of modern measurement and visualization techniques in research of turbulent swirl flow in ventilation systems. In: 15th International Passive House Conference 2011, pp. 579–580 (2011). Passivhaus Institute, Innsbruck, Austria
60. Mattern, P., Sieber, S., Čantrak, Đ., Fröhlig, F., Caglar, S., Gabi, M.: Investigations on the swirl flow caused by an axial fan: A contribution to the revision of ISO 5801. In: Proceedings Fan 2012, International Conference on Fan Noise, Technology and Numerical Methods, paper fan2012-68-MATTERN 11 p. (2012). IMechE, CETIAT, CETIM, Senlis, France
61. Čantrak, Đ., Janković, N., Nedeljković, M., Lečić, M.: Stereo PIV and LDA measurements at the axial fan outlet. In: Proceedings, 15th International Symposium on Flow Visualization, paper ISFV15-072-S16 (2012). A. V. Luikov Heat and Mass Transfer Institute of the National Academy of Sciences of Belarus, Minsk, Belarus
62. Čantrak, Dj.S.: Advanced research in energy systems - Bilateral project Karlsruhe-Belgrade. In: Proceedings, Resources of Danubian Region: The Possibility of Cooperation and Utilization, pp. 55–76 (2013). Humboldt-Club Serbien, Belgrade, Serbia
63. Ćoćić, A.S.: Modeling and numerical simulations of swirling flows. Ph.D. thesis, University of Belgrade, Faculty of Mechanical Engineering, Belgrade, Serbia (2013). (in Serbian)
64. Ćoćić, A., Pritz, B., Gabi, M., Lečić, M.: Numerical simulation of turbulent swirling flows. In: PAMM, pp. 309–310. GAMM, University of Novi Sad, Novi Sad, Serbia (2013)
65. Ćoćić, A.S., Lečić, M.R., Čantrak, S.M.: Numerical analysis of axisymmetric turbulent swirling flow in circular pipe. Therm. Sci. **18**(2), 493–505 (2014)
66. Vukašinović, B.: Turbulent transport processes and problems in its modeling in swirl flows. Mag. thesis, University of Belgrade, Faculty of Mechanical Engineering, Belgrade, Serbia (1996)
67. Burazer, J.M., Lečić, M.R., Čantrak, Đ.S.: On the non-local turbulent transport and non-gradient thermal diffusion phenomena in HVAC systems. FME Trans. **40**(3), 119–126 (2012)
68. Čantrak, Dj., Dušanić, A., Božić, I., Lečić, M.: On the anisotropy of the turbulent viscosity. In: Proceedings of the International Conference Classics and Fashion in Fluid Machinery, pp. 139–148 (2002). University of Belgrade Faculty of Mechanical Engineering, Belgrade, Serbia
69. Novković, Đ.M.: Modelling and numerical calculations of the incompressible flow in a straight conical diffusers. Ph.D. thesis, University of Belgrade, Faculty of Mechanical Engineering, Belgrade, Serbia (2019). (in Serbian)
70. Ilić, D.B.: Swirl flow in conical diffusers. Ph.D. thesis, University of Belgrade, Faculty of Mechanical Engineering, Belgrade, Serbia (2013). (in Serbian)
71. Protić, Z., Benišek, M.: Determination of system characteristic at swirling flow in a duct connected with a fan without straightening vanes. In: Proceedings, 15th Yugoslav Congress of Theoretical and Applied Mechanics, pp. 361–368 (1981). Yugoslav Society of Mechanics, Kupari, Yugoslavia (in Serbian)
72. Benišek, M., Protić, Z.: Investigation of the lost flow energy for swirling flow in long lined circular pipes. In: Proceedings, 15th Yugoslav Congress of Theoretical and Applied Mechanics, pp. 269–276 (1981). Yugoslav Society of Mechanics, Kupari, Yugoslavia (in Serbian)
73. Protić, Z., Benišek, M.: Determination of the flow losses in pipes with inbuilt axial fans without guide vanes. In: Proceedings, World Congress on heating ventilating and air conditioning CLIMA 2000, pp. 1–6 (1985). VVS Kongres, Charlottenlund, Copenhagen, Denmark (in German)

74. Čantrak, Đ., Janković, N., Ristić, S., Ilić, D.: Influence of the axial fan blade angle on the turbulent swirl flow characteristics. Sci. Tech. Rev. **LXIV**(3), 23–30 (2014)
75. Janković, N.Z.: Investigation of the free turbulent swirl jet behind the axial fan. Therm. Sci. **21**(S3), S771–S782 (2017)
76. ISO 5801: Industrial fans – Performance testing using standardized airways. ISO (2017)
77. Strscheletzky, M.: Equilibrium forms of the axisymmetric flows with constant swirl in straight, cylindrical rotation geometries. Voith Forschung und Konstruktion **5**, S.1.1–1.19 (1959)

Organizational Changes in Development Process of Technology Startups

Milan Ž. Okanović(✉), Miloš V. Jevtić, and Tijana D. Stefanović

Faculty of Organizational Sciences, University of Belgrade, Jove Ilića 154, 11000 Belgrade, Serbia
milan.okanovic@fon.bg.ac.rs

Abstract. Organizational growth and development are spurred by changes that arise as a result of business development. These organizational changes are most often explained in theory as a reaction to the resulting crisis, which ends in a transition from one to the other stage of organizational development. While in theory this topic is significantly explored within the traditional organizational forms, the changes that arise in the process of development of new technology startups are insufficiently researched. In the process of reaching business success in a highly turbulent environment, these temporary organizations are looking for a repeatable and scalable business model that will continuously deliver value to a large number of customers. In the process of development of new technology startups, there is a need for continuous changes and alignment within the organizational elements in order to achieve their goals. This paper is focused on the key changes that happen in the early stages of developing technology startups, with the aim of elaborating the most important elements of these changes that should lead to the successful development of technology startups.

Keywords: Organizational development · Organizational elements · Technology startups

1 Introduction

Organizational changes can often happen as a reaction to changes in the environment or as proactive management strategy that aims to make changes within the company. Regardless of the nature of the change, whether it is organizational, behavioral or technological changes, their ultimate goal is to improve business processes in the company, making it more competitive. While changes in traditional organizations have been extensively explained in theory, changes in emerging organizations that happen much more quickly, such as in new ventures, represent a challenge to both theory and practice.

The subject of research in this paper are technology startups and organizational changes that happen as a result of their growth and development. Organizational changes were observed through different stages of development of technology startups. The assumptions for research are based on traditional models of growth and development

N. Mitrovic et al. (Eds.): CNNTech 2020, LNNS 153, pp. 203–219, 2021.
https://doi.org/10.1007/978-3-030-58362-0_13

of organizations, as well as on the changers of organization elements that are most often identified in the organization model theory. The contribution of this paper is in the synthesis of previous research and pointing out the questions that need to be answered in the future.

2 Organizational Models and Their Development

Changes which occur in the process of organizational growth and development can be explained through changes of organizational elements or parts that one organization consists of (Greenwood and Hinings 1988). Regarding the definition of organizational elements, authors usually have conversions, but also a certain differences in opinions is common. Organizational structure as one of the organizational elements is mentioned by all the authors who have written on this subject. Number and size of the organizational units, criteria for grouping activities work division and structural configuration represents inevitable topic in all papers of all researchers who have dealt with the designing the organizations (Mintzberg 1980; Waterman et al. 1980; Daft 2004; Galbraith 2011).

Over time, by adding other elements in the analysis, the scope and importance of the organizational structure has decreased, and the focus is shifted to other elements of the organization and concepts: processes, accountability systems, employees, organizational culture, reward systems, coordination mechanisms, common values and other. Focus shifts from "hard" elements (structure, system of decisions, control systems and processes) to "soft" elements of organizations and concepts (people, rewards, culture and shared values). The number of elements involved in time analysis increases, as the disappointment of researchers and managers in the results of a two/factor approach to design (structure and strategy) grows.

Henry Mintzberg (1980) is one of the most famous authors who dealt with the problem of the organizational design. In his paper "Structure in 5'S …", Mintzberg gave an overview of the "basic elements of structuring the organization" that he considers to be the most important parts of the design process of the organization. Those are:

1. Basic parts of the organization - operational core, top management, middle management, technical and support staff;
2. Basic coordination mechanisms - mutual adaptation, direct monitoring and standardization of processes, outputs and skills;
3. Design parameters - job specialization, formalization of behavior, training and indoctrination, grouping into organizational units, size of organizational units, action planning and performance control systems, inter-entity links, vertical and horizontal decentralization;
4. Situation factors - age and size, technical system, environment and power.

The Weisbord (1976) model looks at the six elements of the organization and their change resulting from the development of a business. (Weisbord 1976) The specialty of this model is that it evaluates the functionality of business operations and processes that trigger growth factors and the success of an organization. The six elements that define the model are the purposes, structure, relationships, rewards, leadership, and helpful mechanisms.

The purpose is to understand the products and services offered to the market, which is our target group, which is our target market, whereby our product and service differs from the competition and the expected obstacles that we can face. Also, this element implies that it is well studied and analyzed by who our direct and indirect competitors may pose a threat in the future. The organizational structure refers to the way in which a business operates, how we produce what we offer and what costs. It defines the hierarchy of responsibilities in the organization's position. Proper structuring of business operations helps to achieve the set goals and ensure long-term competitiveness of the organization.

The third element in the Weisbord model is the relationship between management and employee relations, between employees in the organization, between the organization and external partners such as investors, clients, suppliers and stakeholders. Managing relationships requires the organization's management to take steps to resolve conflicts between either sides. Conflicts will certainly happen, but it is crucial to know how to solve them, because healthy relationships are very important for business development. Personnel rewards are important because employees feel satisfied and give the best results when they are adequately rewarded for the results of their work. Adequate rewarding system should be established from the top of the organization to the lowest positions and aligned with the defined organizational and individual goals and desired behaviors.

Leadership is viewed as an element that most affects the employee's sense of organization and the creation of a positive organizational culture. The model emphasizes the importance of developing leadership skills, and the impact on employee behavior and on the organizational and individual outcomes that are being achieved. The last element concerns the Helpful Mechanisms. It emphasizes the importance of possessing competitive advantages that can relate to raw materials, technological resources. Management plays an important role in tracking market trends and changes taking place in the environment, because mechanisms that were once relevant could become irrelevant in the future.

The model of the 7s organization developed by Peters and Waterman (strategy, structure, systems, shared values, style, staff and skills) arose as a result of dissatisfaction with the conventional way of implementing organizational changes, which mainly involved changes in the organizational structure or restructuring of organizations (Waterman et al. 1980). The proposed model is based on the theory that for the effectiveness of the organization it is important that all seven elements are mutually harmonized and supported mutually. In practice, the model is used to identify the elements that need to be redefined to make the effects greater, or to monitor and maintain compliance (and performance) during the implementation of the changes.

The last model that has been taken into consideration in this paper is Galbraith's Star Model (Galbraith 2011). Model originated from a long-standing consulting practice and scientific research work of the author and over the years it has been modified several times. The Star Model can also be used as a framework for decision-making in the design process and as a point of view from which managers will think about the interaction of the most important elements of the organization. The elements of the organization that make up the model are mainly mentioned in other models of designing organizations. These are: strategy, capabilities, structures, processes, reward systems and people.

The following Table 1 shows the four models of the organization with the elements that make it. Some elements appear in different models under a different name, but with the same or similar description of elements (for example, in Marvin's model, "purpose" is a competitive strategy for organizations, and in the table, this element is referred to as "strategy").

Table 1. Organizational elements in models of organization

	Mintzberg (1980)	Weisbord (1976)	Peters and Waterman (1980)	Galbraith (2011)
Strategy and goals		Purposes	+	+
Organizational structure	Basic elements of organization	+	+	+
Leadership		+	Style	
Systems		Rewards	+	Rewards
Relationships		+		
People			Staff	+
Design parameters	+			
Skills			+	
Shared values			+	
Process				+
Helpful mechanism	Basic coordination mechanisms	+		

All these models are used in practice to analyze the existing way of functioning of the organization, but also to design organizational changes and align elements of the organization. For this work of interest there are changes to the elements that arise as a result of the growth and development of the organization.

Organizational growth and development has been explored in many works in the theory of management. Greiner (1972) described one of the first and most famous models of growth (increasing the number of employees, capacities, capital, etc.) and development (introduction of new products, technologies, competences, etc.). In the business they have to change the organization's elements. He observed the changes that arise in the organization's strategy and goals, organizational structure, leadership, control and reporting system and formalization. Operations on these elements of the organization arise in response to crisis situations or proactive attitude of management that is aware of the fact that the development of business must also lead to changes in the way of organizing (Scott 1987).

In describing the growth and development phases of the organization, Greiner (1998) describes the changes in organizational elements through 5 phases of organizational development. The first stage of an organization's functioning is characterized by objectives that are related to achieving productivity (how to make a product or service), with no clear structure and low formalization of relationships (Greiner 1998). Young organizations are also characterized by a leadership based on the charisma of the owner or founder and a poorly developed control and reporting system. This phase is characterized by a rapid increase in the number of employees and the capacity of the organization. The organization's structure is undergoing transformation. The organizational structure is from the so-called informal and "entrepreneurial structure" changes into a centralized functional structure characterized by the existence of different business functions (Greiner 1998). In the further stages of growth and development there is a decentralization of the structure and the emergence of some form of diversity of organization (territorial, production). Parallel to the change in structure, there are changes in the organization's strategy and goals. The organization's strategy is mainly concerned with defining the product/service portfolio and their positioning on the market. Organizational goals are increasingly linked to achieving the desired profit and return on investment.

Regarding formalization, the growth and development of the organization leads to an increase in the degree to which the rules and roles of individuals in the organization are specified (Daft 2004). The growth and development of the organization changes the informal structure, in which individuals are more valued than positions. Over time, there is a more precise job description, standardization of work procedures and the adoption of general rules and guidelines. Growth and development also lead to a change in the management mode (Colombo et al. 2014). While the initial stage of organizational development is characterized by the charisma of the founder or owner of the organization, and during time roles and relationships are becoming more professional and there is an emphasis of value system, and the leadership style changes to behavioral style.

In this paper, the changes that occur within the elements of the organization in the process of development of technology startups will be analyzed below. The paper will analyze the changes of the four elements that are the subject of analysis in all models of the organization: strategies and goals, structure, leadership and people.

3 Technology Startups

3.1 The Concept and Characteristics of Technology Startups

Gruber (2004) considers that new entrepreneurial ventures can not be seen as smaller formats of large companies. Technology startups are new early-stage ventures that arise in the areas of innovation and the latest technologies, characterized by high dynamics, growth orientation, profit and creation of new market values (Kariv 2013).

Blank and Dorf (2012) define "startup" as a temporary organization, designed to find a repeatable and scalable business model. By "repeatable" it is assumed that the organization searches for a model that will consistently create and deliver value to customers, while "scalable" means the ability to adapt the business model in accordance with the growing customer base (Duening 2014). Ries (2011) thinks that the startup venture is created and guided by people in order to create a new product or service under extremely uncertain conditions. According to Ries (2011) a startup is a human institution designed to deliver a new product or service under conditions of extreme uncertainty. Paternoster and associates (2014) think that besides creating innovating products, technology startups must provide growth with aggressive business expansion on the fast-growing markets.

These startups invest most resources in research and development activities, engaging a large number of engineers (usually developers) and scientists, creating innovative, technologically advanced products, complex configurations, with a brief development and short market life span (Oakey 1988). Technology startups characterize great dynamics and orientation towards the future (Crowne 2002; Graham 2012).

3.2 Phases of Development of Technology Startups

Spinelli and Adams (2012) point out that no new venture is of a static nature, and in the course of transition through the stages of the growth process encounters many limitations and challenges. The development process is not uniform, given that at each stage new ventures meet with periods of ups and downs. Initially, startups in the early stages of development are oriented only to the development of products, while over time, the organizational structure and processes are complicated, resulting in the decline of certain performance (Giardino et al. 2015). According to him new ventures start without any established workflows, grow over time, create and stabilize processes to eventually improve them only when sufficiently mature. Crowne (2002) defines three stages of the development of new ventures: Startup, Stabilization and Growth, while Spinelli and Adams (2012) explain this development through five phases: Research and Development, Startup phase, Accelerated Growth Phase, Maturity Phase and Stability Phase, while Crowne (2002) defines three phases Startup, Stabilization and Growth.

In the Fig. 1, the indicated stages of growth of the new ventures are shown in correlation with time, closed sales and the number of employees, which according to the same authors represents the key characteristics of the transition at different stages of growth. The data relate to performance indicators of new ventures in the United States. Also, these characteristics represent the key challenges faced by founders, who are responsible for the growth and success of entrepreneurial ventures. Analyzing different approaches in defining the phase of growth and development of technology startups, paper further explains the organizational elements that play a significant role in the development of technology startups.

3.3 Organizational Elements in Technology Startups

Goals and Strategies of Technology Startups. The goals and strategies of technology startups are to a large extent related to their development. Unlike firms and companies that have their business history, setting and measuring goals for technology startups this is a major challenge (Van de Ven et al. 1984). If one takes into account some of the traditional indicators, such as sales revenue, this indicator in itself is not entirely a relative measure for a comprehensive assessment of the success of the technology startup (Spiegel et al. 2011; Baum and Silverman 2004; Van de Ven et al. 1984).

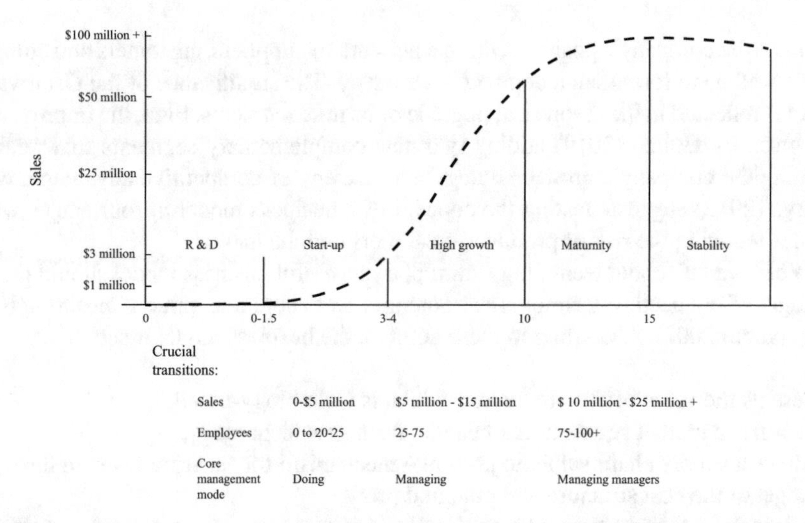

Fig. 1. Stages of venture growth and crucial transitions (Spinelli and Adams, 2012)

One of the main goals of technology startups is the construction and testing of minimum viable product (MVP). The MVP is a product version with minimal features to satisfy need of early customers, and to provide feedback for future product development. Moogk (2012) highlights the efficiency in the process of presenting a minimal viable product to the market in as short a time as possible, in order to test its value and entrepreneurial potential for growth. Testing is done in the form of experiments that measure indicators relevant to growth and allow for accelerated learning to reduce uncertainties that accompany the commercial commercialization of technology. According to Ries (2011), the three indicators that are measured by these "experiments" and are the drivers of growth are: the difference in the rate of providing new customers and the rate of customers who have left the product of the company, the rate of viral product recommendation, the difference in the long-term income of each existing customer in relation at the cost of securing a new customer.

Another important goal of technology startups is establishing a fundable business model. According to Osterwalder and Pigneur (2010), the business model is one of the most important components of new entrepreneurial ventures and it should explain the way in which the organization creates, delivers and provides value. Spinelli (2012) emphasize that the business model must be based on the needs of the market and not exclusively on the need to create inventions or innovations. One of the major revolutions in modern business is switching from traditional to digital business models for achieving higher competitiveness level. (Jovanović et al. 2018). For this reason, it is important to identify the costs, margins and delivery times, the size of the market segment and the characteristics of customers, based on which the positioning strategy is defined. Chesbrough and Rosenbloom (2002) point out that the business model should guide the value creation of the offer, identify the target market and the customer revenue mechanism, define the value chain structure, estimate the cost structure and potential profit, describes the company's position within a network of suppliers, customers and competitors as well as to formulate a competitive strategy. The significance of the Ostarwalder model is reflected in the theoretical upgrade of its nine segments. First, the improvement was given by Hulme (2019) adding two new complementary segments that relate to planning the company's growth strategy and strategy of competitive advantage, while Maurya (2012) suggests testing the concept of a business model in four stages which justifies reducing the risk of products, customers and the market.

When we talk about technology startup, a successful business model should present the logic of connecting technological potential and economic value (Chesbrough and Rosenbloom 2002). According to these authors, the business model must:

- presents the value offered to buyers, which is technology-based,
- identifies a market segment that benefits from new technology,
- defines a supply chain value, to present a mechanism for securing revenue through a budget of the cost structure and potential profit,
- explain the role of the firm in the environment of suppliers, customers and competitors, and
- defines a competitive strategy.

The strategy of technology startups needs to be defined in a dynamic and uncertain environment surrounded by limited resources. The success of a new firm often depends on the ability of entrepreneurs to identify and access sufficient sources and levels of capital (Auken and Neeley 1996). When it comes to financial resources, the development of technology startups can be bootstrapped or funded by investors. Bootstrapping involves highly creative ways of acquiring the use of resources without borrowing money or raising equity financing from traditional sources, such as financial institutions or business angels (Auken and Neeley 1996). Most often it involves the sale of the entrepreneurs, personal property, personal indebtedness, the use of credit cards, and the use of leases to the mix of bootstrap methods employed by business owners. On the other hand, startups can raise capital from venture capitalists, funds or business angels, who at different stages of startup development can make a decision to finance development by taking a certain equity. These startups lose certain control over the management of development strategy, which in some cases can also lead to a change in the original development strategy. Venture capital firms have a growing interest in their startups and they typically increase the skill set of the existing management team in a more proactive way than other financing methods, such as bank loans (Davila 2003).

Organizational Structure in Technology Startups. Organizational structure of technology startup is different from traditional, older and already defined organizations. In structured and formalized organizations, core and non-core parts (Porter 1985) or market-oriented and operational oriented units (Simons 2005) can be identified. The significance of these organizational parts cannot always be easily identified, for example in the case of functionally structured organizations. On schemes of functionally structured organizations, all units in the organizational scheme are displayed in the same way. On the contrary, in technology startup organizations, core units that have resources and convoys to place an entrepreneurial idea are easy to identify (Colombo and Rossi-Lamastra 2013). The role of each individual in technology startup can be easily linked to the elements of organizational design at the top management level and with the wider concept of a concrete tech-startup.

The high-startup organizational design is under great influence of contingency factors because these organizations operate in a turbulent environment characterized by fast changes in demand, competition and technology. Fast changes in the business environment require an organizational structure that will simultaneously fulfill the criteria of flexibility and efficiency (Colombo et al. 2014). Flexibility implies greater ability to accept new ideas and chances and efficiency to faster implementation and exploitation of new ideas (Adler 1999). Under the influence of situational factors there is a change in the formalization of roles and job specialization. This raises the quality of decision-making process through the application of specific knowledge (Huber and McDaniel 1986). Parallel with specialization, decentralization of decision-making takes place (Cudanov et al. 2009). Decentralization of decision-making simultaneously influences the greater involvement of employees in business decision-making, and allows management to focus on the chances of the environment and the strategic decision.

According to the (Colombo et al. 2014) with the startup development, after the initial stages, there is a need for greater formalization, which avoids the initial confusion over who is competent. Between top management members, a clearer division of competencies is established, thus achieving savings due to reduced coordination and lower costs of consensus decision-making (Daft 2004). Task specialization assists all employees, including top management, to focus on an area for which they have specific jurisdiction. Unlike traditional organizations, the high-startup organizational structure must not grow into a bureaucratic structure. Flexibility and speed must remain the main features of high-startup organizations.

Also, the organizational structure of technology startups is characterized by changes in how business functions are organized. Generally speaking, high technology startups have a very simple structure (Scott 1987; Colombo and Grilli 2013). The hierarchy is straightforward and involves mainly two levels: the level of team management and the level of perpetrators. Over time, with the growth of the organization, the number of hierarchical levels can be increased through the creation of middle management (Grimpe 2017). In the structure of business functions, in line with the shift of focus from the product market to the idea market, R&D takes the primary role. R&D function strategically, resourcefully and organizationally becomes the most important for the success of growth and development of high technology startup.

Leadership in Technology Startups. Technology entrepreneurs represent one of the growth drivers of industry growth in general, and in addition to scientific and technological knowledge and skills, the additional development of business knowledge and skills is also required for the individual to become a successful technology entrepreneur (Oakey 1988). According to Brüggemann (2014), a critical success or failure factor towards startups can be the leadership styles of entrepreneurs. In general, leadership can be defined as the process of influencing group members and guiding them to achieve the organizational goals (Esmer and Dayi 2017). However, leadership in technology startups has certain specifics in relation to leadership in companies.

According to Ensley et al. (2006) it is crucial for founding teams to take on leadership roles and lead because in new ventures there are no standard operating procedures of organizational structure to fall back on when you are creating a new company. More established corporations which have more well-defined goals, structure and work process to guide is what differentiates their top management teams from new venture top management teams. Thus, whereas there may be substitutes and/or blockers of leadership in larger and more established organizations, there are, by definition, far less alternatives or impediments to leadership in new ventures.

In startups leadership behavior of founders is likely to have a greater and more direct impact on company performance than in larger and more established organizations. Authors such as Timmons and Spinneli (2003) explore that successful team leader needs to have the ability to accept new knowledge and to teach younger team members, to face the challenges and build entrepreneurial culture and organization. Also Brüggemann (2014) states that the challenge lies in creating an enthusiastic atmosphere for creative, entrepreneurial action, which means that the subordinates just receive the overall assignment and the deadline. The employees are then free in how they structure the task, when they work on it and how to work it out. startups leader considers more how their subordinates can handle the situation more than how to force them to do something. But, too friendly atmosphere might be counterproductive and can lead to unintended consequences (Chen 2007). He suggests that in a new venture's context, entrepreneurial leadership can stimulate entrepreneurial team members to be more creative. The second finding suggests that when lead entrepreneurs have higher risk-taking, pro-activeness and innovativeness, they can stimulate their entrepreneurial team to be more creative during the patent creation process.

As Duchesneau and Gartner (1990) explain, external professionals and mentors are hired by successful startups to solve specific problems during startup phase, and they use advice and information they get from industry individuals, customers and vendors. Internal and external flexibility as well as adaptability are characteristic of successful startups. Policy of rewarding and respecting each success and failure, as well as sharing the revenue and wealth of the team and setting high standards for accomplishing positive performance are adopted by entrepreneurial leaders.

People in Technology Startups. Specifics of technology startups is that they are driven by one or more people who are gathered around a common idea or goal that should be achieved in the future. In theory, the orientation towards the development of startups is justified by the entrepreneurial mindset of the founder, which is defined as an individual characteristic that is considered crucial for the success of the development of startups, and that it is inherited and cannot be learned (Pollard and Wilson 2014). According to the Startup Genome, the success factors that make up the entrepreneurial mindset are: initiation, reflection and patience, breathe, depth and structure (Startup Genome Report 2018).

On the other hand, the entrepreneurial team is represented by two or more individuals who jointly establish a business in which they have equity (financial) interest, and commitment to a venture's future and success (Kamm et al. 1990). Factors that influence the success of the startup include equally internal and external relations of the startup team. Interpersonal relations as well as good communication in the team are important for long-term growth and development of the venture and building a good entrepreneurial team. Lechler (2001) points out that social interaction significantly influences the success of entrepreneurial ventures. This impact of the exam is its model that includes six dimensions including communications, cohesion, work norms, mutual support, coordination and conflict management. The quality of social interaction among team members is important but not the only factor in the success of entrepreneurial ventures.

A special feature of startups by investors is the engagement of new and top talent members and the construction of a quality management team. In this regard, Duchesneau and Gartner (1990) point out that successful start-ups engage external professionals and mentors to solve specific problems during the startup phase, but also use advice and information obtained from individuals from other industries, customers and suppliers. These characteristics of successful startups explain as internal and external flexibility and adaptability.

4 Changes in Organizational Elements of Technology Startups

4.1 Changes in Goals and Strategy of Technology Startups

In the previous part of the paper, important goals for the development of an MVP and an adequate business model were emphasized. During the development process, startups are often forced to make some changes to the MVP. Depending on the results of the MVP testing process with customers, it is often the case that a product feature gets the opportunity to become a new product, which creates easier supply on the market. On the other hand, in phases when the product begins to deliver the value to the customers, it is possible that this product becomes one of the features of a more complex product. Other changes that relate to the MVP relate to changing the technology on which the product is designed to improve its reliability, or the appearance and design of a user interface that makes it easier to use the product. All these changes in the process of developing technology startups are called pivoting or pivot strategies.

Changes that are happening in the context of the technology startup goals relate to the entire business model. The business model of technology startups is changing during the implementation of the initial stages of development, as entrepreneurs face many difficulties. Once the business model is defined, it can be adapted according to new knowledge gained in different stages of development of technology startups. Maurya (2012) defines a business model through nine segments of "lean canvas", and the changes that occur in his model as a result of testing business assumptions through four phases. The realization of these phases should lead to a reduction in the risk of products, customers and markets with maximization of learning and minimal use of resources. Assumptions are tested through understanding the problem, defining the solution, qualitative confirmation, and quantitative confirmation of the solution. The first two phases are called "problem/solution fit", the other two phases are called "product/market fit".

4.2 Changes in Organizational Structure of Technology Startups

Technology startup is characterized by changes in the organizational structure that are in line with the growth and success of the entrepreneurial venture. According to Scott

(1987), the organizational structure in a small business is going through the following phases: unstructured, simple, functional-centralized, functional-decentralized and functional decentralized/product. Certain changes in the organizational structure also occur in cases of tech-startup.

According to (Colombo and Grilli 2013) the establishment of middle management is a crucial milestone in the maturing of startups. Although middle management is associated with a higher degree of formalization and more "mechanistic" and less "organic" organizational structure (Mintzberg 1980), some researchers have pointed to the advantages of formal structure in an environment that implies innovation and flexibility. Sine et al. (2006) indicate that startups with more specialized and formalization in the founding team are better performers than startups with organic structure. Grimpe et al. (2017) extended this perspective by claiming that the R & D function must not be exclusively in the founder's hands, but the responsibility for it can also be extended to middle management.

What precedes management, as a challenge is to maintain initial innovation and flexibility in later phases of technology startup development, in phases of rapid growth, maturity and stability. If innovation is provided by direct involvement of founders (Colombo and Rossi-Lamastra 2013) at the initial stages, and in the rapid growth phase by selective delegation, formalization and inclusion of middle management (Grimpe 2017), the question remains how to provide this innovation at later stages of growth and development organizations. One of the solutions that intuitively impose is the structural separation of the operational part in charge of the realization of ideas and efficiency, from an innovative part that characterizes the speed of response to business opportunities and flexibility (Day 2001). In practice, this is not a rare case of the existence of such structural solutions. What can be said with certainty is that the structural separation of the innovative and operative part is the "foundation for a new house". The functionality and effects of this structured organization must be sought in the design of other elements of an organization: the strategies and goals of the various structural parts of the organization, leadership, performance measurement and reward systems, values that will be promoted etc. (Powell 1992; Sender 1997; Kaplan 2006; Burton 2015). There must be consistency between all the elements analyzed in this paper and the elements that are mentioned in other models of the organization in order to achieve the desired outcomes.

4.3 Changes in Leadership of Technology Startups

In the first few months to a year, startups are characterized by the research and analytical activities of a founder or a smaller team in the development of their business ideas. According to Hyytinen and associates (2015), at this stage, entrepreneurs are most often looking for innovations that will be launched on the market, but such an approach does not necessarily mean survival of the ventures in the early stages of development, especially since the process of establishing a new venture is quite complex in this case (Hyytinen et al. 2015).

In his work Swiercz (2002) proposes two phases of entrepreneurial leadership: the formative growth phase and the institutional growth phase, where in first phase the firm thrives on the free form energy characteristics of an emerging enterprise. Leaders are usually concerned with product or service invention, establishment of market niche, attracting new customers, and manufacturing and marketing the product. (Swiercz et al. 2002). Ensley (2002) emphasize the importance of vertical versus shared leadership which is dependent on the stage in the development or evolution of the organization. For the pre-formation stage of the new venture vertical leadership is important. (Ensley et al. 2006)

According to Brüggemann in the startup phase, role and responsibilities of the team leader must be defined in order for him/her to make decisions. This is due to the fact that this phase has most risks and it is necessary for leader to take full commitment in management process and delegate certain processes to other team members so that they stay consistent with the goals and motivation for work. Brüggemann, H. (2014). With the growth of startups, one person is not able to take on all the leadership roles. Both Ensley (2002) and Swiercz (2002) explore the complexity of the business growth of startups. Ensley states that shared leadership produce superior outcomes, because leaders share responsibilities with skilled and diverse top management. And in the second phase of Swiercz's model explains how daily operations become more systematic once the startup grows. More structure and long-term stability must be focus of leaders while still maintaining the innovative, entrepreneurial spirit that made it successful in the first phase. Certain challenges arise in the process of growth and development of startups. Swiercz (2002) additionally explores how founders need to challenge previous successes through constant innovation and change which usually does not happened because founder is blinded by previous success. Entrepreneurial leaders, therefore need to acquire new leadership competencies while the company is doing well, and focus on subtle warning signs such as undefined roles and responsibilities, lack of coordination, daily short-term crises, high turnover, and management overload which occur much earlier to signal a poorly managed firm.

4.4 Changes in People of Technology Startups

Colombo and Grilli (2010) concluded from their research that founders have a direct and indirect impact on the growth of new ventures. Founders, as intermediaries between investments and their new venture, indirectly positively influence the effect of its growth. Research by these authors has confirmed that investors are more interested in investing in those ventures of individuals and teams of entrepreneurs who have managerial experience and competencies. Also, founders who have an economic and managerial education profile significantly influence the growth of new ventures in relation to entrepreneurs who have different educational profile (Colombo and Grilli 2005). Therefore, the founders during all stages of growth of new ventures must face the change of their own roles from leader to managerial, as the growth of the company is expected to increase human resources at each stage of growth. According to Wang and Wu (2012), this role also entails the trust of other team members, which affects the team's greater commitment during the development of the startups. Crown argues that conflicts often occur between new executives and founders of the company who may be shareholders, and that employees continue to seek product and thought leadership from founders. This might be disruptive in the organization, especially for product development which is what founders are most attached to.

At the startup stage, the strategic orientation of all members of the team and the rights and responsibilities in adopting and challenging strategic decisions are emphasized. Also, entrepreneurial culture and values in an organization play an important role in achieving goals. Within the cohesive teams there is mutual understanding, and all members share the same values, so these teams are more easily faced with challenges because the basic assumptions and goals are not questioned.

Team cohesion is an important feature of new ventures, which contributes to cognitive, constructive conflicts, while on the other hand it reduces affective, destructive conflicts (Ensley et al. 2002) and provides better results for entrepreneurial ventures (Higashide and Birley 2002). Watson et al. (2003) emphasizes the achievement of the synergistic effect of elements of interpersonal processes, such as readiness to resolve conflict, information sharing and coordination in team work on profit and growth of the startup.

5 Conclusion

The aim of this paper was to point out the most important organizational elements of technology startups and to explain their changes that happen during the various stages of development. Also, the work was supposed to enrich the gap in theory in this area using existing organizational models on technology startups.

The paper presents the basic elements of organization and specific analysis from different authors. Furthermore, the focus of the paper was on the elements that are common to most of the considered models (goals and strategy, organizational structure, leadership and people), which are explained in the context of technology startups. These organizational changes, which accelerate fast development, great flexibility and work in a turbulent environment, are aimed at keeping the initial proactivity of the enablers of technology startups as well as the flexibility of the organization itself. As the startup evolves from phase to phase, goals are also adjusted, so while goals are initially focused on product development and testing, the minimum functionality is achieved, so later goals and strategies are related to the business model and financing models. Leadership changes are based on the transition of leadership styles, from a vertical to a horizontal style. Good leadership implies continuous review of success in earlier stages of development of technology startups, as well as timely detection of signals that can lead to crisis. During the development of technology startups, it is necessary to develop entrepreneurial culture and values among human resources in order to achieve cohesion in the team and to face challenges more easily. The development of such culture is very important in the process of recruiting and onboarding new people.

This paper presents a pioneering undertaking explaining the changes of organizational elements in technology startups, which in a comprehensive way presents the most important elements of the organization. Future research in this area should be based on additional systematization of knowledge in this field, as well as on the theoretical foundation of positive and negative practice of development of organizational elements in technology startups.

References

Adler, P.S., Goldoftas, B., Levine, D.I.: Flexibility versus efficiency? A case study of model changeovers in the Toyota production system. Organ. Sci. **10**(1), 43–68 (1999)

Auken, H.E.V., Neeley, L.: Evidence of bootstrap financing among small start-up firms. J. Entrep. Small Bus. Finan. **5**(3), 235–249 (1996)

Baum, J.A., Silverman, B.S.: Picking winners or building them? Alliance, intellectual, and human capital as selection criteria in venture financing and performance of biotechnology startups. J. Bus. Ventur. **19**(3), 411–436 (2004)

Brüggemann, H.: Entrepreneurial leadership styles: a comparative study between startups and mature firms. Bachelor's thesis, University of Twente (2014)

Burton, R.M., Obel, B., Håkonsson, D.D.: Organizational Design: A Step-by-Step Approach. Cambridge University Press, Cambridge (2015)

Chen, M.H.: Entrepreneurial leadership and new ventures: creativity in entrepreneurial teams. Creativity Innov. Manage. **16**(3), 239–249 (2007)

Chesbrough, H., Rosenbloom, R.S.: The role of the business model in capturing value from innovation: evidence from Xerox Corporation's technology spin-off companies. Ind. Corp. Change **11**(3), 529–555 (2002)

Colombo, M.G., Grilli, L.: Founders' human capital and the growth of new technology-based firms: a competence-based view. Res. Policy **34**(6), 795–816 (2005)

Colombo, M.G., Grilli, L.: On growth drivers of high-tech start-ups: exploring the role of founders' human capital and venture capital. J. Bus. Ventur. **25**(6), 610–626 (2010)

Colombo, M.G., Grilli, L.: The creation of a middle-management level by entrepreneurial ventures: testing economic theories of organizational design. J. Econ. Manage. Strategy **22**(2), 390–422 (2013)

Colombo, M.G., Rossi-Lamastra, C.: 21. The organizational design of high-tech start-ups: state of the art and directions for future research. In: Handbook of Economic Organization: Integrating Economic and Organization Theory, p. 400 (2013)

Colombo, M.G., Mohammadi, A., Rossi-Lamastra, C., Foss, N.J., Saebi, T.: Innovative business models for high-tech entrepreneurial ventures: the organizational design challenges. In: Business Model Innovation: The Organizational Dimension, pp. 169–190 (2014)

Crowne, M.: Why software product startups fail and what to do about it. Evolution of software product development in startup companies. In: IEEE International Engineering Management Conference, vol. 1, pp. 338–343. IEEE (2002)

Cudanov, M., Jasko, O., Jevtic, M.: Influence of information and communication technologies on decentralization of organizational structure. Comput. Sci. Inf. Syst. **6**(1), 93–109 (2009)

Daft, R.L.: Organization Theory and Design. South-Western Publishing, Southampton (2004)

Davila, A., Foster, G., Gupta, M.: Venture capital financing and the growth of startup firms. J. Bus. Ventur. **18**(6), 689–708 (2003)

Day, J., Mang, P., Richter, A., Roberts, J.: The innovative organization. McKinsey Q. **2**, 21–31 (2001)

Blank, S., Dorf, B.: The Startup Owner's Manual: The Step-by-Step Guide for Building a Great Company. BookBaby, Pennsauken Township (2012)

Duening, T.N., Hisrich, R.A., Lechter, M.A.: Technology Entrepreneurship: Taking Innovation to the Marketplace. Academic Press, Cambridge (2014)

Duchesneau, D.A., Gartner, W.B.: A profile of new venture success and failure in an emerging industry. J. Bus. Ventur. **5**(5), 297–312 (1990)

Ensley, M.D., Pearson, A.W., Amason, A.C.: Understanding the dynamics of new venture top management teams: cohesion, conflict, and new venture performance. J. Bus. Ventur. **17**(4), 365–386 (2002)

Ensley, M.D., Hmieleski, K.M., Pearce, C.L.: The importance of vertical and shared leadership within new venture top management teams: implications for the performance of startups. Leaders. Q. **17**(3), 217–231 (2006)

Esmer, Y., Dayi, F.: Entrepreneurial leadership: a theoretical framework. J. Mehmet Akif Ersoy Univ. Econ. Admin. Sci. Fac. **4**(2), 112–124 (2017)

Galbraith, J.R.: The Star Model. Galbraith Management Consultants, Colorado (2011)

Giardino, C., Paternoster, N., Unterkalmsteiner, M., Gorschek, T., Abrahamsson, P.: Software development in startup companies: the greenfield startup model. IEEE Trans. Softw. Eng. **42**(6), 585–604 (2015)

Graham, P.: Startup = growth (2012). http://www.paulgraham.com/growth. Accessed 1 Jun 2019

Greenwood, R., Hinings, C.R.: Organizational design types, tracks and the dynamics of strategic change. Organ. Stud. **9**(3), 293–316 (1988)

Greiner, L.E.: Evolution and revolution as organizations grow. Harvard Bus. Rev. **50**(4), 37–46 (1972)

Greiner, L.E.: Evolution and revolution as organizations grow. Harvard Bus. Rev. **76**(3), 55–64 (1998)

Grimpe, C., Murmann, M., Sofka, W.: The organizational design of high-tech startups and product innovation. ZEW-Centre for European Economic Research Discussion Paper, pp. 17–074 (2017)

Gruber, M.: Marketing in new ventures: theory and empirical evidence. Schmalenbach Bus. Rev. **56**(2), 164–199 (2004)

Higashide, H., Birley, S.: The consequences of conflict between the venture capitalist and the entrepreneurial team in the United Kingdom from the perspective of the venture capitalist. J. Bus. Ventur. **17**(1), 59–81 (2002)

Huber, G.P., McDaniel, R.R.: The decision-making paradigm of organizational design. Manage. Sci. **32**(5), 572–589 (1986)

Hulme, T.: Business model framework. http://hackfwd.com/documents/Business%20Model%20Framework.pdf. Accessed 6 May 2019

Hyytinen, A., Pajarinen, M., Rouvinen, P.: Does innovativeness reduce startup survival rates? J. Bus. Ventur. **30**(4), 564–581 (2015)

Jovanović, M., Dlačić, J., Okanović, M.: Digitalization and society's sustainable development–measures and implications. Zbornik radova Ekonomskog fakulteta u Rijeci: časopis za ekonomsku teoriju i praksu **36**(2), 905 (2018)

Maurya, A.: Running Lean: Iterate from Plan A to A Plan that Works. O'Reilly Media, Inc., Sebastopol (2012)

Mintzberg, H.: Structure in 5's: a synthesis of the research on organization design. Manage. Sci. **26**(3), 322–341 (1980)

Moogk, D.R.: Minimum viable product and the importance of experimentation in technology startups. Technol. Innov. Manage. Rev. **2**(3), 23–26 (2012)

Kamm, J.B., Shuman, J.C., Seeger, J.A., Nurick, A.J.: Entrepreneurial teams in new venture creation: a research agenda. Entrep. Theory Pract. **14**(4), 7–17 (1990)

Kaplan, R.S., Norton, D.P.: Alignment: Using the Balanced Scorecard to Create Corporate Synergies. Harvard Business Press, Brighton (2006)

Kariv, D.: Start-up and small business life. In: Encyclopedia of Creativity, Invention, Innovation and Entrepreneurship, pp. 1734–1742 (2013)

Lechler, T.: Social interaction: a determinant of entrepreneurial team venture success. Small Bus. Econ. **16**(4), 263–278 (2001)

Oakey, R.P., Rothwell, R., Cooper, S., Oakey, R.P.: The Management of Innovation in High-Technology Small Firms: Innovation and Regional Development in Britain and the United States. Pinter, London (1988)

Osterwalder, A., Pigneur, Y.: Business Model Generation: A Handbook for Visionaries, Game Changers, and Challengers. Wiley, Hoboken (2010)

Paternoster, N., Giardino, C., Unterkalmsteiner, M., Gorschek, T., Abrahamsson, P.: Software development in startup companies: a systematic mapping study. Inf. Softw. Technol. **56**(10), 1200–1218 (2014)

Powell, T.C.: Organizational alignment as competitive advantage. Strateg. Manage. J. **13**(2), 119–134 (1992)

Pollard, V., Wilson, E.: The "entrepreneurial mindset" in creative and performing arts higher education in Australia. Artivate J. Entrep. Arts **3**(1), 3–22 (2014)

Porter, M.E.: Technology and competitive advantage. J. Bus. Strategy **5**(3), 60–78 (1985)

Ries, E.: The Lean Startup: How Today's Entrepreneurs Use Continuous Innovation to Create Radically Successful Businesses. Crown Books, New York City (2011)

Scott, M., Bruce, R.: Five stages of growth in small business. Long Range Plann. **20**(3), 45–52 (1987)

Sender, S.W.: Systematic agreement: a theory of organizational alignment. Hum. Resour. Dev. Q. **8**(1), 23–40 (1997)

Simons, R.: Levers of Organization Design: How Managers Use Accountability Systems for Greater Performance and Commitment. Harvard Business Scholl Press, Boston (2005)

Sine, W.D., Mitsuhashi, H., Kirsch, D.A.: Revisiting burns and stalker: formal structure and new venture performance in emerging economic sectors. Acad. Manage. J. **49**(1), 121–132 (2006)

Spiegel, O., Abbassi, P., Fischbach, K., Putzke, J., Schoder, D.: Social capital in the ICT sector–a network perspective on executive turnover and startup performance (2011)

Spinelli, S., Adams, R.: New Venture Creation: Entrepreneurship for the 21st Century. McGraw-Hill, New York City (2012)

Swiercz, P.M., Lydon, S.R.: Entrepreneurial leadership in high-tech firms: a field study. Leaders. Organ. Dev. J. **23**(7), 380–389 (2002)

Timmons, E., Spinelli, S.: Entrepreneurship for the 21st century. New Venture Creation **3**, 249–256 (2003)

Van de Ven, A.H., Hudson, R., Schroeder, D.M.: Designing new business startups: entrepreneurial, organizational, and ecological considerations. J. Manage. **10**(1), 87–108 (1984)

Wang, C.J., Wu, L.Y.: Team member commitments and start-up competitiveness. J. Bus. Res. **65**(5), 708–715 (2012)

Waterman Jr., R.H., Peters, T.J., Phillips, J.R.: Structure is not organization. Bus. Horiz. **23**(3), 14–26 (1980)

Watson, W., Stewart Jr., W.H., BarNir, A.: The effects of human capital, organizational demography, and interpersonal processes on venture partner perceptions of firm profit and growth. J. Bus. Ventur. **18**(2), 145–164 (2003)

Weisbord, M.R.: Organizational diagnosis: six places to look for trouble with or without a theory. Group Organ. Stud. **1**(4), 430–447 (1976)

UAV Positioning and Navigation - Review

Dragan M. Raković[1(✉)], Aleksandar Simonović[2], and Aleksandar M. Grbović[2]

[1] Faculty of Mechanical Engineering, University of Belgrade, 11000 Belgrade, Serbia
draganrakovic964@gmail.com

[2] Faculty of Mechanical Engineering, Department for Aeronautics, University of Belgrade, 11000 Belgrade, Serbia

Abstract. The capabilities of UAVs for civil and military operations are being largely improved by the advances made in UAV technology hence allowing UAVs to become a wide spread tool for a wide range of possible applications. This spread increases the need for more efficient and accurate navigation and positioning which means that improvement of existing and innovative new solutions for this challenge is going to remain of high interest to researchers. In this paper some of the most usual positioning and navigation methodologies are presented and comparisons between their advantages and disadvantages are given.

Keywords: Navigation · UAV · Positioning · Drone · Surveillance

1 Introduction

By definition, navigation is the science of determining the exact position of an object, its speed and direction, and positioning relative to the environment as the data needed so that a given path of movement from place A to place B can be successfully realized [1].

Positioning is a technique of determining the exact position of a body in space and is directly related to navigation and navigation systems.

Navigation is the science that studies the guidance of objects, from one point to another on Earth, regardless of whether the object moves on land, air or water. It also applies to spacecraft during space flight. The primary role of navigation is to determine the course, the distance between two places as well as the position of the object. In this way, we can determine the exact value of the distance traveled.

The successful realization of this task as well as its accuracy are directly related to the technology level in various fields.

The easiest way to navigate is to follow the given detailed instructions and to rely on available means. The map, as the basic navigation mean, provide additional information that are related to the topographic characteristics of the terrain, latitude and longitude, etc. In that case, the passenger can determine his position by comparing the terrain he is on with the given map.

Other natural and geographical means can be used as the reference. One such technique is observing fixed stars. While knowing the characteristics of Earth movement and

N. Mitrovic et al. (Eds.): CNNTech 2020, LNNS 153, pp. 220–256, 2021.
https://doi.org/10.1007/978-3-030-58362-0_14

the time of observation, it is possible to determine the position on the Earth by using celestial bodies.

For the long time the biggest problem while navigating on the seas and oceans has been the inability to determine the exact time in order to determine longitude. Latitudinal position was relatively easy to determine using celestial bodies as the reference. However, the determination of the longitudinal positioning was highly influenced by the precision of the time measurement. This problem was overcomed in 18th century when John Harrison's chronometer enabled accurate time measurement on the moving ship [2].

Since that time, there are numerous navigation systems that have emerged as a product of the development of techniques, science and technology.

Navigation of various types can be:

- Landmark-based navigation

Land navigation is the discipline of tracking routes through unfamiliar terrain using terrain, compass, other navigation tools, and landmarks.

- Radio-based navigation;

Radio navigation or radio navigation is the application of radio frequencies to determine the position of an object on Earth;

- Celestial Navigation;

Celestial navigation, also known as astronavigation, is an ancient technique improved by modern approaches and technology which is based on positioning of celestial bodies to determine position. The most commonly used celestial bodies are Stars, Moon.

- Sensor-based navigation;

The use of various sensors to provide adequate data used in the navigation process such.

- Radar-based navigation;

Navigation systems has a very important role from the early age of aviation. The developments of civil and military aviation have been closely connected to the development of navigation where growing importance of navigation has induced a large number of scientific researches. Significant importance of navigation and wide effort involved resulted with its rapid development and various improvements achieved.

Modern systems of navigation have been developed, encouraged by major technological breakthroughs in aviation, especially since the mid-1950s. Then, the Soviet Union successfully launched the first artificial satellite into Earth orbit. In response, the United States launched its APOLO space program, enabling the intensive development of aerospace technology.

The result of this research is new technological concepts that have been applied in solving engineering problems such as the recursive minimum deviation estimator, known as the Kalman-Buci filter, time-domain concepts related to linear algebra, and probabilities, and so on [3].

The research of these areas proved to be extremely useful while enabling the design of optimal flight paths and today they are used in different variations. However, the successful application of new methods in solving problems related to aviation quickly found application in other areas such as medicine, energy production, and management, nuclear technology, etc. [4].

When designing, due to simplicity and efficiency, engineers divide the avionics system into three parts and three systems [5].

1. navigation system;
2. guidance system;
3. flight control system;

The navigation system must ensure the correct position, speed, and direction of flight concerning the reference coordinate system. The data required to perform this function are obtained from sensors (gyroscope and accelerometer).

The guidance provides the flight path that will lead the aircraft to its destination. The guidance system chooses the best path based on the information it receives from the sensor, such as an inertial system, and is based on the closed-loop principle [5].

The flight control system monitors the aircraft during flight and corrects the trajectory based on information obtained from sensors (altimeter, barometer, radar, etc.).

Based on the direction of communication, navigation can be divided into two categories:

1. active navigation system;
2. passive navigation system;

Active navigation systems work based on two-way communication, i.e. there are transmitters and signal receivers in the system. In contrast, passive navigation systems receive a signal from the transmitter, but the receiver does not transmit the signal further. In this way, the passive navigation system is more energy efficient because it consumes less energy and, more importantly, hides its position. Passive navigation systems are Satellite navigation, INS, etc.

Active navigation systems are not used in cases when the position of the receiver must be hidden, such as military operations, etc. However, active navigation systems are easier to implement due to the lower load on the computer system and are therefore more suitable for civilian services such as police, fire, security, and others.

Based on the technique, navigation systems can be divided into the incremental and absolute system.

In an incremental navigation system, such as INS, the initial position of the receiver is known, and then the position of the receiver is updated based on data about the initial position, elapsed time, and incremental movements of the receiver. The accuracy of position estimation is a function of the initial position, accuracy, and precision of incremental displacements. On the other hand, an absolute navigation system, such as Satellite navigation, is the system where the position of the receiver is determinated without any prior data. That is why an absolute navigation system is more practical than an incremental navigation system.

In the field of unmanned aerial vehicle navigation techniques, during their development and application, the following techniques were mainly used:

- Satellite Navigation (SatNav);
- Inertial Navigation System (INS);
- Terrestrial Navigation (TAN);
- Geomagnetic Navigation;
- Vision Navigation System;

2 Satellite Navigation (SatNav)

A satellite navigation system is a system that uses satellites in Earth orbit to position themselves on ground water and in the air. With the help of radio communication, satellites in Earth orbit enable receivers to determine with great precision the location (latitude, longitude and altitude) or tracking the position of the object that the receiver has. The system operates independently of any other radio traffic.

A satellite navigation system with global coverage is called a Global Navigation Satellite System (GNSS). Today, on the world stage, there are the Global Positioning System (GPS) and the Global Navigation Satellite System (GLONASS), which are completely operative. In addition, there is the navigation satellite system BeiDou (BDS), the navigation system Galileo, while the Quasi-Zenith Satellite System (KZSS) and the Navigation with Indian Constellation (NAVIC), and they are not fully operational [6].

The global coverage of each system is mainly achieved by a satellite constellation of 18–30 Earth orbit satellites distributed between multiple orbital planes. Current systems differ in the number of satellites and orbits they use but use orbital inclinations greater than 50° and orbital periods of approximately twelve hours in non-geostationary orbits at a distance of about 20,000 km.

Theoretical Settings and Mathematical Models of Functioning

All satellite systems are collectively known as the Global Navigation Satellite Systems (GNSS).

The GPS navigation system uses the Time-Of-Arrival (TOA) concept to determine a user's position [7–9].

$$TOA = Time\_Instant\_of\_Arrival - Time\_Instant\_of\_Transmission \quad (2.1)$$

In this system, each satellite is equipped with a clock that is synchronized with the other satellites. Each satellite transmits a signal with a specific code and contains accurate weather information about the transmission time of the signal. The positioning system is based on a one-way ranking technique. Time information is retrieved from the signal in the receiver and then scaled at high speed to calculate the distance between the transmitter and the receiver (equation) [6].

$$Distance = Speed\_of\_Light \times Time\_of\_Arrival \quad (2.2)$$

$$D = ct \quad (2.3)$$

where D is the distance between the transmitter and the receiver, c is the speed of light and t is the time difference between the transmission and reception of the signal. Each pair of satellite and receiver positions represents a sphere with a radius equal to the distance between the transmitter and the receiver. The receiver receives additional signals from two other satellites, forming three intersecting spheres. The intersection point of all three spheres determines the position of the receiver. The basic concept of satellite navigation is shown in the Fig. 1.

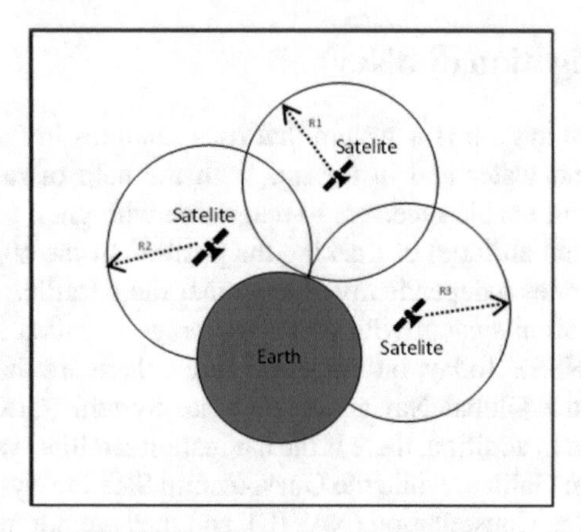

Fig. 1. The basic concept of GPS [6]

The system assumes that the satellites and the receivers are synchronized and thus the three spheres determine the correct position estimate. However, the receiver clock and satellite clock are not synchronized correctly, which adds uncertainty to the position estimation. Therefore, the receiver receives a fourth signal from another satellite and thus provides four spheres to estimate the position of the receiver. Then the receiver generates closed-form equations for solving the intersection of four spheres with the intersection point being the correct position [10].

Satellite navigation is based on the trilateral method of positioning and measuring lengths by the principle of determining the travel time of a radio signal. To measure the travel time of radio signals, the clocks in satellites and receivers must be synchronized with very high accuracy. Since quartz oscillators incorporate much less accuracy than Satellite navigation atomic oscillators in satellites, all time intervals at the time of measurement will have a systematic error. This means that at that moment the distances to all the satellites are equally longer or equally shorter than the actual values, which is why these lengths are called pseudocodes.

The notion that satellite navigation receivers do not measure lengths but pseudocodes length changes the trilateral principle of satellite navigation positioning. Figure 2 illustrates the basic idea.

The receiver from point A simultaneously measures the pseudogapes Pi according to Si satellites, whose positions are known at the time of measurement due to the data

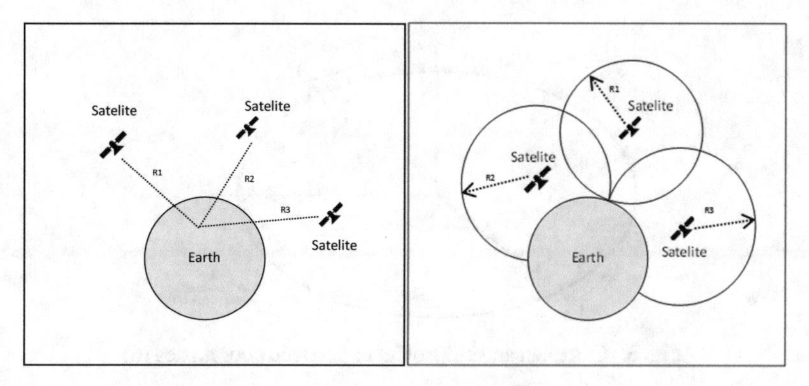

Fig. 2. Satellite navigation trilateration principle [11]

(messages) they emit. Each measured pseudogap defines one sphere whose center is in the corresponding satellite. The three-dimensional position of the unknown point A is determined in principle by the intersection of three such spheres. However, since the measured pseudogapes contain the same synchronization error of the receiver and satellite b, at least four are required for the complete solution. Mathematically, satellite navigation trilateration is used to solve the equation:

$$P_i = (x - x_i) + (y - y_i) + (z - z_i) + b \tag{2.4}$$

by the unknown coordinates (*x, y, z*) of point *A* and the unknown synchronization *error b*, with the known satellite coordinates indicated by the *above index i = 1, 2, 3, 4*.

Satellite navigation equations are formulated after the establishment of a reference coordinate system that applies to both satellites and the receiver. Satellites and receivers are defined by position and velocity vectors in the Cartesian coordinate system [9].

The reference system is a conceptual idea of a coordinate system. The reference framework is the practical realization of the reference system. The frame of reference is a set of coordinates and velocities in the area of interest, together with the estimated error values.

In GPS applications, the position of a point in the coordinate system can be expressed in Fig. 3:

- Cartesian coordinates (x, y, z);
- Ellipsoidal or geodetic coordinates (λ, ϕ, h): λ is the latitude, ϕ is the geographic longitude, and h is the height above the Earth's surface [12].

Satellite navigation uses three coordinate systems to respond to three different requirements [5]: determining satellite orbits, which require an inertial Earth-directed coordinate system, estimating the coordinates of the receiver's position, which is better performed in Earth-directed Earth. a fixed coordinate system and their presentation in a user-friendly form (i.e., in the usual geographical coordinates of latitude, longitude, and altitude).

The Earth-centered inertial (ECI) inertial coordinate system is used to predict the position of artificial satellites in orbit around the Earth (Fig. 4) [9].

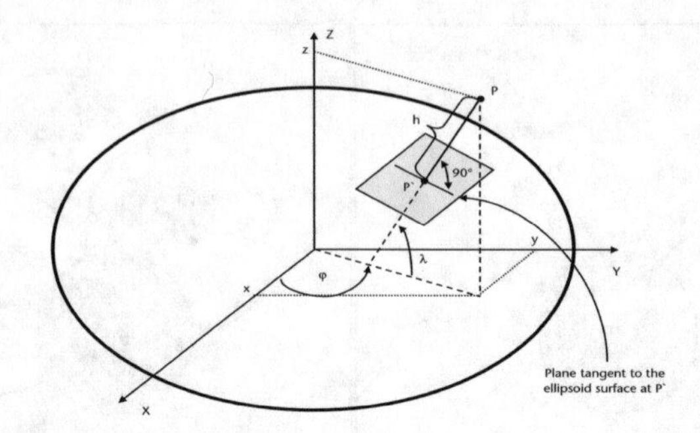

Fig. 3. Cartesian and ellipsoidal or geodetic coordinates [6]

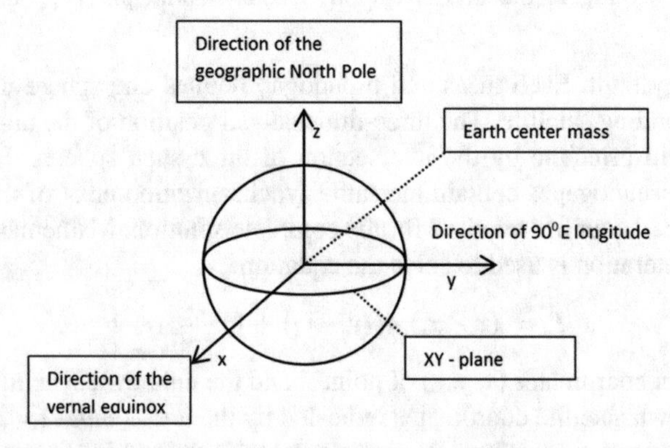

Fig. 4. Earth-centered inertial (ECI) coordinate system [6]

Earth-centered rotational (ECEF) Earth-centered coordinate system is not inertial and can be used to define the three-dimensional position in Cartesian coordinates, which can then be easily transformed into latitude, longitude, and height [13–17] (see Fig. 5).

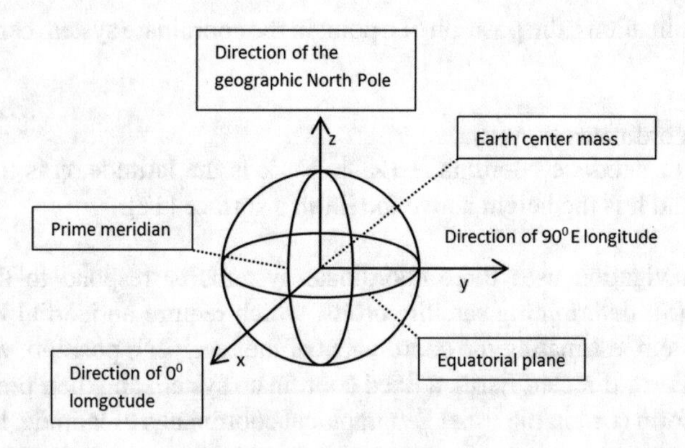

Fig. 5. Earth-centered Earth-fixed (ECEF) coordinate system [6]

The ECEF coordinate system is Cartesian, while satellite navigation receivers display their position in terms of latitude, longitude, and altitude that relate to the physical model of the Earth.

World Geodetic System - 1984 (WGS-84) is the ECEF frame of reference and provides a comprehensive model of the earth and information on gravity irregularities for the calculation of satellite ephemeris [9].

In satellite navigation systems, the concept of trilateration is used to fix the user position in three-dimensional space (Fig. 6) [9].

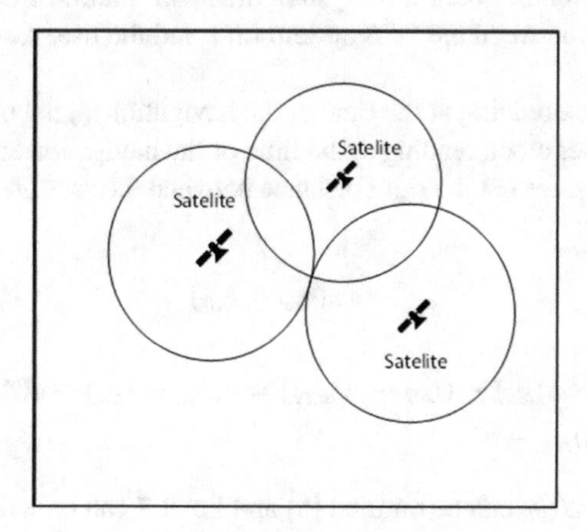

Fig. 6. The concept of trilateration in three-dimensional space [18].

Mathematical equations for determining user position by looking at Fig. 7, where the vector *s* is known from the navigation message, which includes satellite ephemera, *r* is the distance vector estimated by multiplying TOA and the speed of light *c*, and *u* is the unknown vector to be determined [6].

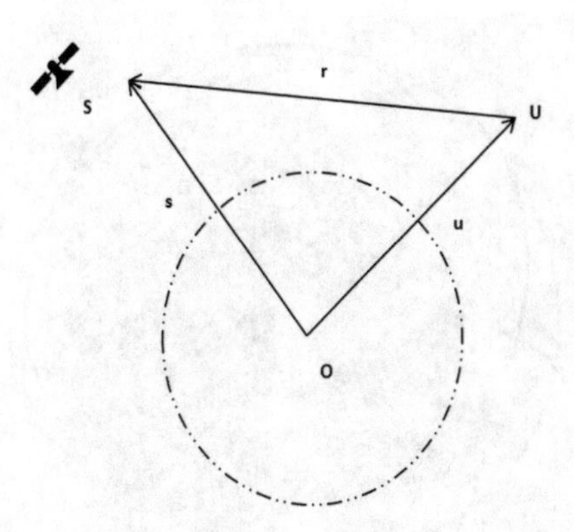

Fig. 7. User position [6]

$$r = \|\underline{r}\| = \|\underline{d} - \underline{u}\| \tag{2.5}$$

If there is no time synchronization, the receiver measures the time interval that contains the deviation between the GPS system time and the satellite clock and the deviation between the GPS system time and the receiver clock, we can define.

- t_{sat} is the moment of the GPS system when the navigation signal was transmitted from the satellite.
- t_{rec} is the timing of the GPS system when the navigation signal arrived at the user's receiver..
- Δt_{sat} is the deviation between GPS system time and satellite clock
- Δt_{rec} is the offset between the GPS system time and the user receiver clock.

The satellite clock reading at the time of the navigation signal transmission is $t_{sat} + \Delta t_{sat}$, and the receiver clock reading at the time of the navigation signal arrival is $t_{rec} + \Delta t_{rec}$ [9]. In this way, we get the right distance between the satellite and the receiver (as the pseudo-range ρ):

$$r = c(t_{rec} - t_{sat}) \tag{2.6}$$

$$\begin{aligned}\rho &= c[(t_{rec} + \Delta t_{rec}) - (t_{sat} + \Delta t_{sat})] = c(t_{rec} - t_{sat}) + c(\Delta t_{rec} - \Delta t_{sat}) \\ &= r + c(\Delta t_{rec} - \Delta t_{sat})\end{aligned} \tag{2.7}$$

The corrections Δt_{sat} can be omitted [5] and Eq. 2.7 can be written:

$$\rho = r + c\Delta t_{rec} = \left\| \underline{s} - \underline{u} \right\| + c\Delta t_{rec} \tag{2.8}$$

Measuring the distance between three or more satellites and the receiver is sufficient to obtain the position of the receiver (see Fig. 8). Thus Eq. 2.8 is expressed over pseudocodes:

Fig. 8. Global Positioning System (GPS) [35]

$$\rho_i = \sqrt{(x_i - x_u)^2 + (y_i - y_u)^2 + (z_i - z_u)^2} + c\Delta t_{rec} \tag{2.9}$$

where (x_i, y_i z_i,) and (x_u, y_u, z_u) denote the three-dimensional position of the satellite and the user, respectively. So we have four unknowns: *x, y, z* и Δt_{rec}. Therefore, at least four independent equations are required to solve the system 2.10.

$$\begin{aligned}
\rho_1 &= \sqrt{(x_1 - x_u)^2 + (y_1 - y_u)^2 + (z_1 - z_u)^2} + c\Delta t_{rec} \\
\rho_2 &= \sqrt{(x_2 - x_u)^2 + (y_2 - y_u)^2 + (z_2 - z_u)^2} + c\Delta t_{rec} \\
\rho_3 &= \sqrt{(x_3 - x_u)^2 + (y_3 - y_u)^2 + (z_3 - z_u)^2} + c\Delta t_{rec} \\
\rho_4 &= \sqrt{(x_4 - x_u)^2 + (y_4 - y_u)^2 + (z_4 - z_u)^2} + c\Delta t_{rec}
\end{aligned} \tag{2.10}$$

This system of nonlinear equations can be solved by various methods: closed-form solutions [19–24]. or iterative solutions based either on linearization [25–29] or on Kalman filtering [30–34]. The least-squares method is most commonly used. In this case, the clock error of the receiver is multiplied by *c*, making it the length denoted by $b = c\Delta t_{rec}$.

2.1 Global Positioning System - GPS

Global Positioning System (GPS) is the Global Navigation System (GNNS placed in Earth's orbit. GPS is part of a satellite navigation system developed by the US Department of Defense as part of its NAVSTAR satellite program [5] (see Fig. 8.) which became fully operational in March 2000.

Satellite navigation originated as an idea in the early 1960s as a requirement to create a high-precision navigation system. The first GPS satellite was launched in 1978 as part of the NAVSTAR GPS program. The first GPS consisted of 12 satellites, while the second generation of the system consisted of 9 satellites launched into orbit by 1990.g [36].

A fully operational GPS consists of 24 or more active satellites arranged in six circular orbits with four or more satellites each. The orbits are tilted at an angle of 55° concerning the equator and are separated from each other by multiple directions of the ascent of 60°. The orbits are non-geostationary and approximately circular, with radii of 26,560 km and half-day orbital periods (11,967 h). Theoretically, three or more GPS satellites will always be visible from most points on the earth's surface, and four or more GPS satellites can be used to determine the position of an observer anywhere on the earth 24 h a day [36].

GPS navigation signal contains navigation information about the GPS-satellite ephemeris, with parameters that allow approximate corrections for the signal propagation delay through the ionosphere. This is suitable for single frequency receivers and shifts time between satellite clock and GPS real-time. Navigation information is transmitted at a speed of 50 Bd (baud) [37].

Although the original purpose of GPS was military, civil use of GPS technology began in 1993 [38]. The signal available to civilian users had a deliberate error (degradation) called selective availability (SA). The accuracy level with selective availability was limited to a horizontal position of 100 m (95% probability) and 140 m vertical position (95% probability) [39]. The reason for implementing selective availability is to prevent potential misuse of GPS systems [40, 41]. Modernization of the system and the creation of technical conditions for the selective limitation of system capabilities have led to the abolition of selective availability and the creation of new GPS applications for widespread use around the world.

2.2 GLONASS - GLObal NAvigation Satellite System

GLONASS arose as an idea on the theoretical basis of the use of radio astronomy for navigation. The demand for a universal navigation system of high accuracy on the entire surface of the Earth led to its creation.

The launch of the GLONASS satellite began in the early 1980s, and the system has been operational since 1993 [42–46]. For civilian use, the system has been available since 1996.

At the beginning of the 2000s, a total of 11 GLONASS satellites were in operation, which is less than the estimated number of satellites during its design and commissioning [6].

However, as of October 2011, a system with 24 satellites placed in orbit provides full global coverage.

The 24-satellite system is located in four orbits and has a mean circular orbit at an altitude of 19,100 km, with an inclination of 64.8° and a period of 11 h and 15 min 44 s.

2.3 GALILEO

The GALILEO global navigation satellite system was developed for the needs of the primary civil sector in the field of geodesy and navigation.

The development of the project started in 1994, and with the launch of the satellite in 2005. The GALILEO system is designed to provide three modes of navigation services:

1. Open Service - open signals, without subscription and other fees, available to all types of consumers.
2. Commercial service - encrypted signal, higher data transfer speed.
3. The commercial service will provide two functions - global high-precision navigation.

The GALILEO satellite navigation system works with 30 satellites (27 plus 3 satellites that are ready to work if one stops working). Their orbit above the Earth is 23,260 km, they move at a speed of 3.6 km/s. An integral part of the system is a network of ground stations that manage the system. This navigation system has the technical ability to determine the position with an accuracy of 4 m.

Galileo performance will gradually be improved and new services will be introduced as further spacecraft is launched.

The planned number of satellites in orbit is expected to be realized during 2020. When all the satellites are in orbit and the system is operational, a phase of full operational capability will be declared.

3 Inertial Navigation System - INS

The functioning of inertial navigation systems is directly related and depends on the laws of classical mechanics formulated in 1687 by the English scientist Isaac Newton. Newton's first law tells us that each body remains in a state of relative rest or even, perpendicular motion, until the action or action of another body forces it to change that state. Also, force will produce a proportional acceleration of the body [47].

Given that we can measure the acceleration of the body, it is possible to calculate the change in velocity and position of the body by performing successive mathematical integrations of acceleration for time. The acceleration can be determined using an accelerometer.

A structurally inertial navigation system is usually made up of three accelerometers, each of which detects acceleration in one direction. The accelerometer shafts are mounted normally on top of each other, i.e. they are at right angles to each other. To move concerning our inertial reference system, it is necessary to follow the direction in which the accelerometers are directed. The rotational motion of the body for the inertial reference system can be detected by gyroscopic sensors and used to determine the orientation of the accelerometers at any time. Based on this information, it is possible to place accelerations in the reference system before the integration process occurs.

Inertial navigation is the process by which the information provided by gyroscopes and accelerometers is used to determine the position of the body in which they are embedded. By combining two datasets it is possible to define the translational motion of a body within an inertial reference system and thus to calculate its position in it.

Inertial systems, unlike other navigation systems, are completely autonomous within the body, i.e. they are not dependent on the transmission of signals from the body or the environment. A very important and important feature of an inertial navigation system is the need to know its exact data at the beginning of navigation. Inertial measurements are then used to obtain estimates of posture changes that occur thereafter during movement. The application of inertial sensors and inertial navigation has had a very dynamic development due to its great technological advances, especially in the field of microelectronics and computer technology. New sensors with state-of-the-art computer technology have dramatically improved the performance of the miniature systems they are built into.

In the recent past, the design and construction of high-precision sensors, such as laser gyroscopes, miniature cameras of extreme resolution, etc., have been won. These devices are still evolving and provide a unique approach to measuring movement and apparent anomalies. As a result, these devices are being deployed on a growing several new applications, including surveying and fundamental physical studies.

The basic principles of operation are the same for all types of inertial navigation systems, but their application may be different. The initial application of inertial navigation systems used stable platforms. In such systems, the sensors were mounted on a stable platform while being mechanically separated from the rotational movement of the body. This system is still used in cases where precise knowledge of navigation of ships and submarines is required.

However, in modern systems, sensors are directly connected to the body structure. This results in reduced system dimensions, lower costs, and greater reliability compared to the platform system. Also, the dramatic increase in the performance of computer technology combined with the development of appropriate sensors has enabled the design of new multisensory navigation systems.

Inertial navigation system INS, is designed to determine the navigation parameters of an object in space using computers and sensors. Changes in the "state vector" of an object in space are detected by measuring its translational motion (accelerometers) and rotation (gyroscopes) using sensors. The INS retrieves data from the sensor, calculates the navigation parameters and compares the results concerning the initial idle state of the object or some new reference state. In this way, it determines the orientation and speed of the object (direction and speed of movement) in real-time. There is no need for a constant reference comparison about one's environment to determine the parameters of the free-body motion in space.

The system works by measuring the kinematical parameters of a moving object by six degrees of freedom, three translations (along three axes) and three rotations (around them). It is used for navigation on moving objects such as ships, planes, submarines, guided missiles and spacecraft.

The inertial guidance system is designed so that the aircraft flies a predetermined path - a trajectory. The aircraft is operated by self-contained automatic devices called accelerometers. Accelerometers are inertial devices that measure acceleration. When operating the aircraft, they measure the vertical, rotational, and longitudinal accelerations of the controlled aircraft. Although there is no contact between take-off points - launching an aircraft after take-off can make corrections to its flight path with very high precision.

During the flight, unpredictable external forces, such as wind, affect the aircraft, causing changes in its movement. These changes are transmitted to the aircraft, which uses its computer to make the necessary calculations and determine the correct position based on new input data. Recently, however, inertial systems have been combined with GPS (Global Position System) to move aircraft more precisely.

Also, there are vibrations that are transmitted throughout the system and for which an adequate system should be designed that could control them [48]. One solution could be to control the free vibrations of smart faces using an adaptive logic controller [49]. It is certainly necessary to consider the vibration problems of the whole structure [50].

However, even with the best inertial systems, aircraft suffer from a phenomenon called "drift" [51]. These are small acceleration and angular velocity errors that integrate into progressively larger velocity errors that are compounded by even greater position errors. The new position of the aircraft is determined - calculated based on the previously calculated position and the measured acceleration and angular velocity, and the errors accumulate in proportion to the time relative to the initial position. Errors are measured in meters per hour. For example, during the iteration of Tomahawk cruise missiles it was found that even with an inertial navigation system, they would have a displacement (error) of 900 m per hour. This means that after one hour of flight, the aircraft could miss its target point by as much as 900 m. When the Tomahawk cruise missiles appeared between the 1970s and 1980s, that error was 2 nautical miles. Now this error is (0.01°/h) [52]. The best accelerometers, with a standard error of 10 μg, have a cumulative error of 50 m within 17 min. Therefore, the position of the aircraft is periodically corrected, the error is compensated by another navigation system.

Cruise missiles are moving at subsonic speeds, so they are more likely to miss the target.

With that said, inertial navigation systems are usually used in conjunction with some other navigation systems. This increases the accuracy of the system about the use of only one navigation system.

An essential feature of a mathematical tool used in navigation is the transformation of kinematics between coordinate systems. In the following presentation, we will summarize the equations that describe the relationship of one system to another and the Carthusian position, velocity, acceleration, and angular relationship between references to an inertial, terrestrial, and local navigation system. We will show the equations for transferring the solution of the navigation from one object to another Cartesian position, velocity, acceleration and angular course directed to the same transformation system are solved by simply applying the transformation of the coordinate matrix [53]:

$$x^{\gamma}_{\beta\alpha} = C^{\gamma}_{\delta} x^{\delta}_{\beta\alpha} \quad x \in r, v, a, \omega \quad \gamma, \beta \in i, e, n, b \tag{3.1}$$

Therefore, these transformations are not explicitly represented for each pair of systems. Coordinate transformation matrices involving the body system - i.e.

$$C^{\beta}_{b}, C^{b}_{\beta} \quad \beta \in i, e, n \tag{3.2}$$

They describe the position of that body against the reference system. The position of the body according to the new reference system can be obtained by simply multiplying the matrix of transformed coordinates between the two reference systems

$$C^{\delta}_{b} = C^{\delta}_{\beta} C^{\beta}_{b} \quad C^{b}_{\delta} = C^{\delta b}_{\beta} C^{\beta}_{\delta} \quad \beta, \delta \in i, e, n \tag{3.3}$$

Euler's transformation, quaternion, or vector rotation into a new reference system is more complex. One solution is to convert the matrix of transformed coordinates, transform the references, and then convert everything back.

The center and z axis of the ECI and ECEF coordinate systems coincide (see Fig. 5 and Fig. 6). The k and i axes coincide at time t_0, and the systems rotate about the z axis to ω_{ie}. So,

$$C^{e}_{i} = \begin{pmatrix} cos\omega_{ie}(t - t_0) & sin\omega_{ie}(t - t_0) & 0 \\ -sin\omega_{ie}(t - t_0) & cos\omega_{ie}(t - t_0) & 0 \\ 0 & 0 & 1 \end{pmatrix}$$

$$C^{i}_{e} = \begin{pmatrix} cos\omega_{ie}(t - t_0) & -sin\omega_{ie}(t - t_0) & 0 \\ sin\omega_{ie}(t - t_0) & cos\omega_{ie}(t - t_0) & 0 \\ 0 & 0 & 1 \end{pmatrix} \tag{3.4}$$

The positions related to the two systems are the same, so only the axes for solving are transformed:

$$r^{e}_{eb} = C^{e}_{i} r^{i}_{ib}, \quad r^{i}_{ib} = C^{i}_{e} r^{e}_{eb} \tag{3.5}$$

The transformation of speed and acceleration is more complex:

$$V^{e}_{eb} = C^{e}_{i}\left(v^{i}_{ib} - \Omega^{i}_{ie} r^{i}_{ib}\right)$$

$$V_{ib}^{i} = C_{e}^{i}\left(v_{eb}^{e} - \Omega_{ie}^{e} r_{eb}^{e}\right) \tag{3.6}$$

$$\begin{aligned} a_{eb}^{e} &= C_{i}^{e}\left(a_{ib}^{i} - 2\Omega_{ie}^{i} v_{ib}^{i} - \Omega_{ie}^{i}\Omega_{ie}^{i} r_{ib}^{i}\right) \\ a_{ib}^{i} &= C_{e}^{i}\left(a_{eb}^{e} + 2\Omega_{ie}^{e} v_{eb}^{e} + \Omega_{ie}^{e}\Omega_{ie}^{e} r_{eb}^{e}\right) \end{aligned} \tag{3.7}$$

The angular displacements are transformed as

$$\omega_{eb}^{e} = C_{i}^{e}\left(\omega_{ib}^{i} - \begin{pmatrix} 0 \\ 0 \\ \omega_{ie}^{i} \end{pmatrix}\right), \quad \omega_{ib}^{i} = C_{e}^{i}\left(\omega_{eb}^{e} - \begin{pmatrix} 0 \\ 0 \\ \omega_{ie}^{i} \end{pmatrix}\right) \tag{3.8}$$

The relative orientation of the Earth and the local navigation systems is determined by the latitude, L_b and longitude, λ_b, for the body system whose center coincides with that of the local navigation system:

$$C_{e}^{n} = \begin{pmatrix} -\sin L_b cos\lambda_b & -\sin L_b sin\lambda_b & \cos L_b \\ -sin\lambda_b & cos\lambda_b & 0 \\ -\cos L_b cos\lambda_b & -\cos L_b sin\lambda_b & -\sin L_b \end{pmatrix}$$

$$C_{n}^{e} = \begin{pmatrix} -\sin L_b cos\lambda_b & -sin\lambda_b & -\cos L_b cos\lambda_b \\ -\sin L_b sin\lambda_b & cos\lambda_b & -\cos L_b sin\lambda_b \\ \cos L_b & 0 & -\sin L_b \end{pmatrix} \tag{3.9}$$

The location, speed and acceleration of the local navigation system are meaningless because the center of the body system coincides with the center of the navigation system. The determination of the axes of position, velocity and acceleration concerning the Earth is done by transformation using (3.1). Angular motion is transformed as

$$\begin{aligned} \omega_{nb}^{e} &= C_{e}^{n}\left(\omega_{eb}^{e} - \omega_{en}^{e}\right), \quad \omega_{eb}^{e} = C_{n}^{e}\left(\omega_{nb}^{n} + \omega_{en}^{n}\right) \\ &= C_{e}^{n}\omega_{eb}^{e} - \omega_{en}^{e} \end{aligned} \tag{3.10}$$

noting that we have a solution for ω_{en}^{n}.

The coordinates of the local inertial transformed navigation systems are obtained by multiplying (3.9) and (3.10)

$$C_{i}^{n} = \begin{pmatrix} -\sin L_b cos(\lambda_b + \omega_{ie}(t-t_0)) & -\sin L_b sin(\lambda_b + \omega_{ie}(t-t_0)) & \cos L_b \\ -sin(\lambda_b + \omega_{ie}(t-t_0)) & cos(\lambda_b + \omega_{ie}(t-t_0)) & 0 \\ -\cos L_b cos(\lambda_b + \omega_{ie}(t-t_0)) & -\cos L_b \sin(\lambda_b + \omega_{ie}(t-t_0)) & -\sin L_b \end{pmatrix}$$

$$C_{n}^{i} = \begin{pmatrix} -\sin L_b cos(\lambda_b + \omega_{ie}(t-t_0)) & -sin(\lambda_b + \omega_{ie}(t-t_0)) & -\cos L_b cos(\lambda_b + \omega_{ie}(t-t_0)) \\ -\sin L_b sin(\lambda_b + \omega_{ie}(t-t_0)) & cos(\lambda_b + \omega_{ie}(t-t_0)) & -\cos L_b \sin(\lambda_b + \omega_{ie}(t-t_0)) \\ -\cos L_b & 0 & -\sin L_b \end{pmatrix} \tag{3.11}$$

Ground speed and acceleration in the axis of the navigation system are transformed into a corresponding inertial reference from the inertial system as

$$\begin{aligned} V_{eb}^{n} &= C_{i}^{n}\left(v_{ib}^{i} - \Omega_{ib}^{i} r_{ib}^{i}\right) \\ V_{ib}^{i} &= C_{n}^{i} v_{eb}^{n} + C_{e}^{i}\Omega_{ie}^{i} r_{eb}^{e} \end{aligned} \tag{3.12}$$

$$a_{eb}^{n} = C_{i}^{n}\left(a_{ib}^{i} - 2\Omega_{ie}^{i}v_{ib}^{i} - \Omega_{ie}^{i}\Omega_{ie}^{i}r_{ib}^{i}\right)$$
$$a_{ib}^{i} = C_{n}^{i}\left(a_{eb}^{n} + 2\Omega_{ie}^{n}v_{eb}^{n} + C_{e}^{i}\Omega_{ie}^{e}\Omega_{ie}^{e}r_{eb}^{e}\right) \tag{3.13}$$

The angular motion is transformed as

$$\begin{aligned}\omega_{nb}^{n} &= C_{i}^{n}\left(\omega_{ib}^{i} - \omega_{in}^{i}\right)\\ &= C_{i}^{n}\left(\omega_{ib}^{i} - \omega_{ie}^{i}\right) - \omega_{en}^{n}\\ \omega_{ib}^{i} &= C_{n}^{i}\left(\omega_{nb}^{n} - \omega_{in}^{n}\right)\\ &= C_{n}^{i}\left(\omega_{nb}^{n} - \omega_{en}^{n}\right) + \omega_{ie}^{i}\end{aligned} \tag{3.14}$$

Sometimes there is a requirement for the body object to transmit a navigation solution from one position to another, for example between the INS and the GPS antenna, between the INS and the center of gravity, or between the reference and alignment INS. We will present equations for transposing position, velocity, and position from describing the b system to describing the B system. Let the orientation of frame B with respect to frame b be given as C_{b}^{B}, and the position of B with respect to the axes of frame b is l_{bB}^{b}, known as the lever arm. It should be borne in mind that the lever is mathematically identical to the Cartesian position with B as the object system and b as the reference and resolution systems. Figure 9 illustrates this:

$$\begin{aligned}C_{\beta}^{B} &= C_{b}^{B}C_{\beta}^{b}\\ C_{B}^{\beta} &= C_{b}^{\beta}C_{B}^{b}\end{aligned} \tag{3.15}$$

We can move the Cartesian position using

$$r_{\beta B}^{\gamma} = r_{\beta b}^{\gamma} + C_{b}^{\gamma}1_{b}^{b} \tag{3.16}$$

Precise transformation of the Earth's latitude, longitude and height requires conversion to the Cartesian position and back. However, if we apply the small-angle

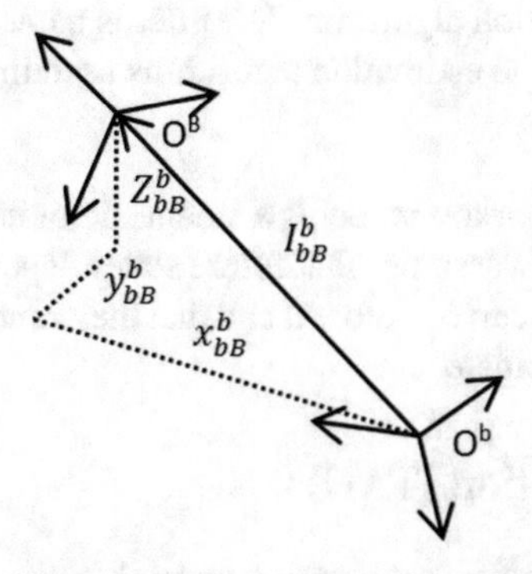

Fig. 9. The lever arm from frame b to frame B [53]

approximation to l/R, where R is the radius of the Earth, a simpler form can be used:

$$\begin{pmatrix} L_B \\ \lambda_B \\ h_H \end{pmatrix} \approx \begin{pmatrix} L_B \\ \lambda_B \\ h_H \end{pmatrix} + \begin{pmatrix} 1/(R_N(L_b)+h_b) & 0 & 0 \\ 0 & 1/[(R_E(L_b)+h_b)cosL_b] & 0 \\ 0 & 0 & -1 \end{pmatrix} C_b^n 1_{bB}^b \tag{3.17}$$

Finally, velocity transposition is obtained by differentiating (3.16) and substituting [54]:

$$v_{\beta B}^{\gamma} = v_{\beta b}^{\gamma} + C_b^{\gamma} C_b^{\gamma\beta} 1_{bB}^b \tag{3.18}$$

assuming $1_{\beta B}^b$ is constant.

$$v_{\beta B}^{\gamma} \;=\; v_{\beta b}^{\gamma} + C_b^{\gamma}(\omega_{\beta b}^b \wedge l_{bB}^b) \tag{3.19}$$

Estimation theory [55] and the Kalman filter provide a theoretical framework for the use of information obtained from different sensors in navigation. One of the most common alternative sensors is a satellite navigation radio signal such as GPS, which can be used for all types of movement, with direct sky visibility. By correctly combining data from INS and other systems (GPS/INS), position and speed errors are stable. Also, the INS can be used as a short-term replacement when GPS signals are not available, for example when passing through a tunnel [56–59].

Estimation theory is a field of statistics that deals with parameter estimation based on measured (empirical) data that have a random component. The parameters describe the basic physical settings so that their value affects the distribution of the measured data. The estimator tries to approximate the value of unknown parameters by measuring.

Thus, radars aim to determine the distance of an object (planes, ships, etc.) by analyzing the two-way transit time of the received rejected signals. Since the reflex impulses are embedded in the electrical noise, their measured values are randomly distributed, so the transit time must be estimated.

For such a model, there must be more statistics to implement the estimators.

Kalman Filter

In statistics and control theory, the Kalman filter, also known as linear quadratic estimation (LKE), is an estimation algorithm. The filter is named after R. E. Kalman and it forms the basis of most of the estimation algorithms used in navigation systems [54].

Markov Model

In probability theory, the Markov model is a stochastic model used to model randomly varying systems. This model assumes that future states depend only on the current state and not on the events that occurred before it and that they cannot be calculated accurately but only predicted - to anticipate.

4 Terrestrial Navigation (TAN)

Ground Navigation (TAN) is a technique for determining the position of an aircraft during flight, by comparing the terrain profile below the aircraft with an altitude map located in the computer's memory. This system has proven to be a good technical navigation solution with great potential for improvement. Considering the long-term stability of the terrain profile, the easy mapping and manipulation of maps, the reliability of the system, and the refinement of computer technology, we can expect long-term use of this technique in navigation systems [60].

TAN is an important part of "integrated navigation systems" that provides corrections to the position of the aircraft to the central navigation system. This is especially important when no other navigation systems such as Global Positioning System (GPS) are available. This technique provides reliable positioning information at low altitudes over a given terrain [61].

Navigation charts can be stored using two techniques. The first uses small high-resolution maps that are used on some sections of the flight route. The second technique uses one small resolution map that covers the entire area of operation.

4.1 Terrain Contour Matching - TERCOM

TERCOM (Terrain Contour Matching) is the most famous TAN system and it was developed to guide cruise missiles.

The TERCOM system significantly enhances the accuracy of missiles compared to inertial navigation systems (INS) and thus makes it possible to reach the target at small issues, making them difficult.

The diagram and concept of TERCOM are shown in Fig. 10.

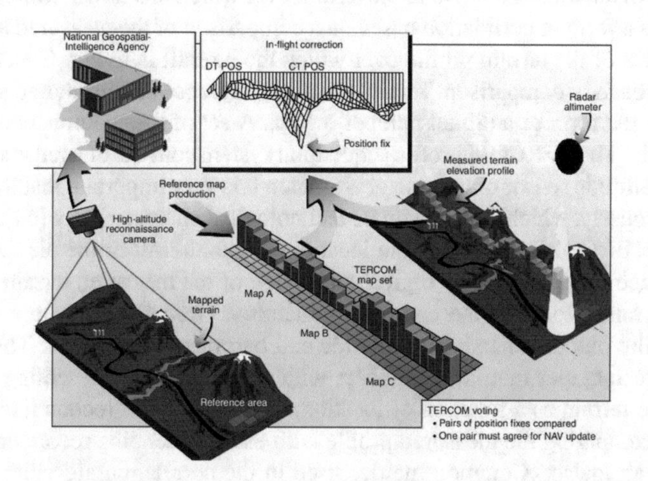

Fig. 10. The TERCOM operational concept [62]

Mission planning has become an important aspect of the TERCOM process. Absolute accuracy, however, is based on the accuracy of radar mapping information and the processor's ability to quickly compare altimeter data with the map as resolution increases. This limited TERCOM first-generation systems to several hundred meters fault. When the advancement of digitalization technology made it possible to store and process them on a computer, the DSMAC (Digital Scene Matching Area Correlator) method was used to guide the aircraft.

The development of Shuttle Radar Topography Mission maps has enabled the new development potential of TERCOM. This new TERCOM route planning capability was implemented in the Tomahawk planning system and started operating in 2007.

TERCOM was last used operationally to guide Tomahawk cruise missiles in 1998, but remains the optional navigation mode for all variations of current Tomahawk missiles, which is a problem as the risk of GPS interference is steadily increasing.

The advantage of the TERCOM system is its accuracy, which does not depend on the flight length. Their absolute accuracy is based on the accuracy of radar mapping information, which is usually within one meter and the processor's ability to quickly compare the altimeter and map data stored in the computer's memory [63].

The disadvantage of this system is the need to plan the entire route of movement and to place it in the memory of the computer, including the starting position. If a rocket launches from an unknown location or flies too far out of its trajectory, it will never fly over control positions and recorded maps. That way it will be "lost". In this case, the INS system can assist, allowing flying in the general area to repair or checkpoint. However, major mistakes cannot be corrected.

This makes the TERCOM system much less flexible than modern systems such as GPS, which can be set up to reach any location, from any location, without searching for pre-recorded data, which means they can be targeted just before launch.

Terrain contour monitoring (TERCOM) can be defined as a technique for determining the position of an aircraft relative to the terrain over which it is flown. More specifically, TERCOM is a form of correlation based on a comparison of the measured and recorded characteristics of the terrain profile over which the aircraft is flying. Generally, terrain height is the basis of comparison. Terrain elevation reference data are stored in the aircraft computer in the form of a digital number matrix. A set of these numbers describes the terrain profile. The TERCOM profile acquisition system consists of a radar altimeter and a reference altitude sensor or barometer altimeter. It is very important that the barometer functions properly, which depends on its technological characteristics [64].

In fact, TERCOM determines the location of the aircraft in the air relative to the terrain through which it flies by digital correlation of the measured terrain profile with the data. terrains stored in the computer's memory. Ground profiles are obtained by combining the outputs of radar for altitude and barometric altimeters. The TERCOM 'map' is a rectangular matrix of numbers with each number representing the average height of the terrain as a function of location. The TERCOM "reconciliation process" consists of comparing the measured profile with each descending record column in the reference map matrix. Common metrics used in the correlation algorithm are: mean-absolute difference or mean-square difference [65].

TERCOM is a stand-alone precision navigation system designed for airplanes, drones, cruise missiles, and the only repair system that works autonomously in a warm environment. It can be used at any time of day or night, in all weather, conditions of strong electronic interference, and low altitude flights.

In recent years, the TERPROM (TERRAIN PROFILE MATCHING) navigation system has been developed as part of TAN technology. It is a high precision computer navigation system that uses radar altimeter data and a high resolution digital map to determine the exact position of the aircraft.

To obtain the best position estimate, the TERCOM system uses an altimeter and a computer to correct measured terrain contours. During the summer, the radar altimeter first measures changes in the country profile. These measured variations are then digitized and entered into a correlator for comparison with in-memory data. The TERCOM system uses digital maps that are stored in the memory of the aircraft's computer. These maps consist of a series of numbered squares representing the change in ground elevation above sea level as a function of location. As the aircraft approaches the area for which there is a digital map in the computer, the altimeter provides elevation information above ground. In this way, the computer compares the received data with the data in memory and determines the actual path and orders autopilot correction if needed (see Fig. 11) [65].

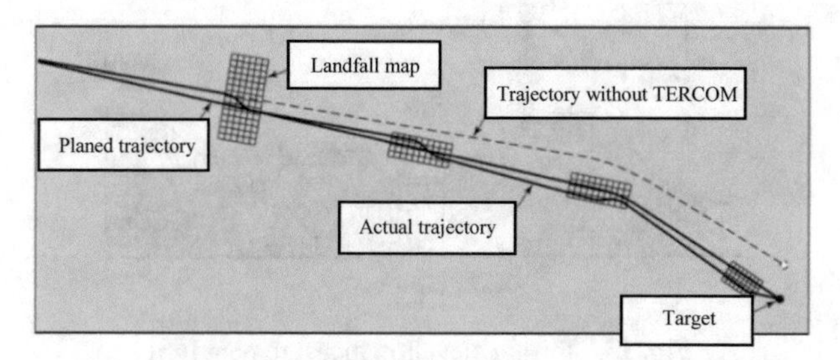

Fig. 11. TERCOM maps in us [65]

The determination of the position of the aircraft can be described in three phases: data preparation, data collection and data correlation, and Fig. 12 illustrates this concept.

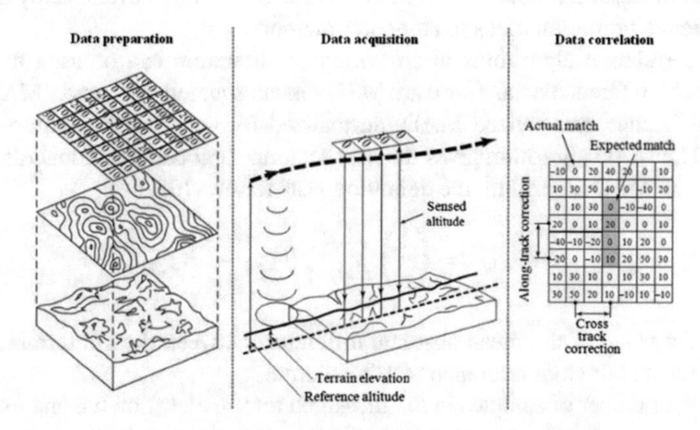

Fig. 12. TERCOM concept [65]

Source material in the form of maps or stereo photographs of the terrain is used to collect the set of heights that represent the reference matrix. The left side of the diagram shows the reference data loop and the right side describes the data collection loop. The radar altimeter provides altitude data. The output of the radar altimeter is different from the reference height of the system. Then various arithmetic operations (e.g. quantization etc.) are performed. Finally, the correlation of the data in the computer's memory and the data collected is done with the MAD (mean absolute difference) function, thus determining the position fixation.

As the aircraft flies over a given area, altitude information is obtained from the radar altimeter (Fig. 13). Radar altitude is measured at equal distances along the path, with at least one altitude measurement being made for each station traveled *d*. However, several measurements are usually made during the transition of each cell of the matrix, and the average of the measurements is kept as the measured radar height (distance from the terrain) for that cell.

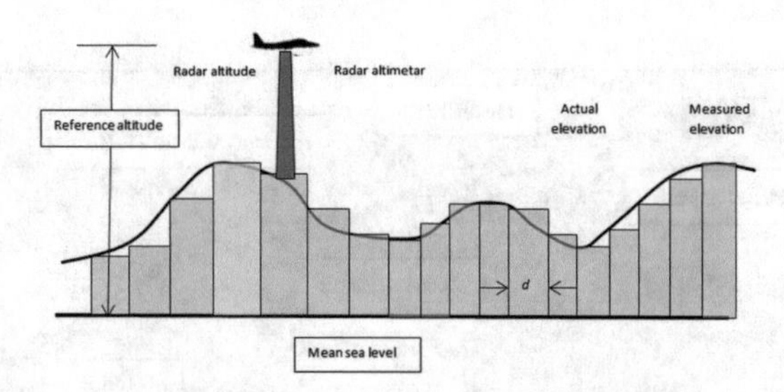

Fig. 13. Terrain elevation measurement [65]

The TERCOM procedure is based on the alignment of the measured terrain contour along the flight path of the aircraft with each column of the reference matrix housed in the memory of the aircraft computer before the flight. Given that the TERCOM system is noiseless, the terrain profile measured in flight will probably never exactly match one of the reference profile matrices in computer memory.

Many correlation algorithms in contemporary literature can be used to correlate measured with reference data. These are MSD - mean squared difference, MAD - mean absolute difference, normalized MAD, normalized MSD, of varying complexity and accuracy. The MAD algorithm gives the best accuracy and computational efficiency in real-time. The MAD algorithm, the definition is as follows [65, 66]:

$$MAD_{k,m} = \left(\frac{1}{n}\right) \sum\nolimits_{i=1}^{N} \left|h_{k,m} - H_{m,n}\right| \tag{4.1}$$

$MAD_{\text{k,m}}$ the value of the mean absolute difference between the kth terrain elevation file and the mth reference matrix column,

N the number of samples in the measured terrain elevation file and usually it is also equal to the number of rows in the reference matrix,

M the number of reference matrix columns,

K the number of measured terrain elevation files used in the correlation process (for the SSLM technique K=1),

$|\,|$ the absolute value of the argument,

n,m,k row, column, and terrain elevation file indices,
$H_{m,n}$ the stored reference matrix data: $1 \leq m \leq M,\ 1 \leq n \leq N,$
$h_{k,m}$ the kth measured terrain elevation file: $1 \leq k \leq K.$

The mathematical expression for the MSD algorithm is

$$MSD_{j,k} = \left(\frac{1}{n}\right)\sum\nolimits_{i=1}^{N}\left(S_{ij} - S_{ik}\right)^2 \tag{4.2}$$

$$S_{iS_{ij}}, S_k = j\text{th and } k\text{th profiles}$$
$$N = length\ of\ each\ profile$$

We can also express the uniformity of the MAD algorithm as in the expression for the MSD algorithm. Thus:

$$MAD_{k,m} = \left(\frac{1}{n}\right)\sum_{i=1}^{N}\left|S_{ij} - S_{ik}\right| \tag{4.3}$$

From (4.2) and (4.3), the ambiguity ξ between any two profiles is defined as the probability P and can be mathematically expressed:

$$\xi_{jk} = \begin{cases} P[C_{jk} < C_{jj}], \ here\ a\ minimum\ of\ C_{jk}\ is\ sought, \\ P[C_{jk} > C_{jj}], \ where\ a\ miaximum\ of\ C_{jk}\ is\ sought \end{cases}$$

For a MAD processor, C_{jk} is given by the following expression:

$$C_{jk} = \left(\frac{1}{N}\right)\sum_{i=1}^{N}\left|S_{ij} - R_{ik}\right|$$

Wherein:

$$S_j = j_{th} \text{ measured profile}$$

$$R_i = k_{th} \text{ measured profile}$$

Terrain correlation system required to update the position of the inertial navigation system for cruise aircraft. The concept of terrain correlation is shown in Fig. 14.

In this system, correlation data is a set of terrain height data. The array is obtained from a radar altimeter and a bar-temperature-inertial system that gives a reference mean altitude. These two measurements, from the radar altimeter and the mean altitude, are subtracted to obtain the variation of flight altitude. This new data is compared to the reference circuit in the aircraft computer. The reference height matrix is called the terrain correlation map. By calculating the best match of the measured and altitude data, the navigation system estimates its position when positioned above the center of the map and then updated [67].

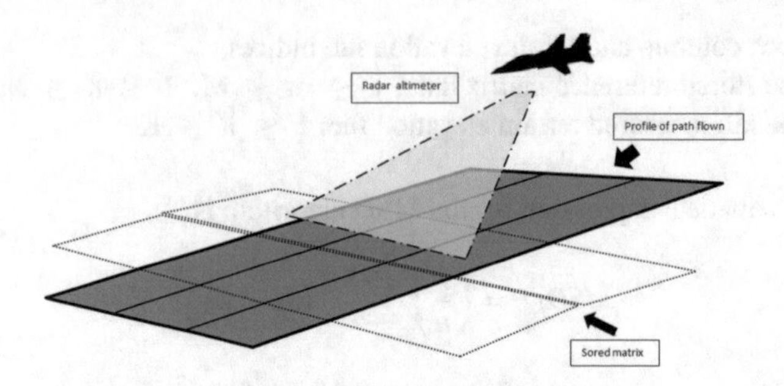

Fig. 14. Terrain correlation concept [65]

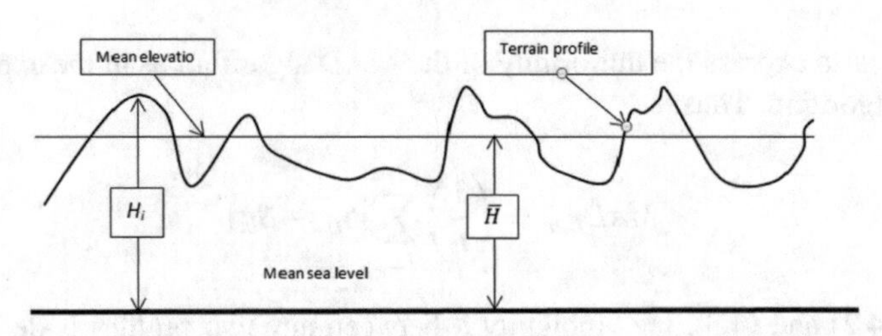

Fig. 15. Terrain standard deviation (σ_T) [65]

The factor used in selecting the update area is the roughness and uniqueness of the terrain. The TERCOM concept will not work on all types of terrain. For example, more diverse terrain is better than monotonous. However, good terrain must be unique i.e. TERCOM folders should not be similar. Terrain roughness is the standard deviation of terrain elevation patterns (see Fig. 15) and is commonly referred to as "*sigma-T*" or σ_T.

$$\sigma_T = \sqrt{\left(\frac{1}{N}\sum_{i=1}^{N}\left(H_i - \bar{H}\right)^2\right)} \tag{4.4}$$

Where $\bar{H} = \left(\frac{1}{N}\right)\sum_{i=1}^{N} H_i.$

Lakes and very flat areas have low *sigma-T* values. Therefore, they are not as suitable as fixing points. However, *sigma-T* is not the only criterion for determining the suitability of TERCOM operations.

There are three parameters used to describe the terrain-related to TERCOM, and their values can indicate the terrain's ability to support a successful TERCOM solution. These parameters are *sigma-T*, *sigma-Z* (σZ), and the X_T terrain correlation length.

The X_T correlation length represents the separation between two rows or columns of the terrain elevation matrix required to reduce their normalized autocorrelation function to the value of e^{-1}.

Sigma-Z is defined as the standard point-to-point deviation of pitch change as shown in Fig. 16. Like sigma-T, the sigma-Z value shows terrain roughness. *Sigma-Z* validly demonstrates the performance of TERCOM. The expression for *sigma-Z* is given by the equation:

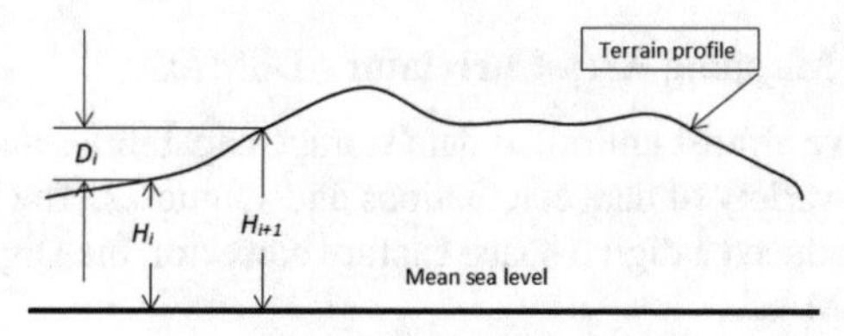

Fig. 16. Definition *sigma-Z* [65]

$$\sigma_z = \sqrt{\frac{1}{(N-1)} \sum_{i=1}^{N} (D_i - D)^2}$$

$$D_i = H_i - H_{i+1}, \ and\ D = \frac{1}{(N-1)} \sum_{i=1}^{N-1} D_i \tag{4.5}$$

Where,

$$D_i = H_i - H_{i+1}$$

$$D = \frac{1}{N-1} \sum_{i-1}^{N-1} D_i$$

The two parameters *sigma-T* and *sigma-Z* refer to the third parameter X_T according to the ratio:

$$\sigma_z^2 = 2\sigma_T^2[1 - exp\left(-\frac{d}{X_T}\right)^2$$

where d is the cell size (or distance between elevation samples.

With TERCOM we have vertical measurement errors that give erroneous height measurements and horizontal errors that induce vertical errors by causing terrain elevation measurements to move horizontally from the desired location.

Vertical measurement errors are due to inaccuracies in the original data, radar altimeter error, barometer altimeter error. Horizontal errors are due to horizontal velocity and slope errors, vertical height errors, horizontal quantization (i.e. cell size), etc.

The concept of using sensor data to correct body position began to be explored in the late 1950s. The goal of the terrain contour matching process is to determine the system's motion error, and the navigation system then uses the measured position error to update its estimate of the actual geographic position of the body. In such situations, a Kalman filter is usually used to reduce the error of the navigation system based on the measured error in body position.

The Kalman filter is used to reduce the speed of abandonment of the aircraft's inertial navigation system. It optimally estimates the internal errors in the inertial system (e.g. platform slope angles in the case of slope systems) based on the error measurement positions calculated from each correction of the terrain correlation position. These estimated internal errors are given to the inertial navigation system as negative feedback. In this way, errors in calculating the current position of the system can be minimized each time [5].

4.2 Digital Scene-Mapping Area Correlator - DSMAC

Modern systems have almost unlimited data storage capabilities and existing programs can perform a wide variety of data calculations and syntheses. The combination of new technologies has produced a digital space capture corrector, the Digital Scene Matching Area Correlator DSMAC.

Digital Scene Correction (DSMAC) is a technique that has proven to be very successful in navigating autonomous drones and is often combined with TERCOM as a guidance system, enabling very precise targeting.

To move accurately along a given path, the aircraft must determine and correct its position on the path. Some different techniques have been developed to accomplish this task. One of them is a technique of comparing optical images above the terrain with images of a reference map in computer memory, which was made during mission planning before take-off and used with TERCOM and DSMAC navigation systems (see Fig. 17).

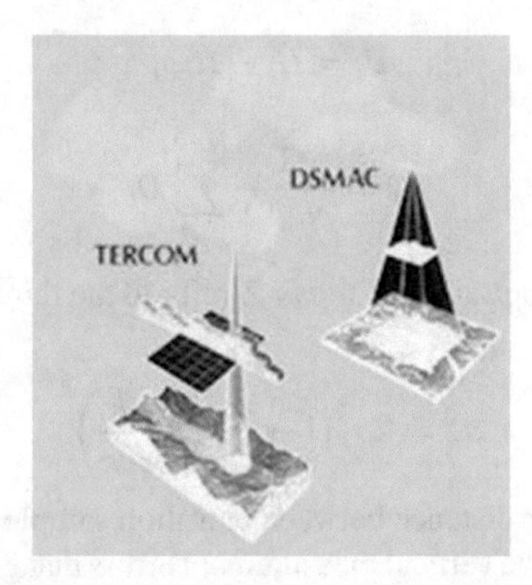

Fig. 17. TERCOM and DSMAC system navigation [68, 69]

The cruise aircraft belongs to the category of autonomously guided aircraft and flies the planned route, starting from launch to destination, guided by an inertial navigation system. Due to the initial errors in the inertial system, which increase during the flight, there is a high probability that the target will be missed if the flight errors are not corrected. To navigate a given path, the cruiser uses three systems to determine its position: terrain altitude matching with the terrain contour matching system (TERCOM), satellite positioning with the Global Positioning System (GPS), and optical scene display with digital scene corrector for DSCAC) [70].

The current operational concept of cruise aircraft consists of cruise and terminal phase i.e. getting to the target. The mission planning system specifies a route with defined control points along the route. Corrective guidance systems TERCOM and DSMAC are used along the flight route (see Fig. 18).

With respect to these systems, the DSMAC provides the most accurate positioning and is therefore used for combat operations to destroy so-called point targets with conventional warheads. In this way, with great precision capability, extensive destruction and collateral damage are avoided.

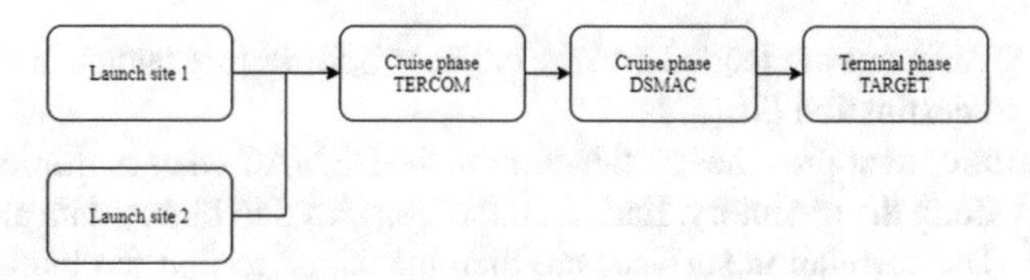

Fig. 18. The current operational concept [71]

DSMAC action planning has two phases. The first stage is to provide the data necessary to update the DSMAC in the summer. The second stage is the planning of the operation with the estimated probability of an accurate update to reach the target. In order to calculate the likelihood of an accurate update during mission planning, the impact of various factors on the performance of DSMACs must be determined. Various models are used for forecasting purposes depending on the time of day or season [72].

The design of the DSMAC system is shown in Fig. 19 with explanations of his work, along with notes on some factors that affect his reliable work.

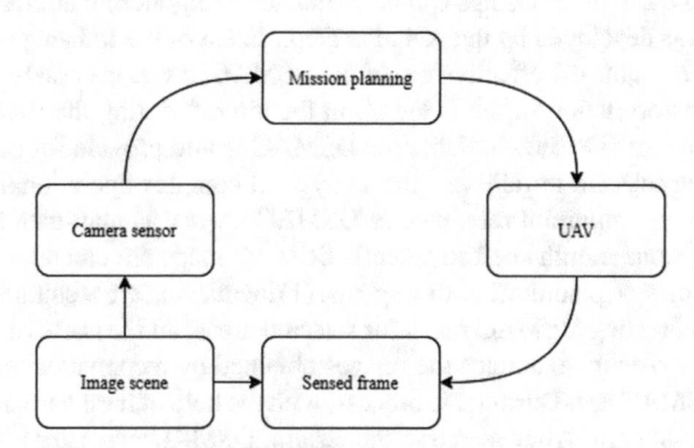

Fig. 19. DSMAC system design [70]

For the terrain selected for the DSMAC update, a gray-scale optical image is prepared, showing a high altitude view and geodetically positioned. That image is converted to a binary folder. Considering that field images are taken days or months before the mission, there are major differences in color, brightness, etc. However, using correlation and image processing, DSMAC determines the exact position and there are large differences between images taken during the flight and stored in the navigation system. In this regard, for the reliable and accurate functioning of the DSMAC, systems of optical modeling, image processing, and correlation modeling are very important.

The system uses analytical and mathematical modeling to reduce the delay in the calculations of image and model analysis. The DSMAC analyst identifies the characteristics in the scout images and predicts the DSMAC's reliability with the Digital Imagery Workstation Suite (DIWS) software and displays it as fake colors applied to the model's binary image. The analyst analyzes these reliability images and adjusts them to the conditions under which the binary reference map provides acceptable reliability. Mission data and DSMAC maps are loaded into the cruise missile before launch and inertial data from launch to destination [70].

When the aircraft approaches certain terrain, the DSMAC takes multiple low-altitude camera shots along the trajectory. Each frame is converted to binary data and correlated using a map. The correlation surfaces are then machined to find the likely position of the best correlation frame. After one or more successful position updates, a target is reached. For better storage of data, images, and prepared maps, digital processing and compression are used. Photogrammetric checkpoints in a geodetic database of stereo images allow DIWS to transform scouting images with the selected resolution.

Near the control terrain, each frame is filtered, converted to binary, and linked to the reference frame. Correlation surfaces are spatially connected series of numbers and images. When DSMAC receives mapped terrain images, the corresponding correlation surface has a single value for the true (matched) position and multiple erroneous ones. The DSMAC should identify the right position, pass it to the inertial navigation system, and reject the false positions.

The true correlation position is related to the flight altitude, speed and position of the aircraft and these relationships can be calculated using inertial aircraft data.

DSMAC was developed by the Aviation Department of the Indianapolis Naval Aviation Center. To enable the effective use of the DSMAC, it was necessary to develop the processing and correlation of the DSMAC in the aircraft during the flight, the prediction of performance (i.e., how reliable the DSMAC would provide for correct position updating for the selected terrain) and the analysis of complex operational models [71].

Time is a very important reference in DSMAC's work as map-data reference processing can be done months before takeoff. DSMAC maps are created during mission planning, and mission planning has three goals: (1) identifying areas suitable for DSMAC operations, (2) creating DSMAC maps for selected areas, and (3) assessing them [71].

The analyst visually examines the images obtained by reconnaissance to find areas suitable for DSMAC use. During this process, with the help of hardware, it improves the contrast and resolution. After selecting the terrain, DSMAC maps are made and based on images obtained from scouts but digitally processed. They have reduced resolution, are digitally filtered to remove local average brightness, and converted from multibit to single-bit (binary) (Fig. 20).

This is due to a reduction in the amount of data, which affects storage space and processing speed. Figure 20 shows a photo that has been processed to create a binary reference map on the left. The simulated frame and the corresponding binary are shown on the right. The green box above the reference map shows the position where this box matches the reference [72].

As the rocket gets each frame, the DSMAC unit correlates the frame with the reference map to determine where exactly the two images match. Correlation compares the frame at each possible location, counting the number of binary pixels that match between the frame and the map. The higher the correlation value, the more likely that position is the exact location of the frame within the map. The highest correlation value obtained is called the correlation peak. The smaller correlation peaks are called lateral layers and they represent noise. The mission planning predicts the probability that the true peak will be the largest, that is, larger than all the lateral ones [70].

Fig. 20. DSMAC operation [72]

The maximum correlation value may appear in the wrong position in the reference map, leading to an error in the measured position. To prevent this possibility, several frames are correlated with the map.

For mission planning, analysts use multiple computer algorithms to be able to perform a DSMAC reliability assessment for correctly updating the position of the aircraft for the selected model. These algorithms model terrains of various characteristics and their impact on DSMAC. Some models simulate calculations in the aircraft, other weather conditions, weather characteristics such as time of day, season, amount of leaves, etc. Instead of considering individual conditions, let us consider the analysis of errors in horizontal geometry.

The differences between the geometric properties of the image in the shot and the geometry of the corresponding DSMAC properties, the map, reduce the correct level of correlation. These geometric differences make scale and rotation errors, as well as a small translation error called a phase. The model is based on a theoretical result, which shows that small geometric errors affect the correlation as if the images were fi altered with low passages [70]. With further development, new models were obtained (see Fig. 21).

For further improvement of DSMAC, special software is being developed for the development and application of prediction algorithms in order to meet the requirement for reliable operation in a wide range of scenes and environments. These improvements require large financial investments for which the question of justification arises [73].

5 Observable Modes and Absolute Navigation Capability for Landmark-Based IMU/Vision Navigation System of UAV

In recent years, the use of drones has increased exponentially. Currently, most drones rely on GPS and Inretial Navigation System (GPS/INS) to obtain their navigation parameters (position, speed, etc.). However, GPS signals can be easily disrupted, i.e. GPS does not

Fig. 21. A gray-scale video frame from DSMAC during flight. (a) Original, uncorrected frame; (b) frame corrected to remove geometric errors [70].

work well in difficult environments such as urban areas, canyons, mountains, forests, etc. Also, the inertial system itself can accumulate error and thus be too sensitive to initial conditions.

To overcome these problems, new sensors such as a video system are being introduced. The navigation system based on the video system is cheaper, works in real time and has great flexibility.

This system is based on landmark recording and is not autonomous. When GPS is not available, landmarks that are sufficiently mapped can be used to estimate navigation parameters, i.e. absolute position, speed, etc. drones. However, this system is effective only when the GPS signal is occasionally available, so it is basically not an autonomous navigation system.

A group of authors [74] analyzed the absolute navigation of an unmanned aerial vehicle using an inertial and video system (INS/VNS) and proposed an analytical method for solving this problem.

They performed the analysis by establishing a mathematical model INS/VNS and forming a matrix obtained based on Lie derivatives. Observation modes with a different number of landmarks are obtained by solving the zero space of the observation matrix. Various analytical methods are used to obtain the navigation capabilities of a drone using a camera.

Their work shows an innovative calculation method for obtaining unique analytical solutions of navigation parameters from the visible mode, and the simulation results confirmed the accuracy of the stated analysis method. These methods can be extended to other systems, such as geomagnetic navigation systems, etc. (Fig. 22).

6 Intelligent GNSS/INS Integrated Navigation System for a Commercial UAV flight Control System

Commercial GPS positioning solutions are satisfactory for drones flying a hundred meters above the ground. However, this positioning is not accurate enough for safe movement in urban areas with a high density of tall buildings. One solution to this problem is to integrate GPS with INS based on a conventional Kalman filter, machine

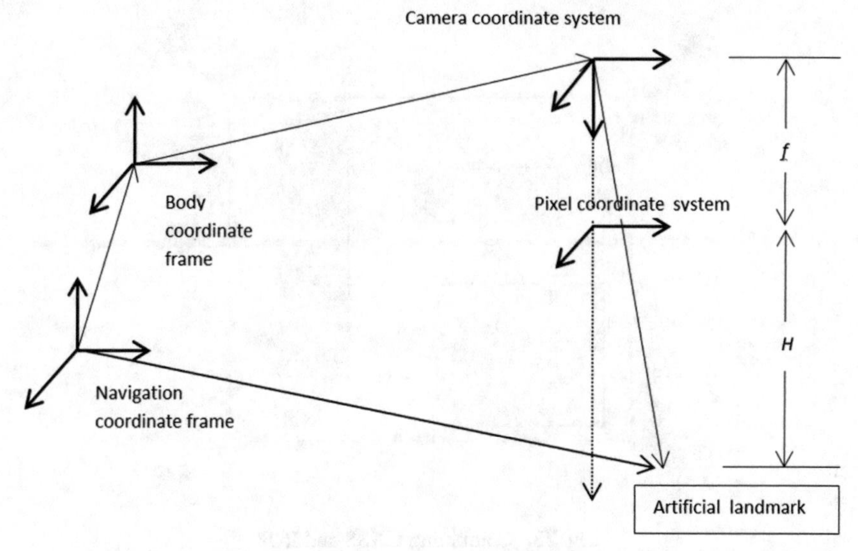

Fig. 22. Definitions of coordinate system in the IMU/Vision navigation system

learning algorithms, and positioning error models in the predicted area. In this way, the GPS error level is efficiently determined and the positioning of the drone is improved.

The group of authors [75] proposes an adaptive Kalman filter for adjusting the noise covariance of GPS measurements with different positioning accuracy as a solution. The authors rely on the interactive learning of the system itself following the data obtained using adequate algorithms and previously designed scenarios.

The results show that the proposed adaptive Kalman filter can achieve a better classification of GPS accuracy compared to others and improve positioning by 50%.

By combining GNSS and INS they improve each other and get a good navigation solution shown in Fig. 23. When the GNSS conditions are good, there is a field of view for several satellites, the GNSS receiver enables the position. When no satellites are seen, the INS determines the position and navigation until the conditions for using GNSS are created.

7 Attitude Estimation for Cooperating UAVs Based on Tight Integration of GNSS and Vision Measurements

Interest in the development of mini and micro drones has been on the rise in recent years, mainly due to advances in new technologies and their miniaturization. Despite the technological revolution, small UAVs are limited in terms of coverage, reliability, and performance. One of the solutions for determining the exact position of the aircraft is the installation of a dual GPS antenna and phase differential processing of the GPS signal.

A group of authors [76] considered a cooperative framework for improving UAV navigation performance, with an emphasis on position assessment. The proposed concept, differential GPS, and video system are tightly integrated with an extended Kalman

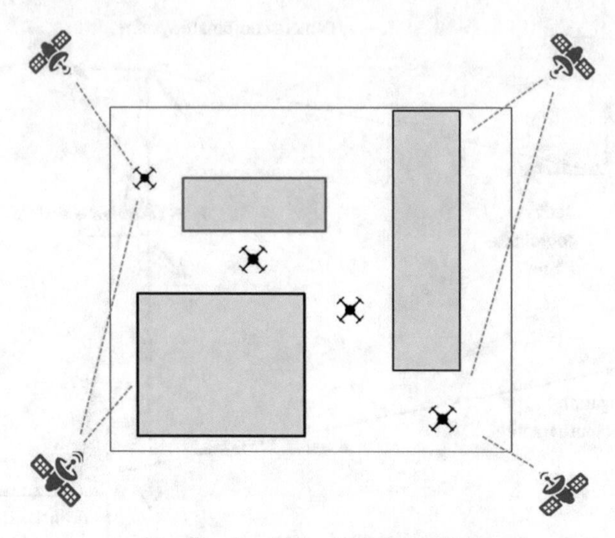

Fig. 23. Combining GNSS and INS

filter that receives measurement data from each drone individually. The approach in the work is based on data fusion that is possible for every unmanned aircraft equipped with a camera for tracking other aircraft, in order to increase the accuracy of its position via its GPS.

Ongoing research activities relate to the implementation of the proposed concept with the latest generation Micro-Electro-Mechanical System (MEMS) inertial system and differential GPS processing. This strategy is consistent with the latest technology-leading trends while reducing costs (Fig. 24).

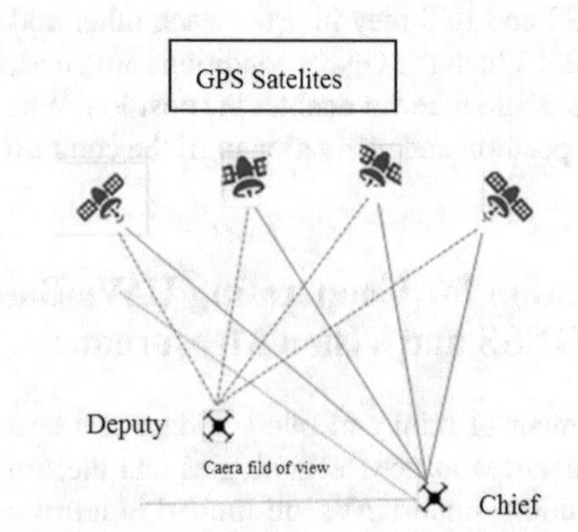

Fig. 24. Cooperating UAVs based on tight integration of GNSS and vision measurements

8 Combined Weightless Neural Network FPGA Architecture Surveillance and Visual Navigation of UAVs

To assess the position of the unmanned aerial vehicle and its navigation, binary images of the observed area are used, among other things, and they are often used for deforestation control and visual navigation. The designed models are estimated during the actual flight above the known environment (rural area) while the monitoring of deforestation is estimated using data (images) e.g. ozone forests.

The authors [77] present a combined weightless neural network architecture (WNS) and evaluate the use of WNS for deforestation control and the visual navigation of drones. The system has the task of processing the recorded data in memory with the actual data. The goal is autonomous visual navigation and monitoring of deforestation. The authors built systems with a programmable hardware implementation of the learning and classification system that make up the Field Programmable Gate Array (FPGA) and WNS.

Further development may include other control tasks, such as fires, burnt areas, and other problems that can be treated in the Boolean domain of neural systems (Fig. 25).

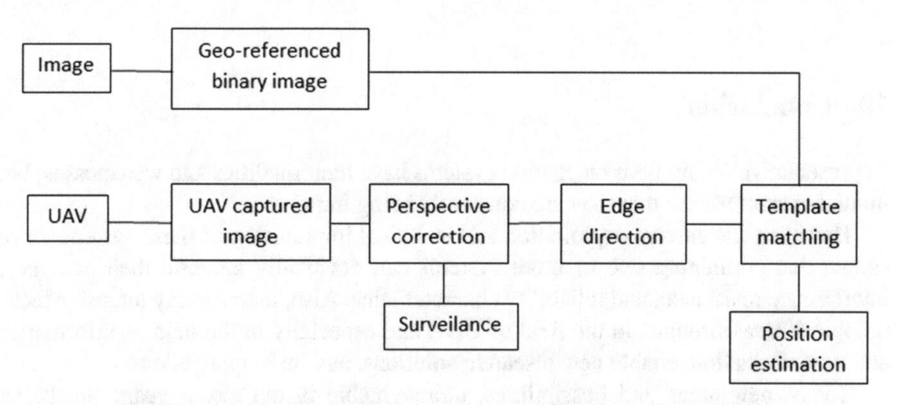

Fig. 25. Combined weightless neural network FPGA architecture [77]

9 The Vector Matching Method in Geomagnetic Aiding Navigation

Geomagnetic matching is a technological tool in navigation technology that can correct an inertial navigation system (INS) by comparing the geomagnetic profile read in flight with a geomagnetic map stored in computer memory. In this way, geomagnetic matching is an ideal choice for autonomous unmanned aerial vehicle navigation for long-term missions.

In that sense, a group of authors [78] developed a navigation method of geomagnetic matching that uses a geomagnetic vector and thus increases the accuracy of positioning. The vector geomagnetic matching method uses a multidimensional geomagnetic element to obtain stable and accurate positioning performance and represents a significant improvement over previous methods based on measuring the geomagnetic field of the vector.

The geomagnetic matching procedure uses all the information of the geomagnetic vector to increase accuracy. In the future, work is planned on several matching regions, a complete long-range route, etc. (Fig. 26).

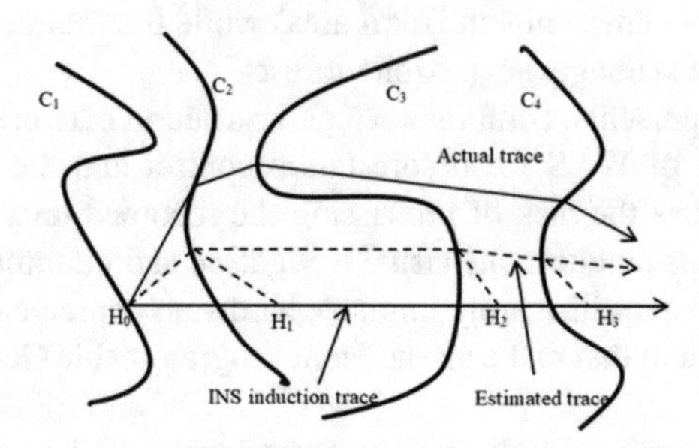

Fig. 26. Illustration of inertial navigation system (INS) indication trace, mEstimated trace, and Real trace [78].

10 Conclusion

Representatives of modern navigation systems have their qualities and weaknesses, but mutual characteristics that they are constantly being improved.

However, the current exploitation and technical imprecision of these systems have shown that combining two or more systems can drastically increase their accuracy, interference resistance, and reliability characteristics. Also, increasingly intensive technological breakthroughs in the field of UAV, and especially in the field of informatics and miniaturization, enable new research, solutions, and their application.

Today, new areas and possibilities, unimaginable in the recent years, are being explored. Artificial intelligence, modern computer technology, miniaturization, cybernetics, heuristics provide broad range of opportunities in the development and application of new ideas and solutions. The solutions are multidisciplinary and the use is diverse.

In the future, major breakthroughs can be expected in the field of artificial intelligence and all other systems in the domain of UAV (optical, navigation, etc.). In that way, development of autonomous systems of this kind with integrated artificial intelligence will enable more complex use in various environments.

References

1. Hofman-Wellenhof, B., Legat, K., Wieser, M.: Navigation: Principles of Positioning and Guidances, pp. 5–6. Springer, Heidelberg (2007). ISBN 978-3-211-00828-7
2. Sobel, D.: Longitude: The True Story of a Lone Genius Who Solved the Greatest Scientific Problem of His Time. Penguin Books, London (1996)

3. Masten, M.K., Stockum, L.A.: Precision stabilization and tracking systems for acquisition, pointing, and control applications. In: Society of Photo-Optical Instrumentation Engineers (SPIE) (1996)
4. Kerry, G.: The multi-state Kalman Filter in medical monitoring. Comput. Methods Programs Biomed. **23**(2), 147–154 (1986)
5. Lin, C.-F.: Modem Navigation, Guidance, and Control Processing. Prentice Hall, Englewood Cliffs (1991)
6. Prasad, R., Ruggieri, M.: Applied Satellite Navigation Using GPS, GALILEO, and Augmentation Systems. Artech House, Norwood (2005)
7. Kaplan, E.D., Christopher, J.H.: Ch. 1–4 Understanding GPS, 2nd edn. Artech House, London (2006)
8. Parkinson, B.W., Spilker Jr., J.J. (eds.): Global Positioning System: Theory and Applications, Progress in Astronautics and Aeronautics, vols. 163 and 164. American Institute of Aeronautics and Astronautics (1996)
9. Kaplan, E.D.: Understanding GPS: Principles and Applications. Artech House, Norwood (1996)
10. Pace, S., Frost, G.P., Lachow, I., Frelinger, D.R., Fossum, D., Wassem, D., Pinto, M.M.: The global positioning system assessing national policies, GPS History. Chronology, and Budgets, Appendix A and Appendix B (1995)
11. El-Rabbany, A.: Introduction to GPS the Global Positioning System. Artech House, Norwood (2002)
12. Tsui, J.B.-Y.: Fundamentals of Global Positioning System Receivers: A Software Approach. Wiley, New York (2000)
13. Gaposchkin, P.: Reference coordinate systems for earth dynamics. In: Proceedings of the 56th Colloquium of the International Astronomical Union, Warsaw, Poland, July 1981
14. Minkler, G., Minkler, J.: Aerospace Coordinate Systems and Transformations, Adelaide. Magellan Book Co., Australia (1990)
15. Wolper, J.S.: Understanding Mathematics for Aircraft Navigation. McGraw-Hill, New York (2001)
16. Sudano, J.J.: An exact conversion from an earth-centered coordinate system to latitude, longitude and altitude. In: Proceedings of the IEEE 1997 National Aerospace and Electronics Conference, NAECON 1997, 14–17 July 1997, vol. 2, pp. 646–650 (1997)
17. Maling, D.H.: Coordinate Systems and Map Projections, 2nd edn. Pergamon Press, New York (1992)
18. https://gisgeography.com/trilateration-triangulation-gps/. Accessed May 2020
19. Abel, J.S., Chaffee, J.W.: Existence and uniqueness of GPS solutions. IEEE Trans. Aerosp. Electron. Syst. **27**(6), 952–956 (1991)
20. Fang, B.T.: Comments on 'existence and uniqueness of GPS solutions' by Abeland, J.S., Chaffee, J.W. IEEE Trans. Aerosp. Electron. Syst. **28**(4), 1163 (1992)
21. Phatak, M., Chansarkar, M., Kohli, S.: Position fix from three GPS satellites and altitude: a direct method. IEEE Trans. Aerosp. Electron. Syst. **35**(1), 350–354 (1999)
22. Hoshen, J.: On the Apollonius solutions to the GPS equations. In: AFRICON 1999, 28 September–1 October 1999, vol. 1, pp. 99–102. IEEE (1999)
23. Leva, J.L.: An alternative closed-form solution to the GPS pseudorange equations. IEEE Trans. Aerosp. Electron. Syst. **32**(4), 1430–1439 (1996)
24. Chaffee, J., Abel, J.: On the exact solutions of pseudorange equations. IEEE Trans. Aerosp. Electron. Syst. **30**(4), 1021–1030 (1994)
25. Hassibi, B., Vikalo, H.: On the expected complexity of integer least-squares problems. In: IEEE International Conference on Acoustics, Speech, and Signal Processing, vol. 2, pp. 1497–1500 (2002)

26. Hassibi, A., Boyd, S.: Integer parameter estimation in linear models with applications to GPS. IEEE Trans. Signal Process. **46**(11), 2938–2952 (1998)
27. Abel, J.S.: A divide and conquer approach to least-squares estimation with application to range-difference-based localization. In: International Conference on Acoustics, Speech, and Signal Processing, 23–26 May 1989, vol. 4, pp. 2144–2147 (1989)
28. Peng, H.M., et al.: Maximum-likelihood-based filtering for attitude determination via GPS carrier phase. In: IEEE Position Location and Navigation Symposium, 13–16 March 2000, pp. 480–487 (2000)
29. Hassibi, B., Vikalo, H.: On the expected complexity of sphere decoding. In: Conference Record of the Thirty-Fifth Asilomar Conference on Signals, Systems and Computers, 4–7 November 2001, vol. 2, pp. 1051–1055 (2001)
30. Chaffee, J.W., Abel, J.S.: The GPS Filtering Problem. In: IEEE PLANS 1992 Position Location and Navigation Symposium, Record 500 Years After Columbus—Navigation Challenges of Tomorrow, 23–27 March 1992, pp. 12–20 (1992)
31. Ponomaryov, V.I., et al.: Increasing the accuracy of differential global positioning system by means of use the Kalman filtering technique. In: Proceedings of the 2000 IEEE International Symposium on Industrial Electronics, 4–8 December 2000, vol. 2, pp. 637–642 (2000)
32. Mao, X., Wada, M., Hashimoto, H.: Investigation on nonlinear filtering algorithms for GPS. In: IEEE Intelligent Vehicle Symposium, 17–21 June 2002, vol. 1, pp. 64–70 (2002)
33. Mao, X., Wada, M., Hashimoto, H.: Nonlinear filtering algorithms for GPS using pseudorange and Doppler shift measurements. In: Proceedings of the 5th IEEE International Conference on Intelligent Transportation Systems, Singapore, pp. 914–919 (2002)
34. Wu, S.C., Melbourne, W.G.: An optimal GPS data processing technique for precise positioning. IEEE Trans. Geosci. Remote Sens. **31**(1), 146–152 (1993)
35. http://www.mountainsafety.co.uk/How-The-GPS-System-Works.aspx. Accessed May 2020
36. Hofmann-Wellenhof, B., Lichtenegger, H., Collins, J.: GPS: Theory and Practice. Springer, Vienna (1997)
37. Logsdon, T.: The NAVSTAR Global Positioning System, pp. 1–90. Van Nostrand Reinhold, New York (1992)
38. Altamimi, Z., Sillard, P., Boucher, C.: ITRF2000: a new release of the International Terrestrial Reference Frame for earth science applications. J. Geophys. Res. **107**(B10), 2214 (2002)
39. McCarthy, D. (ed.): IERS Conventions, IERS Technical Note No. 21, U.S. Naval Observatory, July 1996. http://www.maia.usno.navy.mil/conventions.html
40. Swift, E.R.: Improved WGS 84 coordinates for the DMA and air force GPS tracking sites. In: Proceedings of ION GPS-94, Salt Lake City, UT, September 1994
41. Cunningham, J., Curtis, V.L.: WGS 84 Coordinate Validation and Improvement for the NIMA and Air Force GPS Tracking Stations, NSWCDD/TR-96/201, November 1996
42. Malys, S., Slater, J.A.: Maintenance and enhancement of the world geodetic system 1984. In: Proceedings of ION GPS-94, Salt Lake City, UT, September 1994
43. Merrigan, M.J., et al.: A refinement to the world geodetic system 1984 reference frame. In: Proceedings of ION GPS-2002, The Institute of Navigation, Portland, OR, September 2002
44. True, S.A.: Planning the future of the world geodetic system 1984. In: Proceedings of the Position, Location and Navigation Symposium, Monterey, CA, 26–29 April 2004 (2004)
45. Conventions on International Civil Aviation, Annex 15: Aeronautical Information Services, International Civil Aviation Organization, Montreal, ICAO (2003)
46. World Geodetic System – 1984 (WGS-84) Manual, Doc. 9674, 2nd ed., International Civil Aviation Organization (2002)
47. Newton, I.: Newton's Principia: The Mathematical Principles of Natural Philosophy. Daniel Adee, New York (1846)

48. Jovanović, A.M., Simonović, A.M., Zorić, N.D., Lukić, N.S., Stupar, S.N., Ilić, S.S.: Experimental studies on active vibration control of smart plate using a modified PID controller with optimal orientation of piezoelectric actuator. Smart Mater. Struct. **22**(11), 115038 (2013)
49. Zorić, N.D., Simonović, A.M., Mitrović, Z.S., Stupar, S.N., Obradović, A.M., Lukić, N.S.: Free vibration control of smart composite beams using particle swarm optimized self-tuning fuzzy logic controller. J. Sound Vib. **333**(21), 5244–5268 (2014)
50. Tomović, A., Šalinić, S., Obradović, A., Grbović, A., Milovančević, M.: Closed-form solution for the free axial-bending vibration problem of structures composed of rigid bodies and elastic beam segments. Appl. Math. Model. **77**(2), 1148–1167 (2020)
51. Shukla, K.S., Talpin, J.: Synthesis of Embedded Software: Frameworks and Methodologies for Correctness by Construction, p. 62. Springer, Heidelberg (2010). ISBN 978-1-4419-6400-7
52. Titterton, H.D., Weston, L.J.: Strapdown Inertial Navigation Technology, 2nd edn. The Institution of Electrical Engineers (2004)
53. Grovess, P.D.: Principles of GNSS, Inertial, and Multisensor Integrated Navigation System (2008)
54. Walter, E., Pronzato, L.: Identification of Parametric Models from Experimental Data. Springer, London (1997)
55. Kalman, R.E.: A new approach to linear filtering and prediction problems. J. Basic Eng. **82**, 35–45 (1960). https://doi.org/10.1115/1.3662552
56. Bolstad, W.M.: Introduction to Bayesian Statistics, 2nd edn. Wiley, Hoboken (2007). ISBN 0-471-27020-2
57. Milić, T., Ivanović, B.: Kalmanov filter i primene u finansijama, Vremenske serije i primene u finansijama, Matematički fakultet, Univerzitet u Beogradu (2017)
58. Zouaghi, L., Alexopolous, A., Koslowski, M., Kandil, A., Badreddin, E.: An integrated distributed monitoring for mission-based systems: on the example of an autonomous un manned helicopter. In: 6th IEEE International Conference on Intelligent Systems, pp. 415–420, September 2012
59. Kortunov, V.I., Dybska, I.Y., Proskura, G.A., Kravchuk, A.S.: Integrated mini INS based on MEMS sensors for UAV control. IEEE Aerosp. Electron. Syst. Mag. **24**(1), 41–43 (2009)
60. Johnson, N., Tang, W., Howell, G.: Terrain aided navigation using maximum a posteriori estimation. In: IEEE Position Location and Navigation Symposium (1990)
61. Metzger, J., Wendel, J., Trommer, G.F.: Hybrid terrain referenced navigation system using a Bank of Kalman filters and a comparison technique. In: AIAA, Guidance, Navigation, and Control Conference and Exhibit, Providence, Rhode Island, August 2004
62. Riedel, F.W., Hall, S.M., Barton, J.D., Christ, J.P., Funk, B.K., Milnes, T.D., Neperud, P.E., Star, D.R.: Guidance and navigation in the global engagement department. Johns Hopkins APL Tech. Digest **29**(2), 118–132 (2010)
63. Titterton, D.H., Weston, J.L.: Strapdown Inertial Navigation Technology, 2nd edn. The Institution of Electrical Engineers (2004)
64. Perić, B., Simonović, A., Ivanov, T., Stupar, S., Vorkapić, M., Peković, O., Svorcan, J.: Design and testing characteristics of thin stainless steel diaphragms. Procedia Struct. Integrity **13**, 2196–2201 (2018)
65. Siouris, G.M.: Missile Guidance and Control Systems. Springer, New York (2004)
66. Mobley, M.D., Brown, J.I.: Impact of terrain correlation elevation reference data on Boeing's air launched cruise missile. Institute of navigation National Meeting, Dayton, Ohio, pp. 108–112, March 1980
67. Mobley, M.D.: Air launched cruise missile (ALCM) navigation system development integration test, pp. 1248–1254 (1978)
68. https://fas.org/man/dod-101/sys/smart/bgm-109.htm. Accessed May 2020
69. https://slideplayer.com/slide/10587447/. Accessed May 2020

70. Irani, G.B., Christ, J.P.: Image processing for Tomahawk scene matching. Johns Hopkins APL Tech. Digest **15**(3), 250–264 (1994)
71. Hatch, R.R., Luber, J.L., Walker, J.H.: Fifty years of strike warfare research at the applied physics laboratory. Johns Hopkins APL Tech. Digest **13**(I), 1–8 (1992)
72. Mostafavi, H., Smith, F.W.: Image correlation with geometric distortion part I: acquisition performance. IEEE Trans. Aerosp. Electron. Syst. AES **14**(3), 487–493 (1978)
73. Popović, V.M., Vasić, B.M., Lazović, T.M., Grbović, A.M.: Application of new decision making model based on modified cost-benefit analysis - a case study: Belgrade Tramway transit. Asia-Pacific J. Oper. Res. **29**(06), 1250034 (2012)
74. Huanga, L., Songa, J., Zhanga, C., Caib, G.: Observable modes and absolute navigation capability for landmark-based IMU/Vision Navigation System of UAV. Optik **202**, 163725 (2019)
75. Zhang, G., Hsu, L.-T.: Intelligent GNSS/INS integrated navigation system for a commercial UAV flight control system. Aerosp. Sci. Technol. **80** 368–380 (2018)
76. Vetrella, A.R., Fasano, G., Accardo, D.: Attitude estimation for cooperating UAVs based on tight integration of GNSS and vision measurements. Aerosp. Sci. Technol. **84**, 966–970 (2019)
77. Torresa, V.A.M.F., Jaimesa, B.R.A., Ribeiroa, E.S., Bragaa, M.T., Shiguemor-ic, E.H., Velhod, H.F.C., Torresa, L.C.B., Bragaa, A.P.: Combined weightless neural network FPGA architecture for deforestation surveillance and visual navigation of UAVs. Eng. Appl. Artif. Intell. **87**, 103227 (2020)
78. Song, Z., Zhang, J., Zhu, W., Xi, X.: The vector matching method in geomagnetic aiding navigation. Sensors **16**(7), 1120 (2016)

Experimental and Numerical Analysis of Stress-Strain Field of the Modelled Boiler Element

Milena Rajić[1](✉), Dragoljub Živković[1], Milan Banić[1], Marko Mančić[1], Miloš Milošević[2], Taško Maneski[3], and Nenad Mitrović[3]

[1] Faculty of Mechanical Engineering, University of Nis, Nis, Serbia
milena.rajic@masfak.ni.ac.rs
[2] Innovation Center of Faculty of Mechanical Engineering, University of Belgrade, Belgrade, Serbia
[3] Faculty of Mechanical Engineering, University of Belgrade, Belgrade, Serbia

Abstract. Hot water boilers are most commonly used devices in thermal and industrial plants for heat production. Boilers elements are exposed to high pressures and temperatures during boiler operation. Non-stationary working regimes of boiler may lead to breakdowns and fatigue of boiler elements. There are numerous cases of breakdowns of such units, which lead to reduced reliability and safety of the plant. In order to avoid this state, certain investigation should be made that would examine the influence of different regimes on the boiler structure. Lack of available data concerning the temperature as well as stress-strain field in boiler elements led us to perform such kind of experiment. This paper presents results of numerical analysis performed on the modelled boiler. Validation of the numerical model was performed based on experimental results obtained using DIC (Digital Image Correlation) method by using Aramis system. The character of parameters change, such as strain and stress, occurring in the critical zones of boiler elements can be verified both by experimental and numerical results. In this paper the novel approach of experimental and numerical analysis is presented. The method can be used and conducted in similar units for testing on different loads or working regimes, as well as to improve the safety and reliability of similar pressure vessel units.

Keywords: Hot water boiler · Finite element method · Digital Image Correlation

1 Introduction

Hot water boilers, especially boilers with fire tube where the combustion takes place, are widely used in process industry. Exploitation experience indicates constant and permanent breakdowns occurring as a result of accidental states of individual boiler elements. Studies done before in this field indicate certain critical boiler elements that may have influence on the reliability and safety of the entire unit [1–6].

N. Mitrovic et al. (Eds.): CNNTech 2020, LNNS 153, pp. 257–273, 2021.
https://doi.org/10.1007/978-3-030-58362-0_15

One of the most critical elements is certainly the tube plate, located on the first and second flue reversing chamber. Design calculations of boiler elements are done by using existing norms and standards that impose certain safety factors for stress calculations that is allowed in nominal regime [7–13]. Very often a large value of safety factor can lead to over-dimensions of elements thickness, which not only can increase design costs but also to have an unfavourable impact to the strength of the structure. The paradox in this case is that, when the wall of the tube plate is wider, the lower are the stresses caused by pressure, but the thermal stresses are higher under the same parameters. Then, the wall thickness should be chosen to minimize the total cumulative load - pressure load and thermal load [14]. It should not be allowed to have any non-elastic deformation in these barring element constructions.

Exact application of analytical elasticity method is not possible due to the complex structure of the boiler elements. For boiler structure calculation, Finite Element Method FEM is already used in numerous studies [7–12, 14–21]. FEM represents a numerical method that enables modeling and calculation of complex constructions and problems by dividing the structure into a numerous finite elements with correct geometric form, whose behaviour is possible to describe.

Published studies done in this field, with this specific unit, indicate the lack of available data on measurements conducted on boiler itself or boiler elements. Stress-Strain analysis of hot water boiler done by FEM [7–12, 14–21] are based on data given in standards or in manufacturer documentation, but present general averaged data. There is no evidence of experimental data that can be used to determine the construction temperature field of the boiler's element, neither stress-strain field measured in any part in this type of boiler. One of the limitations of experimental procedure is high cost of experiment that should be performed. Boilers are high expensive pressure vessel unit, and therefore experiments are limited to certain part of the boiler that can be easily repaired and not endangered the essential part of the unit. Not neglecting that, this type of boilers operate in high temperature and pressure regime, therefore security and safety level of the plant and also the personnel involved in experimental procedure is also one of the limitations. One of the reasons for not having this kind of experiments is the limitation of experimental equipment and quality data that would be collected. Conventional methods of measuring temperature (or stress/strains) filed of boiler construction would give us the data in singular points. The distribution of stress field would be also averaged.

Having in mind all the presented limitations, the new approach of an experimental method has been introduced. The suitable model of boiler construction was made, maintaining the rigidity of the elements in the real boiler unit (that are exposed to the real operating loads) and the model (with the maximum allowed loads that can be simulated in the laboratory conditions). The presented method included the laboratory testing where several measurement procedures can be conducted in different loads, measurement conditions etc. The performed measurement includes: temperature measurement of critical part of the model by using thermocouple, strain measurement of the same critical part by strain gauge, the 3D Digital Image Correlation (DIC) method for stress-strain measurement. The numerical calculation of the model with all real time parameters data was performed and presented. Considering a large amount of data that is obtained in this experimental procedure, only the critical parts of the tube plate (as one of the most critical part of this type of boilers) in critical regimes are presented.

The construction model testing or its structure is of high importance especially in cases of prototype testing in solving the constructions design problems. Using the method of experimental analysis, a number of relevant data are usually obtained for assessing the load capacity and stability of the structure. Particularly important are the data related to the effect of local stress concentrations, determining the area of plastic deformation, the fracture mechanism, as well as the impact of mechanical characteristics changes depending on time and temperature. Experimental methods are of the highest importance, especially in those cases where due to the complexity of the problem the theoretical model is almost impossible or very difficult to solve. Experimental methods are particularly important in cases where the significant stress concentrations on the construction occur, which in operating conditions very often lead to the failures and breakdowns.

The 3D Digital Image Correlation (DIC) method represents the novel method of stress-strain measurement [22–34]. It enables full-field stress-strain measurements. During one measurement procedure big datasets are obtained that in conventional experimental method would replace a large number of strain gauges and therefore significantly reduces experimental planning time and preparation and also costs. Also numerical methods such as finite element method enables results of full covered stress-strain field, that can be easily verified by the data obtained from experiment. The main idea in this study was to link the numerical experiment of the model within finite element method to experimental measurement and to have not only the data that is verified by using different measurement methods, but also the interconnection between the numerical and real model testing that can be used also for similar units with the same load nature.

This paper analyzes critical elements of the hot water boiler by experimental data obtained from its model and improved numerical model in order to analyze constrains and loads the model is exposed to and to have fully-covered stress-strain field obtained from experimental data.

2 Design Parameters and Geometrical Data

In order to perform the experimental laboratory study, the suitable model for testing was design. As it is mentioned before, real boiler unit is huge, and as such it is not able to fit the laboratory. It was necessary to scale the object to the appropriate dimensions of the model, which could easily be installed in laboratory. The aim of the model forming (design) was to preserve the rigidity of the model and the real object, since stress-strain analysis of tube plate would be examined under different conditions. All dimensions were scaled 10 times, including the thickness of the tube plates. The model includes the plain tube as the fire tube, instead of the Fox corrugated flame tube, with the appropriate thickness so that the rigidity of that element would be identical in the conditions of the real object. Due to complexity of the layout and the number of flue pipes of the real object, in which there are 208 flue pipes, another approximation was made. Namely, flue pipes scaled 10 times, would give an extremely dense pipe network in the model, which is impossible to provide on the market because of its dimensions. This is the reason where 45 "full tubes" or rods, distributed on the tube plate of the model, were adopted, following the schedule and distribution of the flue pipes on the real object. The dimensions of these rods will not affect the rigidity of the structures, which means that it would remain identical or substantially the same in the already defined critical points of the model and the real object. Geometric model is presented in Fig. 1.

3 Numerical Analysis of the Model

The geometric model was transformed into the discretized FE model with the application of advanced meshing tool capable of creating adaptive discrete model. The discretized model consisted of 438128 nodes, which formed 352921 finite elements. The discretized model of the analyzed boiler is presented in Fig. 2. The finite element mesh has an identical topology for thermal and structural analysis, but these meshes are formed by different types of finite elements. The resulting meshes are automatically unified into a unique hybrid finite element mesh in which the definite topology overlaps the finite elements relevant to individual analyzes. In the analysis, the materials that are used (which correspond to the real boiler construction) have the characteristics [35, 36]. Working loads were also defined, which were simulated in laboratory conditions. By analyzing the structural load of the design model, the loads can be determined as:

1. loads resulting from the weight of the model's elements;
2. loads resulting from the thermal expansion on higher temperatures;
3. real boiler operating loads that should have been simulated.

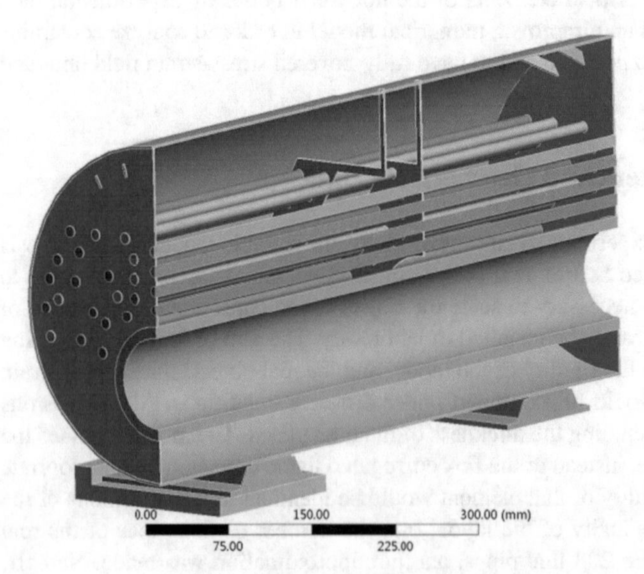

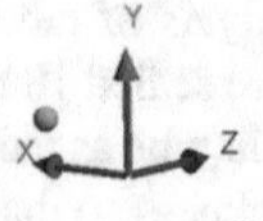

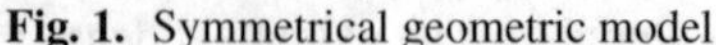

Fig. 1. Symmetrical geometric model

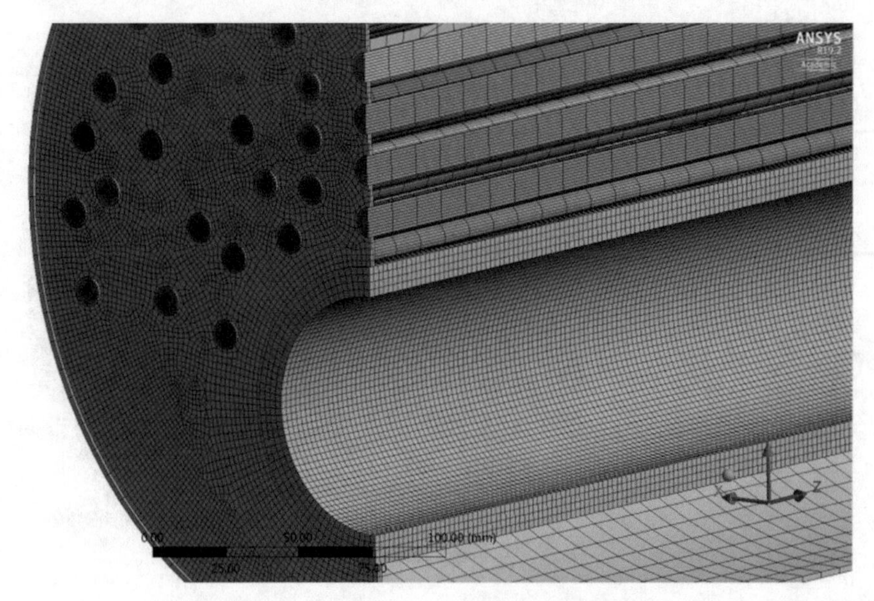

Fig. 2. Numerical model of discretized structure of the model

Beside the thermal and hydraulic pressure loads, other mechanical loads were also accounted during the static structural analysis: the gravity force and the hydrostatic water pressure.

Below (Fig. 3 and 4) the equivalent (Von Misses) stresses of the tube plate element are presented. In Fig. 3 the model was exposed to 5 bar pressure on the water side and without any source of heating (cold phase) and in Fig. 4 the model was under 5 bar pressure on the water side and with thermal load that was simulated with heaters inside the fire tube. The results of analysis show that the maximal stresses occur on the upper part of the tube plate in the case without heating (Fig. 3) and on the part of the tube plate closer to the fire tube outlet (Fig. 4). These are the critical locations, where the particular attention was made. Therefore, on these locations was the main focus in presenting the following experimental results. It should be noted that the boiler, as real object, shows specific sensibility in exactly those specific locations. As is was previously mentioned, one of the critical elements of the boilers is the tube plate and especially the part that is near the fire tube outlet, where combustion takes place and where the maximum temperatures are and where accidents often occur. The major breakdowns of the hot water fire-tube boilers are caused by the damage and failures of the welding joints in these elements [37, 38].

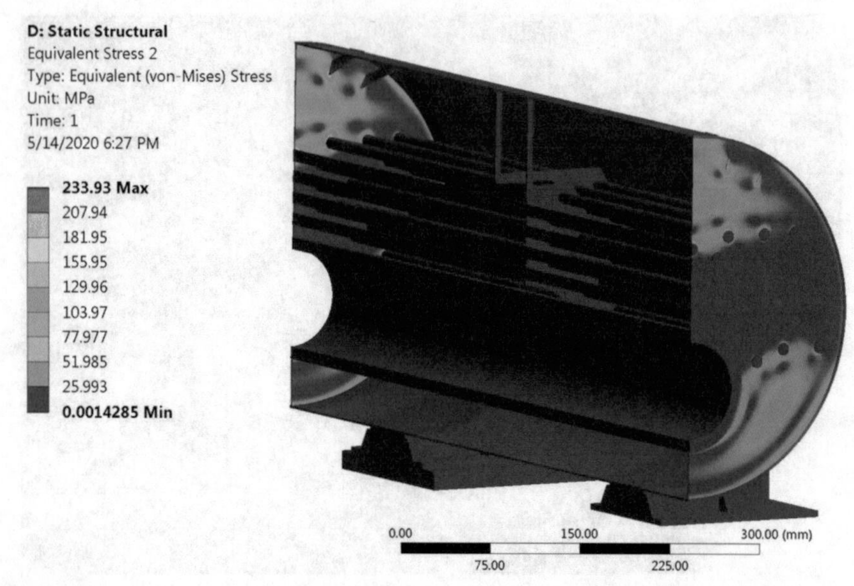

Fig. 3. Equivalent (von-Mises) stresses of the model exposed to 5 bar pressure on the water side-cold state

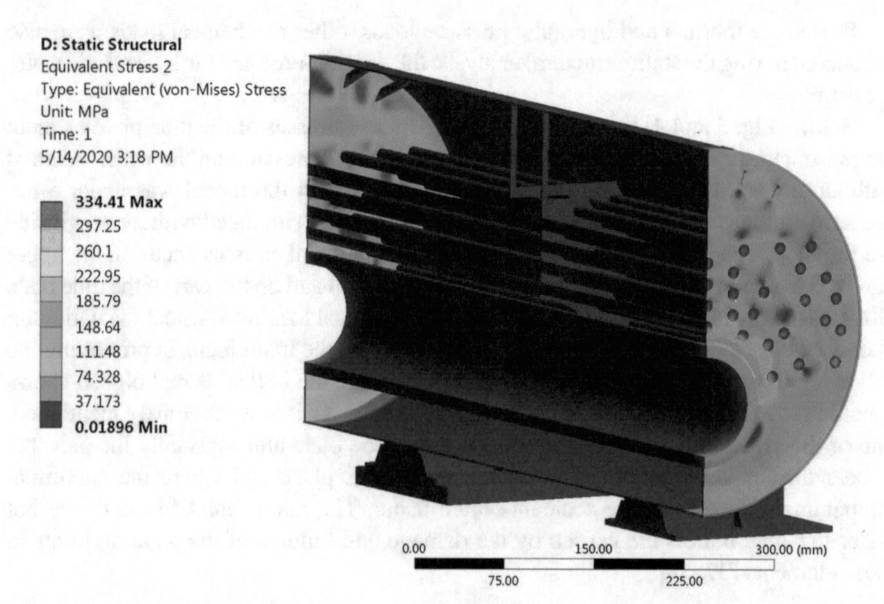

Fig. 4. Equivalent (von-Mises) stresses of the model exposed to 5 bar pressure on the water and with heaters inside the fire tube

4 Experimental Procedure

The aim of experimental testing was to verify the numerical calculation of the model with physically installed model in the laboratory. In this manner, reliability and accuracy of the numerical calculation of the defined model with the given loads would be presented. The similarity of the appearance and character of the change in physical variables (stress-strain) is expected by comparing the results of the experiment on the model and the real object.

The experimental methods used are strain gauge method and methods for contactless measurement of stress and deformation (DIC). The given load is symmetric, so that it allows (along with the already existing symmetry of the construction itself) parallel measurement of the strain gauge and the Digital Image Correlation (DIC) system. Two strain gauges are placed at the selected points on the same half of the structure (i.e. the tube plate), and on the other half they will be used to record correspondence fields with cameras (Fig. 5).

Fig. 5. The strain gauges locations and measuring areas for DIC system

For DIC system for strain measurement, appropriate procedures are defined by producer and developed for experimental testing. The experiment was performed according to following procedure: sample preparation (the measuring surface must have a pattern with contrast to clearly allocate the pixels in camera images), measuring volume selection (selecting the measuring volume is based on the sample size), system calibration (for certain measuring volume, the appropriate calibration panel is used for sensor configuration), sample positioning (the boiler model was positioned in a way that appropriate tube plate was fully exposed to cameras, the one side of the boiler was fixed as it is in real operating conditions), measurement (after calibration, measurement procedure was performed. The pressure on the water side was gradually loaded, up to 5 bars. Digital images were recorded 60 s after the loading), data processing (recorded data is in a form of report, that can be further analyze).

Experiment was performed both in a cold state, i.e. without the installation of heaters and in the warm state, i.e. with the installed heaters. The recordings of each measuring area (6 marked measuring area) were performed in phases, with different loads - pressures on the water side (the boiler was previously filled with water). Filling the boiler with water and setting the pressure was carried out by a manual pump. The pump is connected to the model with the valve to allow constant pressure during the test. The measuring pressure was in interval 1–5 bar with 60 s stabilization time period between the measured pressure set. To perform the experiment as faithfully as possible in laboratory conditions, the idea was to install a heater, inside the fire tube, whose power will be able to be controlled and adjusted during the experiment. For testing purposes, a heater with a power of 5 kW was made, which consists of 6 individual heaters. It was possible to activate the heaters individually, all at once or in the desired combination, in order to achieve equal heating of the structure. The heater also had a potentiometer, which allowed regulating the power of the heaters and maintaining a constant power over time.

5 Results and Discussion

In the experimental procedure, the critical measuring areas were identified, where the stress strain field has its maximum. The first phase of the examination procedure, the installation without heaters, in the cold state, gave us the significant results.

Von Mises strain results are presented for the first measuring area, Fig. 6. The examined condition was under max test pressure of 5 bar. Experimental data is presented graphically as function of section length. Scale in % is given on ordinate of Fig. 6. The 3D Von Mises strain field across the sample surface (the first measuring area) (Fig. 6c and d) is presenting the highest measured strains (orange color). Von Mises strain values are also given as section length function (Fig. 6a and b). Strain stages (0–3) represent pressure increase, where stage 0 represents beginning of the experiment to stage 3 where maximal test pressure is for the experiment.

In (Fig. 6d) it is shown a line which "imitates" virtual strain gauge. The Aramis system software provides ability to determine the distance between any two points in any moment of experiment. The horizontal line here is in the same place as the real strain gauge in order to verify results and compare.

In Table 1, the comparison of obtained results by using strain gauge, the Aramis system and results obtained numerically in Ansys software. The results of certain pressures values are missing form DIC Aramis system because the measurement was conducted in stages of (2 bar, 3.5 bar, 5 bar), also measured results from strain gauge under pressure of 3.5 bar also missing, while the results are only taken for pressure of 4 and 5 bar. The results form experimental and numerical analysis are presented more detailed in [39].

The results from strain gauge and obtained numerically have significantly match. Therefore, the numerical model can be verified under different pressure loads by the results obtained from the measurements of strain gauge. The 3D DIC Aramis system has certain limitation, especially when there are small displacements. There are deviations about 30% in experimental results compared to numerical model or measured strain by

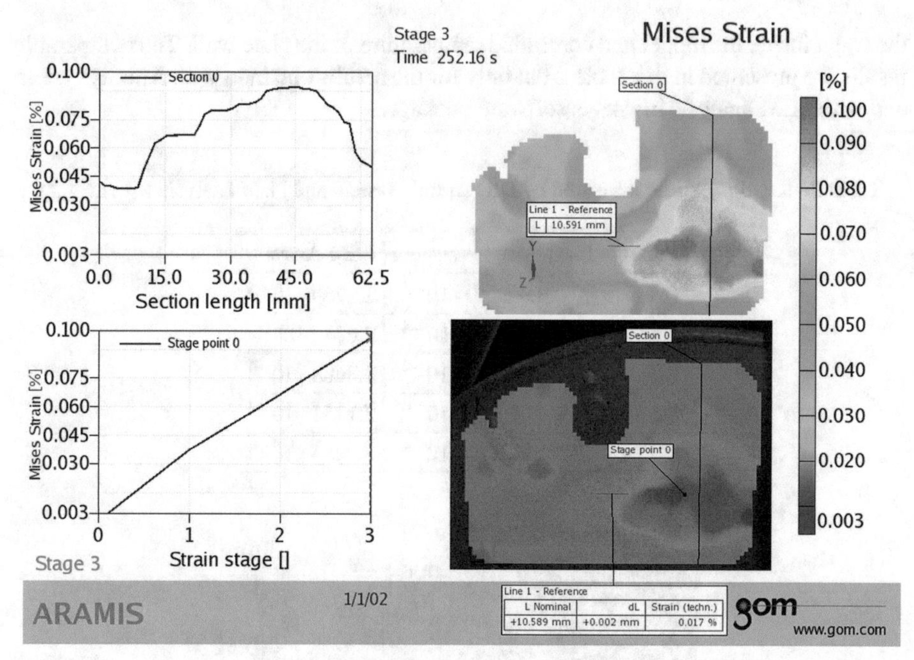

Fig. 6. Experimental results of Von Mises strain for maximal test pressure of 5 bar. a) Von Mises strain as a function of distance for marked Section. b) Von Mises strain as a function of strain stage. c) Von Mises strain field. d) Sample image with overlaying Von Mises strain field (measuring area 1).

Table 1. Resulting strain value measured by strain gauge, DIC Aramis system, results of FEM

Pressure [bar]	Strain gauge	Aramis system	FEM Ansys
1	$3.2931 \cdot 10^{-5}$	/	$3.0189 \cdot 10^{-5}$
2	$5.7071 \cdot 10^{-5}$	$9.44376 \cdot 10^{-5}$	$6.01 \cdot 10^{-5}$
3	$8.5409 \cdot 10^{-5}$	/	$8.9972 \cdot 10^{-5}$
3.5	/	$1.30344 \cdot 10^{-4}$	$1.009 \cdot 10^{-4}$
4	$1.47662 \cdot 10^{-4}$	/	$1.201 \cdot 10^{-4}$

strain gauge. The advantage of this measurement method is the whole measured strain area.

The second phase of experimental procedure includes the thermal strains, simulated with suitable adjustable heaters installed inside the fire tube. The properties of the heaters allow us to set the desired amount of heat that we want to keep during the examination phase. In order to simulate the real boiler conditions in described experiment, the heaters were design to be installed inside the fire tube. The heat transfer mechanisms were similar as the ones in the real hot water boiler. On the other hand, the constant water pressure was maintained by a pump. The measurements were perform on the desired pressure of the water inside the model and controlled temperature of the plate wall. The comparable results are presented in the Table 2 but only for the results obtained with Aramis system and with FEM analysis in Ansys software package.

Table 2. Resulting strain measured by DIC Aramis system and FEM analysis with heating

Pressure [bar]	Aramis system	FEM Ansys
0.5	$9.5840 \cdot 10^{-5}$	$2.3769 \cdot 10^{-5}$
1.5	$1.9168 \cdot 10^{-4}$	$7.624 \cdot 10^{-5}$
2.5	$2.8752 \cdot 10^{-4}$	$1.3688 \cdot 10^{-4}$
3.5	$4.7920 \cdot 10^{-4}$	$2.1151 \cdot 10^{-4}$
5	$8.6256 \cdot 10^{-4}$	$2.9284 \cdot 10^{-4}$

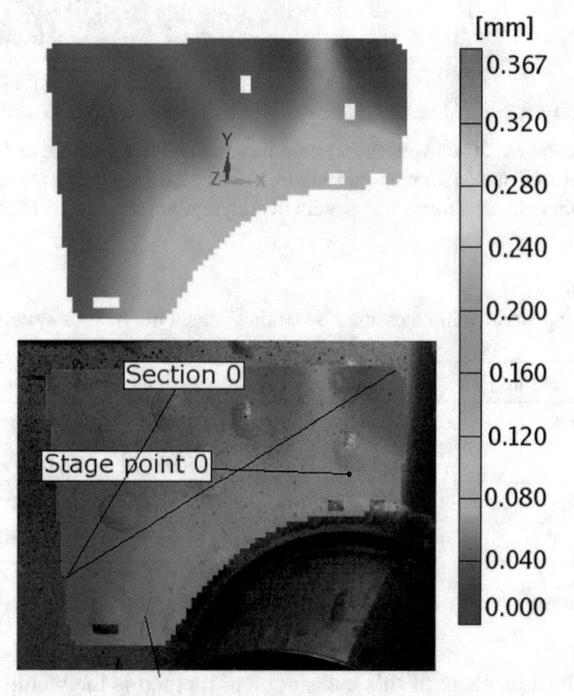

Fig. 7. Results of directional deformation measured by DIC Aramis system - pressure load 1.5 bar

Results obtained by Aramis system and the FEM in Ansys software show significant matches in different loads with a similar temperature field. As already noted, Aramis system has certain limitation, especially in cases of small displacements/deformations, as is the presented case (Table 2).

The results of directional deformation are presented from Aramis system and numerical model done in Ansys software, as it can be seen on Fig. 7 and Fig. 8. The pressure

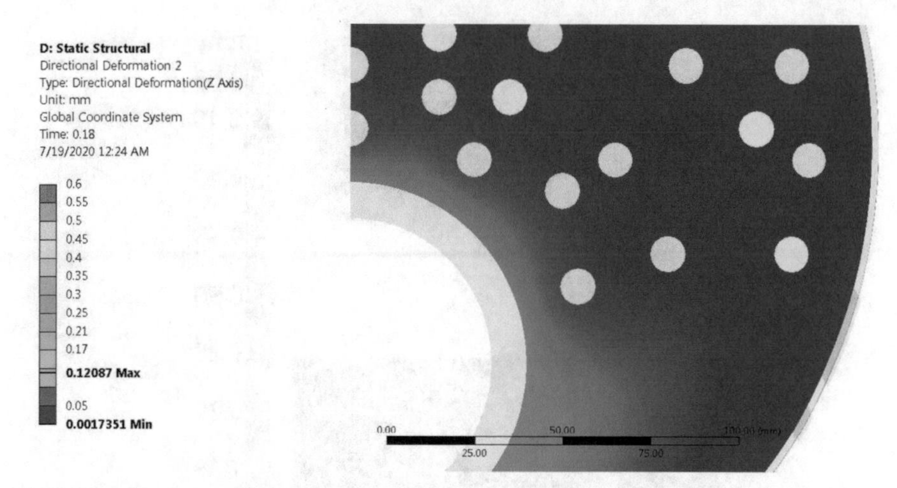

Fig. 8. Results of directional deformation done numerically by Ansys software package - pressure load 1.5 bar

load was 1.5 bar on the water side and the temperature of the inner surface of the fire tube was maintained approx. 90 °C. The distribution of deformation field is quite the same as well as the critical zones.

Results of measured directional deformation by Aramis system and results obtained numerically by finite element method are mostly the same. The critical zones are identified as well as the measured deformations (in critical zone 0.04–0.12 mm). More precisely, the minimal directional deformation under the test pressure of 1.5 bar obtained in Ansys is $1.7351 \cdot 10^{-3}$ mm and maximal is 0.12087 mm.

The next presented experimental stage, was under pressure load of 3.5 bar. The result of directional deformation done by Aramis and numerical method by Ansys is presented in Fig. 9 and 10.

The obtained results from the measured directional deformation by Aramis system and results obtained numerically by finite element method have significant matches. The critical zones are identified as well as the measured deformations (in critical zone 0.15–0.3 mm). The results obtained in Ansys are: the minimal directional deformation under the test pressure of 3.5 bar is $3.3521 \cdot 10^{-3}$ mm and maximal is 0.24365 mm.

In order to analyze one more stage in the conditions with heated surfaces, one more case is presented with the maximal pressure load of 5 bar and it represents the last stage in analysis. These results are presented in Fig. 11 and 12, i.e. the result of directional deformation done by Aramis and numerical method by Ansys, respectively.

The critical zone is now wider in both cases (in critical zones 0.40–0.50 mm). The distribution of directional deformation field has similar position. Minimal deformation in this stage is $7.6147 \cdot 10^{-3}$ mm and maximal 0.51549 mm.

These results are presented in Table 3 as well. It can be seen that in three analyzed cases, there are significant matches with both experimental and numerical results. The deformation field of analyzed plate is quite similar. Having that in mind, the critical parts as well as the critical elements are identified both in numerical and experimental analysis. The experimental procedure on the model had the aim to confirm the phenomenon of

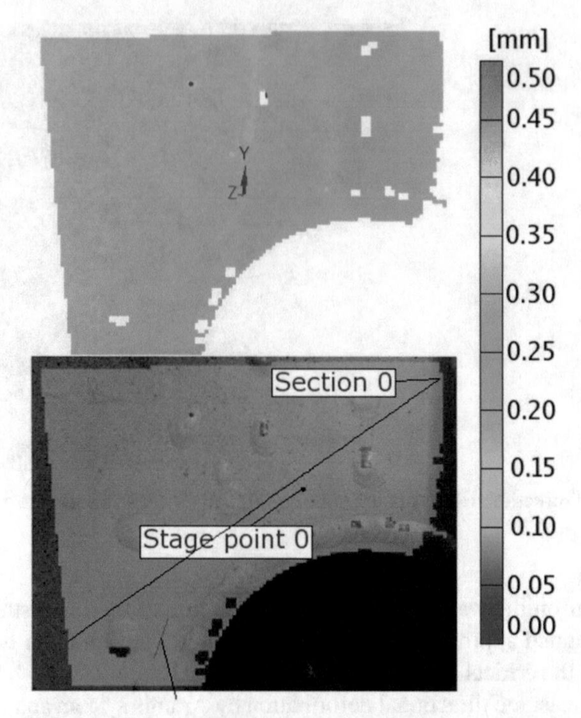

Fig. 9. Results of directional deformation measured by DIC Aramis system - pressure load 3.5 bar

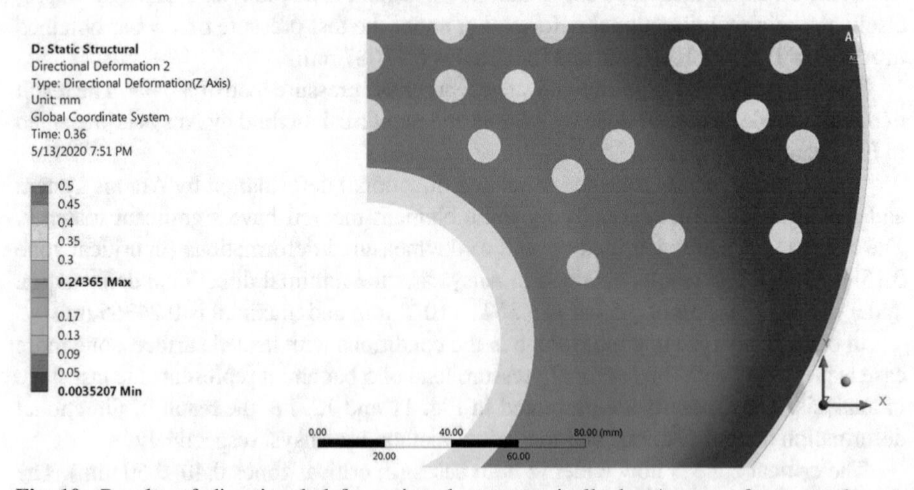

Fig. 10. Results of directional deformation done numerically by Ansys software package - pressure load 3.5 bar

formation and characteristics of deformations of the modeled construction with deformations that occur on a real object. The main goal was to examine the occurrence and characteristics of stresses due to bending of the most sensitive part - the pipe plate of

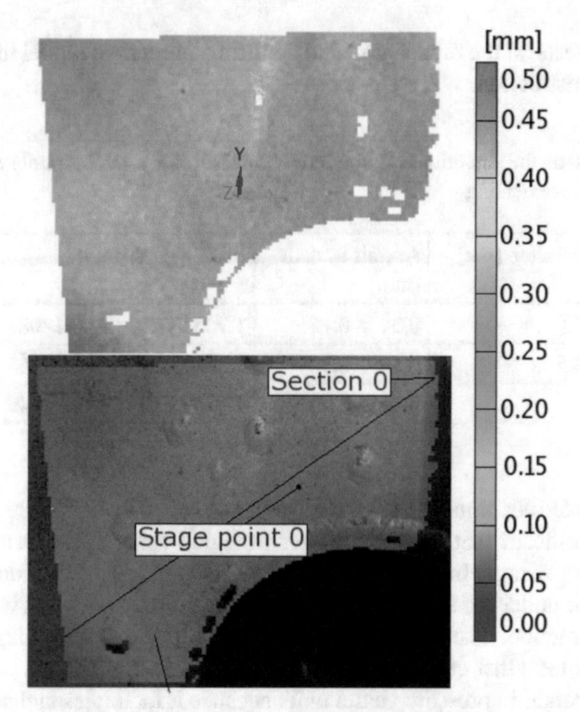

Fig. 11. Results of directional deformation measured by DIC Aramis system - pressure load 5 bar

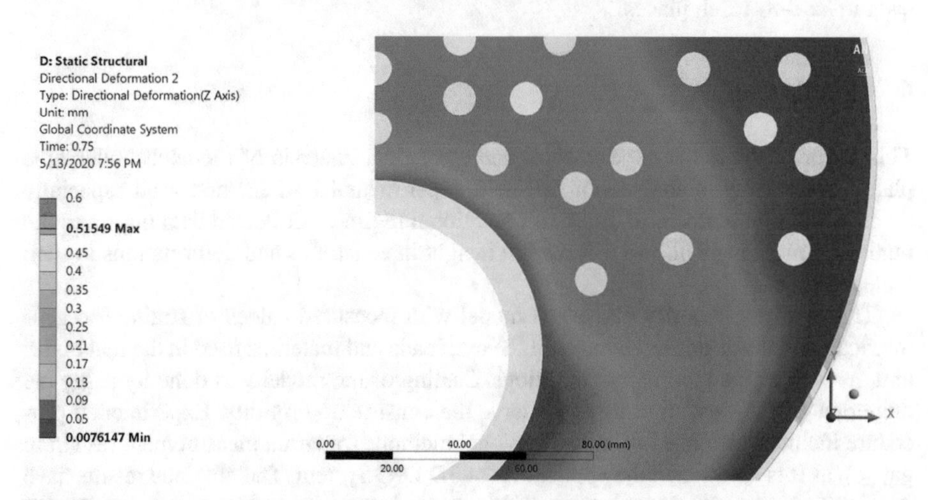

Fig. 12. Results of directional deformation done numerically by Ansys software package - pressure load 5 bar

the first reflecting chamber on the real boiler. The numerical analysis, which included several different stages with different loads and conditions, indicated the occurrence of pipe plate bending, especially in the part between the fire tube and the flue gas pipes.

The presented tests on the model aimed to confirm a numerical model in cold and warm states of the construction.

Table 3. Results of the directional deformations measured by DIC Aramis system and FEM analysis with surface heating

Pressure [bar]	Aramis system [mm]	FEM Ansys [mm]
1.5	$0.04 \div 0.12$	$1.7351 \cdot 10^{-3} \div 0.12087$
3.5	$0.15 \div 0.3$	$3.3521 \cdot 10^{-3} \div 0.24365$
5	$0.40 \div 0.50$	$7.6147 \cdot 10^{-3} \div 0.51549$

In all presented examination stages the values obtained by measured 3D DIC Aramis system has significant matches with values obtained using finite element method by Ansys software package. In this case, it can be said that numerical model has the quite similar behavior under the same loads as the model, both in cold and warm state of the analyzed construction. The verified numerical model can be used with great certainty in other similar tests that can be done on the model. The application of this method is especially important in pressure vessel units because it facilitates and simplifies operational tests, which are non-destructive and without any contacts, that can be applicable even in hard-to-reach places.

6 Conclusion

This paper presents the experimental and numerical analysis of modeled boiler tube plate. The novelty is analysis of full-field experimental DIC method used especially on critical boiler element in different operational regimes. Collected data improves the numerical model and allows to have full insight in constrains and deformations in such boiler elements.

The aim was to verify numerical model with measured values of strain. The geometrical model was designed to fulfill the real loads and materials used in the real boiler unit, as well as the laboratory conditions. Scaling of the model was done by using the numerical experiment in order to preserve the construction rigidity. Experimental procedure includes also one of the conventional methods for strain measurement (as strain gauge), as it is mentioned the application of 3D DIC System. The obtained results have the significant matches in both cases. This strain values are also compared to results conducted by finite element method using Ansys software package. Therefore the formed numerical model is completely verified and can be used for other model testing in order to improve the performances, reliability, safety of the unit and also the design of the boiler with similar geometry.

Acknowledgments. This research was financially supported by the Ministry of Education, Science and Technological Development of the Republic of Serbia.

This paper presents the results of the research conducted within the project "Research and development of new generation machine systems in the function of the technological development of Serbia" funded by the Faculty of Mechanical Engineering, University of Niš, Serbia.

References

1. Taler, J., Dzierwa, P., Taler, D., Harchut, P.: Optimization of the boiler start-up taking into account thermal stresses. Energy **92**(1), 160–170 (2015)
2. Alobaid, F., Karner, K., Belz, J., Epple, B., Kim, H.G.: Numerical and experimental study of a heat recovery steam generator during start-up procedure. Energy **64**, 1057–1070 (2014)
3. Kim, T.S., Lee, D.K., Ro, S.T.: Analysis of thermal stress evolution in the steam drum during start-up of a heat recovery steam generator. Appl. Therm. Eng. **20**(11), 977–992 (2000)
4. Krüger, K., Franke, R., Rode, M.: Optimization of boiler start-up using a nonlinear boiler model and hard constraints. Energy **29**(12–15), 2239–2251 (2004)
5. Dzierwa, P.: Optimum heating of pressure components of steam boilers with regard to thermal stresses. J. Therm. Stress. **39**(7), 874–886 (2016)
6. Dzierwa, P., Taler, D., Taler, J., Trojan, M.: Optimum heating of thick wall pressure components of steam boilers. In: Proceedings of the ASME 2014 Power Conference POWER2014, Baltimore, Maryland, USA, July 28–31 2014. American Society of Mechanical Engineers, Power Division (Publication) POWER (2014)
7. Gaćeša, B., Milošević-Mitić, V., Maneski, T., Kozak, D., Sertić, J.: Numerical and experimental strength analysis of fire-tube boiler construction. Tehnički vjesnik **18**(2), 237–242 (2011)
8. Gaćeša, B., Maneski, T., Milošević-Mitić, V., Nestorović, M., Petrović, A.: Influence of furnace tube shape on thermal strain of fire-tube boilers. Therm. Sci. **18**(Suppl. 1), S29–S47 (2014)
9. Živković, D., Milčić, D., Banić, M., Milosavljević, P.: Thermomechanical finite element analysis of hot water boiler structure. Therm. Sci. **16**(Suppl. 2), 443–456 (2012)
10. Čukić, R., Maneski, T.: Thermomechanical stress analysis of the hot-water boiler by FEM. In: Proceedings of the third International Congress of Thermal Stress 99, Cracow, Poland (1999)
11. Živković, D., Milčić, D., Banić, M., Mijajlović, M.: Numerical method application for thermomechanical analysis of hot water boilers construction. In: Proceedings of the 24th International Conference on Efficiency, Cost, Optimization, Simulation and Environmental Impact of Energy Systems – ECOS 2011, Novi Sad, Serbia, pp. 1351–1362 (2011)
12. Mancic, M., Zivkovic, D., Djordjevic, M., Rajić, M.: Optimization of a polygeneration system for energy demands of a livestock farm. Therm. Sci. **20**(Suppl. 5), S1285–S1300 (2011)
13. EN 12953 - Shell Boilers - General, Materials for pressure parts of boilers and accessories, Design and calculation for pressure parts, Requirements for equipment for the boiler
14. Rajic, M., Banic, M., Zivkovic, D., Tomic, M., Mančić, M.: Construction optimization of hot water fire-tube boiler using thermomechanical finite element analysis. Therm. Sci. **22**(Suppl. 5), 1511–1523 (2018)
15. Tanasic, I., Tihacek-Sojic, Lj., Mitrovic, N., Milic-Lemic, A., Vukadinovic, M., Markovic, A., Milosevic, M.: An attempt to create a standardized (reference) model for experimental investigations on implant's sample. Measurement **72**, 37–42 (2015)
16. Milosevic, M.: Polymerization mechanics of dental composites – advantages and disadvantages. Procedia Eng. **149**, 313–320 (2016). International Conference on Manufacturing Engineering and Materials, ICMEM 2016, 6–10 June 2016, Nový Smokovec, Slovakia, Publisher: Elsevier Ltd.

17. Jovicic, R., Sedmak, A., Colic, K., Milosevic, M., Mitrovic, N.: Evaluation of the local tensile properties of austenite-ferrite welded joint. Chem. Listy **105**, 754–757 (2011)
18. Mitrovic, N., Milosevic, M., Momcilovic, N., Petrovic, A., Miskovic, Z., Sedmak, A., Popovic, P.: Local strain and stress analysis of globe valve housing subjected to external axial loading. In: Key Engineering Materials, vol. 586, 214–217 (2014). Trans Tech Publications
19. Qian, C.F., Yu, H.J., Yao, L.: Finite element analysis and experimental investigation of tubesheet structure. J. Press. Vessel. Technol. **131**(1), 111–114 (2009)
20. Ju, Y.: The analysis of the stress and shift of tube plate and edge of manhole of boiler. J. Dalian Fish. Univ. **1**(1), 71–75 (2000)
21. Wu, G., Zhao, J.: The cause and prevention of the tube plate crack of one gas-fired boiler. Ind. Boil. **1**(1), 54 (2009)
22. Orteu, J.: 3-D computer vision in experimental mechanics. Opt. Lasers Eng. **47**, 282–291 (2009)
23. Pan, B., Wu, D., Yu, L.: Optimization of a three-dimensional digital image correlation system for deformation measurements in extreme environments. Appl. Opt. **51**, 440–449 (2012)
24. Sutton, M., Orteu, J., Hubert, W.: Image Correlation for Shape, Motion and Deformation Measurements: Basic Concepts, Theory and Applications. Springer, Berlin (2009)
25. Mitrovic, N., Petrovic, A., Milosevic, M., Momcilovic, N., Miskovic, Z., Tasko, M., Popovic, P.: Experimental and numerical study of globe valve housing. Chem. Ind. **71**(3), 251–257 (2017)
26. Milosevic, M., Milosevic, N., Sedmak, S., Tatic, U., Mitrovic, N., Hloch, S., Jovicic, R.: Digital image correlation in analysis of stiffness in local zones of welded joints. Tech. Gaz. **23**, 19–24 (2016)
27. Mitrovic, N., Petrovic, A., Milosevic, M.: Strain measurement of pressure equipment components using 3D DIC method. Struct. Integr. Procedia **13**, 1605–1608 (2018). 22nd European Conference on Fracture - ECF22
28. Milosevic, M., Mitrovic, N., Jovicic, R., Sedmak, A., Maneski, T., Petrovic, A., Aburuga, T.: Measurement of local tensile properties of welded joint using Digital Image Correlation method. Chem. Listy **106**, 485–488 (2012)
29. Lezaja, M., Veljovic, Dj., Manojlovic, D., Milosevic, M., Mitrovic, N., Janackovic, Dj., Miletic, V.: Bond strength of restorative materials to hydroxyapatite inserts and dimensional changes of insert-containing restorations during polymerization. Dent. Mater. **31**(2), 171–181 (2015)
30. Milosevic, M., Mitrovic, N., Sedmak, A.: Digital image correlation analysis of biomaterials. In: 15th IEEE International Conference on Intelligent Engineering Systems 2011, pp. 421–425 (2011)
31. Tanasic, I., Tihacek-Sojic, Lj., Milic-Lemic, A., Mitrovic, N., Milosevic, M.: Enhanced in-vivo bone formation by bone marrow differentiated mesenchymal stem cells grown in chitosan scaffold. J. Bioeng. Biomed. Sci. **1**(107), 2 (2011)
32. Mitrovic, A., Mitrovic, N., Maslarevic, A., Adzic, V., Popovic, D., Milosevic, M.: Thermal and mechanical characteristics of dual cure self-etching, self-adhesive resin based cement. Exp. Numer. Investig. Mater. Sci. Eng. **54**, 3–15 (2018)
33. Milosevic, M., Mitrovic, N., Miletic, V., Tatic, U., Ezdenci, A.: Analysis of composite shrinkage stresses on 3D premolar models with different cavity design using finite element method. Key Eng. Mater. **586**, 202–205 (2014)
34. Tanasic, I., Tihacek-Sojic, Lj., Milic-Lemic, A., Mitrovic, N., Mitrovic, R., Milosevic, M., Maneski, T.: Analysing displacement in the posterior mandible using digital image correlation method. J. Biochips Tissue Chips **S1**, 006 (2011)
35. ThyssenKrupp Materials International, Seamless Carbon Steel Pipe for High-Temperature Service. www.s-k-h.com

36. Lucefin Group, Tehnical Card – P235GH (2019). www.lucefin.com
37. Todorović, M., Živković, D., Mančić, M.: Breakdowns of hot water boilers, In: Proceedings of the 17th International Symposium on Thermal Science and Engineering of Serbia SIMTERM 2015, Soko banja, Serbia, 761–769 (2015)
38. Todorovic, M., Zivkovic, D., Mancic, M., Ilic, G.G.: Application of energy and exergy analysis to increase efficiency of a hot water gas fired boiler. Chem. Ind. Chem. Eng. Q. **20**(4), 511–521 (2014)
39. Rajic, M., Zivković, D., Banic, M., Mancic, M., Maneski, T., Milosevic, M., Mitrovic, N.: Experimental and numerical analysis of stresses in the tube plate of the reversing chamber on the model of the boiler. In: Proceedings of the 19th International Conference on Thermal Science and Engineering of Serbia, pp. 439–449 (2019)

Structural Integrity and Life Assessment of Pressure Vessels - Risk Based Approach

Aleksandar Sedmak[1](✉), Snezana Kirin[2], Igor Martic[2], Lazar Jeremic[2], Ivana Vucetic[2], Tamara Golubovic[2], and Simon A. Sedmak[2]

[1] Faculty of Mechanical Engineering, University of Belgrade, Kraljice Marije 16, Belgrade, Serbia
asedmak@mas.bg.ac.rs

[2] Innovation Center of the Faculty of Mechanical Engineering, Kraljice Marije 16, Belgrade, Serbia

Abstract. Risk based approach to assess structural integrity and life of pressure vessel has been presented, starring from EU PED 2014/68 and API 581, ending with use of basic fracture mechanics parameters. In the scope of risk matrix, structural integrity has been tackled by use of the Failure Assessment Diagram (FAD), while structural life was assessed by means of Paris law, using fatigue crack growth rate as relevant parameter. Several case studies have been provided to illustrate this comprehensive approach, including penstock and air storage tanks in Reversable Hydro Power Plant, and two spherical storage tanks, one for vinyl chloride monomer (VCM), the other for ammonia. Besides the assessment of integrity and life, including some important issue like over-pressurizing, roles of engineers and managers in this process is defined and explained.

Keywords: Structural integrity and life · Pressure vessels · Risk based approach

1 Introduction

Prologue (Wikipedia)
The Bhopal disaster was a gas leak incident on the night of 2–3 December 1984 at the Union Carbide India Limited pesticide plant in Bhopal, Madhya Pradesh, India. It is considered to be the world's worst industrial disaster. Over 500,000 people were exposed to methyl isocyanate (MIC) poisonous gas. The highly toxic substance made its way into and around the small towns located near the plant. In 2008, the Government of Madhya Pradesh had paid compensation to the family members of 3,787 victims killed in the gas release, and to 574,366 injured victims. A government affidavit in 2006 stated that the leak caused 558,125 injuries, including 38,478 temporary partial injuries and approximately 3,900 severely and permanently disabling injuries. The big irony of this disaster is the fact that just days after, 4–10 December 1984, the Sixth International Congress on Fracture (ICF6 - the world largest Conference of the kind), was held in New Delhi. As a participant of ICF6, the first author of this paper can witness that nobody was aware of the disaster, going on just 750 km away!

N. Mitrovic et al. (Eds.): CNNTech 2020, LNNS 153, pp. 274–293, 2021.
https://doi.org/10.1007/978-3-030-58362-0_16

Obviously, it is our ultimate task to prevent any pressure vessel (PV) failures, not to mention such disasters. Toward this aim, management of pressure vessel safety is of crucial importance and represents multi complex approach and synergy combination of PED 2014/68/EU, [1], API 581, [2], and structural integrity assessment. One of the aims of the Directive is to ensure that 'Pressure equipment is designed and constructed so that all necessary examinations to ensure safety can be carried out' and that 'means of determining the internal condition of the equipment must be available where this is necessary to ensure the continued safety of the PV, [1]. One can notice that PV categorization according to PED can be used also to estimate the consequence of failure, but in oversimplified manner, since it does not take into account crack-like defects, and also does not assume their presence in welded joints, if not elsewhere.

2 API 581 – Risk Based Inspection

API 581 is designed as the Risk Based Inspection (RBI) document, stating that RBI aims to, [2]:

1) Define and measure the level of risk associated with an item.
2) Evaluate safety, environmental and business interruption risks.
3) Reduce risk of failure by the effective use of inspection resources.

Level of risk is assessed by a quantitative analysis applied after an initial qualitative analysis has selected plant items for further analysis. Qualitative assessment of each plant item locates its position in a 5 × 5 risk matrix. The likelihood of failure is determined from the sum of six weighted factors, [2]:

a) Damage mechanism.
b) Usefulness of inspection.
c) Current equipment condition.
d) Nature of process.
e) Safety design and mechanisms.

The consequence of failure is divided into only two factors, [2]:

a) Fire/Explosion.
b) Toxicity.

Risk is then calculated as the product of consequence and likelihood for each da-ma-ge scenario, risks = CS × FS, where CS is Consequence of scenario, FS Failure frequency. The inspection programme is then developed to reduce the risk, establishing, [2]:

1. What type of damage to look for.
2. Where to look for damage.
3. How to look for damage.
4. When to look for damage.

What and Where is established from reviewing the design data, process data and the equipment history. How to look for the damage is decided by reviewing the damage density and variability, inspection sample validity, sample size, detection capability of method and validity of future prediction based on past observations. When to look for damage is related to the estimated remaining life of the component. This document prescribes methods, with worked examples to obtain an idea of how to assess a system, what constitutes a failure and how to assess the resulting consequences.

3 Risk Ranking

Risk matrices are a useful means of graphically presenting the results from risk analyses of many items of equipment. Risk matrices should, however, not be taken too literally since the scale of the axes is only indicative. The simple matrix below is based on a linear scale of probability and consequence ranging from 1 to 5. The numbers in the cells are the product of the probability and consequence values as shown at Table 1.

Table 1. Risk matrix

5	5	10	15	20	25
4	4	8	12	16	20
3	3	6	9	12	15
2	2	4	6	8	10
1	1	2	3	4	5
Probability/ Consequence	**1**	**2**	**3**	**4**	**5**

This matrix draws attention to risks where the probability and consequence are balanced and to risks where either the probability or the consequence is high. Often matrices will be sectored into regions covering different ranges of risk. As this example shows, the boundaries between regions depend on how the ranges are defined: changing the range of the red region from 21 to 25 to 20 to 25 could have a significant effect. For quantitative analyses risk may be presented as a point a probability/consequence plot.

From these processes of ranking, Duty Holders should be able to identify the items of equipment presenting the greatest risks of failure. For the purposes of inspection planning, equipment may be categorized according to the type of risk. For example:

a) There is a known active deterioration mechanism,
b) There is a high frequency of failure but consequences are low,
c) Consequences of failure are high but the frequency of failure is low.

3.1 Structural Integrity and Life – Fracture Mechanics Approach

When faced with evidence of defects, duty holder (manager) and competent person (engineer) will need to assess the implications of defect in more detail and decide what action should be taken. In making these assessments, risk based principles should apply. Initial considerations towards a decision to accept or reject the defect (corrosion, erosion or crack like defects) will need to take into account the following:

- Type and size of the defect, its cause and mechanism, and the accuracy of NDT data.
- The stress at the location which is affected, i.e. high stressed by areas that are unlikely to tolerate the same degree of deterioration as other areas operating at lower levels of stress.
- The type of material, its strength and fracture properties over the range of operating temperature.
- The safe operating limits associated with each operating condition. These must be considered separately, e.g. a certain corrosion type defect may be acceptable under an operating condition where only pressure is considered. However, should a cyclic operating condition apply then the defect may not be acceptable. Care must be exercised when there are different operating conditions.
- Whether the defect has been present since entry to service, or has initiated during service (due to the contents, environment or operating conditions), and the rate at which it is proceeding.
- Whether the defect is within design allowances (e.g. for corrosion) or fabrication quality control levels.

Even if defects more severe than fabrication 'quality control levels' are revealed by an examination, rejection or repair of the equipment may not always be necessary. Quality control levels are, of necessity, both general and usually very conservative. Decision on whether to reject or accept equipment with defects may be made on the basis of an 'Engineering Critical Assessment (ECA)' [3], using fracture mechanics to assess the criticality of the defects. This may be carried out using before or after the examination. Many standards and recommended practice provide 'Guide on methods for assessing the acceptability of flaws in metallic structures', based on the concept of fitness-for-service, using the Failure Assessment Diagram (FAD) as derived from basic fracture mechanics concepts, [4]. The assessment process positions the flaw within acceptable or unacceptable regions of the FAD.

The flaw lying within the acceptable region of the FAD does not by itself infer an easily quantifiable margin of safety or probability of failure. Conservative input data to the fracture mechanics calculations are necessary to place reliance on the result. If key data are unavailable, (e.g. fracture toughness properties of the weld and parent material), then conservative assumptions should be made. Sensitivity studies are recommended so that the effect of each assumption can be tested. Flaw assessment is a process to which risk based principles may apply. Degrees of uncertainty in the input data (e.g. flaw dimensions, stress, fracture toughness) may be reduced using a lower bound deterministic calculation. The consequences of flaw growth and the possibility of leakage or catastrophic failure may also be factors to consider.

For refinery equipment designed to ASME codes, the American Petroleum Institute has published a recommended practice on fitness for service assessment, API 579, [5], which provided guidance on methods for assessing the acceptability of flaws in fusion welded structures. API 579 is in accordance to API 581, [3]. On the other side, PED regulations in Europe gives instruction in accordance to standard ISO 31000, [6]. This International Standard provides principles and generic guidelines on risk management, but also provides concrete procedures to estimate probability, [6]:

a) The use of relevant historical data to identify events or situations which have occurred in the past and hence be able to extrapolate the probability of their occurrence in the future.
b) Probability forecasts using predictive techniques such as fault tree analysis and event tree analysis. Simulation techniques may be required to generate probability of equipment and structural failures due to ageing and other degradation processes, by calculating the effects of uncertainties.
c) Expert opinion can be used in a systematic and structured process to estimate probability. There are a number of formal methods for eliciting expert judgement which provide an aid to the formulation of appropriate questions.

Obviously, the first option is not appropriate for a serious consideration, especially since it has little or no relevance to any specific case. The second one is based on survey or similar activities, not aimed here, because of its complexity. For the same reason even more complicated methods for risk assessment, e.g. RIMAP [7], based on empirical rules, is not considered here. The third option is actually used here, focused on original, simple methodology, based on risk matrix, as introduced in [8].

As for the consequence, the simplest and most efficient approach is to use PV categorization, according to PED. There are 5 classes in this approach, from 0 to 4, according to p·V value (p stands for pressure, V for volume), similar to those presented in the risk matrix, Fig. 1, as well as to 4 classes, as defined in [2]. Anyhow, one should notice that consequence and probability are not separated in this, somewhat arbitrary and over-simplified approach. Thus, more complicated option is typically needed and used here, based on several parameters: health, safety, environment, business and security, [7–10].

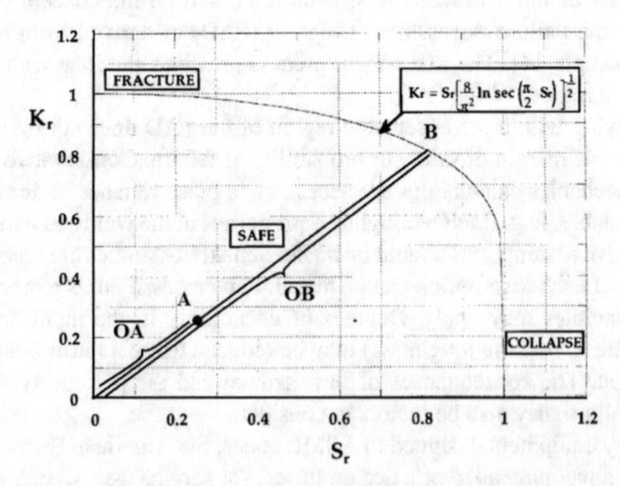

Fig. 1. FAD with service point A and corresponding point B on limit curve, [16]

4 Structural Integrity and Life Assessment

In this chapter, structural integrity and life assessment is described, using recently introduced new concept based on fracture mechanics parameters and laws, [11–17]. This concept goes well with activities of ESIS TC12, [18], especially when so-called "critical" components, such as pressure vessels, are analysed.

It is well known that consequences of failure of pressure vessels can be extremely serious, even catastrophic, as already explained in prologue. This is the major reason to classify pressure vessel as critical components and treat them accordingly. On the other side, frequency of pressure vessel failures is very low, and typically related to welded joints, as the most crack sensitive areas, [13]. Detailed analysis of welded joints in this respect is given in [19, 20], while consequences are analysed in more details in [11–15].

Probability is well known mathematical term, defined as number of events divided by number of possible events. Anyhow, such a definition takes just statistics into account, while much more important factor is missing – state of the component. It is self-evident that the probability of failure of a pressure vessel with a crack is much higher than for a sound pressure vessel, and more or less proportional to defect size, [16].

Therefore, here probability estimation is based on "Expert opinion used in a systematic and structured process to estimate probability", i.e. option c) in Ch. 5.3.4. Likelihood analysis and probability estimation of ISO 31010 standard, as already explained in [16]. In the case of static loading, probability is determined according to the position of the service point in FAD, [8], whereas in the case of fatigue it is taken in accordance with the number of cycles for the given crack length, [17]. Reasoning behind is simple, no defect probability 0, unacceptable defect probability 1, and for all other defects in between, probability is also in between.

This concept can be also applied a design phase, if one assumes existence of a crack, typically in a welded joint. In any case, elastic or elastic-plastic fracture mechanics parameters may be used in the analytical form, or can be evaluated by experimental and numerical methods. Having in mind simplicity, as one of the main features of risk based approach, analytical methods are preferable, but their applicability should be checked by more advanced and precise methods, [16].

4.1 Failure Assessment Diagramme Application

Failure Assessment Diagramme, level 2, as shown Fig. 1, enables one of the most efficient ways to assess structural integrity of a cracked component made of elastic-plastic material, such as High Strength Low Alloyed (HSLA) steels, used for PV. Basically, FAD indicates safe and unsafe position of a point corresponding to a given stress state for a cracked component, divided by so-called limit curve:

$$K_r = S_r\left[\frac{8}{\pi^2} lnsec\left(\frac{\pi}{2}S_r\right)\right]^{-1/2} \tag{1}$$

where $S_r = S_n/S_c$ and $K_r = K_I/K_{Ic}$, S_n stands for stress in net cross section, S_c for the critical stress (Yield Strength, Tensile Strength or any value in-between), K_I for the stress intensity factor and K_{Ic} for its critical value, i.e. fracture toughness.

As already explained, probability of failure is here taken as being proportional to the defect size, so it can be defined as the ratio OĀ/OB, Fig. 1.

4.2 Risk Assessment Based on Remaining Life

The crack growth to its critical size primarily depends on external loads and crack growth rate. Paris equation for metals and alloys, establishes the relationship between fatigue crack growth da/dN and stress intensity factor range ΔK, using the coefficient C and the exponent m:

$$\frac{da}{dN} = C(\Delta K)^m = C\left(Y\left(\frac{a}{W}\right)\Delta\sigma\sqrt{\pi \cdot a}\right)^m \quad (1)$$

In the case considered here, i.e. edge crack growing into depth, one gets:

$$\frac{da}{dN} = C(\Delta K)^m = C\left(Y\left(\frac{a}{W}\right)\Delta\sigma\sqrt{\pi \cdot a}\right)^m \quad (2)$$

where Y(a/W) is the geometry factor depending on crack length. Paris law is then integrated and transformed to calculate the number of cycles from initial, a_0, to final, a_{cr}, crack length:

$$N = \frac{2}{(m-2) \cdot C \cdot (Y(a/W) \cdot \Delta\sigma)^m \cdot \pi^{\frac{m}{2}}}\left(\frac{1}{a_0^{\frac{m-2}{2}}} - \frac{1}{a_{cr}^{\frac{m-2}{2}}}\right) \quad (3)$$

If the total life is known, or certain period of time is analysed, probability of failure can simply be estimated using N, as calculated from Eq. 3.

5 Case Studies

Five case studies are presented, four of them focusing on potential catastrophic events under static loading, whereas the fifth one concerns fatigue crack growth and remaining life of an oil rig drilling pipe. More detailed explanation is given in [8–17]. Some other cases are not considered here, but can be seen in [21–25].

5.1 Penstock RHE Bajina Basta, [25]

Having in mind potential consequence of the penstock failure, as explained in [25], special design procedure has been used, proved with extensive investigation of the full-scale prototype behaviour, Fig. 2, including static, dynamic and fracture mechanics testing. The Sumitomo HSLA steel was used, with nominal Yield Strength (YS) above 700 MPa, and Tensile Strength (TS) cca 800 MPa.

Using experience and results of previous testing of the prototype, [26, 27], two axial surface cracks are analysed here, both with length 90 mm, one with depth 11.75 mm (¼ of penstock thickness, 47 mm), and the other one with depth 23.5 mm (½ of penstock thickness). Two loading cases are analysed, one being 9.02 MPa (design pressure), and the other one 12.05 MPa (hydrostatic proof testing pressure, 33.3% higher).

In the case of surface cracks considered here, reduction of cross-section is negligible, so S_n and S_r are calculated as follows, [25]:

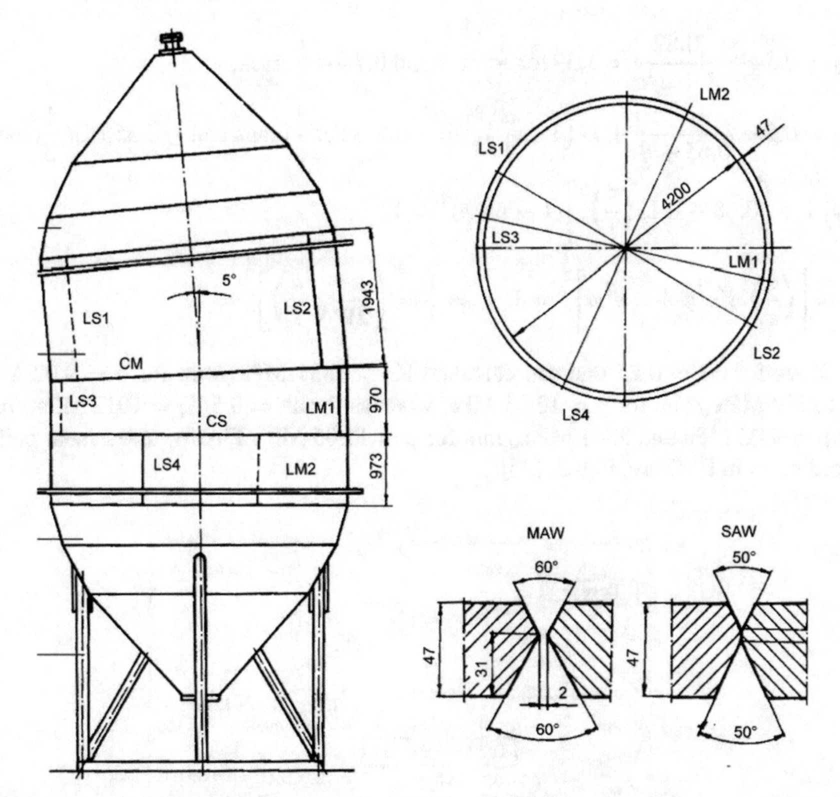

Fig. 2. The full-scale model: L-Longitudinal, C-Circular; MAW – shielded manual arc welding; SAW-submerged arc welding, [25]

- $S_n = pR/t = 9.02 \times 2.1/0.047 =$ **402** MPa, $S_r = S_n/S_c = 402/750 =$ **0.54** for service loading
- $S_n = pR/t = 12.05 \times 2.1/0.047 = 536$ MPa, $S_r = S_n/S_c = 538/750 = 0.72$ for over-loading.

Stress intensity factor for surface edge crack in a plate can be evaluated using the procedure explained in [25]. Plate is taken as better approximation than thin cylinder, since the ratio B/ri is just 47/2100 = 0.023, i.e. much closer to 0 than to 0.1, which is the lowest value for cylinders, [25]. Geometry parameters F and Q are calculated for tensile plate with a surface crack (a/2c = 0.13 for ¼ crack and ≤1) at the point of maximum stress intensity factor (φ = 900) as follows, [25]:

$$Q = 1 + 1.464\left(\frac{c}{a}\right)^{1.65} = 1.08 \text{ for } \frac{1}{4}\text{crack and } 1.22 \text{ for } \frac{1}{2}\text{crack,}$$

$$F = \left[M_1 + M_2\left(\frac{a}{t}\right)^2 + M_3\left(\frac{a}{t}\right)^4\right]gf_\varphi f_w = 1.102 \text{ for } \frac{1}{4}\text{crack and } 1.033 \text{ for } \frac{1}{2}\text{crack,}$$

$$M_1 = \left[1.13 - 0.09\left(\frac{c}{a}\right)\right] = 1.106 \text{ for } \frac{1}{4}\text{crack and } 1.108 \text{ for } \frac{1}{2}\text{crack,}$$

$$M_2 = 0.54 + \frac{0.89}{1 + a/c} = 1.39 \text{ for } \frac{1}{4}\text{crack and } 0.7 \text{ for } \frac{1}{2}\text{crack,}$$

$$M_3 = 0.5 - \frac{1}{0.65 + \frac{a}{c}} + 14\left(1 - \frac{a}{c}\right)^{24} = -0.589 \text{ for } \frac{1}{4}\text{crack and } -0.355 \text{ for } \frac{1}{2}\text{crack,}$$

$$g = 1 + \left[0.08 + 0.15\left(\frac{c}{t}\right)^2\right](1 - cos\varphi)^3 = 1,$$

$$f_\varphi = \left[\left(\frac{c}{a}\right)^2 sin^2\varphi + cos^2\varphi\right]^{\frac{1}{4}} = 1,\ f_w = \left[sec\left(\frac{\pi c}{2W}\sqrt{\frac{a}{t}}\right)\right]^{\frac{1}{2}} = 1.$$

Now, for a/t = 0.25 one can calculate K_I = 2837 MPa$\sqrt{}$mm for p = 9.02 MPa and 3777 MPa$\sqrt{}$mm for p = 12.05 MPa, whereas for a/t = 0.5 K_I = 4012 MPa$\sqrt{}$mm for p = 9.02 MPa and 5341 MPa$\sqrt{}$mm for p = 12.05 MPa. Finally, assessment points coordinates in FAD are, Fig. 3, [25]:

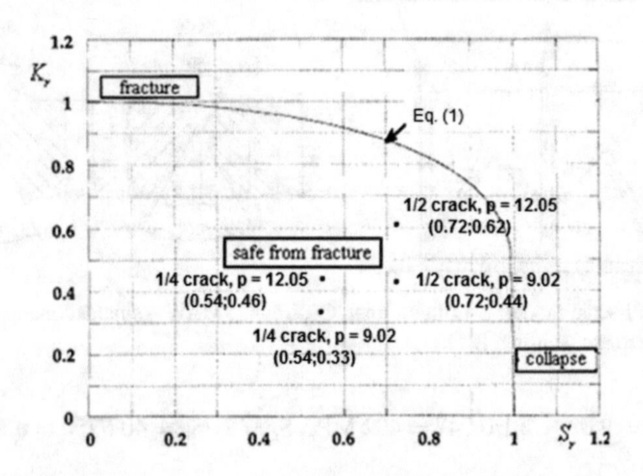

Fig. 3. FAD, including assessment points for surface cracks, [25]

For ¼ crack, p = 9.02 (0.54, 0.33), p = 12.05 (0.54, 0.46). probabilities are 0.58 and 0.62.

For ½ crack, p = 9.02 (0.72, 0.44), p = 12.05 (0.72, 0.62), probabilities are 0.75 and 0.82.

Now, one can get risk matrix in usual way, as presented in Table 2, indicating shift from high (0.58) to very high risk (0.62), for 1/4 crack, or from very high (0.75) to extremely high risk (0.82) for 1/2 crack, clearly demonstrating detrimental effect of over-pressure, [25].

5.2 Case Study – Air Storage Tank in RHPP Bajina Basta, [8]

Another case study in Reversible Hydro Power Plant Bajina Basta is now presented, so-called air storage tanks, which was originally used to establish the risk based procedure,

Table 2. Risk matrix for ¼ crack and ½ crack, p = 9.02 MPa and 12.05 MPa, [25]

Probability category	Consequence category: 1 – very low	2 - low	3 - medium	4 - high	5 - very high	Risk legend
≤0.2 very low						**Very low**
0.2-0.4 low						**Low**
0.4-0.6 medium					1/4 crack p=9.02 MPa	**Medium**
0.6-0.8 high					1/4 crack p=12.05 MPa 1/2 crack p=9.02 MPa	**High**
0.8-1.0 very high					1/2 crack p=12.05 MPa	**Very High**

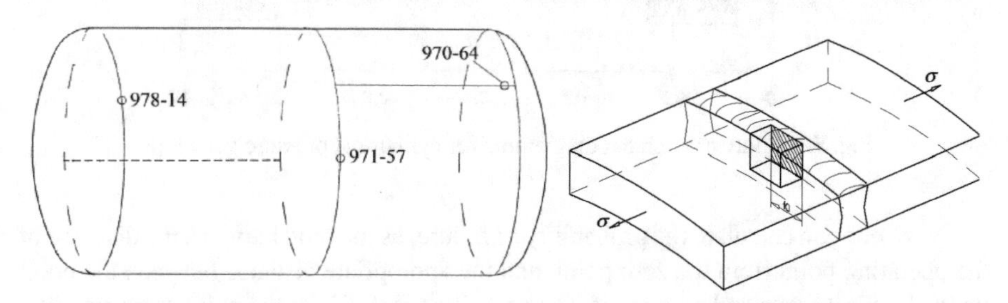

Fig. 4. a) pressure vessel with defects; b) cross section with defect 971-57, [8]

[8]. Eight PVs have been working for many years before the regular NDT (radio-graphy) has revealed 3 non-acceptable defects according to standards, as shown in Fig. 4, [8].

Consequence in this case is simple to evaluate because of the fact that eventual failure would cause catastrophic consequences, [28]. To evaluate probability due to material defects, one of three most dangerous has been selected, defect 971-57, which has been identified as the lack of fusion, and is represented here as a through crack, Fig. 4b, to enable conservative approach and simple 2D analysis, at the same time, using the data:

Geometry: thickness t = 50 mm, diameter D = 2075 mm;

Material: R_{eh} = 500 MPa, R_M = 650 MPa; K_{Ic} = 1580 MPa$\sqrt{}$mm

Crack geometry: length 2a = 10 mm, circumferential weld – lack of fusion;

Loading: maximum pressure p = 8.1 MPa, residual stress σ_R = 175 MPa – maximum value transverse to the weld, in heat-affected-zone (HAZ). There is no data about post-weld-heat-treatment (PWHT), so the maximum possible value of residual stress, according to experience, has been applied, [8].

Curvature effect is negligible (t/D = 50/2075 ≈ 0.025). Having this in mind, the stress intensity factor is: $KI = \sigma\sqrt{\pi a}$; where σ is remote stress ($\sigma = pR/2t + \sigma R$) = 259 MPa: $KI = (pR/2t + \sigma R)\sqrt{\pi a} = 1026\ MPa\sqrt{mm}$, 65% of KIc.

Plastic collapse ratio has been calculated as follows: $S_R = \sigma_n/\sigma_F$; $\sigma_n = p \cdot R/2 \cdot t$ = 84 MPa, no reduction cross-section (negligible), $\sigma_F = (R_{eH} + R_M)/2 = 575$ MPa; S_R = 0.15. Thus, the coordinates are (0.15, 0.65), as shown in Fig. 5, [8].

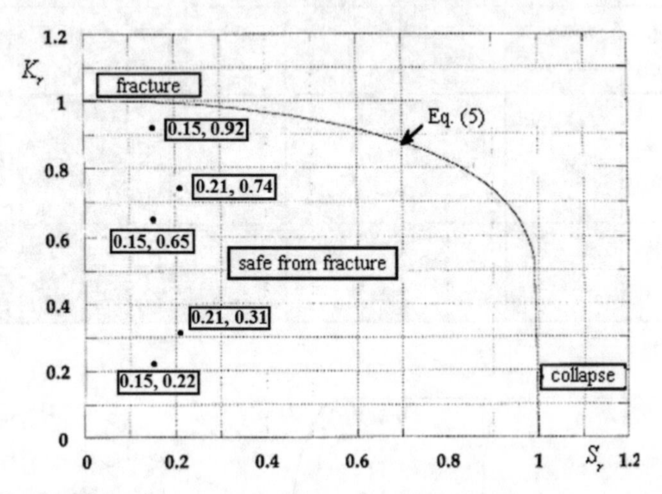

Fig. 5. Failure assessment diagramme for cylindrical pressure vessel, [8]

Now, one can calculate the probability of failure, as the ratio between the distance of the operating point from the zero point, and the appropriate distance between the point on the limiting curve and zero point. The probability is 0.64 for the point corresponding to design pressure. Three important effects now can be followed, [8]:

Common practice in conservative approach is to assume double size of a defect, just for the case that operator has not evaluated the size correctly, leading to the point (0.15, 0.92) and the probability 0.91.

Water proof test pressure (43% higher than the design pressure), raises probability from 0.64 to 0.75.

If one neglects residual stresses, the point in FAD goes to much lower (deeper) part of the safe region, (0.15, 0.22), reducing probability to relatively low value (0.24). Therefore, in this case, water proof is not a problem, because it raises probability up to 0.34, point (0.21, 0.31).

5.3 Case Study – Large Spherical Storage Tank for Ammonia, [14]

The analysis was performed on the spherical storage tanks for ammonia, volume 1000 m^3, diameter $D = 12500$ mm and wall thickness $t = 25$ mm, Fig. 6. The operating pressure was $p = 6$ bar, while the proof test pressure $p = 10$ bar was applied together with non-destructive testing (NDT), [14]. Based on results of previous testing, as explained in [14], $K_{Ic} = 2750\ MPa\sqrt{mm}$ was adopted as the minimum value for fracture toughness in HAZ. Number of cracks were detected before and after repair welding, not uncommon problem for spherical tanks few decades ago, [29]. In order to evaluate its significance, one typical crack was presented as an edge crack with length a = 5 mm, schematically shown in Fig. 6c, as if it was along the whole circumference.

a) spherical storage tank

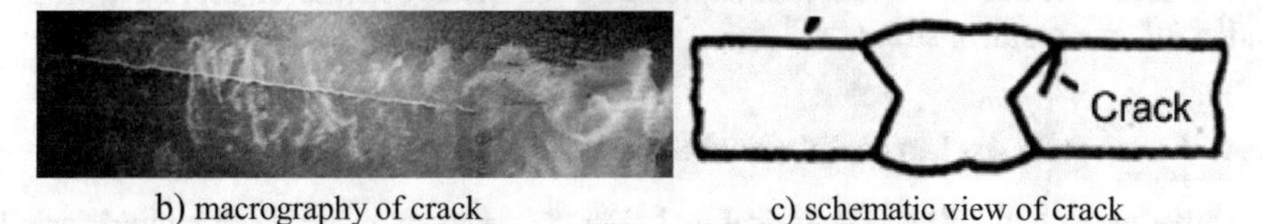

b) macrography of crack c) schematic view of crack

Fig. 6. The spherical tank for ammonia storage, [14]

Therefore, the conservative approach has been applied, with the following data:
PV geometry (thickness t = 25 mm, diameter D = 12500 mm);
St.E460 steel: $R_{eh} = 480$ MPa, $R_M = 680$ MPa; $K_{Ic} = 2750\ MPa\sqrt{mm}$, [14];
crack geometry (edge crack, length 5 mm, ratio length/thickness = 0.2);
loading (max. pressure p = 0.6 MPa, stress $\sigma = p \cdot R/2 \cdot t = 75$ MPa, residual stress $\sigma_R = 196$ MPa - max. value transverse to the weld, no measurements available, no post weld heat treatment, so 40% of the Yield Stress was assumed, [14]);
curvature effect is negligible ($t/R = 25/12500 \approx 0.002$).

The SIF is calculated as: $K_I = 1.12 \cdot (pR/2t + \sigma_R)\sqrt{\pi a} = (75 + 196)\sqrt{\pi \cdot 5} = 1075\ MPa\sqrt{mm}$, resulting in the ratio $K_R = K_I/K_{Ic} = 1075/2750 = 0.39$.

The net stress is $\sigma_n = 1.25 \cdot pR/2t$, due to cross-section reduction for 1/5, critical stress $\sigma_F = (R_{eH} + R_M)/2 = 580$ MPa; resulting in $S_R = (1.25 \cdot 75)/580 = 0.16$. Thus, the coordinates in FAD are $(K_R, S_R) = (0.39, 0.16)$, resulting in probability 0.395, Fig. 7, [14].

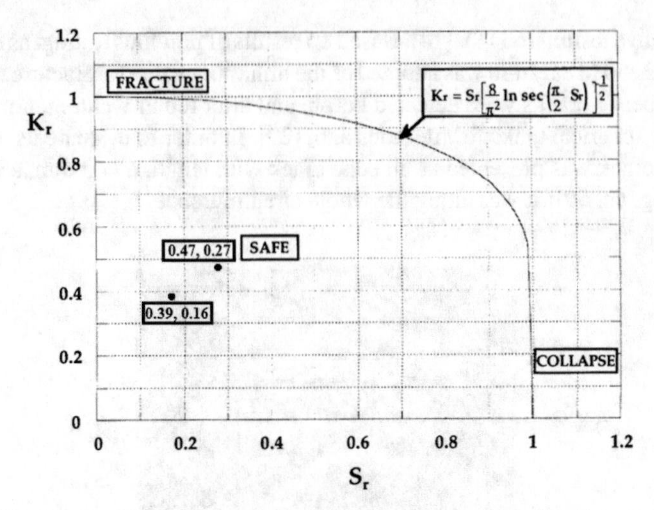

Fig. 7. The FAD for two pressure values, design and proof testing, [14]

Now, the same calculation for the proof testing (pressure p = 1 MPa) leads to the following result: $K_R = K_I/K_{Ic} = 1288/2750 = 0.47$, $S_R = \sigma_n/\sigma_F = 0.27$; the coordinates (0.47, 0.27) and the ratio 0.4, [14]. The FAD is shown in Fig. 6, indicating these two pressure levels, 6 bar (design) and 10 bar (proof test), indicating detrimental role of the proof pressure, but somewhat less expressed than usual, because of overlap with the other effect – residual stress.

5.4 Case Study 4 – Large Spherical Storage Tank for VCM, [13]

Another case study of large spherical tank, Fig. 8a, leakage, was caused by undetected micro-cracks in welded joint, grown through the thickness during proof testing (pressure up to 50% above the design pressure), Fig. 8b–d, [13].

In the case analysed here, it is the large sphere for VCM, 2,000 m^3 in volume, 15.6 m in diameter, made of fine grain, micro-alloyed steel TTSt E-47, Steelworks Jesenice. In regular in-service inspection, many defects, mostly cracks, had been detected in welded joints from the inner side of a sphere. The cracks mostly developed in radial welded joints (RIII, Fig. 8a), in its upper part, at the border of liquid and gaseous phases [13]. The occurrence of cracks was mostly detected in the heat-affected-zone (HAZ), being typical for the micro-alloyed steel TTSt E-47 [13]. Namely, microstructure of TTSt E-47 welded joints is complex one, [13], with some regions in HAZ being sensitive to cracking. This is also clear from the data for fracture toughness, [13]: K_{Ic} (BM) = 4420 MPa$\sqrt{}$mm, K_{Ic} (WM) = 2750 MPa$\sqrt{}$mm, K_{Ic} (HAZ) = 1580 MPa$\sqrt{}$mm. Other data used in this analysis is:

Geometry: thickness t = 20 mm, volume 2,000 m^3, diameter D = 15.6 m; curvature effect negligible (t/R = 20/7800 ≈ 0.025).

Loading: max. pressure p = 0.5 MPa, stress $\sigma = p \cdot R/2 \cdot t = 97.5$ MPa.

Residual stress σR = 196 MPa - max. value transverse to the weld, taken to be 40% of the Yield Stress, R_{eh}, or σ_R = 480 MPa - max. value in longitudinal direction, taken

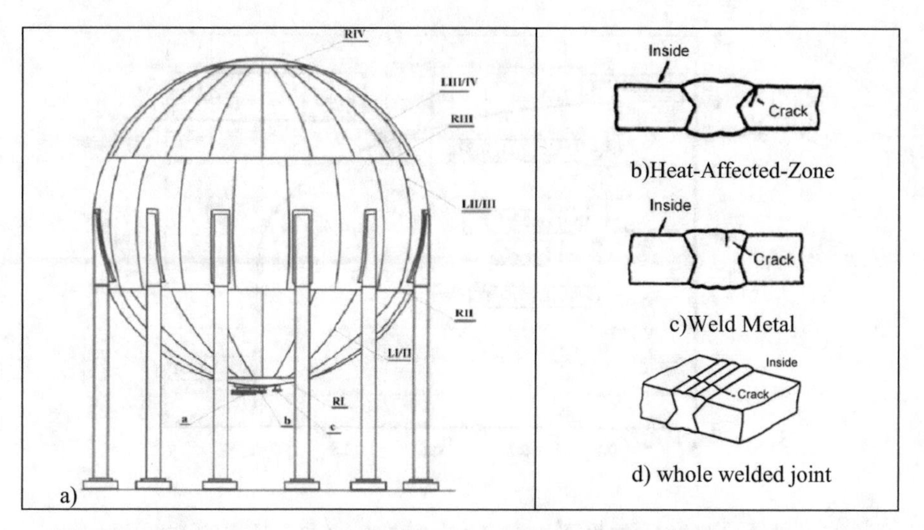

Fig. 8. a) Spherical storage tank, b–d) Cracks on inner wall side of spherical tank, [13]

to be 100% of the Yield Stress, R_{eh}, since no measurements and no record of post weld heat treatment (PWHT) was available, [13]);

All cracks are three-dimensional (3D), i.e. so-called surface cracks, with different lengths (100–200 mm) and depth approximately 5 mm. For cracks of such shape (much longer than deep), it has been shown that they would grow into depth [13], i.e. leakage would precede catastrophic failure. Therefore, the cracks are represented as being 2D edge crack, with length 5 mm (as if they are running all over the circumference, i.e. as they are schematically shown in Fig. 8b–c), enabling conservative and simplified approach to solve the problem, [13]. The stress intensity factor (SIF) is then calculated for two cases, one for longitudinal cracks (HAZ, Fig. 8b, and WM, Fig. 8c), and the other one for the transverse crack (BM, Fig. 8d), [13]:

$K_I = 1.12 \cdot (pR/2t + \sigma_R)\sqrt{\pi a} = 1302.5\ MPa\sqrt{mm}$ for WM and HAZ,

$K_I = 1.12 \cdot (pR/2t + \sigma_R)\sqrt{\pi a} = 2562.8\ MPa\sqrt{mm}$ for BM.

Now, one can calculate ratios $K_R = K_I/K_{Ic}$:

$K_R = K_I/K_{Ic} = 1302.5/2750 = 0.47$ for WM,

$K_R = K_I/K_{Ic} = 1302.5/1580 = 0.82$ for HAZ,

$K_R = K_I/K_{Ic} = 2562.8/4420 = 0.58$ for BM.

The net stress, σn, and the flow stress, σF, is taken as the same for all zones in welded joint: $\sigma_n = 1.33 \cdot pR/2t$, $\sigma_c = (R_{eH} + R_M)/2 = 580$ MPa, $S_R = (1.33 \cdot 97.5)/580 = 0.22$, [13].

The coordinates (K_R, S_R) for WM, HAZ and BM are as follows: (0.22, 0.47), (0.22, 0.82), (0.22, 0.58), respectively, as shown in Fig. 9, with probabilities 0.48 (WM), 0.84 (HAZ) and 0.59 (BM). Based on these results, one can state the following, [13]:

Risk based approach is an engineering tool for assessment of structural integrity, based on the risk matrix presentation as the most suitable for managers to make decisions, even difficult ones.

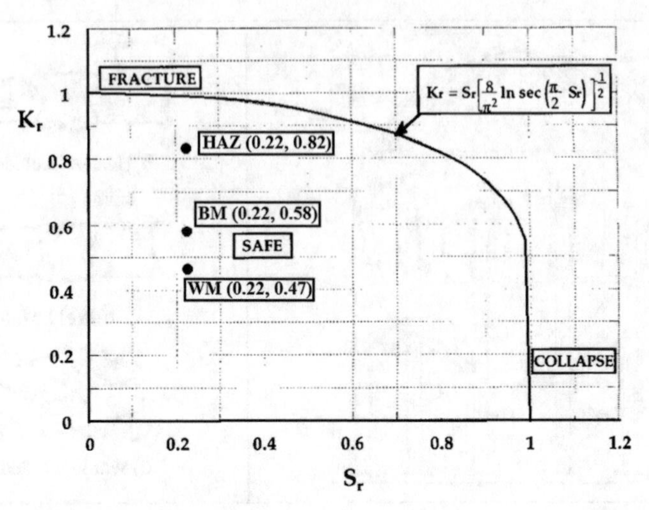

Fig. 9. The FAD for VCM storage tank with cracks in WM, HAZ and BM

Welded joint is a critical region in PV and any welded construction, with either WM or HAZ being the most critical, depending on a combining effect of microstructure (K_{Ic}) and residual stress. In the case study presented here the HAZ is the most critical region.

5.5 Case Study 5 – Static and Fatigue Crack Growth in HF Welded Pipe, [17]

Basic data for HF welded pipe, with an axial crack, made of API J55 steel (YS = 380 MPa, TS = 560 MPa), Fig. 10, is as follows, [17]:

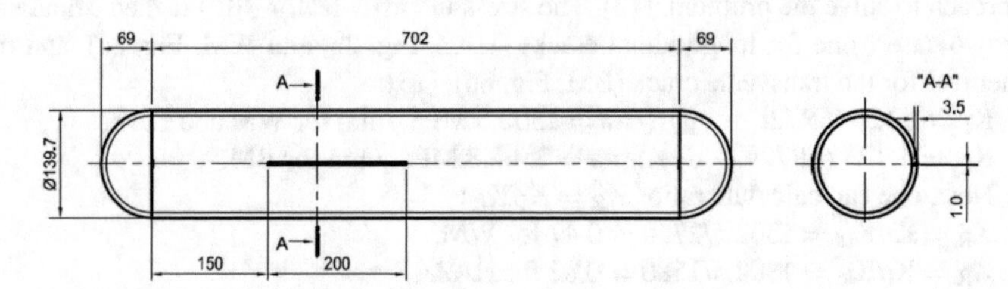

Fig. 10. Oil rig pipe, as tested under pressure, [17]

Diameter 139.7 mm, thickness 6.98 mm, length 702 mm

- Maximum pressure = 10 MPa, testing pressure 22 MPa,
- Number of strokes of pump rod: $n_{PR} = 9.6\ \mathrm{min}^{-1}$

Axial surface crack with dimensions: a = 3,5 mm and 2c = 200 mm, was positioned in the BM, as the weakest point regarding fracture toughness, [17], to test the pipe.

Now one can calculate K_I and S_n to compare with K_{Ic} and S_c, and get the point in FAD: for p = 10 MPa, S = pr/t = 100 MPa, S_n = 200 MPa, S_c = (380 + 560)/2 =

470 MPa, a = 3.5 mm, 2c = 200 mm. $S_{net}/S_c = 200/470 = 0.425$, or 2.2 times more for testing pressure (0.935), [17].

For the conservative assumption, i.e. crack growth into depth in the cross-section "A-A", Fig. 10, one gets (p = 10 MPa): $KI = Y(a/W) \cdot S \cdot \sqrt{\pi a} = 3.52 \cdot 100 \cdot 3.32 = 1169$ MPa$\sqrt{}$mm, KIc = 2850 MPa$\sqrt{}$mm, as given in [17], KI/KIc = 0.41, giving finally the point coordinates (0.425, 0.41), and probability 0.43. For p = 22 MPa, one gets: $KI = 3.52 \cdot 220 \cdot 3.32 = 2571$ MPa$\sqrt{}$mm, KI/KIc = 0.9, point coordinates (0.935, 0.9), which is the UNSAFE region of FAD, [17]. As shown in [17], this is not realistic, because testing at pressure 22 MPa has not caused failure.

Thus, more sophisticated analysis was applied for surface edge crack, [17]:

p = 10 MPa: $K_I = 1.5 \cdot 100 \cdot 3.32 = 498$ MPa$\sqrt{}$mm, $K_I/K_{Ic} = 0.175$, coordinates (0.425, 0.175), Fig. 11, probability 0.43.

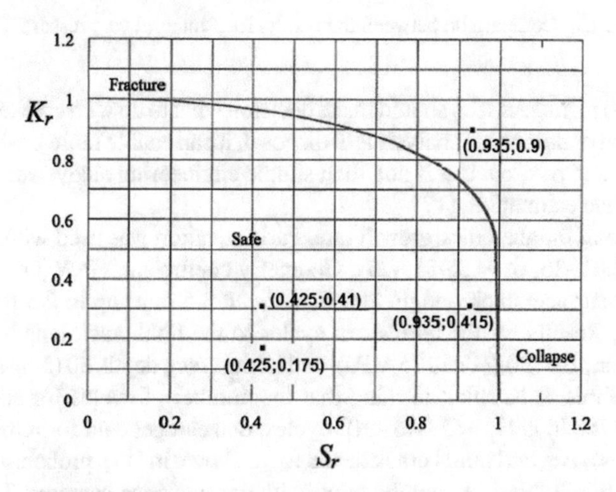

Fig. 11. Failure Assessment Diagram for oil rig pipe, [17]

p = 22 MPa: $K_I = 1.62 \cdot 220 \cdot 3.32 = 1183$ MPa$\sqrt{}$mm, $K_I/K_{Ic} = 0.415$, coordinates (0.935, 0.415), Fig. 11, probability 0.94.

In this case, FEM was used to evaluate J integral via calculated stresses and strain, and compared with direct measurement of J integral, as shown in [17]. Strains were evaluated up to the testing pressure (22 MPa) around the crack, Fig. 12. One can see good agreement between numerical and experimental J integral (16 vs 13 KN/m) for the testing pressure p = 22 MPa. If one takes ratio J/JIc as the probability of failure, significantly lower values are obtained (16/35.8 = 0.45, 13/35.8 = 0.38).

Results for probabilities for surface crack at pressure p = 22 MPa are as follows: analytical 0.94, numerical 0.45, experimental 0.38. In combination with consequence category 3, these probabilities will result in low risk (experimental), medium risk (numerical), and high risk (analytical).

It is clear that risk estimation in this case depends crucially on the method used, since relatively low risk was obtained by more sophisticated and precise estimation (numerical, experimental) and high risk by simple engineering calculation of probability. What shall

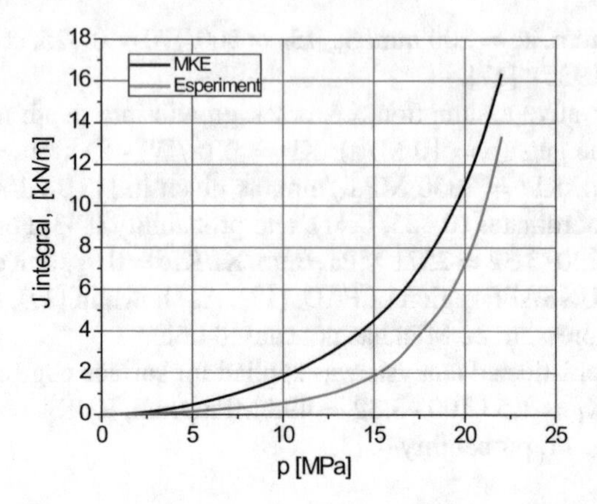

Fig. 12. Comparison between the results for J integral vs. pressure, [17]

be used is up to managers who should make decision what to do with cracked component. If experimental or numerical analysis is at disposal, it can enable more precise estimation of probability and risk, but if it is not, then simple engineering analysis can provide fast and conservative estimation, [17].

In the case of fatigue crack growth rate, Paris equation was used with the following data: C = 1.23E−13, m = 3.931, [17]. Geometry coefficient Y(a/W) was in the range from 1.62 for surface crack (length 200 mm, depth 3.5 mm) up to 2.5 for crack depth 4.8 mm, [17]. Results for the number of cycles to the final crack length (4.8 mm) vs. amplitude stress, $\Delta\sigma$ (20, 21 and 25 MPa), and initial crack depth, a0 (3.5, 4 and 4.5 mm), are given in Table 3. Keeping in mind that the number of cycles for annual working hours, $T_y = 8760$ h, is $N_y = 5.046 \cdot 10^6$ cycles, one can get data for remaining life vs. amplitude stress, $\Delta\sigma$, and initial crack depth, a_0, as shown in [17], probabilities for 1 Year of remaining life, Table 3. In combination with consequence category 3, probabilities given in Table 3 provide low risk for $a_0 = 3.5$ mm with all amplitude stresses, as well as for $a_0 = 4$ mm with $\Delta\sigma = 20$ and 21 MPa, medium risk for $a_0 = 4$ mm with $\Delta\sigma =$ 25 MPa, as well for $a_0 = 4.5$ mm with $\Delta\sigma = 25$ MPa, and finally, high risk for $a_0 =$ 4.5 mm with $\Delta\sigma = 25$ MPa.

Table 3. Remaining life (probability)

$\Delta\sigma/a_0$	3.5 mm	4 mm	4.5 mm
20 MPa	0.11	0.21	0.62
21 MPa	0.14	0.25	0.75
25 MPa	0.27	0.5	1.0

More sophisticated calculation of number of cycles under fatigue loading is provided by extended Finite Element Method (xFEM), as described in [30] and recently applied in few practical problems, [31–34]. Reasonable agreement between numerical and analytical results was obtained with refined mesh, Fig. 13, as generated and used in ANSYS. Anyhow, these results should be treated as initial ones, especially since experiment is not available and practically can't be performed on the pipe itself.

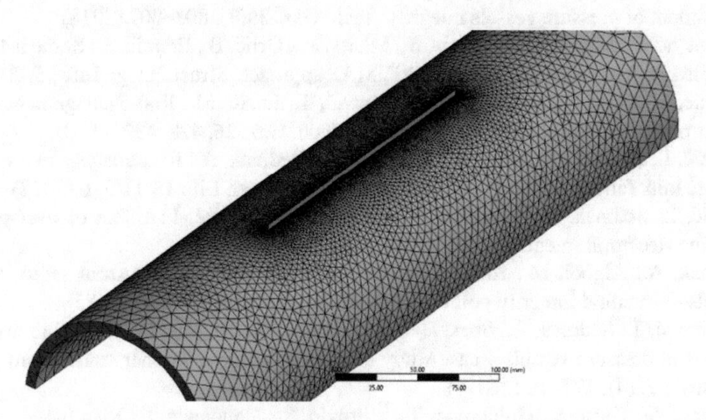

Fig. 13. Fine FE mesh with 414537 nodes, [16]

6 Conclusions

Based on the results shown here, one can conclude the following:

Risk based approach can be useful tool for assessment of structural integrity and life, even if using simple graphical presentation, i.e. the risk matrix. In any case, it can bridge the gap between managers and engineers in decision making process, which is of utmost importance for safe operation of critical components.

Basic structural integrity tools, such as FAD, can be used in combination with the risk based approach to show detrimental effect of proof test in the case of large spherical storage tanks.

More sophisticated methods (experimental, numerical) provide better results, but simple analytical tools can be used in most of cases, even if being over-conservative.

Acknowledgement. The authors are grateful to the Ministry of Education, Science and Technological Development of the Republic of Serbia for its support.

References

1. Directive 2014/68/EC of the European Parliament and of the Council of 15 May 2014
2. API 581 The standard for quantitative Risk Based Inspection. American Petroleum Institute 2010
3. Radu, D., Sedmak, A., Sedmak, S.A., Li, W.: Engineering critical assessment of steel shell structure elements welded joints under high cycle fatigue. Eng. Fail. Anal. **114** (2020). https://doi.org/10.1016/j.engfailanal.2020.104578

4. BS 7910:2019: Guide to methods for assessing the acceptability of flaws in metallic structures
5. API 579: Fitness-for-service assessment of pressure equipment in petrochemical and other industries. American Petroleum Institute (2000)
6. ISO 31000: Risk management – risk assessment techniques (2009)
7. Golubović, T., Sedmak, A., Spasojević Brkić, V., Kirin, S., Rakonjac, I.: Novel risk based assessment of pressure vessels integrity. Tech. Gaz. **25**(3), 803–807 (2018)
8. Stanojević, P., Jovanović, A., Kirin, S., Misita, M., Orlić, B., Eremić, S.: Some achievements in RBIM implementation according to RIMAP approach. Struct. Integr. Life **15**, 79–84 (2015)
9. Vučetić, I., Kirin, S., Sedmak, A., Golubović, T., Lazic, M.: Risk management of a hydro power plant – fracture mechanics approach. Tech. Gaz. **26**, 428–432 (2019)
10. Vučetić, I., Kirin, S., Vučetić, T., Golubović, T., Sedmak, A.: Risk analysis in the case of air storage tank failure at RHPP Bajina Bašta. Struct. Integr. Life **18**(1), 3–6 (2018)
11. Martić, I., Sedmak, A., Mitrović, N., Sedmak, S., Vučetić, I.: Effect of over-pressure on pipeline structural integrity. Tech. Gaz. **26**(3), 852–855 (2019)
12. Sedmak, A., Algool, M., Kirin, S., Rakicevic, B., Bakic, R.: Industrial safety of pressure vessels - structural integrity point of view. Hemijska industrija **70**(6), 685–694 (2016)
13. Golubović, T., Sedmak, A., Spasojević Brkić, V., Kirin, S., Veg, E.: Welded joints as critical regions in pressure vessels – case study of vinyl-chloride monomer storage tank. Hemijska Industrija **72**(4), 177–182 (2018)
14. Sedmak, A., Kirin, S., Golubovic, T., Mitrovic, S., Stanojevic, P.: Risk based approach to integrity assessment of a large spherical pressure vessel. Struct. Integr. Procedia **2**, 3654–3659 (2016). ECF21, Catania
15. Sedmak, A., Kirin, S., Grbovic, A., Lazic Vulicevic, Lj., Stanojevic, P.: Risk based approach to life assessment of drilling pipe. Procedia Struct. Integr. **18**, 379–384 (2019)
16. Kirin, S., Sedmak, A., Zaidi, R., Grbović, A., Šarkočević, Ž.: Comparison of experimental, numerical and analytical risk assessment of oil drilling rig welded pipe based on fracture mechanics parameters. Eng. Fail. Anal. **112** (2020). Article no. 104508
17. Zaidi, R., et al.: Risk assessment of oil drilling rig welded pipe based on structural integrity and life estimation. Eng. Fail. Anal. **112** (2020). Article no. 104508
18. Correia, J., De Jesus, A., Muniz-Calvente, M., Sedmak, A., Moskvichev, V., Calçada, R.: The renewed TC12/ESIS technical committee - risk analysis and safety of large structures and components (Edit.). Eng. Fail. Anal. **105**, 798–802 (2019)
19. Sedmak, S.A., Jovičić, R., Sedmak, A., Aranđelović, M., Đorđević, B.: Influence of multiple defects in welded joints subjected to fatigue loading according to SIST EN ISO 5817:2014. Struct. Integr. Life **18**(1), 77–81 (2018)
20. Gubeljak, N., Predan, J., Rak, I., Kozak, D.: Integrity assessment of HSLA steel welded joint with mis-matched strength. Struct. Integr. Life **9**, 157–164 (2009)
21. Pilić, V., Mihajlović, V., Stanojević, P., Anđelković, A., Baloš, D.: Application of innovative risk assessment methodology for damage mechanisms identification on part of amine regeneration unit. Struct. Integr. Life **19**(1), 29–35 (2019)
22. Pilić, V., Mihajlović, V., Stanojević, P., Baloš, D.: Damage mechanism and barrier identification on hydrogen production unit using innovative methodology for risk assessment. Struct. Integr. Life **19**(2), 131–137 (2019)
23. Čamagić, I., Kirin, S., Sedmak, A., Burzić, Z.: Risk based analysis of temperature and time effects on brittle fracture of A-387 Gr. B welded joint. Procedia Struct. Integrit **18**, 205–213 (2019)
24. Radu, D., Sedmak, A., Băncilă, R.: Determining the crack acceptability in the welded joints of a wind loaded cylindrical steel shell structure. Eng. Fail. Anal. (2018). https://doi.org/10.1016/j.engfailanal.2018.04.032
25. Kirin, S., et al.: Risk based analysis of RHPP penstock structural integrity. Frattura ed Integrita Strutturale. https://doi.org/10.3221/IGF-ESIS.53.27

26. Sedmak, S., Sedmak, A.: Integrity of penstock of hydroelectric power plant. Struct. Integr. Life **5**, 59–70 (2005)
27. Sedmak, S., Sedmak, A.: Experimental investigation into the operational safety of a welded penstock by a fracture mechanics approach. Fatigue Fract. Eng. Mater. Struct. **18**, 527–538 (1995)
28. Vucetic, I., Vucetic, T., Sedmak, A.: Worst case scenario for hydro power plant after pressure vessel failure. Struct. Integr. Life **17**, 159–162 (2017)
29. Hrivnak, I.: Breakdown and repair of large spherical containers for liquefied hydrocarbon gases. In: IIW Doc. IX-1516-88, Bratislava (1988)
30. Jovicic, G., Zivkovic, M., Sedmak, A., Jovicic, N., Milovanovic, D.: Improvement of algorithm for numerical crack modeling. Arch. Civ. Mech. Eng. **10**, 19–35 (2010)
31. Sghayer, A., Grbović, A., Sedmak, A., Dinulović, M., Doncheva, E., Petrovski, B.: Fatigue life analysis of the integral skin-stringer panel using XFEM. Struct. Integr. Life **17**(1), 7–10 (2017)
32. Kraedegh, A., Li, W., Sedmak, A., Grbovic, A., Trišović, N., Mitrović, R., Kirin, S.: Simulation of fatigue crack growth in A2024-T351 T-welded joint. Struct. Integr. Life **17**(1), 3–6 (2017)
33. Sedmak, A.: Computational fracture mechanics: an overview from early efforts to recent achievements. Fatigue Fract. Eng. Mater. Struct. **41**, 2438–2474 (2018). https://doi.org/10.1111/ffe.12912
34. Đurđević, A., Živojinović, D., Grbović, A., Sedmak, A., Rakin, M., Dascau, H., Kirin, S.: Numerical simulation of fatigue crack propagation in friction stir welded joint made of Al 2024-T351 alloy. Eng. Fail. Anal. (2015). https://doi.org/10.1016/j.engfailanal.2015.08.028

Advantages of Using New Technology in Textile and Fashion Design

Maja Milinic Bogdanovic[1](✉) and Nebojsa Bogojevic[2]

[1] Academy of Applied Technical Studies Belgrade, 11000 Belgrade, Serbia
mmbogdanovic@politehnika.edu.rs
[2] Faculty of Mechanical and Civil Engineering in Kraljevo,
University of Kragujevac, Kraljevo 36000, Serbia

Abstract. New technology has transformed traditional approaches to contemporary art and design and has led to a completely new forms in the creative world. 3D technologies enable the production process to accelerate. Digital arts are increasingly being used by artists to express their ideas. With the application of CAD/CAM technology, today, the design solutions of designers are realized faster. Advantage of using additive manufacturing relies on its ability to produce directly based on the CAD model – 3D model, whose file is the source of information for generating the incremental control file. The specific features of additive manufacturing can be developed with digital design, and software based construction of model, which adapts to the capabilities of this technique. Today the Inovation in the textile and fashion design is in an upward trajectory. Additive production is increasingly available especially with FDM Technology (Fused Deposition Modeling). This is production based on the principle of melting deposit modeling by heating and extruding thermoplastic filaments, proved to be very affordable and efficient. Textile and fashion design is developing not only in the aesthetic sense, but also developing software programs in which products of textile design and fashion are designed. Textile design in the industry is necessary to return to a sustainable production model, which involves more localized production, and also allowing small design and production houses to be competitive in the market production. Exploit the new design technics and involve a new technologies, a sustainable production model in textile design is possible to be achieved.

Keywords: Innovation · Additive manufacturing · Textile and fashion design · Fused Deposition Modeling - FDM

1 Introduction to the New Industrial Revolution

1.1 New Industrial Revolution

Technological revolution that will fundamentally alter the way we live, work, and relate to one another is the place where we already standing. 3D printing, additive manufacturing, or rapid prototyping are the terms that are in circulation for new technologies.

N. Mitrovic et al. (Eds.): CNNTech 2020, LNNS 153, pp. 294–309, 2021.
https://doi.org/10.1007/978-3-030-58362-0_17

New industrial revolution is building on the knowledge that we had before, the digital revolution that has been occurring since the middle of the last century. It is characterized by a fusion of technologies that is blurring the lines between the physical, digital, and biological spheres. Until the late 18th century that is, which is when factories, production lines and machines took over this process. Today it appears that once more we are entering into a new era of manufacturing. One where newfangled technologies are pushing our production methods further into the future than we could ever imagine, enabling us to produce more products than ever before, at lower costs and with fewer mistakes [1]. Additive technology, which has been emerging since the 1980s, has developed in various branches of life. With fast, customized, and inexpensive products, 3D printing is poised to bring about a sea change in contemporary culture. Additive Manufacturing is considered by most in the industry as a “disruptive” technology; one that will revolutionize many industrial sectors as it becomes faster and less expensive [2]. It will also fundamentally affect how, when and where heat treatment is performed since sintering will become part of an additive manufacturing cell. As additive manufacturing becomes more sophisticated and as understanding and awareness grows among manufacturers, machine shops as we know them will be fundamentally changed. Digital fabrication technologies, meanwhile, are interacting with the biological world on a daily basis. Engineers, designers, and architects are combining computational design, additive manufacturing, materials engineering, and synthetic biology to pioneer a symbiosis between microorganisms, our bodies, the products we consume, and even the buildings we inhabit. The New Industrial Revolution has the potential to raise global income levels and improve the quality of life for populations around the world. To date, those who have gained the most from it have been consumers able to afford and access the digital world. Technology has made possible new products and services that increase the efficiency and pleasure of our personal lives.

1.2 Precise Term of New Technologies

All known concepts of new technologies, which describe production that differs from classical production methods, are the subject of discussion. To be sure, the terms overlap. They can be used in ways that make them sound like synonyms. But the relationship between them and the difference between them is this: 3D printing is the operation at the heart of additive manufacturing, just as “turning” or “molding” might be the operation at the heart of a conventional manufacturing process. In short, additive manufacturing requires and includes 3D printing, but it also entails more than 3D printing, and it refers to something more rigorous. Additive manufacturing is the term widely used when referring to the process in an industrial context and the debate raised the point that in the industry, additive manufacturing is an established technique that is known for being advanced and highly technical. Furthermore, the argument was raised that some believe there is a material distinction between additive manufacturing and 3D printing, in that additive manufacturing is exclusively to do with metals [3]. It can be recognized that what influenced the use of the term 3D printing came from various media, while the production of additives is an industry-supported term. It seems that additive manufacturing, although the proper term for new technologies, is not so often used, because the term 3D printing is already so networked in the use of media that it cannot be left out of terminology. If we look at the process of making the design of supporting structures, the operation of post-processing and consideration necessary for the realization of process control sufficient for solid repeatability, it can be seen that 3D printing is part of additive production [4]. If the focus is production, then what follows from the consideration of the correct use of the mentioned terms leads to the adoption of the correct term additive production. And seemingly, with each new development in the additive space, that promise becomes clearer. The process has now begun for companies to dedicate themselves to the production of additives in such a way as to increase the entire number of plants oriented around this way of manufacturing production parts. It seems that the debate is not finished when it is lurching foot the right term of the New Technologies.

1.3 Development of New Technologies

In the last 30 years, additive production technology has been developing rapidly. Nowadays, it covers a large number of industries and applications. This technology indicates great potential for future development and an increasing number of industries benefit from the advantages of the additive manufacturing techniques the automotive and medical industries are recognized as leaders, who have introduced new technologies [1]. This is certainly an indication that these technologies are the bearer of the future of the industry and the future. This is a bit of a change for the additive manufacturing industry which is initially be used for prototyping a series of good, but not always technology-justified reasons.

Additive Manufacturing refers to a process in which the raw material is added layer upon layer to create a component part. This is the opposite of machining, often now referred to as “subtractive manufacturing,” which creates a part by removing material from a raw-material form [13, 14]. According to the technologies within additive production, the most used categories are: Inkjet - One of the basic ways of additive production is the upgrade to a classic Inkjet printer. The model is produced layer by layer (of gypsum or resin), and the inkjet disperses the binder in the form of a model to be printed. This technology is the only one with whom is possible to print a prototype in full color. Stereo lithography-SLA (SLA) - In stereo lithography, models are produced by passing a beam of UV (ultra violet) light over a pool with photosensitive liquid. During production, the model is lowered into the tub layer by layer, until the final product is obtained. Another advantage of this type of 3D printing technology is the high level of detail and finishing of the final surface. One of the main advantages of this printing method is its speed.

In this technology, the same models are produced up to 5 times faster than in all other techniques and it is interesting to discover the final product. After the printing of the part is finished, the product is slightly lifted from the tub with the photopolymer solution from which it was made. Production of objects by lamination-LOM (LOM) - The technology of production of the model by lamination in the process implies the use of layers of material (usually paper or plastic), cut with a laser or blade, which is then joined with glue. One of the newest machines of this type is the Mcor Matrix, which uses plain A4 paper and water-based glue. This achieves a much lower cost of making the model, without harmful effects on the environment. Selective laser sintering-SLS (SLS) - the object is produced by successively adding thin horizontal layers which were melted by laser beam. Each of the layers are produced by applying a thin layer of powder above the part top surface that was previously built, which is melted by a laser beam in the form of the next layer.

When the molten part of the powder cools, it connects horizontally (forming a new layer) and vertically (merging with the rest of the part). The shape of each layer is determined by a computer, based on a computer model of the object, and based on the cross sections calculated from 3D model; it controls the process of melting the powder. SLS technology is used for direct production of complex forms and tools. The production of tools with this technology does not introduce restrictions in the geometry of the manufactured designed products, so in this way tools of improved performance are made, optimized from the aspect of their further use. For the production of large batches, any shortening of the production cycle of a single product brings significant savings in time and money for the entire batch. The layers are poured crosswise; each layer is bult at a different angle to the previous one, thus achieving the strength of the final model [15]. The level of the layer produced depends on the material and can be up to 0.3 mm high. To positive design results and with a lower value for layer height of 0.02 mm, this is an extremely delicate design. It is possible to use several different printing materials, with different characteristics, in strength and temperature properties.

The first commercial machine for the production of molten material (FDM) was patented on June 9, 1992 in the United States (US patent 5121329). It involves a process in which a thread of polymeric material is wound on a spool and which passes through an extrusion nozzle. The nozzle is heated where the molted plastics is extruded trough it. The whole process of applying the layers takes place in a chamber in which the temperature is maintained only slightly lower than the melting temperature of the polymeric material.

FDM technology was developed more than 20 years ago by Scott Crump. Different materials can be used with FDM technology to produce parts. In some applications, a support structure is needed, which is also printed as a temporary structure, which helps maintain the desired model, and is removed after the process is completed. Immediately after removing the support, the model is ready for use. By using printers that can use two different filaments at the same time, it is possible to produce a model with a support that decomposes after printing with water [9].

The advantages of FDM technology are that the FDM process is clean, simple and suitable for working in a small space. Thermoplastic parts can withstand exposure to heat, chemicals, wet or dry conditions and mechanical stress. During their development, additive technologies were not adapted to mass production at the very beginning. This aspect of production can still be improved. This unadjusted parameter was conditioned by the fact that the new technology was primarily used for the release of product prototypes. At this point, it has progressed to its sustainability, but as noted, further progress is needed in this area, for high-volume manufacturing. The prime reason was the available materials the test samples were suitable and the purpose of 3D printing, but not necessarily useful for end-use parts. At the outset, many of the early prototyping was focused on testing the simple geometry and simple shape of an object, rather than its intended function, which can have very demanding design solutions that can mimic the complex construction of biform.

1.4 Advantages of New Technologies

Deemed among the most disruptive technologies of our time, additive manufacturing offers exciting possibilities for future development and is already considered to be at the forefront of the fourth industrial revolution. Small batch of unique or complex parts can be produced quickly and at low cost. Unlike Metal Injection Molding (MIM), casting or forging, no expensive molds are required [13]. This reduces time to market, a very valuable commodity today. Also, it is clear that shrinkage is significantly less than that of MIM-produced parts, increasing accuracy and repeatability. One description of Additive manufacturing is that it is a MIM process without the distortion. Additive manufacturing has the ability to pursue new innovations without extending the design cycle. This allows for many generations of design changes in the time that it would normally take to make a single change using conventional technologies. This might be the most revolutionary aspect of the technology. A different lattice structure designs are possible which reducing the part weight while maintaining or even increasing strength. Additive manufacturing offers the ability to make on-the-fly changes. If there's one thing design engineers can count on, it is customer revisions and design changes. With these technologies, the designer simply makes a change in the 3D digital model, which may be relatively quick, be prepared for manufacturing. Highly complex parts can be produced that would be literally impossible with any other technology [1]. There are some shapes and intricate features that cannot be cast, molded or machined but can be additively manufactured. This opens up new possibilities for designers. Regarding that a special mold tools are not used in the porches of the production a high degree of customization is possible without added cost. Additive manufacturing technology allows the manufacture of one-of-a-kind designs like medical implants that are custom made to fit a specific individual. Additive manufacturing generates no waste. Since it is an additive technology, only the material that is needed is actually used. When printing very expensive metals such as titanium or gold, this makes a huge difference in the price of the finished product and the feasibility of the project. In some research, by 2025, the global market for 3D printing and additive manufacturing services is expected to grow to almost US$50 billion, and by 2030 we are expected to see a monumental shift in the global additive manufacturing market, from prototyping to the mass production of parts and accessories. In the automotive industry the building of tooling components and patterns for metal castings using additive manufacturing – which currently accounts for around nine percent of global additive manufacturing spending – will also be an area that will see incredible growth due to continued advances in the technology. Additive technology has always been an attractive choice when production volumes are low, changes are frequent and complexity is high. As build speed increases and costs come down, applications from the additive technology will expand to include more mainstream component parts. Machine shops and in-house manufacturing departments will then be able to choose the most cost-effective technology, with sintering being performed as part of the additive technology manufacturing cell as opposed to a heat-treatment department or an outsource location. The essence of manufacturing is to serially produce items that are always within specified tolerances for various engineering and geometric dimensions.

New additive manufacturing equipment is capable of producing objects in materials actually suitable for end-use applications, such as the many different metal alloys, as well as the new high temperature materials.

With the discovery of these material capabilities, some industries are now producing end-use parts. But by doing so they must then perform the required manufacturing quality control checks. And the additive manufacturing machines being used generally have not included that capability, having inherited much of their design from the previous world of prototyping. This is changing to.

2 Revolution in Design and Art

2.1 Revolution in Contemporary Design

In contemporary design, digital arts are increasingly being used by artists to express their ideas. With the application of computer programs, today, the design solutions of designers and artists are realized faster. Ever since the 1970s, there have been attempts to define the term digital art as well as the terms computer art and multimedia art [7].

Today, we interpret this term in a broader term that is defined in English as a new media art. In addition to the progress that has taken place in the way of expressing contemporary design through the digitization of conceptual solutions, the application of additive production has also improved the production processes of the finished product. Additive production is increasingly available, especially with the development of Fused Fusion Technology (Fused Deposition Modeling) - a model of frozen deposits and printers that are more affordable to their market prices and easy to use. When it comes to this, a smart design, which draws its stronghold in its perfect nature and its billions of years of research and perfected survival solutions, comes to the brilliant solutions that have won the field of architecture, furniture design, fashion design, textiles in recent years.

The development of digital art is flowing along with the development of computer technology. In digital painting, new results and effects are generated through computers and using specific algorithms, as well as using traditional painting techniques. Digital visual art consists of 2D visual information displayed on an electronic visual display of information that is mathematically translated into 3D information that is viewed through a perspective projection on an electronic visual display.

When we look at a design process, we can see that in addition to classical drawing on paper, the work is even more digitized and where from 2D visualizations, which can be displayed in electronic form, it is necessary to translate them into 3D objects, in information that visualizes the desired design in a software program. Digital technology has transformed traditional approaches in contemporary design and art and has led to completely new forms.

With the application of computer programs, today, the design solutions of designers and artists are realized faster. Some of them are Cura, Repetier, 3D StudioMax, Shoemaster, Adobe, Corel Draw, Maya and many others. Indeed, today contemporary design, and all the rest, cannot do without the knowledge of some of the programs that we can be helpful in developing solutions design or art [9]. Certainly the traditional approach to developing conceptual designs is recommended, but if the hand-drawn motifs imported into design software programs realization repeating motif is rapid, precise and assist the work of designers and artists.

Additive production is growing more and more. 3D technologies enable the production process to accelerate. It is necessary to digitally prepare the product itself in some of the software that supports the drawing of 3D models, and digital art and additive production are closely linked.

The basic characteristic of additive production is that objects are formed by adding materials, in contrast to traditional technologies in which objects are produced by subtracting materials (scraping, milling, cutting) or forming materials (pressing, casting, forging). Building objects with additive technologies, unlike traditional technologies, does not require the use of tools [2, 13]. As a result, additive technologies are the fastest possible prototypes of new products.

Additive production is increasingly available, especially with the development of Fused Fusion Technology (Fused Deposition Modeling) - a model of frozen deposits and printers that are more affordable to their market prices and easy to use [4]. When it comes to this, a smart design, which draws its stronghold in its perfect nature and its billions of years of research and perfected survival solutions, comes to the brilliant solutions that have won the field of architecture, furniture design, fashion design, textiles in recent years.

2.2 CAD Modeling

Designers are increasingly creating virtual 3D CAD models as part of their final project deliverables. These digital models also routinely allow designers to use affordable 3D printers (which have become standard equipment at many industry and design schools) to create physical prototypes.

In an advanced computer applications elective at designing process, creators were challenged to reverse design a plastic housing component from a handheld electronic product; for example, a computer mouse or remote control. Designers had to study the fine nuances of form through digital means such as digital photos and measurements. When their designs have to be recreated in the products, they often realized that the products were more complex in terms of geometry than they had initially anticipated.

They also had to print the part on the departments Dimension 3D printer and try to reassemble the printed prototype back into the original product [11]. Designers spent four to five weeks on this process and most were unable to replicate the original parts perfectly.

All creators gained a better appreciation for formal complexity. During the development of an industrial or design product, it is necessary to gradually work through detailed design steps [14]. From idea to realization is a complex process. In addition to sketching, elaboration of ideas, transfer from 2D drawings to a 3D object, the process of searching and making prototypes, this can also be determined in the experimental part of the paper. By correcting possible errors, the prototype is improved and closer to the finished product that would go into mass production. Prototyping is one of the key applications for 3D printing; virtually every 3D printer ever made has produced at least one prototype—unless it hasn't been powered up. Through constant experimental work, it is possible to better master the best practical parameters of prototypes in any field of design, but also industry.

That wasn't always the case, as in the past prototypes were frequently made by hand, due to the cost of setting up manufacturing equipment to make only a single unit. There are considerable efforts required to work with other people during these iterations to more deeply understand the real problems being solved through the implementation of the design.

Prototypes in the further process of elaboration of all parameters are subjected to detailed examination. This part can be entrusted to real users who will be in direct contact with the prototype to see all the parameters that are expected for the good functioning of the finished product.

2.3 "Design of Textile Flexible Materials with Application of Additive Manufacturing"

The results of this research work, related to the new technologies and their application in textile and fashion design, showed that the application of new technologies in this field is possible. In our space, so far, this segment of design has stood still and has not progressed, if we look at technological progress. The conditions have been reached for the use of 3D printers to start in Serbia in the field of textile design. There may be resistance to the application of new technologies by textile designers who continue to create in the traditional way, but it is necessary to understand that the influence of new technologies in this field of design does not conflict the interests of new and traditional, but gives the textile industry a chance a very difficult and unenviable situation and in which she found herself. New technologies, and as already mentioned, speed up procedures, reduce production costs, primarily by avoiding errors, and their rapid elimination through prototyping, also reduces the consumption of materials, and if it refers to natural materials, it further enters the world environmental protection. It is inspired with Word of Fractals, Biomimicry, the Beautiful Nature and solutions that it creates. These natural elements were applied here in the context of contemporary textile and fashion design. Design is developing not only in the aesthetic sense, but also with implement the solutions in software programs. Modern manufacturing methods suggest that it will be available to anyone that has a design product for its custom. Additive manufacturing continues to take an increasing share in the world of design and art. As the additive manufacturing gives a lot of freedom in creativity there is an almost endless array of possibilities and as the technique is developing more solutions will be possible. Especially FDM Technology (Fused Deposition Modeling) is proved to be very affordable and efficient. Digital technology has transformed traditional approaches in contemporary design and art and will bring it to completely new forms. Biomimicry, as the practice of designing based on the nature, is the inexhaustible inspiration source of contemporary design that applied using additive manufacturing in all branches of design. Through numerous examples of 3D printing, it can be seen that a design inspired by biomimicry creates stronger structures, smarter technology and creative aesthetics. The flexibility of textile surfaces produced by additive production is achievable. When this will be reached the additive production will take an increasing share in this field of design. Textile plays an important role in the development of multidisciplinary areas that arise in the processes of pursuing and applying new technologies and the research and reconsideration of the artistic and aesthetic values of structures of textile surfaces. The aim of this work is the affirmation of Additive Manufacturing in textile design, by creating wearable materials with use of FDM technology. Especially FDM technology, production on the principle of melting deposit modeling, which is in the form of coils with winding material in the form of threads, has proved to be very affordable and efficient. Used materials are different and can be biodegradable or not, flexible or solid [11, 12]. This technology is based on model design using solid materials on the principle of extrusion through the nozzle. Basically, the plastic fiber is constantly coming through a small diameter nozzle. The nozzle was heated and the material was melted and applied in layers. During the application of the material, the nozzle is moved in the X-Y plane by pushing the material evenly. Upon completion of the application of one layer, the work table performs a shift along the Z-axis and in this way begins the application of the next layer. The FDM procedure is clean, simple and suitable for work in a small space. Thermoplastic parts can withstand heat, chemicals, wet or dry conditions and mechanical stress. The material used to produce the collection's items was the flexible TPU material NinjaFlex, which can be tricky to be extruded.

Kerry Stevenson, who has been writing posts for the Fabbaloo - 3D printing news, since 2007, wrote the following about the mentioned work on November 19, 2018, through an interview: "This flexible material was required because the stresses caused when wearing clothing require repeated movement and elongation; more rigid materials would likely crack, even if they do bend somewhat when printed at small thicknesses. The most surprising aspect of the project was the machine on which the collection was 3D printed: a basic Printrbot Play. This is a small, inexpensive desktop device that one might suspect could not produce a comprehensive project like this collection. The desktop of this 3D printer allows the development of the models in dimensions 100 × 100 × 100 mm. Producing smaller pieces, which were then joined together after 3D printing, it was possible to complete the larger works, which were done in black and white colors" (Fig. 1).

Also in the interview author sad that the subject of this designed work is the research into new aesthetics of flexible structures of surfaces in textile design. The surfaces of textile, used for clothing, were created with imaginative employment of biodegradable materials, with application of additive manufacture. In the research and the reconsideration of artistic and aesthetic values of the structures of textile surfaces, various kinds of fabric were produced with contemporary technical and technological procedures, starting with a selection of authentic elements created by the nature, as an everlasting source of inspiration for artists. These natural elements were applied here in the context of contemporary textile. Unparalleled is the ability of nature to harmonize beauty, economy and functionality, so it is no coincidence that the great inventions in history have their sources in analogous natural elements. The solutions offered by nature can contribute to the creative process using analogy and at the same time its geometric and mathematical model [8]. For example, permanent mathematical proportions in the physical structure of humans and animals, as well as plant structures, can be observed. This geometry in nature is often combined with the concepts of aesthetics, harmony and balance, making

Fig. 1. "Phantasm" - veil, Maja Milinic Bogdanovic, FDM technology, Ninja Flex material, size 38. Exhibited at Jade Gallery of Museum of Applied Art in Belgrade, Serbia, 2018 [11].

up the true aspect of beauty. The concept of biomimicry, widely accessed in contemporary science and art, consists of analyzing the principles of natural elements and their transfer to specific technology solutions [5]. Nature was a source of inspiration to people, inspiring art, design and architecture. Nature has found numerous solutions to the problems that we are still struggling with. Modern designers have realized this, by using nature; they apply its principles in their design solutions (Fig. 2).

Textile plays an important role in the development and the correlation of multidisciplinary areas that arise in the processes of pursuing and application of new technologies [6]. In the future, the development of design will take part in the advancement and linking of multidisciplinary areas by monitoring new technologies through the concept of work and the contemporary term, which are key elements with a starting point.

The goal of artistic thinking is to support the united position that in the field of applied and fine arts, the influence of new technologies has its justification. Digital art is a new form of expression in contemporary art of the 20th century. This term describes various works of art created by the use of digital technology. Depending on the application of technical means, software or hardware, there are also different types of digital art that continue to develop with the development of computer technology.

Lot of digital technology has transformed traditional approaches to contemporary art and design and has led to a completely new form. The artistic-research contribution has the potential in terms of perceiving innovative patterns for the development of textile design. Technological development is characterized by a series of new technologies, a fusion of science, design and art, it connects the digital and biological worlds, and it will certainly have an impact on the textile industry and the position of designers

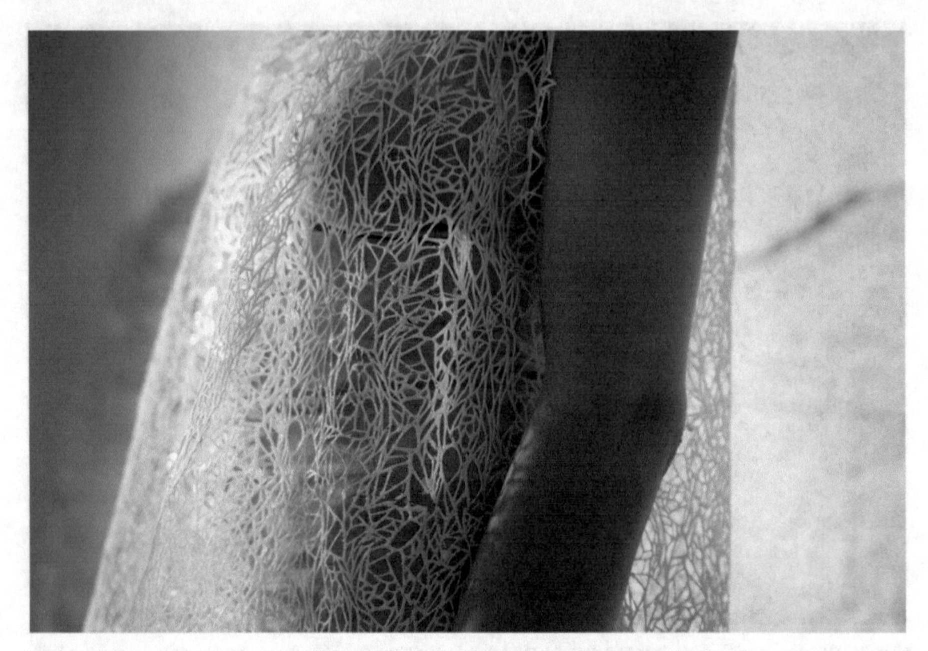

Fig. 2. "Phantasm" - blouse, Maja Milinic Bogdanovic, FDM technology, Ninja Flex material, size 38. Exhibited at Jade Gallery of Museum of Applied Art in Belgrade, Serbia, 2018 [11].

in today's work environment. Additive production can improve the use of waste that greatly pollutes our environment. It enables the creation of designer products that are environmentally friendly. Unlike many traditional manufacturing processes, which have different limitations, additive manufacturing allows greater flexibility for manufacturers to optimize, quickly and efficiently, the design for production. Waste is eliminated by such production. Additive production achieves the possibility of topological optimization of the design, and thus increases the functionality of the product, which further reduces the amount of energy consumed and other natural resources necessary for the functioning of a production process.

Ergonomics is an important part of the fashion design. It involves the science in designing products, so that the products are best adapted to the human body. Today, it cannot be ignored that human psychological and social limitations, needs, requirements can also be limited in the use of a product and that they should be taken into account when designing a technical device or technical system. The emergence of ergonomics is associated with the rapid development of technology and technical means that were more and more perfect and efficient, but then man appeared as one who, with his limitations, became a limiting factor in his development. Technically, taking the most perfect tool is not ergonomic if a person with his bio-psycho-social characteristics cannot use it efficiently, and that is today a limiting factor of technical and technological development. Ergonomics is practiced today by biologists, anthropologists, psychologists, sociologists, but all the knowledge of each of the listed and unlisted professions means nothing if they are not integrated into a set of unique and harmonized requirements that must be set before the designer of the technical device. Movement is thus essential for the body, and therefore extremely important in this work. The desire is to achieve optimal comfort and undisturbed movement by combining form, structure and materials together in a new way (Fig. 3).

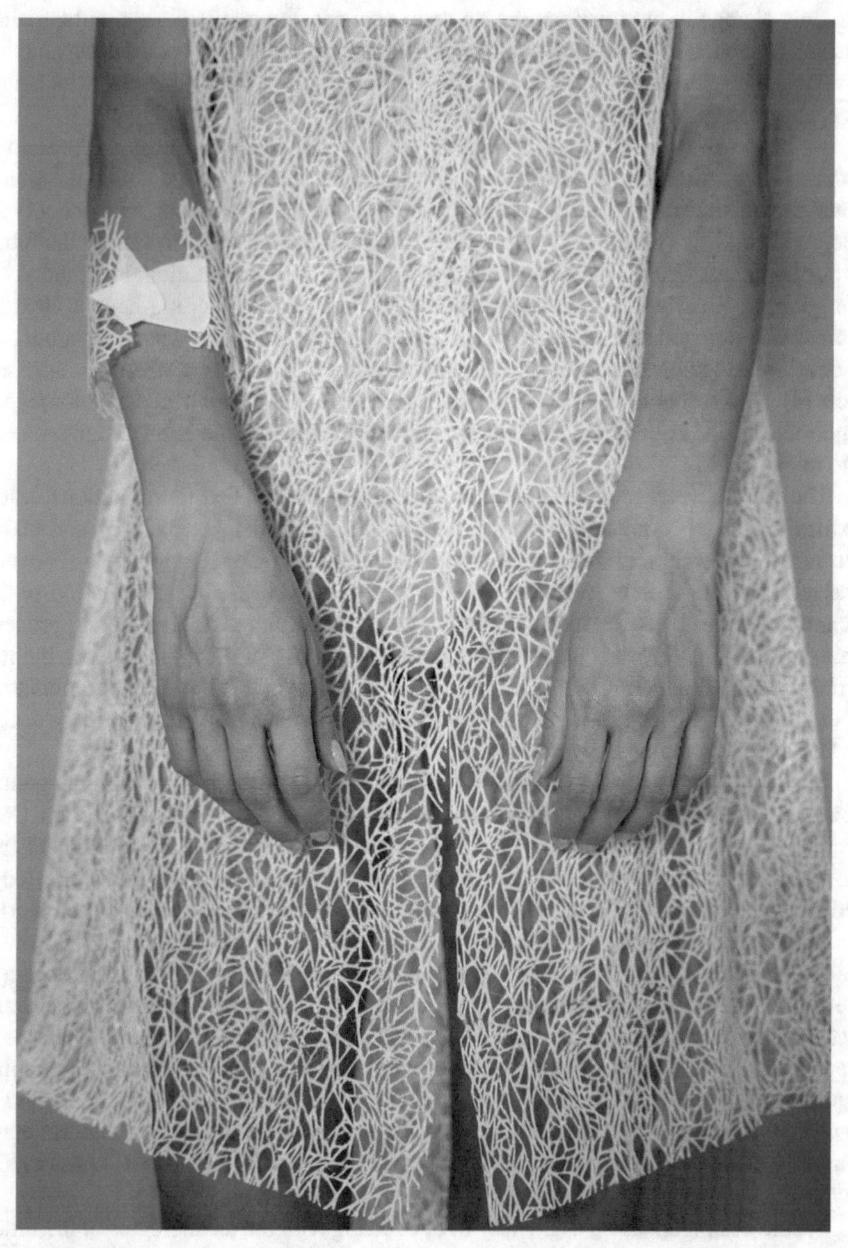

Fig. 3. "Phantasm" - coat, Maja Milinic Bogdanovic, FDM technology, Ninja Flex material, size 38. Exhibited at Jade Gallery of Museum of Applied Art in Belgrade, Serbia, 2018 [11].

Textile surfaces made by additive production can be applied in the design of modern clothing, but also in interior design. With the further development of these technologies, a new freedom of design is achieved, which will open new borders or, more precisely, the borders will be broken. One of the conclusions of this paper is that the design will certainly change in the future, and with these changes, the textile design will certainly change. By conquering a new world of materials and a modern way of thinking and thus new concepts of creativity, new qualities of textile surfaces will be achieved, which must be accompanied by scientific achievements.

As the work further recognizes one of the key factors for progress in the field of textiles is education. The desire for innovation can also be recognized within the educational system, because more and more academies, vocational schools, high schools are integrating additive production within their study programs. That is certainly the future of our humanity. Within the design, textiles are still based on traditional methodology. For good progress in the field of printed textiles, but also in other segments of textiles such as weaving and tapestry, it is necessary to preserve traditional knowledge, but also to develop new knowledge and skills in the use of modern means of making textile surfaces [4]. New technologies allow textile designers to expand their achievements beyond traditional design boundaries, thus turning some of the most challenging design concepts into reality.

The evolution is recognized through traditional methods of textile production towards textiles that have completely grown out of digital technologies. The idea that textile education now embraces these innovative technologies is an encouraging step forward. The education of a new generation of designers, as well as the introduction of 3D technologies and software in textile design is a naturally gradual and lasting process. This may explain why the change in the textile industry is, at first glance, slow, but it is certain that the situation in this field of design will change in the future. The combination of digital technologies has transformed traditional approaches in contemporary design and art and led to completely new forms [10].

Like other areas of life and industry, fashion and textile design will change and improve. How fast progress will be made in these areas depends on many factors. Connecting multidisciplinary areas is one of the steps mentioned, but the key elements are for the designer profession, of these areas, must initiate and improve their knowledge and approach to work, by mastering the use of new technologies for creating their works.

The concept of additive production has not yet reached the mass market, however, the implications of solutions for such production are being considered. Textile design in the industry is necessary to return to a sustainable production model, which implies more localized production, enabling small design and production houses to be competitive in the market. The transformation of the perception of textile design and modern clothing continues. What will be the consequences of the introduction of additive production in design, textile design, and modern costume is a question that will find its answers in the future. In a world of constant technological advances, textiles can also be considered a key means of showing the great possibilities of 3D production.

Textile design opens up a new way of connecting with the technologies of the future and enables greater participation of this branch of design in its progress [12]. The huge flexibility of additive manufacturing is another important advantage for the industry. It is now available to create machines so that they are perfectly adapted to a certain size and ergonomics of each part of the body, enabling real personalization (Fig. 4). Thus, the role of 3D printing in textiles is constantly evolving, with a significant increase in awareness and interest related to the introduction of new technologies by designers. The growth of curiosity and curiosity comes from the full spectrum of the design profession.

Fig. 4. "Phantasm" - hood, Maja Milinic Bogdanovic, FDM technology, Ninja Flex material, size 38. Exhibited at Jade Gallery of Museum of Applied Art in Belgrade, Serbia, 2018 [11].

3 Conclusion

What will influence the future development of textile design, as well as fashion design, is based on the accelerated application of innovations. Using additive technologies, which are a source of constant surprise and constant progress in the development of 3D machines as well as materials and improving the parameters of existing materials used in additive manufacturing, we can expect a change in the approach to work in these industries. Great impact on the good performance of companies that follow new trends, especially to keep pace with the fourth industrial revolution, which is considered, there is precisely the use of additive technology. There are four main effects that the New Industrial Revolution has on the development of the textile industry and fashion design, which are recognized in customer expectations, product improvement, innovation in producer-consumer cooperation, as well as in organizational forms of production. Today, the textile industry in this area is in a position doesn't prosper, but stagnate. The reason for this can be found in the lack of understanding, by textile designers themselves. It is necessary for designers to change their attitude as well as to improve their knowledge, through learning CAD modeling, as well as through learning how to use new technologies

[16]. The view that nothing should be changed in the process of making textile products will make stagnation in the industry of textile. When it comes to improving customer service, it is important to monitor product development according to different consumer requirements. It is new technologies that can improve product quality very quickly and efficiently.

Textile products produced with additive technology are more durable, resistant and produced during the user experience, and in the process of prototyping everything can be analytically based on data from all product performances. There for it can be an improvement in production in an extremely short period. Organizational forms that are recognized through new business platforms will find their application through the fourth industrial revolution and the innovations must be by combining modern technologies [17]. The third industrial revolution brought the digitization and within the current industrial revolution, fourth, new technologies will take precedence over conventional production. Many industries, as well as the textile industry, are seeing the introduction of new technologies that create entirely new ways of serving existing needs and significantly disrupt existing industry value chains.

An important role in the process of improvement is played by business managers who, with their attitudes, must initiate and continuously innovate of the production within their companies. We should thus grasp the opportunity and power we have to shape the New Industrial Revolution, with Additive manufacturing, and direct it toward a future that reflects our common objectives and values.

Acknowledgments. The authors would like to acknowledge support of the Serbian Ministry of Education, Science and Technology Development through contract no: 451-03-68/2020-14/200108.

References

1. Croccolo, D., De Agostinis, M., Fini, S., Olmi, G., Robusto, F., Ćirić-Kostić, S., Morača, S., Bogojević, N.: Sensitivity of direct metal laser sintering Maraging steel fatigue strength to build orientation and allowance for machining. Fatigue Fract. Eng. Mater. Struct. (2018). https://doi.org/10.1111/ffe.12917
2. Vranić, A., Bogojević, N., Ćirić Kostić, S., Croccolo, D., Olmi, G.: Advantages and drawbacks of additive manufacturing. IMK-14 Res. Dev. Heavy Mach. **23**(2), EN57–62. UDC 621 (2017). ISSN 0354-6829
3. Barnatt, C.: 3D Printing-second edition, Explaining TheFuture.com (2014)
4. Croccolo, D., De Agostinis, M., Fini, S., Olmi, G., Bogojevic, N., Ciric-Kostic, S.: Effects of build orientation and thickness of allowance on the fatigue behaviour of 15–5 PH stainless steel manufactured by DMLS. Fatigue Fract. Eng. Mater. Struct. **41**(4) (2017). https://doi.org/10.1111/ffe.12737
5. Clarke, S.E., Braddock, J.H.: Digital Visions for Fashion and Textiles: Made in Code, 1st edn. Thames & Hudson, London (2012)
6. Clarke, S.E., Braddock, M.O.: Techno Textiles 2: Revolutionary Fabrics for Fashion and Design. Thames & Hudson, London (2008)
7. Horvath, J.: Mastering 3D Printing. Heinz Weinheimer, New York (2014)
8. Kapsali, V.: Biomimicry for Designers. Thames & Hudson, London (2016)

9. Lipson, H., Melba, K.: Fabricated-The New World of 3D Printing. Indianapolis, Wiely (2013)
10. McCue, T.J.: 3D Printing In The Home: 1 In 3 Americans Ready For 3D Printer. Forbes (2014)
11. Milinic Bogdanovic, M.: Design of textile flexible materials with application of additive manufacturing, Ph.D. work, Faculty of Applied Arts in Belgrade, University of Science Arts in Belgrade (2018)
12. Milinic Bogdanovic, M.: Additive manufacturing in design inspired by biomimicry. Full paper for Conference Published in at Graz University of Technology, Gratz, Austria (2018)
13. Croccolo, D., De Agostinis, M., Fini, S., Olmi, G., Robusto, F., Ćirić Kostić, S., Vranić, A., Bogojević, N.: Fatigue response of as-built DMLS maraging steel and effects of aging, machining, and peening treatments. Metals **8**(7), 505 (2018). https://doi.org/10.3390/met8070505
14. Rashid, K.: Digipop. Taschen, London (2004)
15. Campione, I., Brugo, T.M., Minak, G., Tomić, J.J., Bogojević, N., Ćirić Kostić, S.: Investigation by digital image correlation of mixed mode i and ii fracture behavior of metallic IASCB specimens with additive manufactured crack-like notch. Metals **10**, 400 (2020). https://doi.org/10.3390/met10030400
16. Ritland, M.: 3D Printing with Skech, Up. Packed Publishing Open Source, Birmingham, Mumbai (2014)
17. Wapner, C.: Progress in the Making: 3D Printing Policy Considerations through the Library Lens, no. 3. American Library Association (2015)

Effects of Dispersion and Particle-Matrix Interactions on Mechanical and Thermal Properties of HNT/Epoxy Nanocomposite Materials

Aleksandra Jelić[1(✉)], Aleksandra Božić[2], Marina Stamenović[2], Milica Sekulić[3], Slavica Porobić[3], Stefan Dikić[4], and Slaviša Putić[4]

[1] Innovation Centre, Faculty of Technology and Metallurgy, University of Belgrade, 11000 Belgrade, Serbia
jelicalexandra@gmail.com
[2] Belgrade Polytechnic, 11000 Belgrade, Serbia
[3] Vinča Institute of Nuclear Sciences, University of Belgrade, 11000 Belgrade, Serbia
[4] Faculty of Technology and Metallurgy, 11000 Belgrade, Serbia

Abstract. Halloysite nanotubes (HNTs), naturally occurring as aluminosilicate nanoclay mineral, have recently emerged as a possible nanomaterial for countless applications due to their specific chemical structure, tubular shape, high aspect ratio, biocompatibility and low toxicity. In this study, HNTs were incorporated into the epoxy resin matrix to improve its mechanical properties and thermal stability. However, heterogeneous size, surface charge and surface hydrogen bond formation, result in aggregation of HNTs in epoxies to a certain extent. Three specific techniques were used to integrate HNTs into neat epoxy resin (NE). The structure and morphology of the embedded nanotubes were confirmed by Fourier-transform infrared (FTIR) spectroscopy and X-ray diffraction (XRD). Tensile testing was carried out and the fractured surface of the tested specimen was analysed using scanning electron microscopy (SEM). The thermal stability of the prepared nanocomposite materials was investigated by thermogravimetric (TG) and derivative thermogravimetry (DTG) studies. The obtained results indicated that improved properties of HNTs/epoxy nanocomposite materials were related to the unique properties of well-dispersed HNTs, agglomerate scale, and reduced void presence, and could be controlled by the manufacturing processes.

Keywords: Halloysite · Nanocomposites · Mechanical properties · TGA · SEM

1 Introduction

Due to their specific properties and potential applications in all areas of human life, polymer nanocomposite materials have attracted great attention over the past three decades [1–3]. In most cases, the addition of nanofiller results in improvement in mechanical, thermal, optical, and electrical properties [3–6], as well as barrier [7], antimicrobial [8], and

N. Mitrovic et al. (Eds.): CNNTech 2020, LNNS 153, pp. 310–325, 2021.
https://doi.org/10.1007/978-3-030-58362-0_18

antioxidant properties [9]. However, the efficiency of polymer nanocomposite materials is influenced by the nanofiller characteristics (including aspect ratio, volume fraction, size and shape, and basic surface area) and, the phase interactions [3, 5, 9–13]. One of the key explanations for the usage of nanoparticles is the high surface-to - volume ratio, which reduces the amount of particle-matrix contacts, thereby enhancing the impact on the overall material properties [14–16]. Based on dimensions, nanofillers are categorized as: two-dimensional layered (layered silicates, graphene, MXene) [17], one-dimensional (carbon nanofibers and nanotubes, montmorillonite nanoclay, halloysite nanotubes) fibrous [17–20], and zero-dimensional (spherical silica, semiconductor nanoclusters, quantum dots) [21, 22].

One of the biggest challenges in preparation of nanocomposite materials is the nanoparticle dispersion in polymer matrices [4]. In the cases of well-dispersed nanofillers composite materials with more desirable properties are obtained. However, the main problem is the tendency of nanofillers to agglomerate or cluster, because of the presence of intermolecular Van der Waals interactions causing decreased material performance. The formation of agglomerated nanofillers is connected to the formation of voids that represent favorable crack initiation and failure sites [23]. Another important factor is the influence of nanofillers on polymer structure and properties (density, stiffness, elongation, etc.), depending on the structure of the polymer and the nanoparticle surface charge, covalent bonds, ionic bonds, and affinity effect interactions among the nanofillers and the polymer matrix [3, 23, 24].

Halloysites (HNTs) are novel nanoparticles with chemical structure similar to kaolinite, but with higher water content and different shapes: tubular, semi rolled forms, platy, and spherical [14, 21, 22, 25–27] depending on the environment. The structure of HNTs was described as silicon oxide with aluminium oxide structure beneath, chemical formula $Al_2Si_2O_5(OH)_4{\cdot}nH_2O$ [28]. Despite the belief that HNTs are filled nanotubes, it has been shown that the halloysite nanotubes are hollow in structure. They consist of several layers of porous aluminum oxide and silicone oxide walls between which are located water molecules that can be extracted by vacuum or heating (Fig. 1) [29]. Prior studies have shown that halloysite nanotubes consist of SiO_4 tetrahedral and $AlO_2(OH)_4$ octahedral layers bound to the tetrahedral layer by apical oxygen [30]. The tetrahedral layer comprises an SiO tetrahedron bound to three neighboring tetrahedrons by exchanging one basic oxygen atom, which creates an endless two-dimensional layer of hexagonal gaps in the ground. The octahedral layer, on the other side, comprises octahedrons of AlO interconnected by edges that form layers of hexagonal and pseudohexagonal symmetry. The units of tetrahedral and octahedral layers are interconnected by the tetrahedron's apical oxygen, so that one surface of the unit layer consists solely of the tetrahedral layer 's basic oxygen atoms, and the other surface consists of octahedral layer OH groups. The stacking of layers of units is in the c-direction. In hydrated HNTs, a layer of water molecules separates the unit layers and the layer distance is 10 Å [30–33]. The dehydrated HNTs form 7 Å type and cannot be rehydrated to 10 Å. The outer nanotube diameter varies from 20 to 190 nm, and inner diameter from 10 to 100 nm and length ranging from 500 nm to 1.2 μm [5, 26, 34].

This non-toxic, biocompatible material [2, 35] with strong physical and electrochemical properties due to its unique shape, low price and quality [36] has been used as a drug

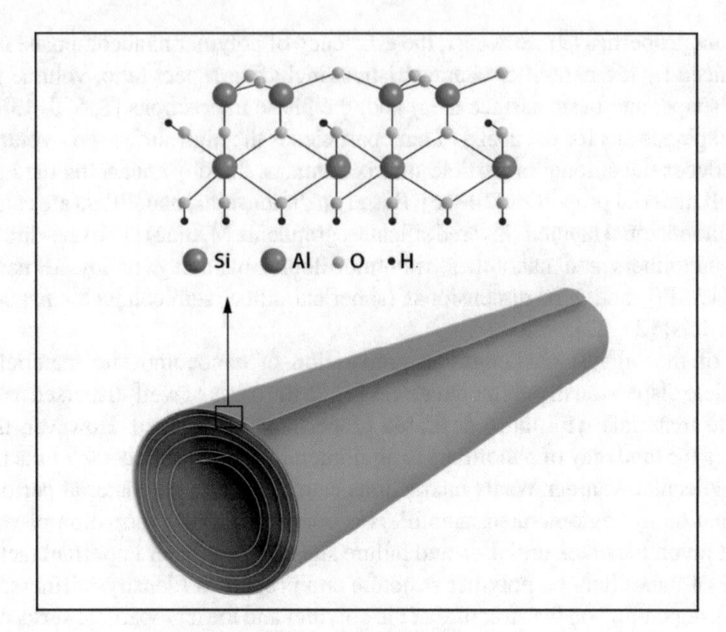

Fig. 1. HNT structure

carrier [37–42], to store energy [43–46], as an adsorbent for remediation of pollutants [47–50], or like filler in composite materials [32, 41, 43, 44, 46, 51–69]. High specific surface area, high aspect ratio and good mechanical properties contribute wide application of HNTs in polymer composites [32]. Halloysite nanotubes affect mechanical properties (tensile strength, impact strength, flexural properties, etc.), thermal stability and flame retardancy of nanocomposite materials, and on the crystallization behavior of semi-crystalline polymers. HNTs possess high flexibility against mechanical loading and high impact energy could lead to plastic deformation of HNTs. Modulus of elasticity of HNTs varies from 130 to 340 GPa. Halloysite has Mohs hardness of 2.5. Thermal properties of HNTs have been much investigated. However, the specific heat, thermal diffusivity, and thermal conductivity were investigated up to 280°. It was concluded that the water loss at 100 °C influenced a sharp decrease of thermal conductivity and diffusivity. Since the values of thermal conductivity and diffusivity of HNTs are very low, they are used as additives for thermal insulation. The loss of water at higher temperatures enabled the application of HNTs as halogen free flame retardant. Nevertheless, the tubular structure maintained intact at up to 970 °C. Due to low coefficient of thermal expansion (CTE) of HNTs, it is used to reduce the CTE of polymers. Hollow tube structure indicates 20% of empty volume. Average density of HNTs varies from 2.14 to 2.59%.

Due to wider usage of nanostructured materials, it is important to understand morphological changes and degradation under different operating conditions, the stability of HNTs and transformation. It is possible to mix HNTs with plastics, rubbers, coatings using basic mixing methods: two-roll mill and screw extruders depending on polymer type, the nanofiller dispersion, and improved interfacial bonding.

However, it is known that overall performance of HNTs-reinforced polymer composites depends on the degree of HNTs dispersion and the interfacial interaction between the nanotubes and the polymer matrix [32, 43]. Behavior of HNTs in polymer composites is related to the surface charge of halloysite. Regardless the medium polarity, HNTs disperse with or without the dispersants and coupling agents. However, lower degree of HNTs dispersion influences lower impact resistance. Better HNTs dispersion leads to easier processing and higher filler load.

Li et al. grafted polymers onto HNTs via ATRP deriving the corresponding polymeric and derivative nanotubes modified with polyacrylonitrile and polystyrene [70]. Deng et al. used ball mill homogenization and potassium acetate for treating HNTs in order to reduce the size of halloysite agglomerates in epoxy matrix. Nonetheless, cetyl trimethylammonium chloride and silane were not suitable for epoxy resin matrix due to the increased nanofiller agglomeration capabilities [71]. Studies reveal positive impact of surface-modified HNT on properties of polymer composites [10, 70, 72–78]. Using melt mixing, solution mixing and melt-cum-solution mixing techniques Ganguli et al. have developed inorganic-organic polymer nanocomposite material with improved mechanical and thermal properties [79], while Ismail et al. used mechanical and solution mixing to prepare HNTs/natural rubber nanocomposite materials with improved properties [80]. Halloysite nanotubes provide the right range of surface size and dispersibility. Because it's the longest dimensions in the 1–2 μ range, the particles are too small to act as voltage concentrations. This means that resistance to impact is maintained, or even improved. Elongation at break maintains at high level or is enhanced in some cases compared to unfilled materials. The incorporation of 2% HNTs strengthens the composite material significantly while leaving values of density largely unchanged.

In this paper different methods of preparation of HNTs/epoxy nanocomposite materials have been developed in order to improve mechanical and thermal properties of the already known materials. High aspect ratio enables HNTs to be used as reinforcement of plastics, elastomers, and coatings. High surface area influences the remediation process of heavy metals. Hollow structure allows controlled release, thermal insulation and influences light weight of HNTs. Due to the presence of water molecules between the layers HNTs are fit to be used as flame retardants, temperature indicators, and foaming agents. The purpose of this paper is to enhance HNTs dispersion and interfacial interaction among HNT and epoxy resin matrix varying basic parameters of nanocomposite preparation and their effect on the mechanical and thermal properties of newly synthesized nanocomposite material.

2 Experimental

2.1 Materials

Halloysite nanotubular clay was obtained from Aldrich Chemistry (30–70 nm × 1–3 μm, nanotube). Medium-viscosity epoxy resin based on bisphenol-A (Araldite GY 250) with epoxy equivalent weight of 183–189 g/Eq, monofunctional, aliphatic, reactive diluent for epoxy resins Araldite DY-E with EEW of 278–317 g/Eq and modified cycloaliphatic polyamine Aradur 2963-1 with total amine value of 325–350 mg KOH/gS were obtained from Huntsman Advanced Materials. Bisphenol A diglycidyl ether was provided by Epoksan, Čačak.

2.2 Preparation of HNTs/Epoxy Nanocomposite Materials

Neat epoxy (NE) was prepared by means of mechanical mixing. First, epoxy components were stirred to obtain homogenous mixture at room temperature and left for 30 min to eliminate the bubbles. Curing agent (Aradur 2963-1) was added to the mixture and stirred for 2 min. The mixture was poured into the mold and cured at room temperature for 2 h. After temperature was raised to 60 °C and samples were cured for 6 h.

HNTs/epoxy nanocomposite materials were prepared by mixing Araldite GY-250 and Araldite DY-E. The homogenous mixture was left for 30 min, and after that 5% of HNTs were added and stirred until the nanotubes were well dispersed. The mixture was poured into the mold and samples were cured. Depending on the mixture, some samples were cured at room or elevated temperature, and degassed. Table 1. describes the experimental design carried out for HNTs/epoxy nanocomposite materials.

Table 1. Experimental design of HNTs/epoxy nanocomposite materials

ID-experiment	HNTs ratio, %	Curing time/temperature	Degassing time
NE	–	2 h/room temperature 6 h/60 °C	–
1HNTs/NE	5	2 h/room temperature 6 h/60 °C	–
2HNTs/NE	5	6 h/room temperature	6 h
3HNTs/NE	5	12 h/room temperature	12 h

2.3 Characterization Methods

The Fourier transformation infrared spectroscopy (FTIR) was performed to confirm the structure of HNTs using Nicolet is 10 spectrometer (Thermo Scientific) with the wavelength range from 3000–500 cm^{-1}.

In order to confirm the crystallographic phase of HNTs, X-ray diffraction (XRD) analysis were performed on Rigaku SmartLab system Cu Kα radiation generated at 30 mA and 40 kV in the 2θ range from 10° to 90°.

The prepared materials were submitted on tensile testing using Universal Testing Machine, AG-XPlus (Shimadzu, Japan) equipped with 1 kN force load cell. All tensile tests were performed with a loading rate of 0.5 mm/min at room temperature and according to the American Society for Testing and Materials (ASTM) standard D 882.

The morphology, size, distribution of HNTs, as well as the presence of voids, and fracture surface were analyzed using field emission scanning electron microscope (SEM) Tescan MIRA 3 XMU operating at 20 kV. To avoid electrostatic charge, all samples were sputter-coated with gold using a POLARON SC502 sputter coater prior to examination.

The thermal stability of the samples was investigated by the TGA and DTG techniques using a Setaram Setsys Evolution 1750 instrument (France). The samples were heated from 30 to 600 °C in a flow rate of $\varphi = 20$ cm^3 min^{-1}, under the pure argon (Ar) with a heating rate of $\beta = 10$ min^{-1}. The average mass of the samples was about 7 mg.

3 Results and Discussion

3.1 Characterization of HNT

The FTIR spectrum of HNTs (Fig. 2) confirmed O-H stretching of inner hydroxyl groups at 3694 cm^{-1} and 3626 cm^{-1}. The bands at 3553 cm^{-1} and 1657 cm^{-1} confirmed the O-H deformation and water stretching. Perpendicular Si-O-Si were attributed to the peak at 1129 cm^{-1}. Si-O stretching and Al_2OH bending appeared at 1015 cm^{-1} and 903 cm^{-1}, respectively. Peaks at 750 cm^{-1} and 674 cm^{-1} were assigned to the perpendicular Si-O stretching.

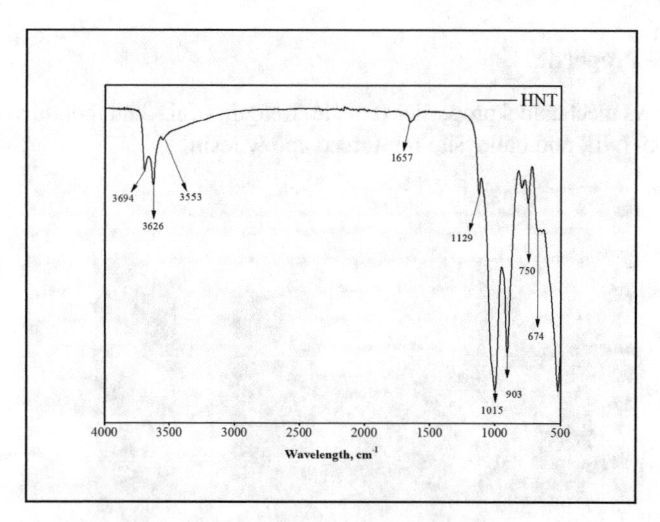

Fig. 2. FTIR spectrum of HNTs

Since kaolinite and halloysite are usually found together, it is necessary to identify HNTs purity using XRD method. The XRD pattern of HNTs (Fig. 3) showed (001) basal reflection at diffraction angle (2θ) at 11.60°, (020)/(110) at 19.85° and, (002) at 24.80° that confirmed the dehydrated state of HNTs [55]. The mineral form SiO_2, quartz, was present at 2θ is 26.52° [81].

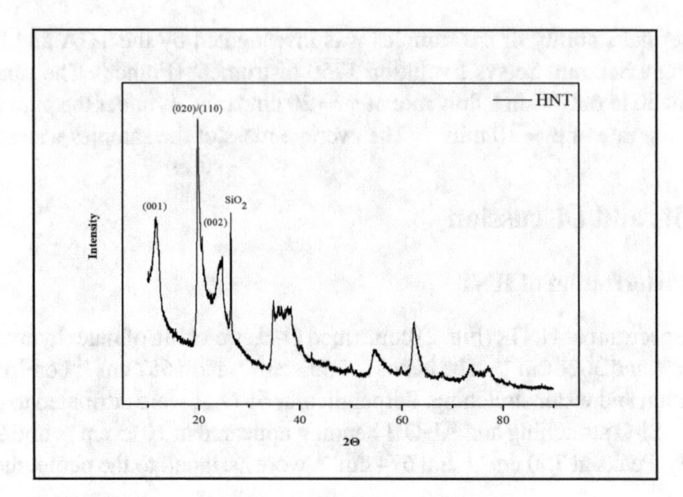

Fig. 3. XRD pattern of HNTs

3.2 Tensile Properties

Figure 4 shows mechanical properties (tensile strength, strain, and modulus of elasticity) of neat epoxy (NE) and halloysite reinforced epoxy resin.

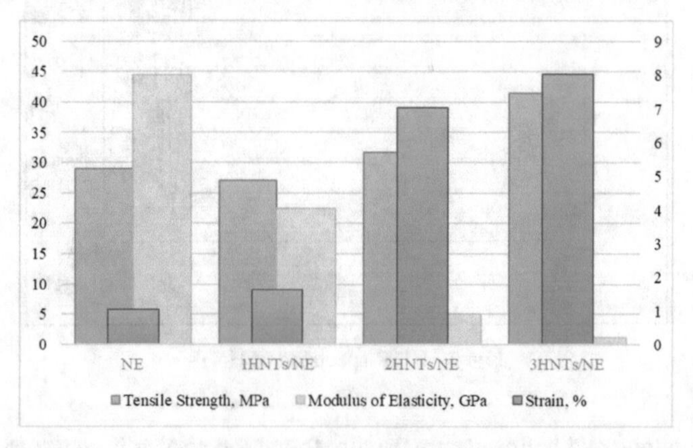

Fig. 4. Mechanical properties of NE and HNTs/epoxy nanocomposite materials

It was noticeable increasement of the values of tensile strength for 2HNTs/NE and 3HNTs/NE compared to NE by 9.1% and 43.18%, respectively, was observed. Values of strain, one of the process parameters that define deformation [82], of HNTs/epoxy nanocomposite materials have significantly increased for 800%, 700%, and 200% for 3HNTs/NE, 2HNTs/NE, and 1HNTs/NE, respectively. However, the values of modulus of elasticity of all HNTs/epoxy nanocomposites have significantly decreased.

3.3 SEM Analysis

The SEM micrographs of neat epoxy (Fig. 5) showed smooth surface that was a result of low resistance to crack propagation and consequently low mechanical strength.

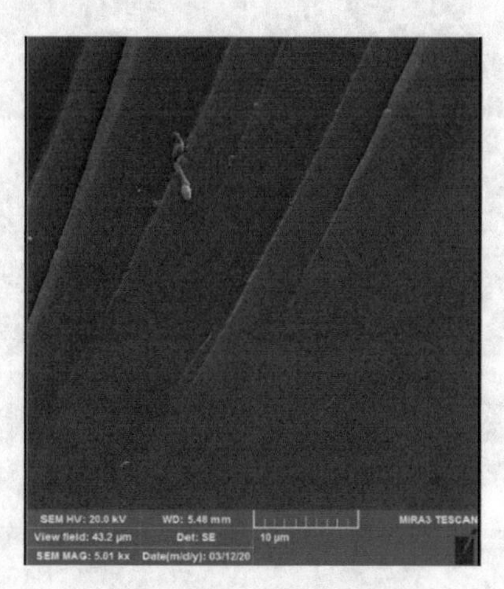

Fig. 5. SEM micrograph of NE

Nevertheless, SEM micrographs of 1HNTs/NE (Fig. 6a,b) showed agglomerated nanoparticles and large voids. One of the important factors is the matrix-reinforcement connection in the material ("interface"). Poor bond between the nanoparticles and epoxy resin matrix caused the brittle deformation of the prepared nanocomposite material. Therefore, the adhesion between the matrix and the reinforcement is granted considerable importance. Bigger adhesion causes better structural resilience. The stronger the material's hardness the more energy is required to sever the connection between the matrix and the reinforcement [16].

HNTs form smaller agglomerates homogenously dispersed comparing to the ones shown in Fig. 6a and b. SEM images of 3HNTs/NE SEM micrographs (Fig. 6e and f) show presence of homogenously dispersed and smaller agglomerates of HNTs. Due to the reduced presence of cavities, the bond between HNTs and epoxy resin was improved. Better nanoparticle dispersion and low presence of voids resulted in higher tensile strength of the prepared 3HNTs/NE nanocomposite material.

3.4 Thermogravimetric (TG) and Derivative Thermogravimetry (DTG) Studies

The results of thermal gravimetric analysis (TGA) and derivative thermogravimetry (DTG) are summarized in Table 2 and presented at the Fig. 7. It can be observed (Fig. 7) that neat epoxy and HNTs reinforced epoxy nanocomposite materials showed the same decomposition trend in all 4 curves, and they became restrained from 400 °C to 600 °C. The main degradation stage consisted of one weight loss step in range of 280° to 430 °C [83].

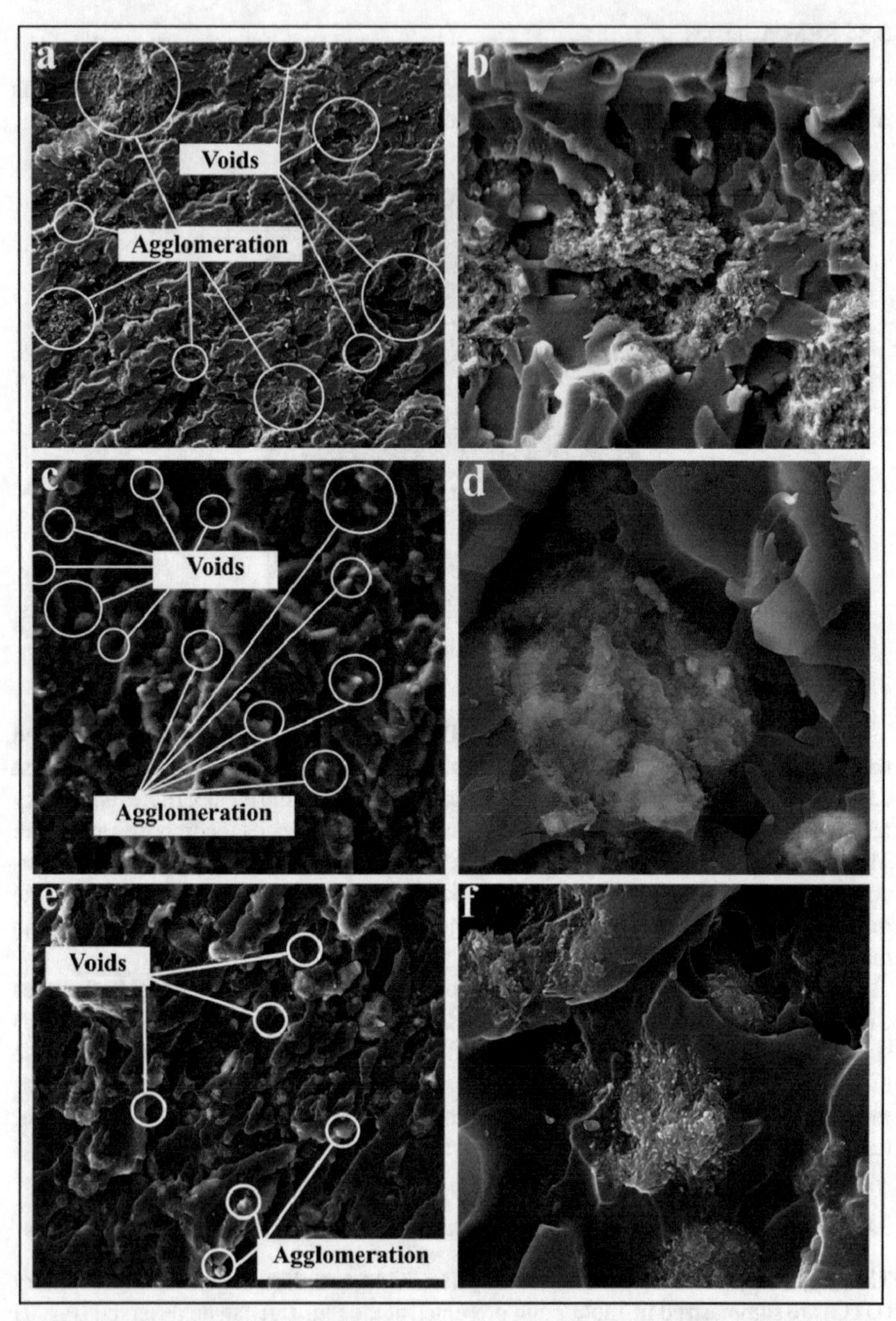

Fig. 6. SEM micrographs of a) 1HNTs/NE at 1.01 kx b) 1HNTs/NE at 5.00 kx c) 2HNTs/NE at 1.00 kx d) 2HNTs/NE at 5.00 kx e) 3HNTs/NE at 1.00 kx f) 3HNTs/NE at 10.00 kx

Table 2. TGA and DTG results for NE and HNTs/epoxy nanocomposite materials

Specimen	T_5 (°C)	T_{max} (°C)	T_f (°C)	Residue (%)
NE	188.94	371	476.97	9.98
1HNTs/NE	177.31	365.86	476.97	15.43
2HNTs/NE	180.14	369.9	476.97	14.24
3HNTs/NE	175.82	365.67	476.97	14.86

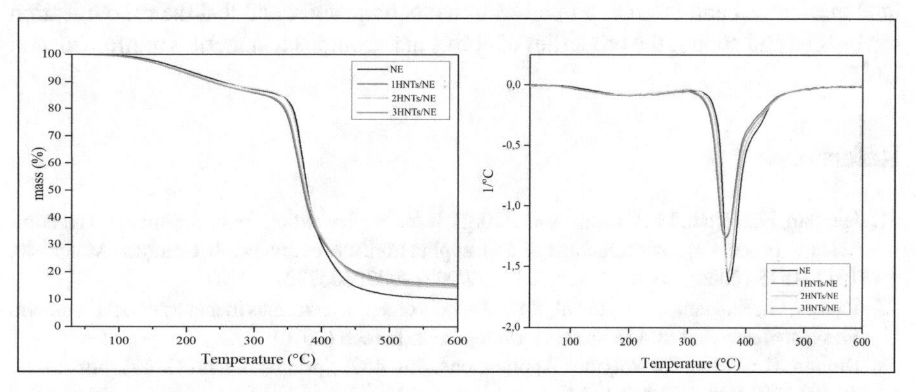

Fig. 7. TGA and DTG curves for NE and HNTs/epoxy nanocomposite materials

The 5% weight loss for HNTs reinforced epoxy nanocomposites at 175 °C was 10 °C lower than that of the NE. The addition of HNTs decreased the initial thermal stability of epoxy resin, however, 15% weight loss temperature for NE and 2HNTs/NE and, 1HNTs/NE and 3HNTs/NE specimen was 337 °C and 314 °C, respectively. The 75% weight loss temperature for 1HNTs/NE, 2HNTs/NE, and 3HNTs/NE reached 424 °C, unlike NE with a 75% weight loss temperature of 409 °C.

The remaining char of the HNTs reinforced neat epoxy was ~5% higher than the neat epoxy resin. The decomposition of aromatic groups and curing agent of epoxy network, and HNTs degradation resulted in weight loss. The T_f has not increased for any of the prepared nanocomposite materials.

4 Conclusions

The incorporation of HNTs into the epoxy resin matrix significantly increased the tensile strength of 2HNTs/NE and 3HNTs/NE up to 9.1% and 43.18%, respectively. The tensile strength value for 1HNTs/NE specimen was reduced for 4.3%. Values of strain have increased for 800%, 700%, and 200% for 3HNTs/NE, 2HNTs/NE, and 1HNTs/NE, respectively. Modulus of elasticity values have significantly decreased for all prepared HNTs reinforced epoxy nanocomposite materials.

HNTs formed agglomerates due to the Van der Waals forces and influenced the mechanical properties of newly synthesized nanocomposite materials. The processing procedures had a great deal of importance with respect to the final properties of nanocomposite products. Based on SEM micrographs, it was concluded that the presence of cavities and the degassing time influenced the bond between the nanotubes and epoxy resin.

Based on results of thermal gravimetric analysis and derivative thermogravimetry it was concluded that better dispersion of nanotubes led to improved char formation and enhanced thermal stability of nanocomposite materials.

This study demonstrated that the production processes are very important for the final mechanical and thermal properties of nanocomposites, and that their modification can change and control the properties of epoxy nanocomposite materials reinforced with HNTs.

References

1. Hussain, F., Hojjati, M., Okamoto, M., Gorga, R.E.: Review article: polymer-matrix nanocomposites, processing, manufacturing, and application: an overview. J. Compos. Mater. **40**, 1511–1575 (2006). https://doi.org/10.1177/0021998306067321
2. Pandey, G., Rawtani, D., Agrawal, Y.K.: Aspects of nanoelectronics in materials development. In: Nanoelectronics and Materials Development. InTech (2016)
3. Rothon, R.: Fillers for Polymer Applications, vol. 489. Springer, Cham (2017). https://doi.org/10.1007/978-3-319-28117-9
4. Ojijo, V., Sinha Ray, S.: Introduction to Nanomaterials and Polymer Nanocomposite Processing. Springer, Cham (2018)
5. Fu, S., Sun, Z., Huang, P., Li, Y., Hu, N.: Some basic aspects of polymer nanocomposites: a critical review. Nano Mater. Sci. **1**, 2–30 (2019). https://doi.org/10.1016/j.nanoms.2019.02.006
6. Godara, S.S., Mahato, P.K.: Potential applications of hybrid nanocomposites. Mater. Today Proc. **18**, 5327–5331 (2019). https://doi.org/10.1016/j.matpr.2019.07.557
7. Dolić, S.D., Jovanović, D.J., Smits, K., Babić, B., Marinović-Cincović, M., Porobić, S., Dramićanin, M.D.: A comparative study of photocatalytically active nanocrystalline tetragonal zyrcon-type and monoclinic scheelite-type bismuth vanadate. Ceram. Int. **44**, 17953–17961 (2018). https://doi.org/10.1016/j.ceramint.2018.06.272
8. Davidović, S., Lazić, V., Miljković, M., Gordić, M., Sekulić, M., Marinović-Cincović, M., Ratnayake, I.S., Ahrenkiel, S.P., Nedeljković, J.M.: Antibacterial ability of immobilized silver nanoparticles in agar-agar films co-doped with magnesium ions. Carbohydr. Polym. **224** (2019). https://doi.org/10.1016/j.carbpol.2019.115187
9. Jamróz, E., Kulawik, P., Kopel, P.: The effect of nanofillers on the functional properties of biopolymer-based films: a review. Polymers (Basel) **11**, 1–42 (2019). https://doi.org/10.3390/polym11040675
10. Kothmann, M.H., de Anda, A.R., Köppel, A., Zeiler, R., Tauer, G., Zhang, Z., Altstädt, V.: The Effect of dispersion and particle–matrix interactions on the fatigue behavior of novel epoxy/halloysite nanocomposites. In: Processing of Polymer Nanocomposites, pp. 121–155 (2019)
11. Shukla, M.K., Sharma, K.: Effect of carbon nanofillers on the mechanical and interfacial properties of epoxy based nanocomposites: a review. Polym. Sci. Ser. A **61**, 439–460 (2019). https://doi.org/10.1134/S0965545X19040096

12. Khostavan, S., Fazli, M., Ahangari, M.G., Rostamiyan, Y.: The effect of interaction between nanofillers and epoxy on mechanical and thermal properties of nanocomposites: theoretical prediction and experimental analysis. Adv. Polym. Technol. **2019** (2019). https://doi.org/10.1155/2019/8156718
13. Fizir, L., Richa, A., He, H., Touil, S., Brada, M.: A mini review on molecularly imprinted polymer based halloysite nanotubes composites: innovative materials for analytical and environmental applications. Rev. Environ. Sci. Bio/Technol. (2020). https://doi.org/10.1007/s11157-020-09537-x
14. Sudo, T.: Shapes of halloysite particles in Japanese clays. Clays Clay Miner. **4**, 67–79 (1955). https://doi.org/10.1346/ccmn.1955.0040110
15. Putić, S., Bajčeta, B., Vitković, D., Stamenović, M., Pavićević, V.: The interlaminar strength of the glass fiber polyester composite. Chem. Ind. Chem. Eng. Q. **15**, 45–48 (2009)
16. Stamenović, M., Putić, S., Rakin, M., Medjo, B., Čikara, D.: Effect of alkaline and acidic solutions on the tensile properties of glass-polyester pipes. Mater. Des. **32**, 2456–2461 (2011). https://doi.org/10.1016/j.matdes.2010.11.023
17. Attia, N.F., Mohamed, G.G., Ismail, M.M., Abdou, T.T.: Influence of organic modifier structures of 2D clay layers on thermal stability, flammability and mechanical properties of their rubber nanocomposites. J. Nanostruct. Chem. **10**, 161–168 (2020). https://doi.org/10.1007/s40097-020-00338-w
18. Sanusi, O.M., Benelfellah, A., Aït Hocine, N.: Clays and carbon nanotubes as hybrid nanofillers in thermoplastic-based nanocomposites – a review. Appl. Clay Sci. **185**, 105408 (2020). https://doi.org/10.1016/j.clay.2019.105408
19. Tasić, A., Rusmirović, J.D., Nikolić, J., Božić, A., Pavlović, V., Marinković, A.D., Uskoković, P.S.: Effect of the vinyl modification of multi-walled carbon nanotubes on the performances of waste poly(ethylene terephthalate)-based nanocomposites. J. Compos. Mater. **51**, 491–505 (2017). https://doi.org/10.1177/0021998316648757
20. Imtiaz, S., Siddiq, M., Kausar, A., Muntha, S.T., Ambreen, J., Bibi, I.: A review featuring fabrication, properties and applications of carbon nanotubes (CNTs) reinforced polymer and epoxy nanocomposites. Chin. J. Polym. Sci. (English Edition) **36**, 445–461 (2018). https://doi.org/10.1007/s10118-018-2045-7
21. Yismaw, S., Kohns, R., Schneider, D., Poppitz, D., Ebbinghaus, S.G., Gläser, R., Tallarek, U., Enke, D.: Particle size control of monodispersed spherical nanoparticles with MCM-48-type mesostructure via novel rapid synthesis procedure. J. Nanopart. Res. **21** (2019). https://doi.org/10.1007/s11051-019-4699-7
22. Gholami, T., Salavati-Niasari, M., Bazarganipour, M., Noori, E.: Synthesis and characterization of spherical silica nanoparticles by modified Stöber process assisted by organic ligand. Superlattices Microstruct. **61**, 33–41 (2013). https://doi.org/10.1016/j.spmi.2013.06.004
23. Šupová, M., Martynková, G.S., Barabaszová, K.: Effect of nanofillers dispersion in polymer matrices: a review. Sci. Adv. Mater. **3**, 1–25 (2011). https://doi.org/10.1166/sam.2011.1136
24. Camargo, P.H.C., Satyanarayana, K.G., Wypych, F.: Nanocomposites: synthesis, structure, properties and new application opportunities. Mater. Res. **12**, 1–39 (2009). https://doi.org/10.1590/S1516-14392009000100002
25. Churchman, G.J., Aldridge, L.P., Carr, R.M.: The relationship between the hydrated and dehydrated states of an halloysite. Clays Clay Miner. **20**, 241–246 (1972). https://doi.org/10.1016/j.jacc.2016.10.047
26. Joussein, E., Petit, S., Churchman, J., Theng, B., Righi, D., Delvaux, B.: Halloysite clay minerals—a review. Clay Miner. **40**, 383–426 (2005). https://doi.org/10.1180/0009855054040180
27. Slansky, E.: Interstratification of 10-Å and 7-Å layers in halloyste: Allegra's mixing function for random and partially ordered stacking. Clays Clay Miner. **33**, 261–264 (1985)

28. Roy, K., Debnath, S.C., Pongwisuthiruchte, A., Potiyaraj, P.: Up-to-date review on the development of high performance rubber composites based on halloysite nanotube. Appl. Clay Sci. **183**, 105300 (2019). https://doi.org/10.1016/j.clay.2019.105300
29. Tharmavaram, M., Pandey, G., Rawtani, D.: Surface modified halloysite nanotubes: a flexible interface for biological, environmental and catalytic applications. Adv. Colloid Interface Sci. **261**, 82–101 (2018). https://doi.org/10.1016/j.cis.2018.09.001
30. Yuan, P., Tan, D., Annabi-Bergaya, F.: Properties and applications of halloysite nanotubes: recent research advances and future prospects. Appl. Clay Sci. **112–113**, 75–93 (2015). https://doi.org/10.1016/j.clay.2015.05.001
31. Mandegarian, S., Taheri-Behrooz, F.: A general energy based fatigue failure criterion for the carbon epoxy composites. Compos. Struct. **235**, 111804 (2020). https://doi.org/10.1016/j.compstruct.2019.111804
32. Tan, D., Yuan, P., Liu, D., Du, P.: Surface modifications of halloysite (2016)
33. Chao, C., Liu, J., Wang, J., Zhang, Y., Zhang, B., Zhang, Y., Xiang, X., Chen, R.: Surface modification of halloysite nanotubes with dopamine for enzyme immobilization. ACS Appl. Mater. Interfaces **5**, 10559–10564 (2013). https://doi.org/10.1021/am4022973
34. Kamble, R., Ghag, M., Gaikawad, S., Panda, B.K.: Review article halloysite nanotubes and applications: a review. J. Adv. Sci. Res. **3**, 25–29 (2012)
35. Yendluri, R., Otto, D.P., De Villiers, M.M., Vinokurov, V., Lvov, Y.M.: Application of halloysite clay nanotubes as a pharmaceutical excipient. Int. J. Pharm. **521**, 267–273 (2017). https://doi.org/10.1016/j.ijpharm.2017.02.055
36. Yuan, P., Bergaya, F., Thill, A.: General introduction. In: Developments in Clay Science, pp. 1–10. Elsevier Ltd., Amsterdam (2016)
37. White, R.D., Bavykin, D.V., Walsh, F.C.: The stability of halloysite nanotubes in acidic and alkaline aqueous suspensions. Nanotechnology **23** (2012). https://doi.org/10.1088/0957-4484/23/6/065705
38. De Silva, R.T., Pasbakhsh, P., Lee, S.M., Kit, A.Y.: ZnO deposited/encapsulated halloysite-poly (lactic acid) (PLA) nanocomposites for high performance packaging films with improved mechanical and antimicrobial properties. Appl. Clay Sci. **111**, 10–20 (2015). https://doi.org/10.1016/j.clay.2015.03.024
39. Lvov, Y.M., DeVilliers, M.M., Fakhrullin, R.F.: The application of halloysite tubule nanoclay in drug delivery. Expert Opin. Drug Deliv. **13**, 977–986 (2016). https://doi.org/10.1517/17425247.2016.1169271
40. Bediako, E.G., Nyankson, E., Dodoo-Arhin, D., Agyei-Tuffour, B., Łukowiec, D., Tomiczek, B., Yaya, A., Efavi, J.K.: Modified halloysite nanoclay as a vehicle for sustained drug delivery. Heliyon **4** (2018). https://doi.org/10.1016/j.heliyon.2018.e00689
41. Clifton, S., Thimmappa, B.H.S., Selvam, R., Shivamurthy, B.: Polymer nanocomposites for high-velocity impact applications-a review. Compos. Commun. **17**, 72–86 (2020). https://doi.org/10.1016/j.coco.2019.11.013
42. Rapacz-Kmita, A., Foster, K., Mikołajczyk, M., Gajek, M., Stodolak-Zych, E., Dudek, M.: Functionalized halloysite nanotubes as a novel efficient carrier for gentamicin. Mater. Lett. **243**, 13–16 (2019). https://doi.org/10.1016/j.matlet.2019.02.015
43. Du, M., Guo, B., Jia, D.: Newly emerging applications of halloysite nanotubes: a review. Polym. Int. **59**, 574–582 (2010). https://doi.org/10.1002/pi.2754
44. Kamble, R., Ghag, M., Gaikawad, S., Panda, B.K.: Halloysite nanotubes and applications: a review. J. Adv. Sci. Res. **3**(2), 25–29 (2016)
45. Anastopoulos, I., Mittal, A., Usman, M., Mittal, J., Yu, G., Núñez-Delgado, A., Kornaros, M.: A review on halloysite-based adsorbents to remove pollutants in water and wastewater. J. Mol. Liq. **269**, 855–868 (2018). https://doi.org/10.1016/j.molliq.2018.08.104
46. Lvov, Y., Guo, B., Fakhrullin, R.F.: Functional Polymer Composites with Nanoclays. The Royal Society of Chemistry, Cambridge (2016)

47. Su, C.: Environmental implications and applications of engineered nanoscale magnetite and its hybrid nanocomposites: a review of recent literature. J. Hazard. Mater. **322**, 48–84 (2017). https://doi.org/10.1016/j.jhazmat.2016.06.060
48. Ramadass, K., Singh, G., Lakhi, K.S., Benzigar, M.R., Yang, J.H., Kim, S., Almajid, A.M., Belperio, T., Vinu, A.: Halloysite nanotubes: novel and eco-friendly adsorbents for high-pressure CO_2 capture. Micropor. Mesopor. Mater. **277**, 229–236 (2019). https://doi.org/10.1016/j.micromeso.2018.10.035
49. Kim, J., Rubino, I., Lee, J.Y., Choi, H.J.: Application of halloysite nanotubes for carbon dioxide capture. Mater. Res. Express **3**, (2016). https://doi.org/10.1088/2053-1591/3/4/045019
50. Jana, S., Das, S., Ghosh, C., Maity, A., Pradhan, M.: Halloysite nanotubes capturing isotope selective atmospheric CO 2. Sci. Rep. **5**, 1–8 (2015). https://doi.org/10.1038/srep08711
51. Ralph, C., Silberstein, M., Thakre, P.R., Singh, R. (eds.): Mechanics of Composite and Multifunctional Materials: Proceedings of the 2015 Annual Conference on Experimental and Applied Mechanics, pp. 141–148. Springer, Switzerland (2015)
52. Lecouvet, B., Sclavons, M., Bourbigot, S., Bailly, C.: Thermal and flammability properties of polyethersulfone/halloysite nanocomposites prepared by melt compounding. Polym. Degrad. Stab. **98**, 1993–2004 (2013). https://doi.org/10.1016/j.polymdegradstab.2013.07.013
53. Yah, W.O., Takahara, A., Lvov, Y.M.: Selective modification of halloysite lumen with octadecylphosphonic acid: new inorganic tubular micelle. J. Am. Chem. Soc. **134**, 1853–1859 (2012). https://doi.org/10.1021/ja210258y
54. He, Y., Kong, W., Wang, W., Liu, T., Liu, Y., Gong, Q., Gao, J.: Modified natural halloysite/potato starch composite films. Carbohydr. Polym. **87**, 2706–2711 (2012). https://doi.org/10.1016/j.carbpol.2011.11.057
55. Khunová, V., Kelnar, I., Kristóf, J., Dybal, J., Kratochvíl, J., Kaprálková, L.: The effect of urea and urea-modified halloysite on performance of PCL. J. Therm. Anal. Calorim. **120**, 1283–1291 (2015). https://doi.org/10.1007/s10973-015-4448-9
56. Albdiry, M.T., Ku, H., Yousif, B.F.: Impact fracture behaviour of silane-treated halloysite nanotubes-reinforced unsaturated polyester. Eng. Fail. Anal. **35**, 718–725 (2013). https://doi.org/10.1016/j.engfailanal.2013.06.027
57. Kim, T., Kim, S., Lee, D.K., Seo, B., Lim, C.S.: Surface treatment of halloysite nanotubes with sol-gel reaction for the preparation of epoxy composites. RSC Adv. **7**, 47636–47642 (2017). https://doi.org/10.1039/c7ra09084f
58. Chen, S., Yang, Z., Wang, F.: Investigation on the properties of PMMA/reactive halloysite nanocomposites based on halloysite with double bonds. Polymers (Basel) **10**, (2018). https://doi.org/10.3390/polym10080919
59. Ma, W., Wu, H., Higaki, Y., Takahara, A.: Halloysite nanotubes: green nanomaterial for functional organic-inorganic nanohybrids. Chem. Rec. **18**, 986–999 (2018). https://doi.org/10.1002/tcr.201700093
60. Bischoff, E., Daitx, T., Simon, D.A., Schrekker, H.S., Liberman, S.A., Mauler, R.S.: Organosilane-functionalized halloysite for high performance halloysite/heterophasic ethylene-propylene copolymer nanocomposites. Appl. Clay Sci. **112–113**, 68–74 (2015). https://doi.org/10.1016/j.clay.2015.04.020
61. Vahedi, V., Pasbakhsh, P.: Instrumented impact properties and fracture behaviour of epoxy/modified halloysite nanocomposites. Polym. Test. **39**, 101–114 (2014). https://doi.org/10.1016/j.polymertesting.2014.07.017
62. Mu, B., Zhao, M., Liu, P.: Halloysite nanotubes grafted hyperbranched (co)polymers via surface-initiated self-condensing vinyl (co)polymerization. J. Nanopart. Res. **10**, 831–838 (2008). https://doi.org/10.1007/s11051-007-9319-2

63. Li, C., Liu, J., Qu, X., Yang, Z.: State: a general synthesis approach toward halloysite-based composite nanotube. J. Appl. Polym. Sci. **116**, 2647–2655 (2009). https://doi.org/10.1002/app.29652
64. Jia, S., Fan, M.: Silanization of heat-treated halloysite nanotubes using γ-aminopropyltriethoxysilane. Appl. Clay Sci. **180**, 105204 (2019). https://doi.org/10.1016/j.clay.2019.105204
65. Naveen Kumar, G., Suresh Kumar, C., Seshagiri Rao, G.V.R.: An experimental investigation on mechanical properties of hybrid polymer nanocomposites. Mater. Today Proc. **19**, 691–699 (2019). https://doi.org/10.1016/j.matpr.2019.07.755
66. Zhang, J., Jia, Z., Jia, D., Zhang, D., Zhang, A.: Chemical functionalization for improving dispersion and interfacial bonding of halloysite nanotubes in epoxy nanocomposites. High Perform. Polym. **26**, 734–743 (2014). https://doi.org/10.1177/0954008314528226
67. Eser, N., Önal, M., Çelik, M., Pekdemir, A.D., Sarıkaya, Y.: Preparation and characterization of polymethacrylamide/halloysite composites. Polym. Compos. **41**, 893–899 (2020). https://doi.org/10.1002/pc.25420
68. Liu, M., Guo, B., Lei, Y., Du, M., Jia, D.: Benzothiazole sulfide compatibilized polypropylene/halloysite nanotubes composites. Appl. Surf. Sci. **255**, 4961–4969 (2009). https://doi.org/10.1016/j.apsusc.2008.12.044
69. Jia, Z., Luo, Y., Guo, B., Yang, B., Du, M., Jia, D.: Reinforcing and flame-retardant effects of halloysite nanotubes on LLDPE. Polym. Plast. Technol. Eng. **48**, 607–613 (2009). https://doi.org/10.1080/03602550902824440
70. Li, C., Liu, J., Qu, X., Yang, Z.: Polymer-modified halloysite composite nanotubes. J. Appl. Polym. Sci. **110**, 3638–3646 (2008). https://doi.org/10.1002/app.28879
71. Deng, S., Zhang, J., Ye, L.: Halloysite-epoxy nanocomposites with improved particle dispersion through ball mill homogenisation and chemical treatments. Compos. Sci. Technol. **69**, 2497–2505 (2009). https://doi.org/10.1016/j.compscitech.2009.07.001
72. Sabatini, V., Taroni, T., Rampazzo, R., Bompieri, M., Maggioni, D., Meroni, D., Ortenzi, M.A., Ardizzone, S.: PA6 and halloysite nanotubes composites with improved hydrothermal ageing resistance: role of filler physicochemical properties, functionalization and dispersion technique. Polymers (Basel) **12** (2020). https://doi.org/10.3390/polym12010211
73. Peixoto, A.F., Fernandes, A.C., Pereira, C., Pires, J., Freire, C.: Physicochemical characterization of organosilylated halloysite clay nanotubes. Micropor. Mesopor. Mater. **219**, 145–154 (2016). https://doi.org/10.1016/j.micromeso.2015.08.002
74. Pasbakhsh, P., Ismail, H., Fauzi, M.N.A., Bakar, A.A.: EPDM/modified halloysite nanocomposites. Appl. Clay Sci. **48**, 405–413 (2010). https://doi.org/10.1016/j.clay.2010.01.015
75. Krishnaiah, P., Ratnam, C.T., Manickam, S.: Development of silane grafted halloysite nanotube reinforced polylactide nanocomposites for the enhancement of mechanical, thermal and dynamic-mechanical properties. Appl. Clay Sci. **135**, 583–595 (2017). https://doi.org/10.1016/j.clay.2016.10.046
76. Lim, J.S., Noda, I., Im, S.S.: Effect of hydrogen bonding on the crystallization behavior of poly(3-hydroxybutyrate-co-3-hydroxyhexanoate)/silica hybrid composites. Polymer (Guildf) **48**, 2745–2754 (2007). https://doi.org/10.1016/j.polymer.2007.03.034
77. Saif, M.J., Asif, H.M., Naveed, M.: Properties and modification methods of halloysite nanotubes: a state-of-the-art review. J. Chil. Chem. Soc. **63**, 4109–4125 (2018). https://doi.org/10.4067/s0717-97072018000304109
78. Rosas-Aburto, A., Gabaldón-Saucedo, I.A., Espinosa-Magaña, F., Ochoa-Lara, M.T., Roquero-Tejeda, P., Hernández-Luna, M., Revilla-Vázquez, J.: Intercalation of poly(3,4-ethylenedioxythiophene) within halloysite nanotubes: synthesis of composites with improved thermal and electrical properties. Micropor. Mesopor. Mater. **218**, 118–129 (2015). https://doi.org/10.1016/j.micromeso.2015.06.032

79. Ganguly, S., Bhawal, P., Choudhury, A., Mondal, S., Das, P., Das, N.C.: Preparation and properties of halloysite nanotubes/poly(ethylene methyl acrylate)-based nanocomposites by variation of mixing methods. Polym. Plast. Technol. Eng. **57**, 997–1014 (2018). https://doi.org/10.1080/03602559.2017.1370106
80. Ismail, H., Salleh, S.Z., Ahmad, Z.: Properties of halloysite nanotubes-filled natural rubber prepared using different mixing methods. Mater. Des. **50**, 790–797 (2013). https://doi.org/10.1016/j.matdes.2013.03.038
81. Bai, S., Sun, X., Wu, M., Shi, X., Chen, X., Yu, X., Zhang, Q.: Effects of pure and intercalated halloysites on thermal properties of phthalonitrile resin nanocomposites. Polym. Degrad. Stab. 109192 (2020). https://doi.org/10.1016/j.polymdegradstab.2020.109192
82. Radović, N., Vukićević, G., Glišić, D., Dikić, S.: Some aspects of physical metallurgy of microalloyed steels. Metall. Mater. Eng. **25**, 247–263 (2019). https://doi.org/10.30544/468
83. Gaaz, T.S., Sulong, A.B., Kadhum, A.A.H., Al-Amiery, A.A., Nassir, M.H., Jaaz, A.H.: The impact of halloysite on the thermo-mechanical properties of polymer composites. Molecules **22**, 13–15 (2017). https://doi.org/10.3390/molecules22050838

Bridging Nature-Art-Engineering with Generative Design

Dusko Radakovic(✉)

Academy of Applied Technical Studies, Belgrade Polytechnic, Belgrade, Serbia
dradakovic@politehnika.edu.rs

Abstract. Since their evolution, human beings designed things by observing and learning from nature. Still, there's a lot to learn from nature when designing advanced components, with the prevalence of materials development and engineering design at the forefront of additive manufacturing. Hence, innovative product design is an increasingly demanding task as more and more technology, science, and art are integrated into the product. Collaboration between designers and engineers increases as more participants get involved in the development process. Winning the race to the market means shorter deadlines. Overwhelmed by the multitude of responsibilities, engineers hardly have time to find a feasible design. Even less to improve it. Designers find an appealing use of novel software tools (organic design, organic form, organic structure) that even more complicates the entire design cycle. Unfortunately, most designers lack the knowledge to address specific functional and technical requirements, resulting in beautiful though impractical designs. Such a gap creates difficulties in later product development. The application of generative design could provide some optimism in this respect. Using topological optimization, morphogenesis, and biomimicry, or/and other algorithms found in nature, enables generating several alternative designs. Consequently, more alternatives are considered than previously possible.

This paper aims to present some design issues, accelerate product development, and shorten the time to market. The review covers the limitations of today's engineers and their effects on design quality. Research on generative design presents some technical considerations and details of use, as well as how to apply such algorithms to design form, structure, or/and function.

Keywords: Generative design · Biomimicry · 3D modeling

1 Introduction

Every design project is an act of balancing between deadlines, budget, targeted material costs, and functional requirements, on one hand, and aspiration to excel in order to create more innovative and competitive products on the other hand. In recent years, several trends have driven this act of balancing toward a more conservative design, which hinders innovation efforts. One factor that drives engineers to be conservative is the reality that new development is fraught with risk. Errors in concept design phase (even

N. Mitrovic et al. (Eds.): CNNTech 2020, LNNS 153, pp. 326–343, 2021.
https://doi.org/10.1007/978-3-030-58362-0_19

in the detailed design phase) that pass along with the design approval can have serious consequences for the whole project, especially for the engineers. Just changes in work orders that bring jobs back to table generate emergencies, pile-up of waste, reworking the designs, and failed prototypes. In addition, design error further affects time, energy, and resources. In fact, over 60% of those surveyed in the Simulation-Driven Design Study missed project deadlines due to failed prototypes [1].

Another factor that hinders design towards a conservative approach is the increasing complexity of technologies being incorporated into today's products. Trends in electronics, including constant miniaturization, low energy requirements, and the need for greater heat dissipation, make it difficult to accommodate the increasing demand for computing power in products. The increasing incorporation of software into products raises integration problems as these applications must seamlessly work with electronic hardware and other systems in products. With the advent of the Internet of Things (IoT), development becomes even more complex as companies have to figure out how to equip their products with the right sensors, how to gather the right data, how to channel that data into proper memory, and then what to do with that data. All this converges with engineering design to improve complexity, but also increase system integration difficulties.

Another thing to consider is the requirement that engineers collaborate more and with more participants. In order to stay on top of the competition, manufacturers need to integrate the latest technologies or lose market competitiveness. As a result, engineers have to collaborate with subject matter experts in a narrower field. In addition, it is not enough to find a design that meets the specifications of form, structure and function. Designers, especially engineers, have to do more. Today's products have significant operational and business constraints that affect the design solution. As such, feedback from an increasing number of participants, including procurement services, suppliers, subcontractors, manufacturing, customers, services and more, is needed. According to data from the Hardware Design Engineers Study conducted by LifeCycleInsights, engineers are overloaded by about 57% of total product development responsibilities [2]. Making design decisions is just one of those responsibilities.

Design processes tend to be highly path-dependent, i.e. early decisions tend to determine and limit the possibilities of the design in later stages. As the design process develops in time, decisions are made as they are challenged. Making a critical decision often limits the space of exploration in later stages. All of these limitations make the design process extremely difficult, and impose an unlikeliness that any single design process will achieve a true optimal design.

2 Designing in Nature

Design problems are inherently multidimensional. Though the outcome of a design process is typically a physical object in three-dimensional space, the design process can itself be thought of as a massively high-dimensional problem space, where each individual decision taken during the process forms one of its dimensions.

As human designers, what can we learn from nature? One way to answer this question is to first consider our own limitations as designers/engineers—the aspects of the design process that are particularly difficult to humans. Then, we can study how design occurs in nature, and the process that natural evolution takes in arriving at its final design solutions. Finally, we can consider how we might develop similar strategies within our design processes, which can augment our own abilities and make us better human designers.

Organisms found in nature represent a great variety of novel formal solutions way beyond the imagination of even the most creative human designer. At the same time, each of these unique organisms is also uniquely adapted (evolved) to the functional requirements of its environment. It seems that nature is able to escape the curse of dimensionality, and produce an endless variety of forms that are both novel and high performing, with an ability to solve difficult problems in novel and beautiful ways. This has always been inspiring for designers. However, up to this point inspiration from nature has been limited to 'biomimicry', or the reproduction of nature's physical forms in new designs. The key to nature's design is the evolutionary process [3]. This process operates at the level of a species, which is a kind of model which encodes all of the unique properties and abilities of its individual members. While each individual member of a species is unique, all members of the same species share common characteristics, most important of which is their ability to reproduce and create new members. Over time, the reproductive process improves the species through adaptation and interaction with other species, as well as the environment. This is known as natural selection. More or less, the process can be coarsely summed within three steps:

1. *Selection* – only members of a species that are best adapted to the environment will survive;
2. *Breeding* – those survivors reproduce new individuals that share characteristics;
3. *Mutation* – some characteristics of new individuals are "randomly" changed.

Within the Nature's design system there are some critical components. For example, the relationship between genotype and phenotype. The genotype of an organism is its DNA (it encodes all information and is unique to the organism). The phenotype is the physical expression of the organism (influenced by its genotype and interaction with the environment). In the evolutionary process, breeding and mutation operate on the genotype, while competition and selection befall at the level of phenotype. The process of growth and development (resulting in the phenotype) is known as morphogenesis. This is the process by which form and structure emerges from nature.

Another critical component of natural design is the ability to weigh the trade-off between exploitation and exploration. Here, exploitation refers to the use of existing knowledge in order to produce the best possible performance. On the other hand, exploration refers to random exploration of unknown system in order to develop new knowledge. When a new, promising strategy is discovered there is a desire to exploit it to get better results. This is the trade-off, because such exploitation may stop further exploration – so there would be no knowledge whether better strategies might develop. In a usual design-production unit, this is a well-known trade-off. Very often the designer/engineer has a limited set of resources resulting in very little time for exploring the unknown. Unfortunately, engineers don't have enough time even to improve the design. In the nature's evolutionary process, selection and breeding are focused primarily on exploitation, while mutation is the step where novel designs are explored. These novelties contribute to the diversity of the species, which in turn allows it to adapt to a changing environment making it survive at the long run.

An engineering optimization process can easily relate to this process of evolution. For example, for a given model of species, the natural selection will try to produce new individuals that better fit the objectives and constraints of the environment. In mathematical and engineering terms, most optimization problems can be expressed in the generic form

$$\text{minimize}\, \boldsymbol{f}_i(\boldsymbol{x}), x \in \mathbb{R}^n, (i = 1, 2, \ldots, M) \tag{1}$$

$$\text{subject to}\, \boldsymbol{h}_j(\boldsymbol{x}) = 0, (j = 1, 2, \ldots, J), \tag{2}$$

$$\boldsymbol{g}_k(\boldsymbol{x}) \leq 0, (k = 1, 2, \ldots, K), \tag{3}$$

where $\boldsymbol{f}_i(\boldsymbol{x})$, $\boldsymbol{h}_j(\boldsymbol{x})$, and $\boldsymbol{g}_k(\boldsymbol{x})$ are functions of the design vector

$$x = (x_1, x_2, \ldots, x_n)^T \tag{4}$$

The goal of the optimization is to find the combinations of input data that best satisfy the objectives and constraints of the given system. Simple objectives and constraints give easy solutions through functional analysis. But, when functions tend to be complex or when there are certain unknown elements, more stochastic methods have to be utilized since they can target optimum solutions by testing many configurations and by learning the patterns of the system. The field of computer science has developed many different tools and algorithms for solving such problems.

If we want to incorporate a process within design methodologies that is very much alike natural evolution, it is necessary to reframe the design problems of optimization and then use the tools of optimization to solve them. This leads to generative design which can be composed of three steps (alike nature's evolution process):

1. Generate – a design space is delineated as a closed system which can generate all possible solutions to the given design problem
2. Evaluate – measures are developed to judge the performance of every design iteration
3. Evolve – searching the design space by evolutionary algorithms to find unique high-performance design solutions

3 Generative Design

Generative design (GD) is a technology that mimics nature's evolutionary approach to design. It starts with design goals and then explores all possible permutations of a solution to find the best option. Using cloud computing, generative design software quickly cycles through thousands—or even millions—of design choices, testing configurations and learning from each iteration what works and what doesn't. This process lets designers generate brand new options, beyond what a human alone could create, to arrive at the most effective design. More or less, it can be characterized by:

- *Focus on components*: At the moment GD is applied to part level. Some major CAD platforms have claimed integrating an optimization methodology simultaneously focusing on more than one component – on the whole assembly.
- *Autonomous execution*: Once executed, GD works autonomously, searching for a design with respect to given requirements and data. In this fashion, it performs similarly as an ability of a gradient structured optimization.
- *Goal driven*: The decision algorithm the CAD uses works toward an explicitly defined goal or an implicit idea. For example, it can be minimizing deformation under load or minimizing weight.
- *Limitations (constraints) relation*: This decision algorithm is also tied by limitations defined by the designer/engineer. Range of limitations can vary from solution to solution. For example, a design limitation by engineering physics (maximum strain). Besides that, limitations can be imposed by the manufacturing process (producing geometry that can be manufactured by molding).

The complete process of generating design by AI comes down to the computer. The computer is a machine with the ability to accurately calculate complex mathematical problems in a split second. In order for a computer to generate a new design, it is necessary that a programmer writes a program of what the computer should do and how. This allows the algorithm. As said, generative design integrates algorithms that mimic the evolutionary approach found in nature. The whole process comes down to achieving the desired design through exploration of possible variations of the solution in order to reach the best possible design [4].

In the application of generative design, decision algorithms are used to determine the final shape. Within generative design, algorithms can be applied in the manufacturing industry, infrastructure, and basically anything that requires a seamless combination of form and function [5]. The diagram on Fig. 1 helps to explain the general principle behind generative design.

3.1 Topology Optimization

Topology optimization is considered to be a subtractive method of generative design, because it removes material in a progressive manner. This term is specific to the study of forms, structures and solids. It is most often associated with projects in the automotive or aerospace industries in the context of lightweight projects. In such projects, engineers and designers try to find ways to reduce the material without compromising design requirements. Every topology optimization problem is comprised of three main characteristics: the model to represent the physics, the optimization problem itself, and the optimization routine used to find an optimized design [6]. For example, how to solve the partial differential equations that represent the physical problem. Or, what is the design objective, and what are the constraints that are taken into account? Also, what to consider as design variables, and how do they describe the design? What topology optimization method should be used? Finally, what choices have to be made regarding the solution methods used to solve the previously formulated optimization problem. These aspects generally depend on each other, i.e. a certain model description may be more suitable for

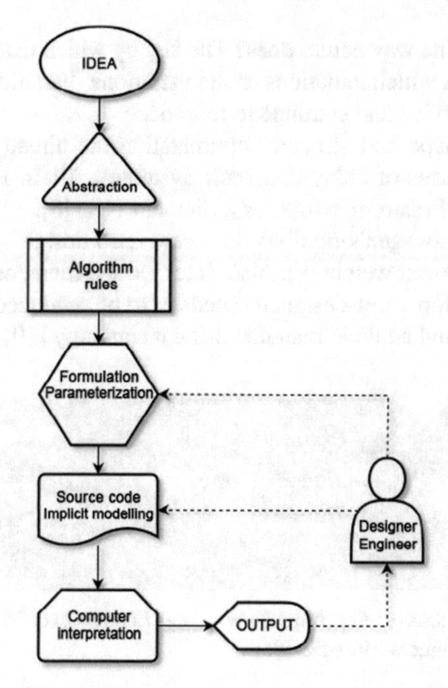

Fig. 1. A general interpretation of the generative design process

a particular topology optimization method, and a certain way of describing the problem (continuous/discrete variables) may be more suitable for particular optimizers.

Besides these aspects, in general, topology optimization approaches can be categorized into element-based, discrete and combined, depending on the different algorithms they use [7]. These types of algorithms usually involve:

- *Shape and structure optimization* – The algorithm was developed in the mid-1980s and is based on structural engineering. The existing part, defined by the designer/engineer, is analyzed and then the material that does not carry significant loads is removed. This procedure is performed iteratively, whereby the material is removed several times during the process. Achieving the settled goal completes the optimization, which ultimately gives the final design.
- *Biomimicry* – Recently established, this algorithm mimics organisms and systems in nature, such as the multiplication of bacterial colony growth, root growth and tree branching, or the evolution of bone structures to optimize the weight-to-strength ratio.
- *Morphogenesis*: This algorithm uses research results related to the way cell groups react to their environment. Cells that are actively loaded are strengthened, while cells that are not are rejected.

Biomimicry and morphogenesis have always been the inspiration of every designer. Finding forms in nature, extracting the underlying principles that can solve functional, structural and technological problems in design is limited by biomimicry. The question is, can people design the way nature does? The key by which nature designs lies in the process of evolution in which mutations create variations. Variations that are adapted to the new conditions survive and continue to reproduce.

Algorithms for shape and structure optimization are already built into the most advanced CAD platforms of today (Dassault Systémes, Altair, ParaMatters, ANSYS, Autodesk, PTC Creo, Frustrum, nTopology, Siemens NX) [8].

For example, Volkswagen's minibus, in a new retro design with wheels and other elements, has a much lower weight [9]. Also, Hackrod and Siemens have joined forces in a joint project to develop a self-designed speedster to be produced using virtual reality, artificial intelligence and additive manufacturing technology [10] (Figs. 2 and 3).

Fig. 2. Volkswagen minibus (source: https://www.core77.com/posts/89318/Where-Does-Generative-Design-Belong-Designers-Must-Decide)

Fig. 3. Self-designed speedster (source: https://all3dp.com/self-driving-cars-yesterday-self-designing-cars-future-thanks-hackrod/)

When it comes to design, an idea is always a solution to some kind of problem. When the designer/engineer has an idea, it does not necessarily mean that they have found a way to solve the problem at hand. Sometimes, an idea can simply be an insight into the nature of the problem itself—the solution often remains opaque or prohibitively complex. The designer starts with an objective (or a set of objectives) and runs them through an algorithm. While they are the architects of the system, it is the system which produces the final result. For example, a shoe designer wants to create a running shoe capable of providing comfortable cushioning and maximum weight support of 130 kg; or an architect who wants to build a structure with massive solar energy conversion without sacrificing aesthetics. In conventional design, the designer or architect would sketch their ideas, create blueprints, make a 3D model or a prototype, and work out how the final product should look and be made. Changes would be made by iterating these steps, until a design is reached that satisfies all of the project goals.

All these processes are called "explicit design". If they put their ideas or problems into computer algorithms and let the computer create design iterations for them, what they do is called generative design. A computer can perform infinitely many iterations much, much faster than any human being could. A good example of how the technology works was set by the works of Lightning Motorcycle shown in Fig. 4, while in Fig. 5 is presented and optimization realized with ParaMatters' CogniCAD 2.

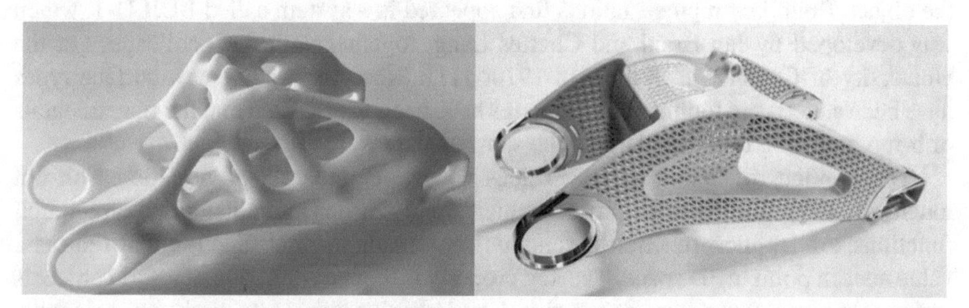

Fig. 4. Topology optimized (left) and prototype (right) (source-left: https://media.wired.com/photos/59547bc45578bd7594c46270/master/w_700,c_limit/LighningMotorcycles_swing_arm2.jpg, source-right: https://blogs-images.forbes.com/bruceupbin/files/2014/09/0909_materials-autodesk-lightning_1024x576-e1410316198993.jpg)

3.2 Implicit Modelling Approach

Current CAD systems (NX Siemens, Dassault Systeme, PTC, and others) all use boundary representations (b-reps) to express the shapes of solid objects. As the name implies, a boundary representation is a collection of faces that form the boundary (the outer skin) of the object. The faces are glued together by "topology" information that describes connectivity, such as which edges lie on each face, which faces meet at each vertex, and so on.

In CAD systems, the surfaces are usually curved; they might be cylinders, or cones, or NURBS surfaces, for example. In computer graphics applications, on the other hand, users often use boundary representations in which all faces are planar. These are much simpler, of course, but the principle is the same – the object is again represented by its external skin. Current systems assume that the interior of the object is homogeneous, in which case the boundary alone does provide enough information to fully describe

Fig. 5. Designed in CogniCAD 2.0, ParaMatters' second-generation holistic and agnostic generative design solution (Source: https://www.autonomousvehicletech.com/ext/resources/AVT/2019/Janunary/20190107-ParaMatters-image-1_web.jpg?1546884976)

the object. Boundary representations first appeared in a system called BUILD-1, which was developed by Ian Braid and Charles Lang, together with their colleagues at the University of Cambridge in the mid-1970s [11]. Since then, some new surface types have been added, and topological structures have been generalized, but the fundamentals of b-rep modeling today are largely the same as they were 40 years ago.

The implicit modelling approach does not use boundary representations (or, not much, anyway). Instead, it's a process which simply means modeling based on implicit functions. An implicit function (i-function) is a mathematical function that returns a value at each point in 3D space. The key property is that an i-function of a solid object is negative at points that are inside the object, and positive outside. In mathematical terms, if F is an i-function of a solid object S, then

$$F(x, y) = S \Rightarrow \begin{cases} F(P) < 0, \textit{point } P \textit{ is inside } S \\ F(P) = 0, \textit{point } P \textit{ is on the boundary of } S \\ F(P) > 0, \textit{point } P \textit{ is outside } S \end{cases} \tag{5}$$

A simple 2D example might serve to make the implicit modeling idea clearer. Suppose we wanted to store some description of a 2D rectangular region. The b-rep approach would be to store the outside boundary of the rectangle, which consists of four lines. The implicit modeling approach is to store an i-function that is negative inside the rectangular region, and positive outside. To illustrate the process, consider the rectangular region whose opposite corners are at the points $(-2, -1)$ and $(2, 1)$ (Fig. 6).

In a b-rep, the four lines shown in the left-hand image in Fig. 7 would be stored. The implicit approach is shown in the right-hand image. A point lies in the rectangle if it lies in both the yellow region and the red region. A point (x, y) lies in the yellow region if $y^2 - 1 < 0$, and it lies in the red region if $x^2 - 4 < 0$. So it lies in the rectangle if both $y^2 - 1$ and $x^2 - 4$ are less than zero, which means that $\max\{x^2 - 4, y^2 - 1\}$ is less than zero. A possible i-function for the rectangle could be defined as

$$F(x, y) = \max\left\{x^2 - 4, y^2 - 1\right\} \tag{6}$$

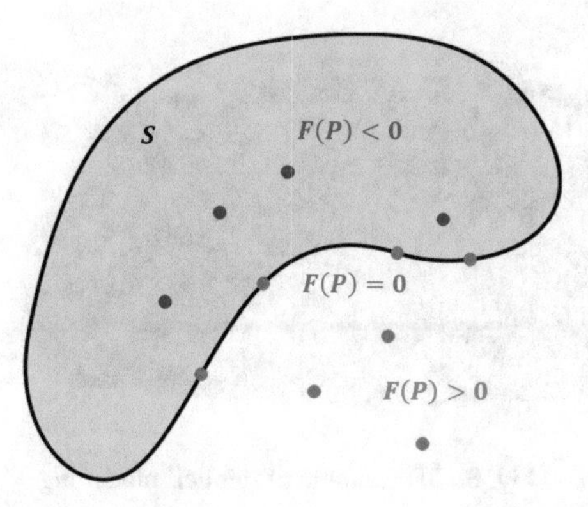

Fig. 6. Implicit function interpretation on an arbitrary 2D space

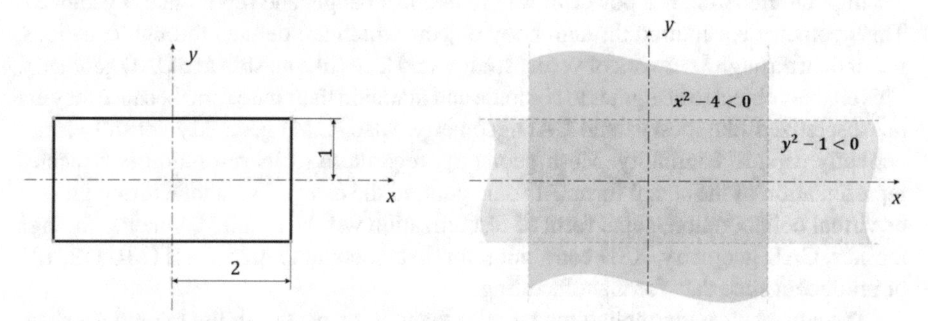

Fig. 7. General difference in interpretation between B-Rep and the implicit modeling approach

Interesting to notice is that i-functions are not unique: the red region above could be represented by any of the i-functions $x^2 - 4$, $|x| - 2$, $x^4 - 16$, or numerous others. Actually, among the listed, the function $|x| - 2$ is the best choice, since it accurately measures signed distance for the red region.

The b-rep has no explicit representation of the interior of the rectangle, and the implicit approach involves no explicit representation of the four edges. The two representation schemes are completely different.

A simple 3D example is a spherical solid of radius 3 centered at the origin, given by

$$F(x, y, z) = x^2 + y^2 + z^2 - 9 \tag{7}$$

In the example on Fig. 8, additionally the sphere is cut parallel to *x-y* plane at $z = 0$ and $z = 2$. In this case a point $P(x,y,z)$ is defined inside the object S by

$$\forall P(x, y, z) \in S \Leftrightarrow \left\{ \sqrt{x^2 + y^2 + z^2} < 3 \wedge z < 2 \wedge z > 0 \right\} \tag{8}$$

Hence, a possible i-function would be

$$F(x, y, z) = \max\left\{ x^2 + y^2 + z^2 - 9, -z, z < 2 \right\} \tag{9}$$

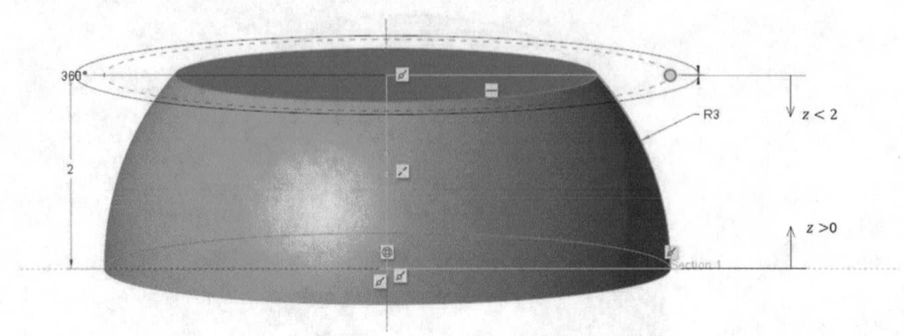

Fig. 8. 3D example of implicit modelling

Implicit modeling is a powerful way to define, change, and represent 3D geometry. This geometry is captured through body objects which are defined through equations, rather than through a network of vertices, edge, and faces like meshes and CAD geometry. This enables object to be lighter to compute and maintain their pure form because they are not discretized like meshes and CAD geometry. Also, CAD geometry doesn't always perfectly capture continuity. Mesh geometry, regardless of its resolution is a faceted representation of the actual form. At some point in the design to manufacturing process or virtual collaboration, some form of discretization will be required, whether through meshes, CAD geometry, X3D conversion for distant collaboration via HTML [12, 13], or producing slice data for manufacturing.

The advantage with implicit modeling is that users have the ability to produce slices from implicit bodies. This means discretization occurs at the very end of the process rather than at the beginning or throughout. Also, outgoing manufacturing data is more precise. Implicit bodies, being orders of magnitude faster to compute, also result in super lightweight files as only a minimal amount of information is needed. Implicit modeling offers a new set of tools that overcome many limitations of traditional techniques for CAD geometry and meshes, which are becoming increasingly problematic in advanced manufacturing and generative design. For example, CAD and mesh modelers are unable to or have difficulty to perform routine operations such as offsetting, rounding, drafting, and even simple Booleans with sufficient reliability. In addition, they cannot handle the complexity of 3D printed models, manually or in automated workflows, let alone describe parts with varying material properties.

The math behind implicit modeling guarantees that operations like Booleans, offsets, rounds, and drafts never fail.

Material structures such as lattices, cellular, periodic structures, architected materials and so on, are easy to construct and can be very diverse. Basically, these are hybrid structures defined by solid material and empty space, but with a precise arrangement that determines the mechanical properties. The amount of material is also significant and is measured by relative density - the mass of the network divided by the mass of the full block. Some typical assumptions in the cellular world include honeycomb structures, foams, gratings, and continuous shell-like structures, such as gyroids, each of which has its advantages and disadvantages. Each structure can be used in different ways:

uniformly as shown in Fig. 9, as a filling in a shell construction, aligned across a surface, or spatially variable in a complex domain.

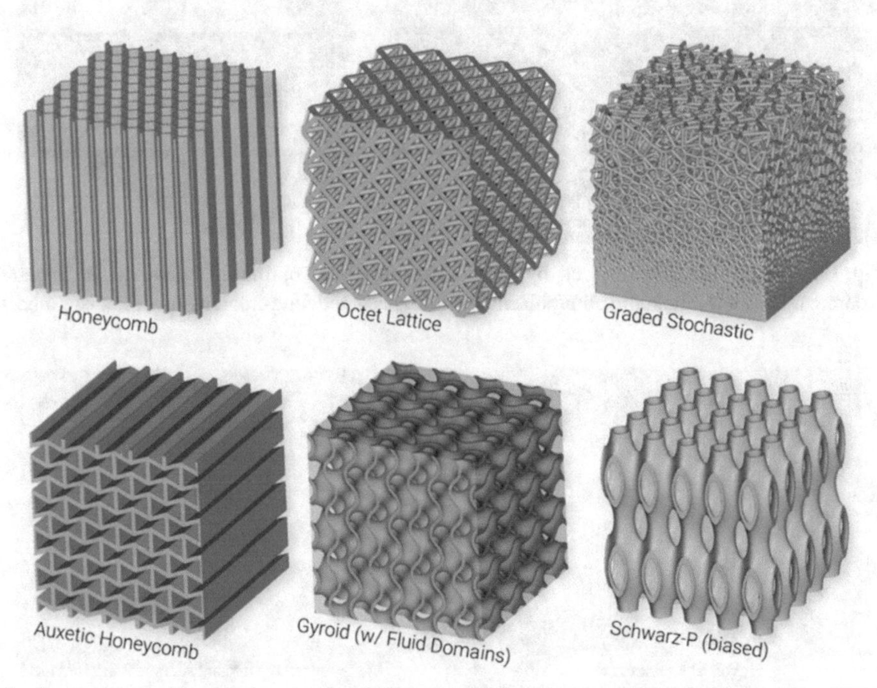

Fig. 9. Architected materials using implicit functions (source: https://ntopology.com/blog/2019/10/18/5-techniques-for-lightweighting-doing-more-with-less/)

Generative design encourages innovation. Designers, especially engineers, can now use applications to create products they would have considered impossible a few years ago. When these tools are combined with metal additive production, an outstanding blend of aesthetics and functionality can be created [14].

Software companies that follow the trend of generative design and additive manufacturing enable product design that correlates with the specificity of design in biological systems. One of the basic principles of added value of additive production is in the field of medicine (shown in Fig. 10). The ability to simulate the morphological properties of bone is the "holy grail" of design. In these applications, the focus is on structural requirements. By improving the computational design program, future orthopedic implant engineers will progressively rely on these tools to create new generations of advanced structures tuned to the structures and loads in biological systems. In addition, the biological characteristics of the patient (size, weight, morphological condition of the bones, etc.) play a huge role in how each part (implant) will be loaded. For example, the design of an hip implant stem whose structure should compensate for loads in patients with a body weight range of 80 kg up to weights over 150 kg will lead to non-optimal outcomes if the structure does not correspond to the given loads (as shown in Fig. 11) [15, 16].

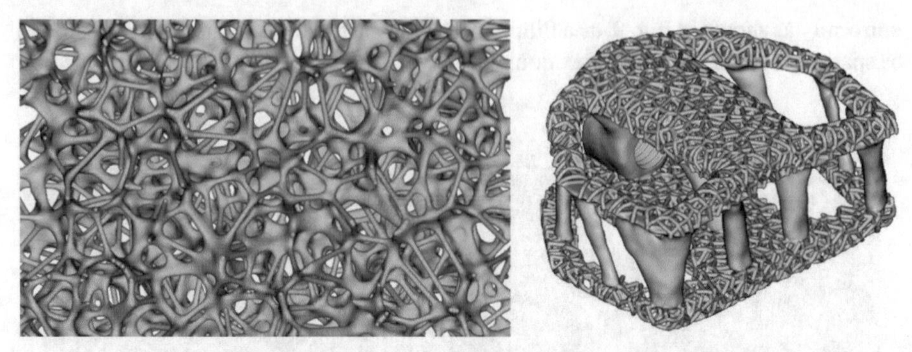

Fig. 10. Lattice structure inspired by bone structure (right: optimized spinal ALIF implant) (source: https://www.linkedin.com/pulse/true-biological-modeling-implants-matthew-shomper)

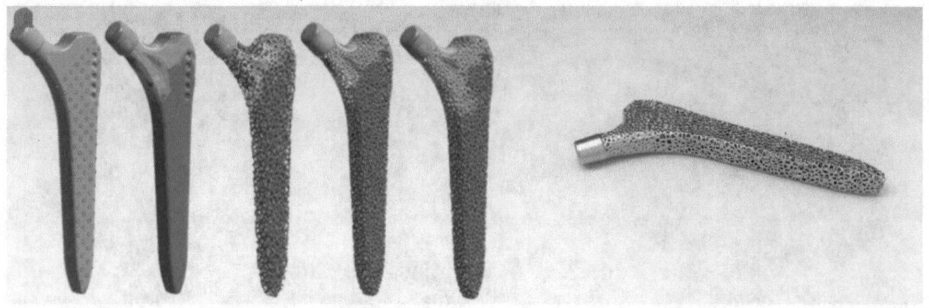

a) Von Misses stress at same scale (from initial design to optimized strut thickness

b) Optimized hip implant stem, AM in Ti-6Al-4V

Fig. 11. Design and optimization of a patient-specific additively manufactured hip implant stem (source: https://ntopology.com/blog/2020/04/27/how-to-design-and-optimize-a-patient-specific-additively-manufactured-hip-implant-stem/)

4 Generative Design in Product Development

4.1 Concept Design

During concept design phase, engineers/designers develop a range of ideas that could potentially meet the requirements of form, structure and function. In practice, once the first feasible design is determined, engineers move on to the next aspect of the design development in order to meet the shorter deadlines defined by the development plans. The disadvantage is that the opportunity to explore different design solutions that would possibly better meet the given requirements is missed. Potential design solutions can vary widely. Some use the Top-Down approach to fit into the already defined spaces of individual components. Others use 2D and 3D sketches created over curves, lines, surfaces and other simple geometries. At this stage, these are not complete and detailed 3D models.

Generative design is very applicable in the concept design phase. Engineers/designers have the flexibility to explore alternative solutions to their ideas. Moreover, they are able to conduct comparative studies of the pros and cons. This provides better insight into the relationships of key variables and desired performance. The resulting generative design will be a surface, faceted geometry whose practical feasibility must be examined. In the concept design phase, facet modeling is very useful for manipulation, as it avoids the conversion of geometry into a b-rep, which is specific to parametric and/or direct modeling. The obtained facet model of the model is the starting point for further development and must be converted into b-rep geometry.

Generative design has a great application on products where aesthetics is important, but not imperative. Figure 12 shows the first commercial chair created by generative design. French designer Phillipe Starck, in collaboration with Autodesk and the Cartell Company, using artificial intelligence and generative design, created the first chair that was designed beyond the reach of the human mind. The essence is in obtaining good aesthetics, innovative design and optimal use of materials (Fig. 12).

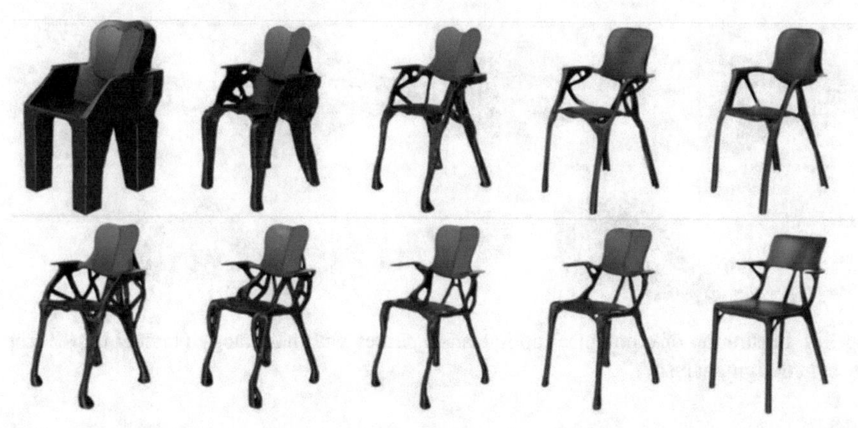

Fig. 12. First commercial product created with generative design (source: https://www.fastcompany.com/90334218/this-is-the-first-commercial-product-made-using-generative-design)

4.2 Detailed Design

In product development, the tested and selected design concept is further elaborated in detail with the aim of launching the design for production, meeting all the requirements of shape, structure and functionality. To do this it is necessary to explore options for different aspects of the design in order to improve performance. This is especially true for finding a balance between conflicting requirements such as mass, structural load, cost, and natural frequencies.

The digital geometric representation of the design at this stage is a fully detailed 3D model. These models are often built using parametric and direct modeling, resulting in a smooth, rounded geometry, but can also be defined by implicit modeling approach. At this point, there is a great opportunity to go beyond the first, feasible design. Experimenting on different detailed geometry and with different configurations of size parameters can drastically affect product performance as well as cost and workability. As in the previous phase, experimenting with variables and design requirements will provide new insights into their relationship. Sometimes it happens that generative design suggests options that the engineer/designer has never considered. This is a very valuable tool for perfecting the design in order to find a solution with a good balance between the given requirements. This allows engineers to fine-tune the detailed design to better meet the requirements.

Integrating generative design solutions with detailed design models is imperative. Finally, engineers must launch detailed design in procurement or manufacturing departments. Surface modeling can greatly affect productivity with the ability to smooth the transition from faceting to b-rep geometry. In some cases, it is necessary to manually remodel a smooth surface based on a faceted model, and corrections in faceted geometry are possible. This mostly applies to organizations engaged in additive manufacturing, while relying on faceted geometry, which allows models to be sent directly to 3D printing (Fig. 13).

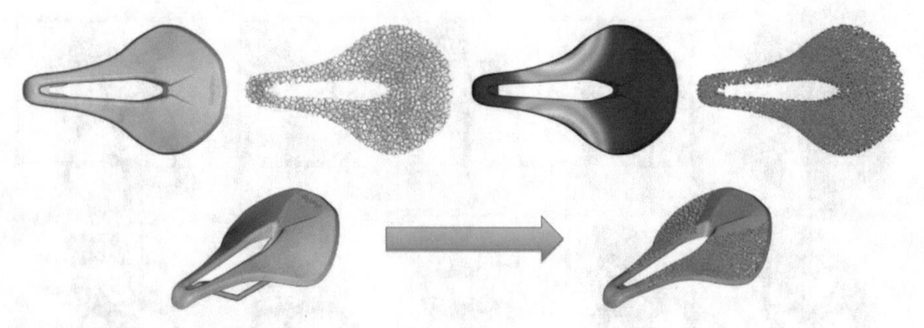

Fig. 13. Illustration of topological optimization phases with nTopology (source: https://ntopology.com/design-analysis/)

5 Potential Solutions

5.1 Using Two or More Applications

In concept design and detailed design, generative design is a very powerful tool in the product development process. However, traditional technologies that support generative design, as a non-integrated set of software applications, inherently have great compatibility issues within the digital workflow.

Basically, traditional geometry modeling can be parametric or direct. Parametric modeling enables the creation of a feature-by-feature model through parametric controls. Direct modeling allows geometry to be shaped by dragging and/or pressing. Both of these approaches work with b-rep, where the geometry is displayed on flat or smooth curved surfaces.

In contrast, the grid geometry contains a cloud of points that represent the outer surface. Some CAD applications convert this cloud into the geometry of solids by creating triangular or trapezoidal surfaces, which are bound at the edges into a homogeneous volume. Facet modeling allows engineers to “tweak” the quality of the resulting grid, as well as to modify the geometry by subtracting or adding material. There are cases where it is necessary to develop smooth rounded geometry as well as mesh geometry. In the concept design phase, it is necessary to work with the sketches and the allocated space along with the network geometry of the scanned components. In detailed design, it is necessary to create detailed 3D models taking into account the network geometry.

Since most CAD applications do not have the ability to work in parallel with mesh geometry and parametric/direct modeling, designers and engineers must solve the problem by applying two or more applications. Some standalone applications (such as 3D scanning applications) offer a CAD-like application that allows facet modeling. Theoretically, two such applications can be used, but in practice there are serious problems.

5.2 Using One Application

In the past few years, some CAD applications have expanded their capabilities to parametric, direct and faceted modeling. Some have even introduced lattice modeling and optimizations. The implications for generative design are crucial.

When it is necessary to make b-rep geometry from the resulting generative design, the workflow becomes simpler. All modeling capabilities are located in one environment, which means that users always have access to the appropriate tool for the current situation.

The second case is interesting in that it is not necessary to transform the generative design into b-rep geometry. Facet modeling provides tools to change the design without additional time-intensive steps. This especially refers to the components that will be produced by additive technology, which already depends on the mesh geometry.

Important in all these scenarios are the activities that this new type of CAD application enables engineers. That is, the exchange of design data is avoided. Because all of these capabilities exist in a single environment, there is no need to exchange 3D data, network geometry, or contour presentations between different software applications. All work can be done in one environment. No more time is wasted on fixing the geometry due to digital incompatibility. Instead, further work can focus on design.

6 Conclusion

Today, engineers are under great pressure in product development. Design is basically an unstable and unpredictable process, because mistakes can cause delays and delays in further development. Increasingly complex technology is being integrated into products. Engineers need to work with more suppliers than ever before. Deadlines are getting shorter. With a multitude of responsibilities, engineers often only have time to find the first feasible option, instead of the better one.

Generative design is the ability of CAD applications that autonomously generates numerous design alternatives for a given number of constraints and operating conditions. Upon completion, engineers can decide which design is more fully explored. This speeds up the design without the detailed attention of engineers. However, engineers must be aware that the result of generative design is actually network geometry, which can only be manipulated by facet modeling. This is especially important if taken into account that the resulting design will be used during the rest of the development process.

In concept design, generative design can be used for exceptional purposes. Designers /engineers can apply it to explore a wide range of design alternatives at the earliest stage when requirements are most flexible. In detailed design, generative design can be used to precisely adapt a design to meet the right balance between conflicting requirements, such as weight and construction load, cost, and natural frequencies.

References

1. Jackson, C.: Enabling simulation driven design: aligning people, processes, and technology, Interal Audit., pp. 1–10, February 2017
2. Hardware Design Engineer Study | Lifecycle Insights, https://www.lifecycleinsights.com/study/hardware-design-engineer-study/. Accessed 16 Oct 2019
3. Darwin, C.: On the Origin of Species, from the first edition, John Murray, London, 1859, released through Project Gutenberg. https://gutenberg.org/files/1228/1228-h/1228-h.htm. Accessed 10 Nov 2019
4. Nagy, D.: The problem of learning (2017). https://medium.com/generative-design/generative-design-introduction-64fb2db38e1. Accessed 14 Oct 2019
5. Wong, K.: Generative design: advice from algorithms. Digit. Eng. **247** (2018). https://www.digitalengineering247.com/article/design-advice-algorithms. Accessed 10 Feb 2019
6. Verbart, A., van Keulen, F., Langelaar, M.: Topology optimization with stress constraints, dissertation, TU Delft (2015)
7. Tyflopoulos, E., Flem, D.T., Steinert, M., Olsen, A.: State of the art of generative design and topology optimization and potential research needs, DS 91. In: Proceedings of NordDesign 2018, Linköping, Sweden, 14th–17th August 2018, ISBN: 978-91-7685-185-2. https://www.designsociety.org/publication/40924/State+of+the+art+of+generative+design+and+topology+optimization+and+potential+research+needs. Accessed Jun 2019
8. Cole Reports, Generative Design Market Applications, Types and Industry Analysis Including Growth Trends and Forecasts to 2027, Cole Market Research, 12 May 2020. https://coleofduty.com/news/2020/05/12/generative-design-market-applications-types-and-industry-analysis-including-growth-trends-and-forecasts-to-2027/
9. Noe, R.: Where does generative design belong? Designers must decide, digital fabrication (2019). https://www.core77.com/posts/89318/Where-Does-Generative-Design-Belong-Designers-Must-Decide. Accessed 10 Nov 2019
10. Freier, A.: La Bandita: hackrod and siemens partner to develop self-designing car (2018). https://all3dp.com/self-driving-cars-yesterday-self-designing-cars-future-thanks-hackrod/
11. Crotty, R.: The Impact of Building Information Modelling: Transforming Construction. Routledge, London (2013). https://doi.org/10.4324/9780203836019
12. Kostić, Z, Cvetković, D., Jevremović, A., Radaković, D., Popović, R., Marković, D.: The development of assembly constraints within a virtual laboratory for collaborative learning in industrial design. J. Tehnički vjesnik – Tech. Gaz., **20**(5), pp. 747–753 (2013). https://hrcak.srce.hr/109781. Accessed 8 May 2020
13. Kostic, Z., Radakovic, D., Cvetkovic, D., Jevremovic, A., Markovic, D., Kocareva Ranisavljev, M.: Web-based laboratory for collaborative and concurrent CAD designing, assembling, and practical exercising on distance. J. Tehnički vjesnik – Tech. Gaz. **22**(3), 591–597 (2015). https://doi.org/10.17559/TV-20140211115630
14. Harris, J.: 5 Techniques for lightweighting: doing more with less, nTopology (2019). https://ntopology.com/blog/2019/10/18/5-techniques-for-lightweighting-doing-more-with-less/

15. Shomper, M.: True biological modeling for implants (2019). https://www.linkedin.com/pulse/true-biological-modeling-implants-matthew-shomper/
16. du Plessisa, A., Broeckhoven, C., Yadroitsava, I., Yadroitsev, I., Hands, C.H., Kunju, R., Bhate, D.: Beautiful and functional: a review of biomimetic design in additive manufacturing. Addit. Manuf. **27**, 408–427 (2019). https://doi.org/10.1016/j.addma.2019.03.033

Automation of Cup Filling Machine by Inserting PLC Control Unit for Educational Purpose

Nebojsa Miljevic, Nada Ratkovic Kovačevic, and Djordje Dihovicni(✉)

The Academy of Applied Technical Studies Belgrade, Blv. Zorana Djindjica 152 a, 11070 New Belgrade, Serbia
ddihovicni@gmail.com

Abstract. In this paper the automation of cup filling machine by inserting PLC unit for educational purpose is presented. Small filling machine has been constructed, and it has ability to recognize two cup sizes and sort them after filling. The machine charger works independently at 220 V, the sizes of the glasses it fills are 0.2 l and 0.1 l, and it has its own liquid tank with level sensor and compressor and bottle with pressure sensors and automatic on and off button. The conditions for starting the machine, are that the air pressure is above 2 bar, and the level of liquid in the vessel is sufficient to fill one whole glass after signaling the disappearance of liquid. The required width of the conveyor belt is 10 cm, and it is driven by a 12 V electric motor. The cups are sorted using a pneumatic piston, and the liquid circulation pump is of constant flow. In the operation of the machine, it is necessary to provide protection against incompletely filled cups due to liquid loss, and protection against incorrect sorting due to loss of air pressure, and light alarm in case of loss of conditions. The alteration of machine's control units is shown by replacing specific components with the programmable logic controller SIEMENS S7 300. A detailed description of the installation wiring and programming of PLC is described, as well as the program execution using PLC SIM software. The final phase involved program testing, and testing of particular system operations.

Keywords: Automation · PLC controller · Filling machine · Mechatronics

1 Introduction

Modern industrial production imposes a constant tendency to rise economy and labor productivity. There are increasingly stringent requirements regarding the flexibility of production facilities. The speed of work increases, the number of failures and scraps decreases to a minimum (the tendency is that there are no scraps, no failures, no waste of materials), the quality of workmanship tends to be higher, sorting, packaging, and other operations are automated and accelerated. The introduction of automation in industrial existence represents a significant improvement of each of the factors of production, with a reduction in the participation of human power. Control systems aim to replace the man in the control, speed up the production and operation process, while reducing possible errors and failures.

N. Mitrovic et al. (Eds.): CNNTech 2020, LNNS 153, pp. 344–361, 2021.
https://doi.org/10.1007/978-3-030-58362-0_20

The beginning of the automation of industrial plants was based on relay technology, and the development of computers significantly contributed to the progress in the design and capabilities of control systems. The most often case for consideration of control systems, is system behavior on infinite interval, which in real cases has only academic importance, in spite of practical stability, which has significance in real life [1]. Increasing efforts are being made to implement and integrate computers into technical systems, in order to monitor and surveillance their work, record relevant process data, calculate and implement processing and management in real-time, data mining and knowledge extraction. During process of analyzing and synthesis of the control systems fundamental problem is stability [2]. Sometimes taking into account the complexity of the model, and the fact that it is described by partial different equations, it is important to develop the program support based on numerical methods [3]. Computers also tend to take complete control of equipment and facilities over time, monitoring the condition of assemblies and machines (predictive maintenance), collecting data, and providing communication between machines, with automated decision making. Various mathematical models are developed for control systems, and some complex systems are described by using partial differential equations. Combining time delay component and as well distributed parameter component, would open the field for discussion of applying adequate approaches for solving these kind of problems [4].

The factories of the future are based on flexible automation, and on Computer Integrated Manufacturing (CIM). The basis of these production technologies are CNC (Computer Numerical Control) machines, industrial computers and PLC (Programmable Logic Controller) devices, industrial robots and specialized SBC computers (Single Board Computer).

Modern factories are already based on the integration of production using computers, with the fact that the equipment itself has built-in computers, which can provide communication between machines and people, but also between the machines themselves. The workforce of the future will be of a different composition than the traditional one. First of all, it means highly educated teams and crews, which will do the preparation, monitoring of production and commissioning, so that the factory can function with as few number of people - operators, as necessary. As a creator, man moves into the field of design, monitoring, planning and realization of all tasks related to engineering. Changes in the industry are accelerating, and the education of engineers must follow them, in order to provide experts, appropriate professions, specialties and competencies.

The methodologies applied in this research include: analysis, synthesis, abstraction, concretization, specification, deconstruction, definition, division, deduction, analogy, experiment. After the process of design and development, a functional machine is constructed, an automatic filler, which is a scaled model of an industrial machine for educational purpose. A control algorithm is specially developed, which is both translated and transferred to the control unit, and then applied to the machine.

2 Study of Requirements of the Control System

The aim of this paper is to show how to make a machine for filling cups, which is a model of decreased scale of an industrial machine, and it has the ability to recognize two sizes of glasses - cups and to sort them after filling, according to their size.

The conditions for starting the machine are:

1. air pressure in the pneumatic subsystem is above 2 bar;
2. level of liquid in the vessel is enough to fill one whole cup, after signaling that the liquid is soon to run-out (to disappear).

Other details of the terms of reference specification are:

- the charger works independently, supplied with alternating voltage of 220 V;
- the sizes of the cups they fill are 0.2 l and 0.1 l;
- has a liquid tank with a level sensor;
- has a compressor and a bottle with pressure sensors and automatic on and off, depending on the pressure value;
- the required width of the conveyor belt is 10 cm;
- the conveyor belt is driven by a 12 V DC electric motor;
- sorting of cups is done using a pneumatic piston;
- the fluid circulation pump is of constant flow;
- the protection against incompletely filled cups, due to loss of liquid should be provided;
- the protection against incorrect sorting, due to loss of air pressure in the pneumatic piston subsystem;
- secure the end of the tap with a pressure valve to prevent dripping or slight leakage of liquid while the cup is not under the tap;
- activate the light alarm in case of loss of conditions, either 1 or 2;
- sound and light alarm for reaching the set number of filled glasses;
- reset the alarm for number of filled glasses.

3 Choice of I/O Ports and PLC Size

Choosing I/O ports and connections means physically counting the necessary sensors and actuators, to know exactly how many inputs and how many outputs is needed. In order for the filler to be able to perform the task, it is necessary to have two conditions fulfilled. The first condition is the level of liquid in the vessel from which the pump drops and pours into cups. Without this condition, the operation of the filler is not possible, due to the fact that when the pump that fills the liquid in the installed container fails, the filler will not recognize whether there is liquid in the cup that passed the pump, so it is possible for an empty cup to reach the imaginary packaging line. To prevent this from happening, the filler will stop at any time when the liquid level sensor responds and detects a lack of liquid. As the level sensor it is chosen the liquid level regulator or floating switch of the Italian company CEME type RKS, as it is presented in Fig. 1.

This type of floating switch is also suitable for fluids that are dirty, because the connection of the floating part with the body is achieved by a flexible Teflon connection.

Technical characteristics of the chosen liquid level regulator:

- body (material) PP/PTFE;
- electrical characteristics: max 50 W/50 VA 1A;
- fluids include all liquid fluids that may come into contact with the body;
- operating temperature in interval −25 °C–+125 °C;
- protection is Ip67.

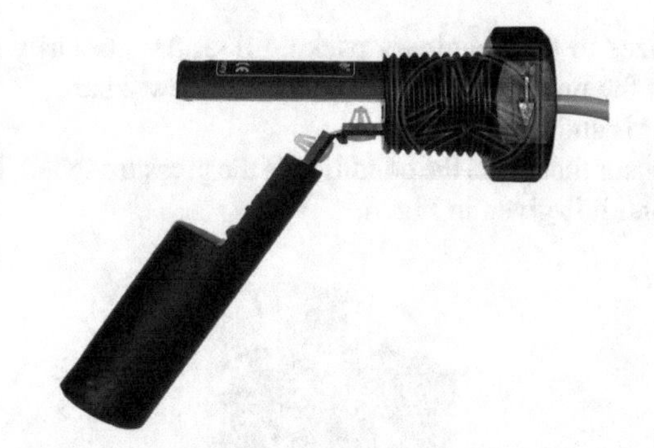

Fig. 1. Regulator-indicator of liquid levels type RKS.

Installed dimensions of the level regulator are shown in Fig. 2.

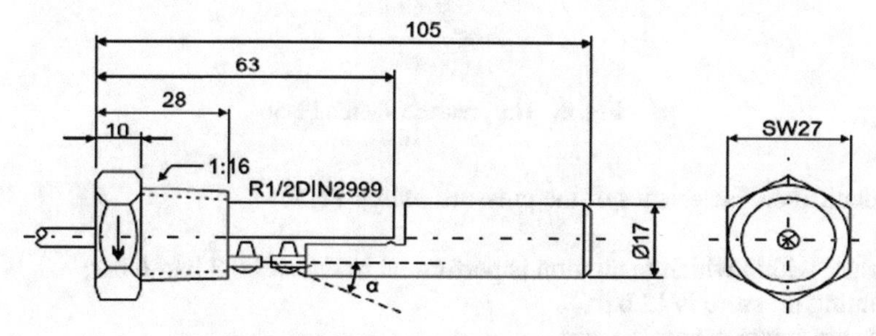

Fig. 2. Installed dimensions of the level regulator

For the second condition, the air pressure in the bottle is set. Using a pneumatic cylinder and piston and a bi-stable solenoid, after filling the filler, the lever rectifier – sorter sorts the cups.

In the case of failure of any pneumatic element and a drop in cylinder pressure below 2 bar, the filler/sorter will not be able to select the cups to a certain side, so it is possible

Fig. 3. The compressor

to send both cup sizes to the imaginary packing line. As a security measure, the filler will stop as soon as the pressure in the bottle drops below 2 bar.

The compressor is shown in Fig. 3.

The pressure sensor that gives the condition is the pressure switch PN56 of the Italian company CEME, and it is given in Fig. 4.

Fig. 4. The pressure switch PN56

Technical characteristics of the pressure switch PN56:

- pressure within which regulation is performed in interval 0.2 bar–6 bar;
- mounting pressure is 12 bar;
- fluids are steam, water, oil, air;
- fluid contact parts are stainless steel diaphragm AISI301-connectors, copper;
- declassified life is 100000 cycles.

The pressure sensor that controls the operation of the compressor is a PT5 pressure switch from an Italian company ITALTECNICA, which is shown in Fig. 5.

Fig. 5. The pressure switch PT5

When both conditions are met, the filler is ready for operation and starts the conveyor belt that carries the cups. The conveyor belt is manufactured by VULKAN COMMERCE Srbobran, from Serbia, dimensions 1.2 mm × 1000 mm × 40 mm, and it is presented in Fig. 6.

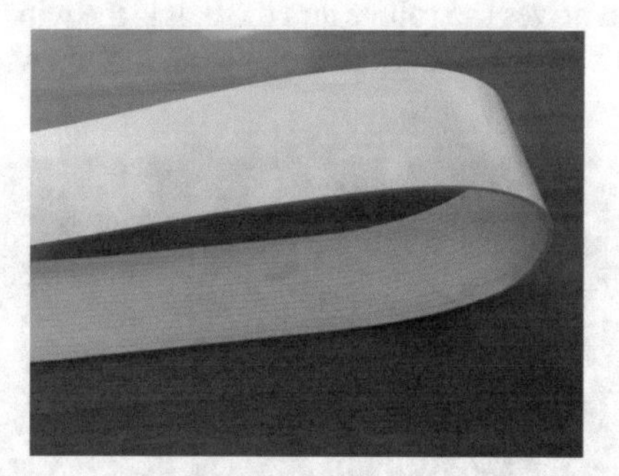

Fig. 6. The conveyor belt

The rollers that carry the conveyor belt are made by cutting from a large roller intended for installation in certain agricultural machines, and the shaft on which the rollers rotate is the shaft of the bicycle. The roller with shaft is presented in Fig. 7.

Fig. 7. The roller with shaft

The drive motor that drives the rollers and the belt is the wiper motor with the rollers, which is connected via two sprockets and a chain, as it is shown in Fig. 8.

Fig. 8. The drive motor

The recognition of the position and height of the glass is done by sensors from the Italian company MICRO DETECTORS given in Fig. 9.

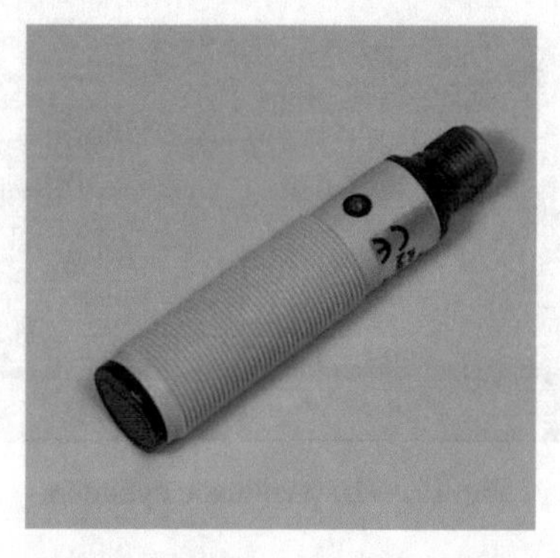

Fig. 9. Optoelectronic sensor from MICRO DETECTORS, Italy

The pump that pours liquid into glasses is the BRUSHLESS DC PUMP Model QR30E pump, and it is presented in Fig. 10.

Fig. 10. The pump that pours liquid into glasses

Technical characteristics of the brushless DC pump model QR30E:

- operating voltage 12 V;
- current 350 mA;
- maximum flow 4 l/min;
- potential used fluids are water, acids.

The actuator that sorts the glasses in relation to the signal obtained from the sensor is a pneumatic cylinder with a piston and a bi-stable solenoid valve from Westing firm from Trstenik, Serbia, and it is given in Fig. 11.

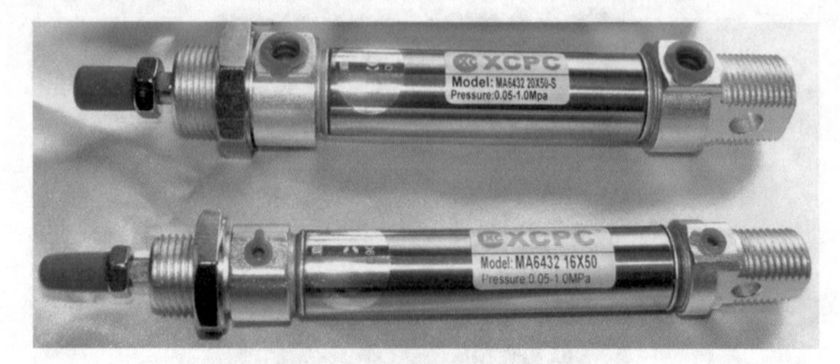

Fig. 11. The pneumatic cylinders

The bi-stable solenoid valve is presented in Fig. 12.

Fig. 12. The bi-stable solenoid valve

A graphical representation of the principle of operation and the method of connecting the solenoid valve and the pneumatic piston is given in Fig. 13.

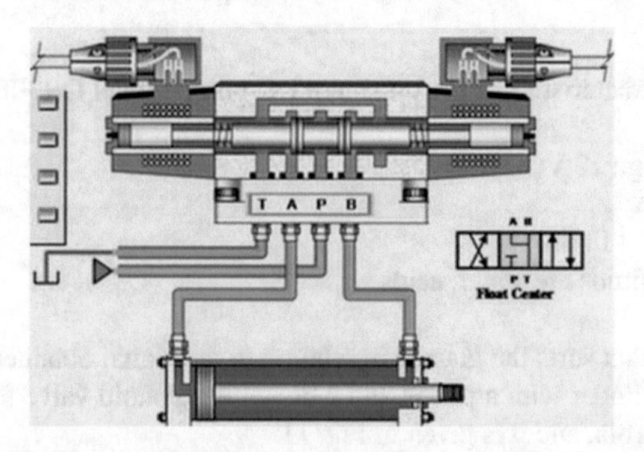

Fig. 13. The graphical representation of the principle of operation

4 Translation of a Block Diagram into a Ladder Diagram

STEP 7 is an engineering tool for SIMATIC controllers, running on the standard Microsoft Windows platform [5, 6]. Block diagram translation is the writing of program code based on a block diagram in one of the standard languages for PLC programming, Ladder logic (LAD), Function block diagram (FBD) or Statement list (STL).

STEP 7 Professional integrates PLCSIM simulation software for testing the user program in offline mode. Using the programming languages LAD and FBD, programming is done by connecting predefined blocks that represent logic circuits. To write a program using these two programming languages, it is necessary to know the basic symbols of automation and their function. STL programming language is more often chosen by those who have previously had points of contact with any type of programming.

In the processing of automation system data, the following basic procedure is applied:

Creating a New Project
The first step in creating a project. The name is assigned and the storage location is determined.

Hardware Configuration
After the unit is added, the hardware settings are specified. Based on the request, a communication bus is installed and modules are added to it. If necessary, the module parameters or the IP addresses of the signal modules are set. Sometimes, an additional communication bus can be added for expansion, to which modules are also added.

Communication Network Configuration
The project can contain several units. Configuring the communication network creates connections between units that enable data exchange.

Assigning Names to Signals
Symbolic names are assigned to signals before writing the program.

Writing the Program
The program consists of separate parts, called blocks. The blocks are programmed using the mentioned programming languages, so the order of execution of the blocks can be determined.

The block diagram with the control flow describing the program that the PLC will be executing is shown in Fig. 14.

The hardware configuration is edited using HW Config in offline mode without the necessary presence of the PLC itself. In order to work with HW Config, it is necessary to open a project with the selected PLC unit.

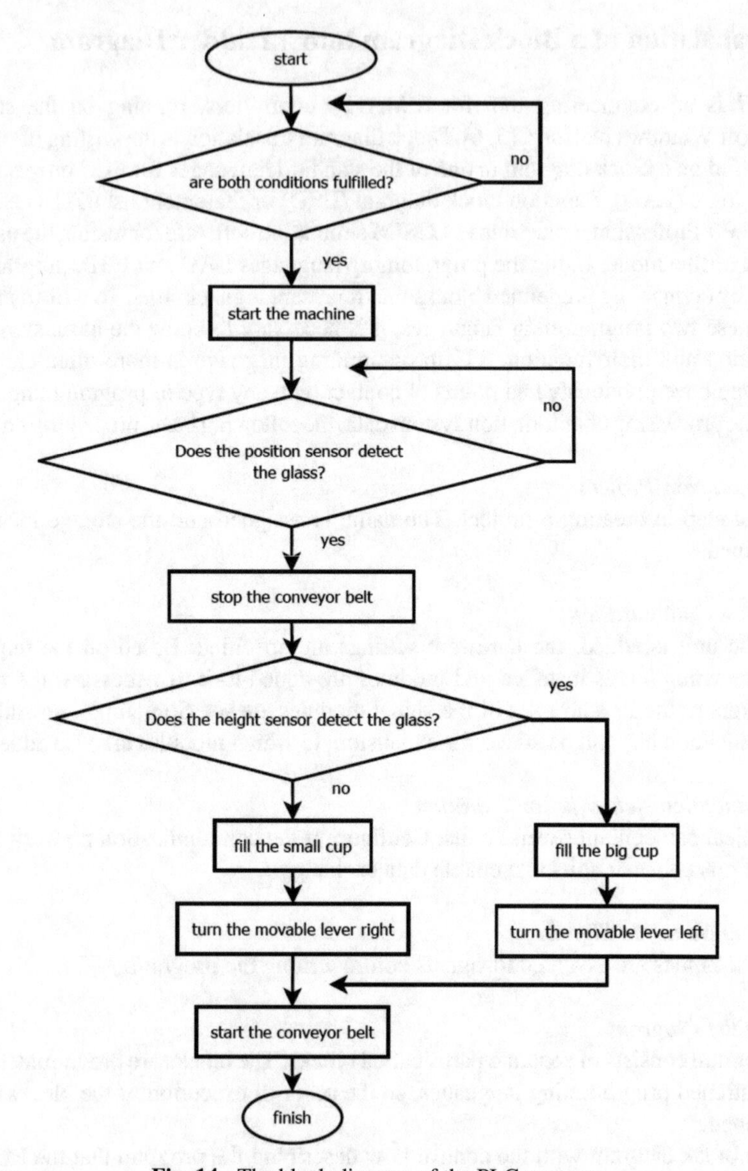

Fig. 14. The block diagram of the PLC program

After the hardware configuration, a realistic picture of the future controlled automatic system is obtained. By selecting the desired module from the module catalog using the Drag & Drop method, the chosen control system is placed in the desired location. When installing the module, it is necessary to take into account the schedule of proper filling of slots. Some of the slots are predefined for certain modules so that other modules cannot reach that slot. For example, the third slot is reserved for the communication module and no other can be installed except it. If not needed, the slot remains empty.

To find the desired module, it is important to go through the folder structure. When the selected module is clicked with the mouse, its most important characteristics necessary for creating the program are displayed.

The selected module switches to the factory settings slot and the program automatically assigns IP addresses.

The hardware configuration is presented in Fig. 15.

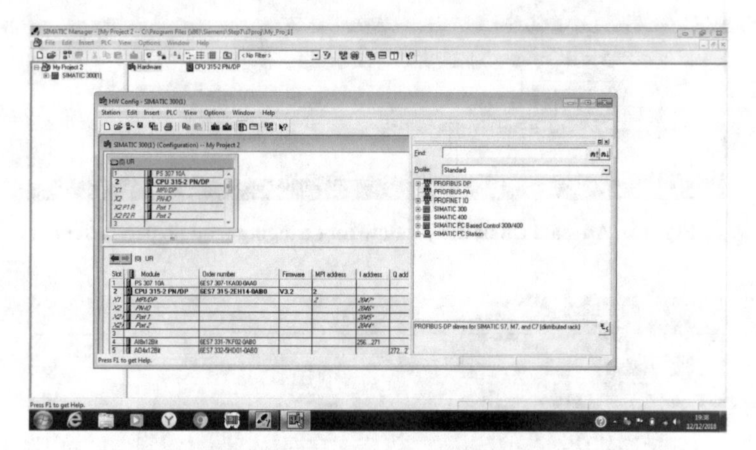

Fig. 15. The hardware configuration

The module settings can be changed by the user as needed. Double-clicking on the selected module or selecting the Edit → Object Properties function opens a dialog box for changing parameters.

Table 1. Input signals and Output signals.

Input signals	Output signals
– bottle air pressure	– belt drive motor
– liquid level in the tank	– liquid dispensing pump
– sensor of the position	– left position of pneumatic piston
– sensor of the height	– right position of pneumatic piston

Assigning names to signals is done for easier navigation in the program. Large programs containing tens of thousands of command lines would be difficult to understand, if the signals are not named. To the absolute address assigned during writing the program, the user also assigns a name that will direct him to what is connected at that input. Table 1 lists input signals and output signals.

Appearance of the window for changing module parameters is shown in the Fig. 16, and the Symbol Editor window with the assigned names is presented in the Fig. 17.

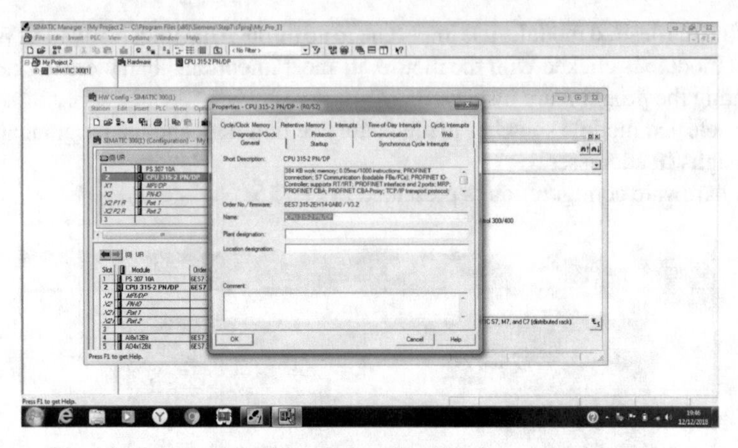

Fig. 16. Appearance of the window for changing module parameters

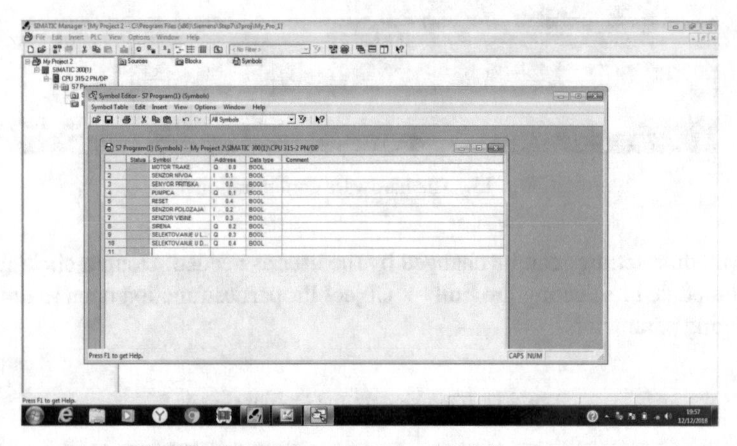

Fig. 17. Appearance of the Symbol Editor window

5 Writing a Control Program

Programming a block begins by opening it by double-clicking on the desired block in the project window or manually selecting the File → Open command in the program editor. After selecting one of the programming languages, in this case LAD, the user starts writing the program code.

On the left side of the window there is the Program Elements from which the user selects the program elements and connects them with the mouse, by simply connecting the inputs and outputs, i.e. the ends of the program elements.

The request of the control system requires that the machine has two conditions, i.e. in order for the machine to work, i.e. the belt to move, it is necessary that the conditions are met. In case of disappearance of one of the conditions, the machine stops and signals the lack of conditions. In case of disappearance of one of the conditions during the pouring of the liquid into the glass, the machine will finish the started cycle and only then stop.

From the Program elements folder of the Bit logic folder by double-clicking or one-clicking from the Quick menu, the circuit symbol is moved to a specific place in the network. Above the symbol, in the place of red questionnaires, the symbol mark is entered and the program itself pulls its name from the symbol database to make it easier for the user to navigate.

In Fig. 18. the Normally Open (NO) contact is shown.

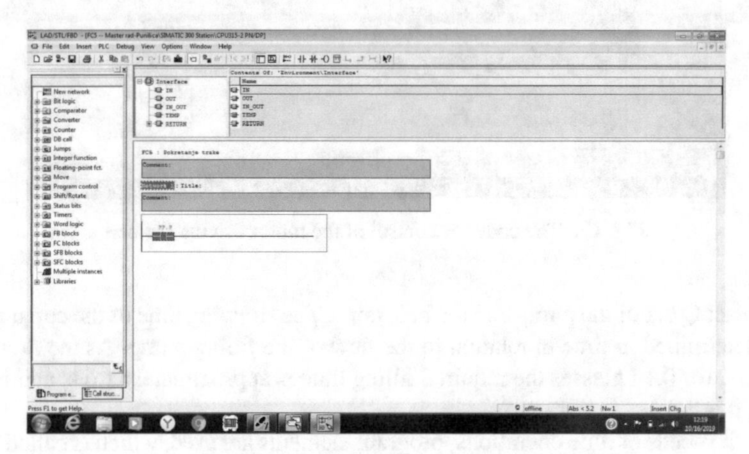

Fig. 18. Normally Open (NO) contact

The serial connection of two NO (normally open) contacts, liquid level sensor I 0.1 and air pressure I 0.0 and two NC (normally closed) contacts, pump Q 0.1 and timer T3 sets all the conditions for starting and stopping the engine.

The final element is the Output coil as the motor for starting the belt. By closing the normally open contacts, i.e. the transition of the sensor contacts to the working position (both conditions are met), the belt motor starts working. The belt runs until one of the 4 conditions changes.

Stopping the belt motor is required in the following conditions:

- in case of liquid loss in the vessel;
- if air pressure drops below 2 bar;
- while the pump is running;
- while the time relay T3 (FC4) is running.

The code for control of the motor driving conveyor belt is presented in Fig. 19.

The belt moves until the position sensor detects that the cup has reached the position under the tap. At that moment, the belt stops. By closing the NO contact of the position sensor, the pump starts with the operation as in Figs. 19. Starting the pump opens the

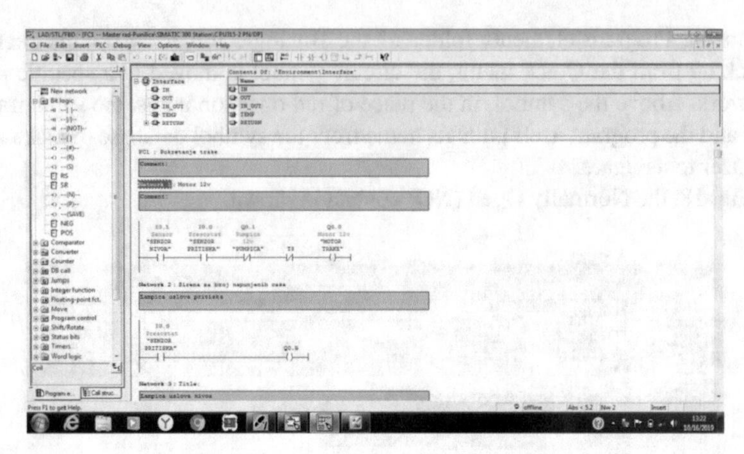

Fig. 19. The code for control of the motor driving the belt

NC contact Q 0.1 of the pump and the belt stops. The stopping time of the cup under the tap is determined in time in relation to the flow of the filling pump. As the flow pump is 4 l/min, for 0.1 l glasses the required filling time is approximately 1.6 s, and for 0.2 l glasses it is 3.3 s.

For the needs of time operations, program elements are used, which is called timers. From the Program elements menu of the Timers folder by double-clicking on the selected S_PULSE timer, the timer symbol is moved to a specific place in the network.

The condition for starting the tapping time measurement is obtained from the position sensor and the height sensor, depending on the size of the cup on the strip.

Two timers are required for two lengths of time interval for filling the glasses.

In Fig. 20 the S_PULSE timer with determined count time is given.

The NO contact of the position sensor is connected to the input S of the first timer and the NO contact of the height sensor is connected to the S input of the second timer. The required counting time is entered at the TV input. A simple coil element is placed on the timer outputs and the pump address is assigned to it.

While the time is measured by either of the two timers, the pump filling cups should be running.

At the output Q, a logical unit is obtained for each counting time, and when the timer does not count, a logical zero is obtained. After recognizing the cup size, the cup selector should be moved to a specific position, bringing the signal to one side or the other of the bi-stable solenoid valve.

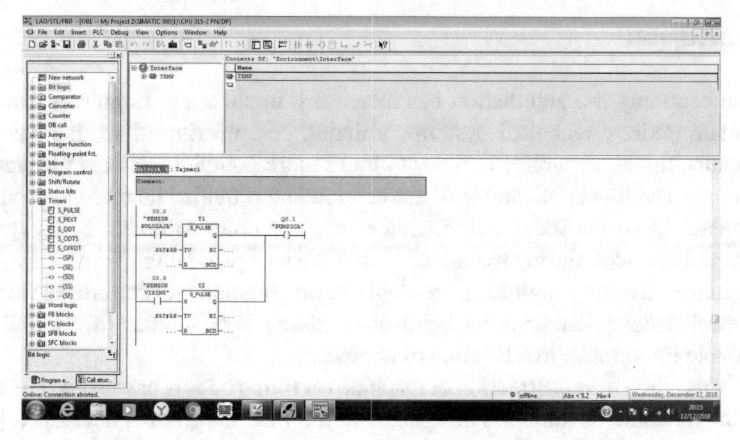

Fig. 20. The S_PULSE timer with determined count time

Fuzzy logic uses the theory of fuzzy sets and provides an analytical modeling algorithm, for the statements whose truth-value can be part of a continuous transition from exact to false [7]. The series connection and the combination of NO and NC contacts of the position sensor and the height sensor determine to which input of the bi-stable solenoid valve the logic unit is fed, i.e. to which side the pneumatic piston will place the selector.

By sequentially connecting the NO contact of the position sensor and the NC contact of the height sensor, the selection of the cups to the left is obtained. By sequentially connecting the NO contact of the position sensor and the NO contact of the height sensor, the selection of the cups to the right is obtained. In other words, if the position sensor detects and the height sensor does not detect the cup, a signal is sent to the solenoid valve which actuates the pneumatic piston to place the selector to the left, and both the position sensor and the height sensor detect the cup, and the solenoid valve actuates the pneumatic piston to place the selector to the right.

Selection of the cups is presented it the Fig. 21.

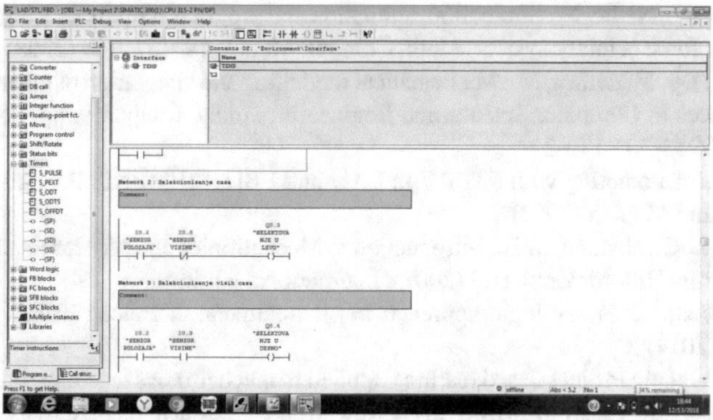

Fig. 21. The selection of the cups

6 Conclusion

In global economy the automation has increasing importance. Engineers are striving to automate entirely technical systems, utilizing combination of mathematical tools with organizational resources, in order to build more complex systems for plethora of applications. The first applications of automation had provided increase in productivity and decrease in costs. Today automation is used in order to obtain better quality of products and services, during manufacturing or service providing.

Teaching or learning engineering through the projects and practice is used worldwide, e.g. in mechatronics, either in bachelor or in master level studies [8, 9], and it is of considerable importance in education of engineers.

In this research the construction of machine for filling cups is presented. The automation of the machine operation is designed and the PLC program is developed [9]. The pneumatic and hydraulic sub-systems of the machine and as well the conveyor belt of the machine, are designed [9–11]. The main contribution of this paper is that cup filling machine is used for educational purpose, and experts will apply their knowledge in mechatronic systems, automation and PLCs, pneumatics and hydraulics, transport and conveyors, in practice.

The machine is built from new and used parts and components, in a real world environment. The machine described here represents physical model of an industrial machine, fully operational, and it can be used as educational aid. Knowledge acquired during engineering practice can be applied and tested in real conditions, similar to some industrial processes or plants. The machine can be used furthermore as an educational aid, for experiments and learning through practice.

References

1. Dihovicni, Dj., Asonja, A., Radivojevic, N., Cvijanovic, D., Skrbic, S.: Stability issues and program support for time delay systems in state over finite time interval. Phys. A Stat. Mech. Appl. **538**, 122815 (2020)
2. Dihovicni, Dj.: Decision making and fundamental matrix approach in process safety. Int. J. Comput. Intell. Syst. **6**(4), 658–668 (2013)
3. Dihovicni, Dj., Nedic, N.: Simulation, animation and program support for a high performance pneumatic force actuator system. Math. Comput. Model. **48**, 761–768 (2008)
4. Dihovicni, Dj., Medenica, M.: Mathematical modelling and simulation of pneumatic systems. In: Advances in Computer Science and Engineering, India, Chapter 9, pp. 161–186 (2011). ISBN 978-953-307-173-2
5. Berger, H.: Automating with STEP 7 in LAD and FBD, SIEMENS. Publicis MCD Verlag, Erlangen and Munich (2003)
6. Alciatore, D.G., Histand, M.B.: Introduction to Mechatronics and Measurement Systems, 3rd edn. McGraw Hill, McGraw Hill (2007). International edition
7. Dihovicni, Dj.N.: Fuzzy logic approach in oil treatment. J. Balcan Tribol. Assoc. **20**(4), 606–614 (2014)
8. Noriega, A. et al.: Project-based learning applied to mechatronics teaching. In: García-Prada, J.C., Castejón, C. (eds.) ISEMMS 2017, New Trends in Educational Activity in the Field of Mechanisms and Machine Theory 2014–2017, MMS, vol. 64, pp. 49–56. Springer, Heidelberg (2019)

9. Miljević, N.: Automation of the cup filling machine utilizing PLC. Master theses in vocational studies - in preparation, The Academy of Applied Technical Studies Belgrade, Technical College Beograd, Belgrade (2020). (in Serbian)
10. Suvajdžić, S.: Design, construction and building the pneumatic sub-system of the device for filling cups with liquid. Master theses in vocational studies-in preparation, The Academy of Applied Technical Studies Belgrade, Technical College Beograd, Belgrade (2020). (in Serbian)
11. Ivetić, N.: Design, construction and making the conveyor belt as a sub-system of the device for filling cups with liquid. Master theses in vocational studies-in preparation, The Academy of Applied Technical Studies Belgrade, Technical College Beograd, Belgrade (2020). (in Serbian)

Process Analysis of a Tank Management System

Jasmina Perišić[1], Marina Milovanović[1], Marko Ristić[2](✉), Jelena Vidaković[3], and Ljiljana Radovanović[4]

[1] Faculty of Entrepreneurial Business and Real Estate Management, UNION "Nikola Tesla" University, Cara Dušana 62-64, Belgrade, Serbia
[2] Institute Mihajlo Pupin, Railway Department, University of Belgrade, Volgina 15, 11000 Belgrade, Serbia
marko.ristic986@gmail.com
[3] Lola Insititute, Kneza Višeslava 70a, 11000 Belgrade, Serbia
[4] Technical Faculty "Mihajlo Pupin", University of Novi Sad, Đure Đakovica Bb, Zrenjanin, Serbia

Abstract. This research focuses on a comprehensive approach to operational processes in the domain of tank management systems which is important in running tank farm safely, smoothly and successfully. The main objective is to provide the necessary information about resources, stakeholders and relationships of all relevant processes related to tank management systems. The acquired information represents a basis for a detailed analysis and process visualization followed by a discussion. We used a BPMN 2.0 notation to create a process model, describe the most important aspects related to tank management systems and connect defined subprocesses. Furthermore, different complex subprocesses, such as loading operations and storage operations, were integrated and enriched with new details. In that way, we developed a standardized process model to meet the evolving demands of the industries to use precise and reliable data for business decisions. In addition, we conducted a process analysis to detect critical points and provide some solutions.

Keywords: Tank management system · Process modeling · Process analysis · BPMN · SCADA · Custody transfer

1 Introduction

In order to increase the reliability of a tank management system and ensure economic feasibility, it is necessary to effectively address all technical, safety, organizational and environmental challenges and ensure compliance with the latest standards in this field. Tank management systems should be designed according to the highest international standards of safety so staff, plants, and the environment can be protected. It should ensure the highest reliability and accuracy of all subprocesses within the system (custody transfer level and flow measurements, liquid and gas analysis, temperature measurements, etc.) and enable plant running in accordance with international laws and regulations. For

N. Mitrovic et al. (Eds.): CNNTech 2020, LNNS 153, pp. 362–377, 2021.
https://doi.org/10.1007/978-3-030-58362-0_21

all tank management systems, it is very important to determine the quality of petroleum products such as the water and sediment content [1] and API gravity and sulfur content of crude oil and petroleum fractions [2]. Another important aspect is to enable highly accurate tank measurement with custody transfer approved level, temperature, and pressure instruments. Finally, a smooth process running enables collaboration and good business decisions among different stakeholders within the process which has an impact on increasing productivity and reducing inventory costs.

Business process modeling can be used to improve business procedures of tank management systems, consider the necessary resources and define responsibilities of participants. It can serve as a basis for selecting the right instrumentation for the precise measurement of various types of liquids, choosing suitable software and resources, as well as defining proper procedures in the early stages of system development. During the following phases, errors made at the beginning of the process can propagate, which makes them increasingly expensive and more difficult to correct. That is the reason to implement business process modelling as a useful tool for identifying and removing these errors [3]. The goal of process analysis is to define all relevant subprocesses, resources, participants and their relations within the overall process, detect critical points, conditions and limitations, and propose mitigation strategies and different ways to improve the process. In combination with a risk analysis, business process modelling and process analysis have a great influence on achieving significant savings, avoiding system failures, increasing reliability and optimizing subprocesses, both individually and integrally.

In this paper, we focused on describing, modelling and analyzing of tank management system processes, which allow optimization and improved management and planning. We defined a process model that depicts identified subprocesses and their interdependence. All the stakeholders and resources in these subprocesses were also captured which enables the following process analysis where it was pointed out on critical aspects and situations, such as measurement conditions, operational considerations, and constraints. In this way, tank management system processes were represented in a more transparent way, with a lot of attention focused on understanding and evaluation of critical aspects, necessary resources and relations between stakeholders in different stages.

As tools for data collection, we implemented questionaries', site visits, analysis, interviews, and discussions with a wide range of participants in the process, such as mechanical, instrumentation and control engineers, financial experts, technologists, tank farm managers and operators, maintenance and transport companies. In that way, we obtained a broad range of information about tank management system related to stakeholders, instrumentation, activities, procedures, necessary resources, and typical problems. This information was built into the process model which represents the starting point in process analysis. A process analysis can be used for deeper insight into a general system and its subprocesses, and finding places it can be enhanced by improving efficiency, supply chain and procedures applied in performing various activities. The final result is that all process parts, stakeholders, limitations and critical points are defined, so it is possible to optimize the process in terms of prevention of equipment failure and inaccurate measurements, avoiding personal injuries, increasing productivity, reducing waste and improving planning activities. We implemented the Business Process

Model and Notation (BPMN) as a graphical tool to represent process workflow with its stakeholders and mutual relations.

2 Materials and Methods

2.1 Process Modeling

Since the tank management system is a very complex system, different subprocesses, participants, technical and economic aspects were engaged in the conducted process modeling. Each aspect required different processes and resources to be processed, which was reflected in the very process of modeling and process analysis. Different technical, manipulative and administrative procedures affect the operation of tank management systems. Therefore, they must comply with international standards in this area, such as IEC 61508, SIL2/SIL3, OIML R85, and API MPMS. Also, all the technical and organizational conditions and criteria must ensure the safe, efficient and continuous operation of a tank management system.

Within the tank management system, there are a number of various stakeholders, resources, relations, and connections between them. Almost all companies have organizational units which are not isolated and there are numerous participants from different fields with different requirements [4], which affect the process flow and complexity of the overall process. This is the reason why it is necessary to carry in-depth identification of the key and supporting stakeholders and their interactions, the consequences of their actions, and the necessary resources for the execution of operations. This procedure serves as a basis for process mapping and the following development of the model which contains all important details about activities and relations between participants, process flow, resources, critical points, etc.

Business process modeling (BPM) is the basis for conducting and improving the process, and it can be utilized in different kinds of real-life scenarios [5, 6]. Because of its expressiveness, descriptiveness, and formality, we implemented Business Process Model and Notation (BPMN) 2.0. modeling language for process description. Additionally, it is understandable to end users, but not restricted to use only by experts [7]. These characteristics are reflected in the fact that BPMN is the most recognized language for specifying process workflows at the early design steps [8, 9]. Through the Business Process Diagram, which is provided with BPMN, it is possible to describe and understand both internal and external business procedures. Process workflow and information are defined with the following basic elements: tasks and connecting elements, elements of an organization, and data elements of artifacts [10]. 'Pool', 'Lane' and 'Milestone' are deployed for the organization of basic elements. Implementation of BPMN enables the representation of complex process semantics understandable to both for technical (developers responsible for implementing the technology that will perform those processes) and business users (business analysts and business people who will manage and monitor those processes) which can enhance business process management.

The goal of process modeling is to identify different stakeholders, resources, interactions and connections between them and to provide a basis related to interactions of involved parties and their activities. The obtained process model will enable users to specify complex process scenarios with all necessary data for further analysis of the critical points in the process, defining execution procedures, and process simulation [11]. We can define three main process layers of the tank management system according to the type of activities undertaken. The organizational layer includes planning and optimization processes which have an impact on the interaction within and between different processes. Management layer includes terminal management, custody transfer and, supply chain management and it covers processes that allow efficient resource and personnel management. Processes from these two layers represent supporting processes which provide requirements necessary for the accurate, safe and continuous operation of the tank management system. In other words, they affect all technical, operational and administrative procedures and their purpose is to synchronize them, provide necessary resources and define appropriate strategies for the processes from the third layer. These strategies are based on information processing in order to facilitate reconciliation and consolidation of stocks.

The third layer includes operational processes such as fuel transport, yard control and logistics, loading operations, production, storage operations, and distribution. These are the processes we analyzed in our work. There are several important aspects the process model should cover such as safety, environment protection, accurate measurements, customer satisfaction, etc. Tank management system performance depends to a great extent on the way different subprocesses are taking place, so it is necessary to organize and execute manipulative activities, resource and personnel management, maintenance, and calibration according to process model which contains complete hierarchy, relationships, participants, and resources of the process.

2.2 BPMN Graphical Concepts

BPMN is designed in order to be easy to interpret by all business participants regardless of their profession and skills. This is the reason it has become an important open standard graphic notation for drawing and modelling business processes [12–14]. We created diagrams by the BPMN specification using Microsoft Visio 2016 software. A set of patterns that can be imported into Microsoft Visio software is also included in this software. BPMN diagram or business process diagram (BPD) is made by importing BPMN graphical tools into Visio workspace. Complete diagram can be saved in different formats: JPEG, XML drawing, AutoCAD drawing, Web page, Windows Bitmap and other. In BPMN, a process is represented with a graph of flow objects. These objects consist of other activities and the controls that sequence them. A Business Process Model can be defined using a network of graphical objects showing activities and their interconnections and flow controls that define the order of performance. The goal of process modelling is to develop simple diagrams that will look familiar to most business stakeholders. This is realized using four basic categories of graphical elements (Fig. 1): Flow objects (Events, Activities and Gateways), Connecting objects, Swimlanes and Artifacts.

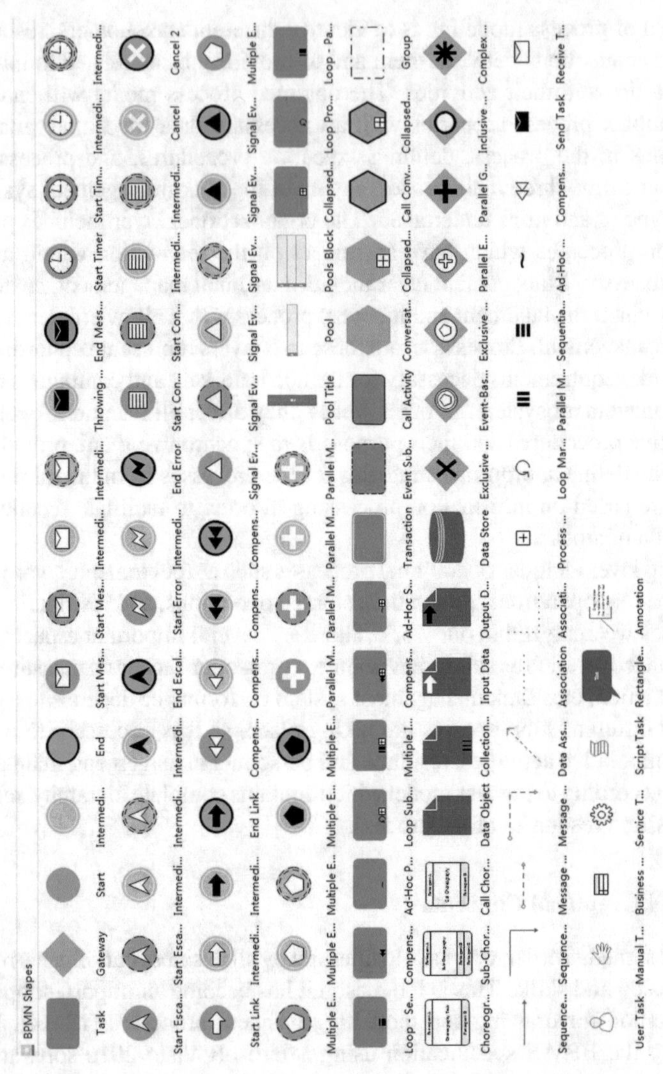

Fig. 1. Graphical representation of BPMN concepts using MS VISIO tool

Flow Objects

Events are circular objects used to show that something is supposed to happen during a business process. There are three categories of events depending on a time when they affect the process flow: start, intermediate and end events. The beginning and the end of a process are indicated using the start event and end event. Intermediate events that interrupt activities (tasks and sub-processes) can also be modeled with BPMN. They refer to situations when an exception occurs in the execution of an activity and are positioned at the boundary of the interrupted activity. On the other side, start and intermediate events are connected with activities using normal sequence flows. We can distinguish catching events (events with a defined trigger which take place once the trigger has been activated) and throwing events (they trigger themselves instead of reacting to a trigger). Each event has a trigger (considered as a cause of the event) and an effect (a result of the event). Activities (processes, sub-processes and tasks) are rectangular objects which represent work that is performed within a business process. Tasks represent elementary activities within a process. They can be marked as loops, multiple instances, or compensations. A subprocesses describes a detailed sequence within a process flow, but it takes no more space in the diagram of the parent process than a task. It is practically used to help with the expanding/collapsing view. Finally, a process is a set of graphical objects. Gateways are objects with a rhomb shape which define all the types of business process sequence flow behavior. There are four categories of gateways: exclusive gateways (XOR) which are used to describe process flow which can only be done under certain circumstances, inclusive gateways (OR) which are used in situations when the process can take place in only one of two directions, parallel gateways (AND) which describe activities which take place simultaneously, and event-based gateways which are used to model decisions based on which event has occurred, not which condition has been met.

Connecting Objects
Connecting objects (sequence flow, message flow and association) are used to connect two objects and define the flow progress through a process. Sequence flow (normal, conditional and default sequence flow) shows the order of flow elements. Message flow is used to show the flow of messages or elements between pools. Association is used to connect BPMN's artifacts with other BPMN elements and it is usually used to link a Text Annotation or a Data Object. If it is necessary, it can show the direction of flow. Solid line with an arrowhead graphically represents a sequence flow, dotted line with an arrowhead represents message flow and a dotted line represents an association.

Swimlanes Swimlanes (pools and lanes) are elements that allow distinguishing participants and responsibilities in a business process diagram. In this way, it is possible to display different functional capacities or organizational roles, i.e. organize the diagram in activities. Participants in the process are represented with pools which can be displayed as "white boxes" with all details shown, or as a "black boxes" with hidden details. Organization and categorization of activities within a pool are presented with Lanes which represent sub-partitions within a Pool.

Artifacts
Some of the process information (inputs and outputs of activities, comments and notes, information about how the tasks are organized) is not directly related to the sequence flow or message flow, and this information can be shown with Artifacts (data objects, text annotations and groups).

During the model developing we followed the next recommendations [15]: decompose the model if it is too large, use an as small number of elements as possible, reduce the number of branching, structure the model as much as possible, the existence of starting and finishing event, avoid OR gateway and use a verb-object activity tags.

3 Tank Management System Processes Model

In the very beginning of the process modeling, it is important to define process levels (Fig. 2). In this way, tank management system processes are divided into smaller subprocesses with accurate sequential and logical connections between them. Standardization of terms and comprehensive representation of the process are the main benefits of this hierarchy. The first level is represented by project phases which have a very long duration, especially compared to sub-processes and elementary processes. The tank management system process model comprises six overlapping project phases: planning, technical design, optimization, construction, commissioning, and operation. Related and dependent cluster processes with an unspecified order of execution, such as operation, stand still and periodical service, are placed on the second level of the hierarchy and they include different organizational, supporting and operation processes. The execution of processes from other project phases are affected by these processes.

Processes at levels 3 (main processes) and 4 (sub-processes) are essentially different in terms of duration, and time and logical sequences of their execution. The main processes are characterized by specified duration and standardized and interruptible implementation with time or logical sequence which is not implicit. Also, some of these processes cannot be executed simultaneously. Level 3 consists of the following processes: planning, preparation, manipulation, validation, and optimization. On the other side, sub-processes at level 4 are temporally and conditionally related. The fuel transport and yard control and logistics take place at the beginning of the sequence and they are connected with external companies. Loading operations, production, and storage operations are all internal sub-processes which include various work procedures with different stakeholders and resources involved. Distribution is the final sub-process of execution main process. The most important information about tank management system is represented by elementary processes at level 5 which provide further process description. Concrete actions, connections, stakeholders, sequence flows and constraints are defined in this part of the process model. The entire organization and management of the overall process are based on the classification of leading and supporting processes, and necessary resources, liabilities, and requirements of different participants in the process. Our model contains a detailed description of the custody transfer elementary process (level 5) within loading operation sub-process and manipulation main process.

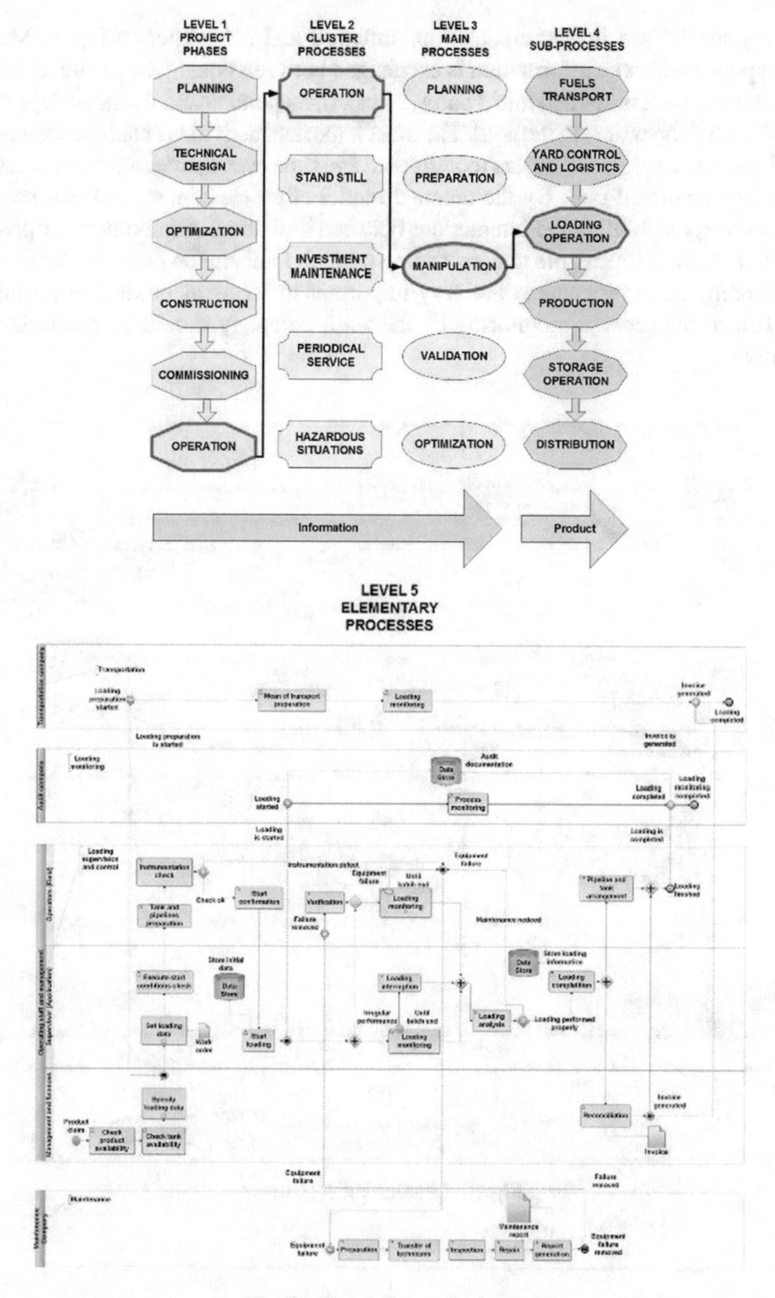

Fig. 2. Detailed process levels

Figure 3 shows the elementary process of "loading operation". There are 4 pools (organizational units) in this process: Operational staff and management, transportation company, audit company specialized in operation supervision, and maintenance company specialized in instrumentation, software and infrastructure repair. Message flows represent the way information is exchanged between pools. "Operational staff and management" pool is divided into 3 parts (lanes): management and financial department, supervisor, and operators in the field. The most important activities or sub-processes (presented with rectangles) and events (conditional and time events, messages) are conducted in this organizational unit. By the detailed model of an elementary process, complete activities, responsibilities and interactions between individual stakeholders are precisely described. Also, it is possible to detect some critical points in the process. For example, sample preparation procedures are very important in terms of producing meaningful data [16], so the process monitoring by the audit company should be conducted very carefully.

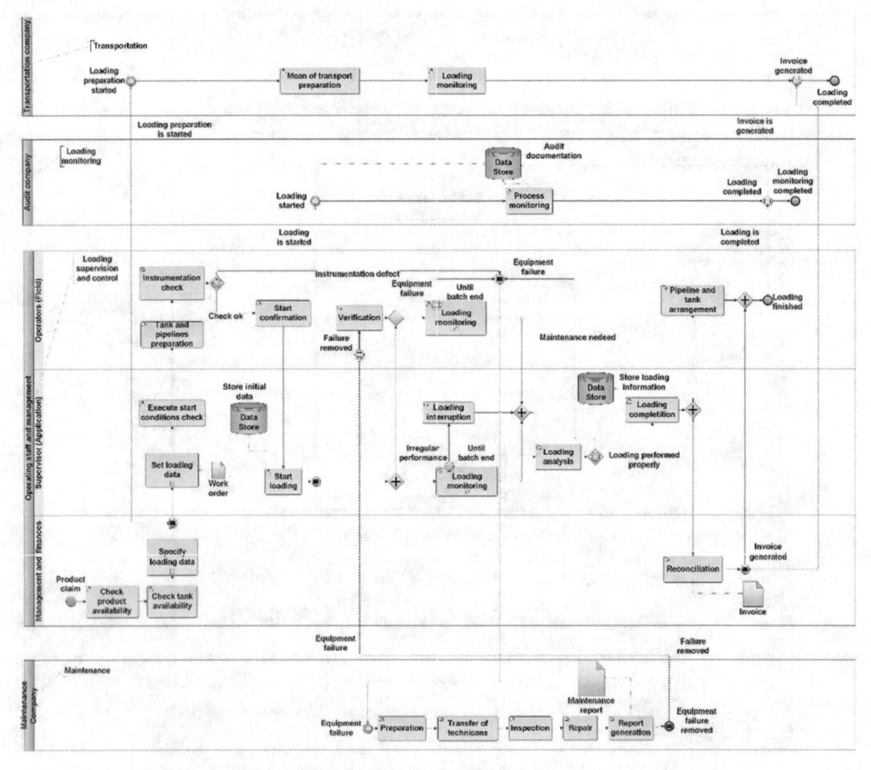

Fig. 3. Loading process flow

4 Process Analysis

Operational processes within a tank management system require a focus on safety, good planning, and efficiency. They should be executed in such a way that they provide accurate measurement data, enable overfill protection, leak detection and alarming, protect the tank farm's personal and environment, and prevent possible disasters. For the smooth running of the system, it is necessary to provide accurate and secure inventory data at all times. This is important for plant management and planning activities. Finally, in order to increase efficiency, reduce costs and improve profits, tank management systems should rely on precise data collected from the field which are integrated and centralized so they represent a good basis for inventory and supply chain management. Figure 4 shows a solution of the SCADA application connected with the central financial database which is used for continuous monitoring and precise control of the tank management system. This solution supports the exchange of data that can be used for financial analysis, logistics planning and process organization.

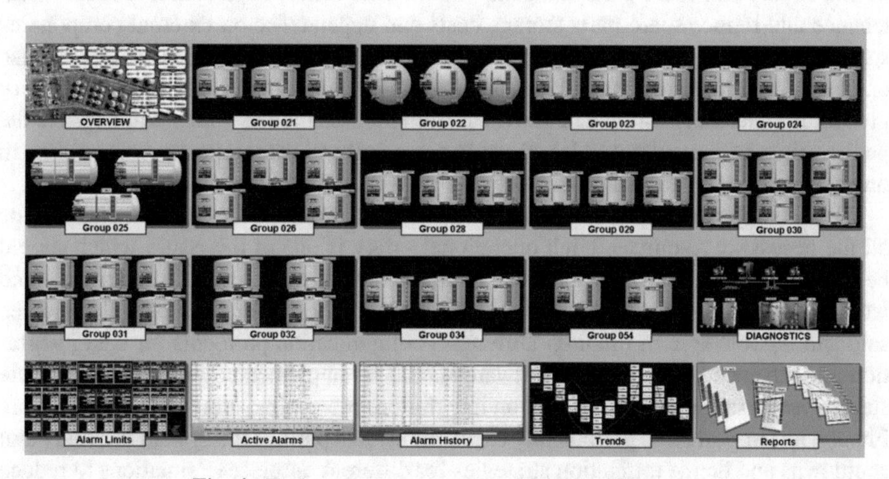

Fig. 4. Tank management system SCADA application

During the planning and optimization phases, it is necessary to provide different functionalities for the future processes of a tank management system. It is very important to enable product gain or loss balance and facilitate the reconciliation of stocks by tracking inflows and outflows in the tank farm. Avoiding product run-outs or emergency deliveries is also important because customer satisfaction relies on delivery performance. This is why it is necessary to optimize a company's supply chain so reactions to inconstancy can be fast and efficient. Due to lack of global vision of a system during design phase of a control system, there are many difficulties related to specification, debugging, testing and verification of tank management control systems [17]. This is the reason why it is necessary to have complete process model. Also, data integration in inventory control systems will reduce inventory management costs, increase productivity and measurement accuracy, and accelerate information exchange, which will altogether improve planning capabilities and cooperation with external companies. The most important stakeholders in these processes are financial analysts, IT experts, and managers. As we can see from the model, these two phases represent supporting processes the system operation will depend on to a great extent.

Commissioning, maintenance and running operations are influenced by various factors (such as instrumentation and its characteristics, requirements related to safety and environmental protection, maintenance needs, necessary resources), so it is very important to consider them during process analysis. These operations should be executed to secure maximum plant availability, reduce complexity, and save time and costs of tank management system supervision and control. One way to accomplish this is to introduce standardized handling and procedures. This is the reason why process modeling and analysis is very important. Another way to increase plant availability is to use modular device concept for spare parts and components. This concept will enable flexible maintenance and reduce spare parts storage costs and dependence on external components suppliers. In these processes, synchronization and good organization of technicians, operators, maintenance department and tank farm managers are required. Different work procedures and checklists can be provided from process models which will improve the performance of the system. Also, there is a necessity to automatically backup data in order to enable easy exchange of electronic systems without any recalibration.

For all subprocesses, it is important to coordinate different stakeholders and provide all the necessary resources. High operational safety is one of the issues which should be considered within process analysis. Instrumentation, installation, infrastructure and working procedures should comply with relevant industry standards which will guarantee safe and flexible system running. Different communication protocols for data integration and advanced management software should be implemented in order to facilitate stable system running and information integration, processing, storage, and protection. Finally, it is important to periodically check the critical parts of the system to avoid plant shutdowns and define mitigation strategies for different unforeseen situations to reduce the consequences and accelerate recovery. Elementary processes are the most important because it is possible to define the smallest details related to the process and analyze them later to detect weak points, necessary resources and participants, conditions, and all other important factors. This the reason we created models of individual elementary processes which we will show on the representative example of the "Transaction preparation" subprocesses and its subprocesses. Transaction preparation represents one of the complex operational transactions which is important because it ensures the safe, accurate, credible and economical crude oil transportation. It consists of 8 subprocesses (logging, network checking, calculating storage capacity, checking instrumentation, water draining, sampler parameterization, alarms/simulation configuration and pump parameterization) and it can be canceled, delayed, stopped and regularly finished. Figure 5 and Fig. 6 represent BPMN models of Transaction preparation and its subprocesses.

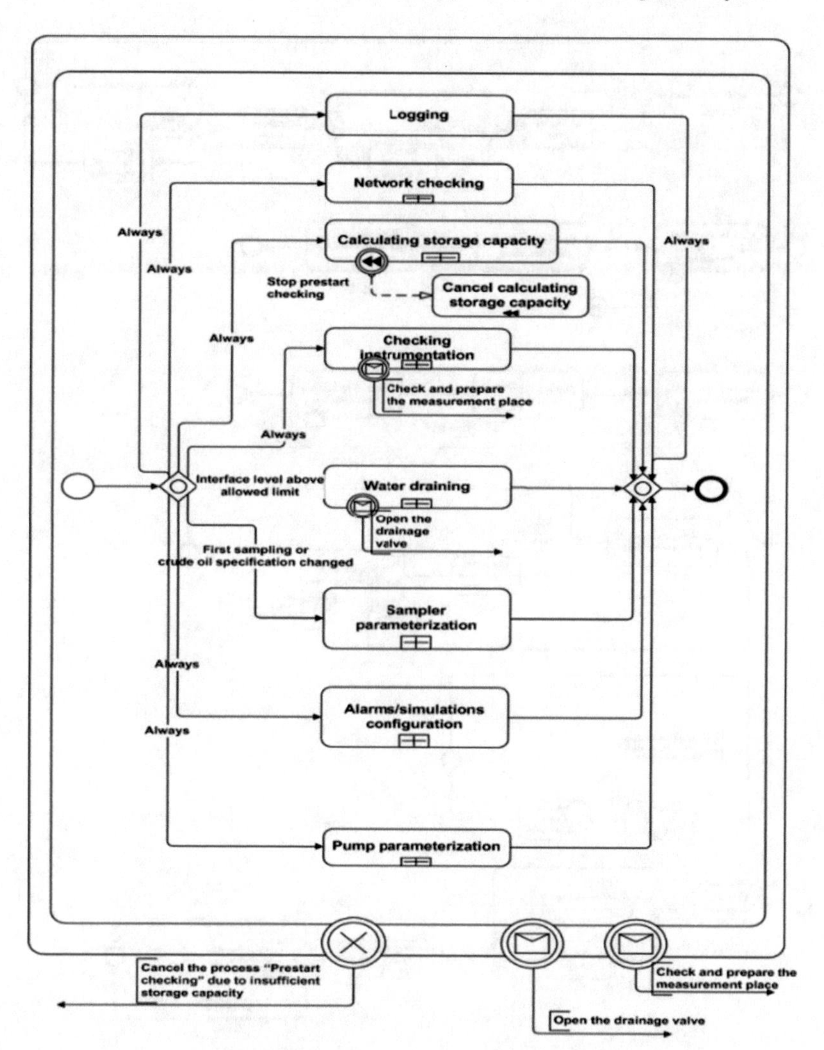

Fig. 5. Pre-start checking BPMN diagram

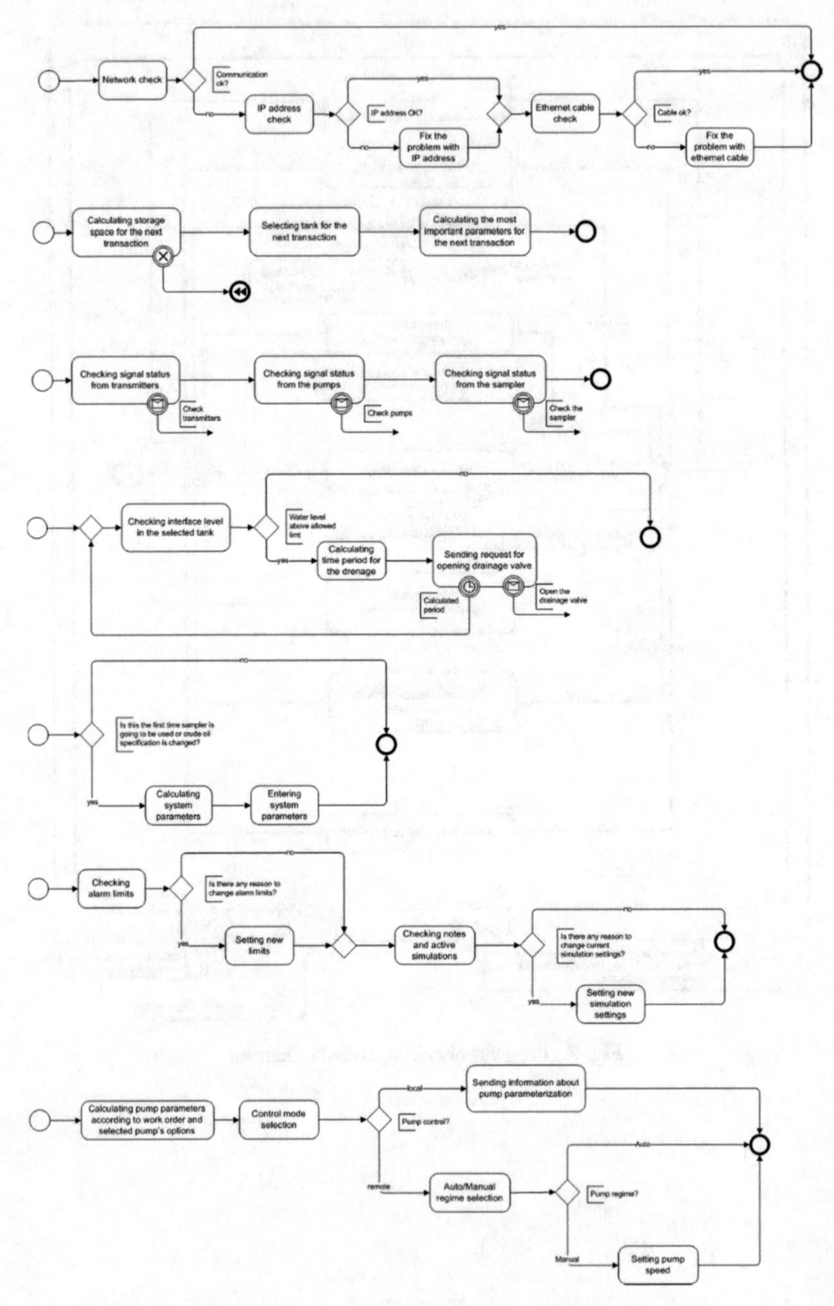

Fig. 6. Transaction preparation subprocesses

5 Discussion and Conclusion

In this study, we represented the process modeling and analysis related to tank management systems. We outlined how the business process modeling provides information about process participants, relationships between them, different tasks and their organization within overall process, and communication flow. The focus of process analysis is on how to improve custody transfer and quality measurement systems, enhance relations between companies, accelerate return on investment and promote management and operation processes. It can serve as a basis for further process simulations and risk analysis, i.e. detection of errors and critical points in these processes, which will reduce measurement, operational and management errors.

Our intention was to make a model assessment, provide accurate and reliable data about the processes, and improve processes in question where possible. Also, the business process modeling and analysis conducted in our study represent support to all managers and analysts at the organizational, financial, operational, and service levels. Implementation of process modeling contributes to a better understanding of the process, perceiving the system as a whole and also as a set of individual processes, and defining the communication modes between participants in the process. It provides detailed guidance for the execution of all manipulative, organizational, and supporting activities regarding critical points, resources, procedures, and stakeholders' responsibilities [18, 19]. Based on the model and analysis, some of the guidelines for process improvement have emerged. Better system reliability and sustainability can be achieved by combining different working tasks and different responsible parties (monthly and investment maintenance, process monitoring, periodic tests, etc.…). [20]. The accuracy and safety can be improved by installing additional instrumentation (flow switches, level transmitters) and protection devices, as well as the purchase of spare parts to make downtime as short as possible [21]. Data stored in numerous data bases should be used for planning optimization, determine optimal working procedures and defining appropriate support for plant personnel such as training for operating staff, service technicians, financial experts and managers, and different documentation (checklists, instruction manuals, protocols, etc.…) [22, 23]. These measures should reduce transaction delays, waste, a risk to the environment and staff.

6 Future Research

The process represented by the obtained model can be implemented as a good basis for further discussion, risk analysis, and model simulation. Simulation can provide information about modeling errors due to inactive paths or events, which will refer to altering the model and affect some modifications of technical or organizational parts of the process [24, 25]. In order to define the complete functionality of the tank management system, our model should be extended with all parts of the entire process. Further steps to be implemented are also to propose, discuss, and analyze different directions of the process flow and define necessary resources, involved participants and other important aspects in order to improve process execution. Various instrumentation, software platforms, and work procedures should be reflected by modeling and process analysis, which will, in the end, lead to the improvement of the system.

Acknowledgments. The research described in this paper are financed from Ministry of Education, science and technological development Republic of Serbia.

References

1. Biazon, C.L., Jesus, V.C.B.M., De Oliveira, E.C.: Metrological analysis by measurement uncertainty of water and sediment in crude oil. Pet. Sci. Technol. **33**(3), 344–352 (2015)
2. Demirbas, A., Alidrisi, H., Balubaid, M.A.: API gravity, sulfur content, and desulfurization of crude oil. Pet. Sci. Technol. **33**(1), 93–101 (2015)
3. Pillat, R.M., Oliveira, T.C., Alencar, P.S., Cowan, D.D.: BPMNt: A BPMN extension for specifying software process tailoring. Inf. Softw. Technol. **57**, 95–115 (2015)
4. Gröner, G., Bošković, M., Parreiras, F.S., Gašević, D.: Modeling and validation of business process families. Inf. Syst. **38**(5), 709–726 (2013)
5. Russo, V., Ciampi, M., Esposito, M.: A business process model for integrated home care. Procedia Comput. Sci. **63**, 300–307 (2015)
6. zur Muehlen, M., Ho, D.T.: Service process innovation: a case study of BPMN in practice. In: Proceedings of the 41st Annual Hawaii International Conference on System Sciences (HICSS 2008), pp. 372–372. IEEE, January 2008
7. Chinosi, M., Trombetta, A.: BPMN: an introduction to the standard. Comput. Stand. Interfaces **34**(1), 124–134 (2012)
8. Decker, G., Kopp, O., Leymann, F., Pfitzner, K., Weske, M.: Modeling service choreographies using BPMN and BPEL4Chor. In: International Conference on Advanced Information Systems Engineering, pp. 79–93. Springer, Heidelberg, June 2008
9. Mauser, S., Bergenthum, R., Desel, J., Klett, A.: An approach to business process modeling emphasizing the early design phases. In: Proceedings of the Workshop Algorithmen und Werkzeuge für Petrinetze. CEUR Workshop Proceedings, vol. 501, pp. 41–56. September 2009
10. Wong, P.Y., Gibbons, J.: Formalisations and applications of BPMN. Sci. Comput. Program. **76**(8), 633–650 (2011)
11. Zur Muehlen, M., Recker, J.: How much language is enough? Theoretical and practical use of the business process modeling notation. In: Seminal Contributions to Information Systems Engineering, pp. 429–443. Springer, Heidelberg (2013)
12. Grosskopf, A., Decker, G., Weske, M.: The Process: Business Process Modeling Using BPMN. Meghan Kiffer Press (2009)
13. White, S.A., Miers, D.: BPMN Modeling and Reference Guide: Understanding and Using BPMN. Future Strategies Inc., Lighthouse Point (2008)
14. Weilkiens, T., Weiss, C., Grass, A.: Modeling business processes using BPMN. In: OCEB Certification Guide, Chapter 6, pp. 77–121 (2011). ISBN-10: 0123869854
15. Mendling, J., Reijers, H.A., van der Aalst, W.M.: Seven process modeling guidelines (7PMG). Inf. Softw. Technol. **52**(2), 127–136 (2010)
16. Cowles, R.J.: Particle characterization for oil sand processing. III. Sample preparation. Pet. Sci. Technol. **21**(7–8), 1241–1252 (2003)
17. Firoozshahi, A.: Innovative tank management system based on DCS. In: Proceedings ELMAR-2010, pp. 323–328. IEEE, September 2010
18. Radovanovic, L., Perisic, J., Milovanovic, M., Speight, J.G., Bozilović, Z., Momcilovic, O., Obucinski, D.: Modeling of petroleum products sampling processes. Pet. Sci. Technol. **36**(23), 2003–2010 (2018)

19. Ristic, M., Radovanovic, Lj., Prokic-Cvetkovic, R., Otic, G., Perisic, J., Vasovic, I.: Increase energy efficiency of thermal power plant Kostolac B by revitalization ventilation mills. Energy Sources Part B Econ. Plann. Pol. **12**(2), 191–197 (2017)
20. Perisic, J., Radovanovic, Lj., Milovanovic, M., Petrovic, I., Ristic, M., Bugarcic, M., Perisic, V.: A brine mixing mobile unit in oil and gas industry - an example of a cost-effective, efficient and environmentally justified technical solution. Energy Sources Part A Recovery, Util. Environ. Effects **38**(23), 3470–3477 (2016)
21. Kvrgić, V., Vidaković, J., Lutovac, M., Ferenc, G., Cvijanović, V.: A control algorithm for a centrifuge motion simulator. Robot. Comput.-Integr. Manufact. **30**(4), 399–412 (2014)
22. Ferenc, G., Dimić, Z., Lutovac, M., Vidaković, J., Kvrgić, V.: Open architecture platforms for the control of robotic systems and a proposed reference architecture model. Trans. FAMENA **37**(1), 89–100 (2013)
23. Perisic, J., Milovanovic, M., Petrovic, I., Radovanovic, L., Ristic, M., Perisic, V., Vrbanac, M.: Modeling and risk analysis of brine mixing mobile unit operation processes. Energy Sources Part B Econ. Plann. Policy **12**(7), 646–653 (2017)
24. Kovacevic, M., Lambic, M., Radovanovic, Lj., Kucora, I., Ristic, M.: Measures for increasing consumption of natural gas. Energy Sources Part B Econ. Plann. Policy **12**(7), 443–451 (2017)
25. Perišić, J., Milovanović, M., Petrović, I., Radovanović, Lj., Ristić, M., Speight, G.J., Perišić, V.: Application of a master meter system to assure crude oil and natural gas quality during transportation. Pet. Sci. Technol. **36**, 1222–1228 (2018)

Experimental Evaluation of Correlations of Evaporation Rates from Free Water Surfaces of Indoor Swimming Pools

Marko Mančić[1(✉)], Dragoljub Živković[1], Mirjana Laković Paunović[1], Milena Mančić[2], and Milena Rajic[1]

[1] Faculty of Mechanical Engineering, University of Niš, Aleksandra Medvedeva 14, 18000 Niš, Serbia
marko.mancic@masfak.ni.ac.rs

[2] Faculty of Occupational Safety, University of Niš, Čarnojevića 10a, 18000 Niš, Serbia

Abstract. Indoor swimming pool buildings can be considered as significant energy consuming public buildings. Most energy is consumed to heat swimming pool water and maintain the desired thermal comfort conditions in the swimming pool hall. Water evaporation from the swimming pool water surface increases humidity in the pool hall air, and therefore the consumption of energy for heating and ventilation of the swimming pool and the hall increases, especially in the scenario of more strict humidity control of the pool hall air. Mathematical correlations for predicting the evaporation rates from free water surfaces can be found in literature, but not all of them were designed specifically for indoor swimming pools. In this paper, the properties of indoor swimming pool water, pool hall air and evaporation rates are measured in a real indoor swimming pool building and measurement results are presented in the paper. The measured results are confronted to the evaporation rates calculated by applying literature mathematical correlations for determination of evaporation rates. A large discrepancy of results from the literature is determined. Based on the original measured results, a mathematical correlation for estimation of evaporation rates of indoor swimming pools is created by application of the least square method.

Keywords: Evaporation · Measurement · Free water surface · Swimming pool

1 Introduction

Buildings with indoor swimming pool consume significant amount of energy for heating and ventilation. Energy in swimming pool building facilities is used for maintaining the thermal comfort conditions in the swimming pool hall, as well as for maintaining the pool water at desired temperature. In indoor swimming pool buildings 45% of energy is used for pool hall ventilation, and 33% for heating pool water. Heating and ventilation of the rest of the building accounts for around 10%, 9% of energy is used for lighting and equipment and sanitary hot water accounts for 3% of total energy consumption [1]. The highest thermal loads in indoor swimming pool buildings, often originates from water evaporation from the pool water surface [2].

N. Mitrovic et al. (Eds.): CNNTech 2020, LNNS 153, pp. 378–393, 2021.
https://doi.org/10.1007/978-3-030-58362-0_22

1.1 Water Evaporation in Indoor Swimming Pools

There are many mathematical models representing efforts to describe the physical phenomena of water evaporation from a free water surface [4–22] and most of them are empirical or semi-empirical, heavily relying on the results from experiments on either real objects or laboratory installations. Simultaneous transport of heat, mass and momentum occurring as well as the fact that this is phase flow process, make it's modelling difficult. In a general case, water evaporation rate from a surface depends on the air flow above the water surface, and the gradient of partial pressures of water vapor at water level and water vapor in the air above the surface of the water. Correlations for predicting water evaporation rate from a free water surface to both still and moving air can be found in literature [4–22]. Bowen provided a general solution for determination of heat transfer by convection and evaporation from a water surface of an element of volume for three different conditions [15]. In order to do this, he first determined a model of vapor diffusion from a unit area. Most of the correlations are based on the Dalton's theory, but there are also attempts of creating correlations based on analogy between heat and mass transfer [7, 15], where a ratio between conduction heat loss and evaporation heat loss is determined. Many evaporation models are analyzed or reviewed in literature [7–22], but even for detailed numerical simulations [16, 20] it is necessary to first determine the evaporation rate from the water surface and the evaporation rate coefficient.

Since the air velocity in indoor swimming pools originates from heating and ventilation equipment operation, it is usually kept at low levels. Heating and ventilation equipment of swimming pool halls usually consists of supply and return ducts spread on several locations through-out the hall, which cause complex air movement in the hall and contributes to forced evaporation. The correlations obtained by experiments for predicting evaporation rates from a water surface of indoor and outdoor swimming pools represent a function of air velocity above the water surface [7]. Air velocity over the indoor swimming pool water surface is not uniform, but extremely complex. The intensity and direction of the air velocity vector show large variations over time. This can be seen in results of a CFD simulation of a public swimming pool [20], where air temperature and humidity at the air return intake was measured and compared to the simulation results. Most of the simulated air velocities were up to 0.2 m/s, but there were also some slightly higher values. In addition, apart from air velocity, the nature of air flow over the water surface in indoor swimming pools has a strong influence on evaporation, which is determined by the value of Reynolds number and Sherwood numbers [15].

Evaporation Rate from Free Water Surface Correlations

The first equation of evaporation from free water surface was given by Dalton (1802) [6], when it was found to be proportional to the partial pressure difference of water vapor near the boundary surface p_a and away from the surface p_{sw}:

$$-dE = K(p_{sw} - p_a)df_{sw} \tag{1}$$

Where df_{sw} is an element of evaporation surface, E is evaporation rate per unit time, and K is a coefficient affected by properties of air flow over the boundary water surface.

According to Lewis the evaporation rate may be determined as:

$$-dE = K_E(x_{sw} - x)df_{sw} \quad (2)$$

Here, K_E is the evaporation rate coefficient given in (kg/s m^2), which is again affected by the properties of air flow over the boundary water surface and is usually given as a linear function of air velocity:

$$K = A + BV_a \quad (3)$$

Where, V_a is the velocity of air above the free water surface, A and B are correlation constants. Values of this coefficient from literature are given in Table 1.

Table 1. Literature evaporation rate prediction correlations

Author	Correlation coefficients of Eq. (3)	Application
McMilan	A = 0.0360; B = 0.250	Lakes
Chernecky	A = 0.05053; B = 0.06638	Swimming pools
Carrier	A = 0.088403; B = 0.001296	Solar pond
Hahne and Kübler	A = 0.0850 B = 0.0508	Outdoor swimming pools
Rohwer	A = 0.0803; B = 0.0583	Laboratory model
Smith et al.	A = 0.0888; B = 0.0583	Swimming pool
Smith et al.	A = 0.0638; B = 0.0669	Outdoor swimming pool
Himus and Hinchey	A = 0.1538; B = 0.06898	General
Lurie and Michailoff	A = 0.109; B = 0.0859	General

A review and comparison of mathematical models for predicting evaporation rates by Sartori [7], indicated a large scattering of results obtained using investigated literature correlation models. Some of the models neglect the influence of relative humidity of the air. Sartori provided a correlation model, where evaporation is a coefficient of air velocity to the power of 0.8, and length of the water surface, instead of actual water surface. Shah proposed a model where evaporation rate coefficient equals to 0.00005 [21, 22]. Asdrubali et al., determined values of the correlation coefficients for air velocity values of 0.05 m/s, 0.08 m/s and 0.17 m/s, equal to 3.4×10^{-8}, 4.2×10^{-8} and 5.2×10^{-8} respectfully, based on results from a laboratory indoor swimming pool scale model [5].

A comparison of the evaporation rates obtained using the correlations in Table 1, are given in Fig. 1. It is clear, that a significant discrepancy of the results can be found.

2 Determination of Evaporation Rates and Water and Air Properties by Real Object Measurement

Having in mind that empirical equations strongly depend on the experimental conditions, as well as the results they are based on. According to Sartori, one empirical evaporation prediction equation is necessary for each class of these conditions [7]. There are

Fig. 1. Apparatus used for measuring on a real swimming pool

differences of height of the point above water level where air velocity is measured, which usually ranges from 0.3–10 m, however most of the heights range from 0.5–2 m for which the differences in results are not significant and are considered not to affect results [3, 5, 7, 10–14]. Smith et al. performed measurements on an outdoor swimming pool, where water body temperature was kept at 28.9 °C, which was monitored using thermocouples. Pool water level was monitored using a Microtector gauge, whereas wind speed was monitored using a rotating cup anemometer at 0.3 m above the pool. Tang and Etzon, compared water evaporation from free water surface and wetted water surface, for which they constructed two identical water "ponds", and found that the evaporation rate of from a free water surface is proportional to the difference in partial pressure of water vapour at water temperature and air above it to the power of 0.82 [4]. Ruiz and Martinez measured relevant parameters of water and air at the border of an out-door pool, at 0.5 m above water level, and water temperature was measured 1 m below water surface, but evaporation rate was not measured, instead it was calculated using correlations from the literature which lead to difference in results [8]. Asdrubali created a scale model of an indoor swimming pool, and measured evaporation rates for air velocities of 0.05 m/s, 0.08 m/s, 0.17 m/s and relative humidity 0.5, 0.6 and 0.7 [2], but the air flow in the laboratory model could be considered uniform and the air flow originating from a typical duct ventilation system in a pool hall is extremely complex. Smith et al. [10] used shallow aluminum floating evaporation pans with diameter of 20 cm, to determine short term evaporation from an out-door swimming pool, but such small pan evaporation surface and the air flow above it are either affected by pan walls, or permit water penetration, thus affecting reliability of the results.

Results presented in this paper were acquired by measuring air flow velocities at 5 points along the pool border, at 0.65 m above water level in the direction of the air flow coming from the ventilation ducts. Average air velocities ranged from 0.01 m/s to 0.07 m/s at the opposite ends of the pool border, and from 0.15 m/s to 0.34 m/s at the middle of the pool border. Water temperature was measured 1 cm below water level using a thermo K type couple probe. Relative humidity and temperatures were measured 1 cm above the water level and 90 cm above water level. Evaporation rate was measured hourly using an evaporation pan with a needle, with a diameter of 0.8 m immersed in the pool water 1.3 m from the pool border. The diameter of the evaporation pan was chosen so it would be big enough to account for the affect of the complex air flow above the water surface in the pool hall. Evaporated water is measured with precision of 1 g, hourly.

Results are measured and collected in a the hall of active indoor swimming pools, in the Sport and Recreation Center Dubočica, Leskovac, in the period from May to August. The following apparatuses were used (Fig. 1):

1. TESTO 454 with 0420 relative humidity and air temperature probe,
2. Cole Palmer 37950-12 relative humidity sensor with air temperature probe,
3. AIRFLOW TA5 anemometer with thermometer,
4. K type thermocouple probe,
5. Evaporation pan with a needle.

Some of the evaporation rate results are discarded, since they were affected by activity of the swimmers causing water penetration into the evaporation pan. Evaporation rate E as function of air velocity Va is given in Fig. 1.

In Fig. 2, measured evaporation rates are presented as a function of the difference of partial pressures of saturated water vapour at water surface temperature and partial pressure of water vapour of air above water level (Eq. 4), calculated in Pa according to IAPWS Industrial Formulation for the Thermodynamic Properties of Water and Steam.

It can be observed from Fig. 2 and Fig. 3 that the evaporation rates mostly rise with the increase of air flow velocity above the water surface, and generally rise with the increase of difference of water pressures of water saturated water vapour at water surface and water vapour in the air above the water surface.

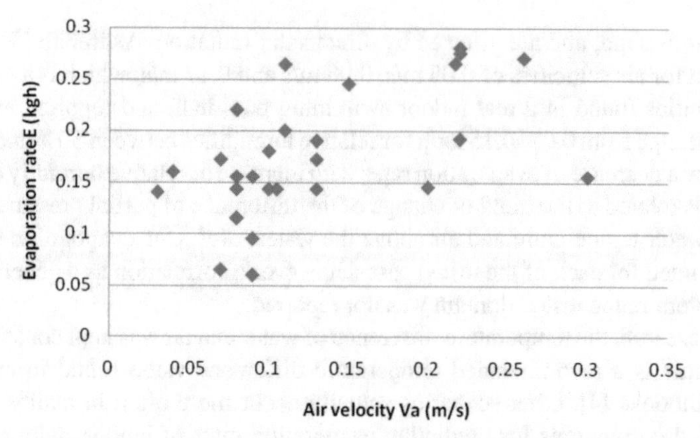

Fig. 2. Measured evaporation rates as a function of measured air velocity

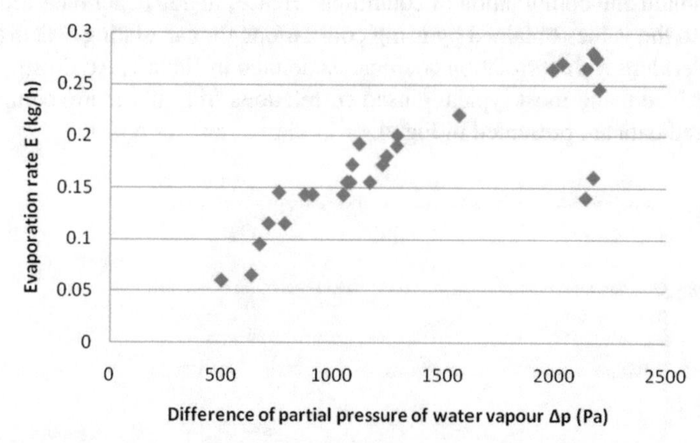

Fig. 3. Measured evaporation rates as a function of difference of partial pressures of water vapour

This is kind of relationship is in accordance with the majority of correlations found in literature. The measured results also showed a strong dependence on the air flow velocity. A significant drop of air velocity caused by the swathed off ventilation system, caused a drop of air velocity to near zero values (Fig. 2), which led to low values of evaporation rates despite relatively high partial pressure difference gradient. This can be observed in Fig. 3.

3 Comparison of Measured Results and Literature Correlation Calculated Results

Obtained measurement results are compared to the values of evaporation rates, obtained for the same air and water parameter values by calculations using correlations from the literature (Table 1). Most of the evaluated results from literature correspond to outdoor swimming pools [3, 5, 8, 10] representing usually higher air flow velocities which originate from wind, and are affected by direct solar radiation. Asdrubali [5] measured evaporation for air velocities of 0.05 m/s, 0.08 m/s and 0.17 m/s, which is a good match to air velocities found in a real indoor swimming pool hall, and reported evaporation rates in the range from 0.07–0.15 kg/h for relative humidifies between 50% and 70%. His results show a decrease of evaporation rates with increase of relative humidity of ambient air, which is related to the trend of change of the difference of partial pressures of water vapour at water temperature and air above the water level. One evaporation coefficient was determined for each of the tested cases, however a correlation as a general solution of the problem in the tested domain was not reported.

In his research, the temperature difference of water and air was kept constant at 2 °C. Although this is a recommended temperature difference value found in engineering design handbooks [4], other scenarios actually occur more often in reality. Shah [21, 22] provided correlations for predicting evaporation rates of indoor swimming pools, but he used data from literature to fit correlation curves, which is again limited by the covered domain and combination of conditions. Hence, in this paper measured data are compared to the values obtained by using correlations for calculating evaporation rates found in literature with correlation coefficients defined in Table 1. An illustration of the results obtained using most typically used correlations from literature, compared with the measured data are presented in Fig. 4.

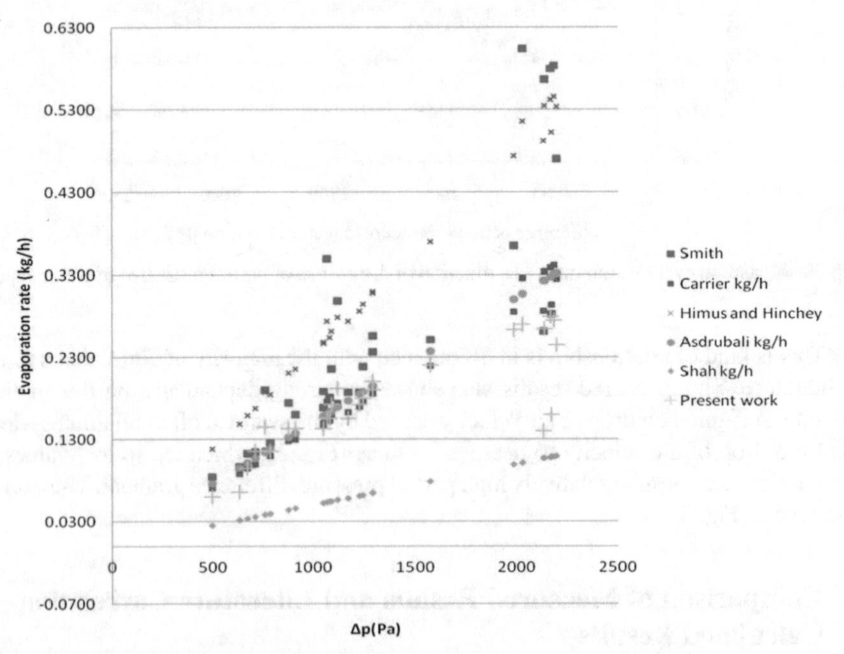

Fig. 4. Measured evaporation rates as a function of difference of partial pressures of water vapour compared to literature correlation results

It can be observed that some of the correlations, such as Smith's and Himus and Hinchey's significantly overestimate the evaporation rate for indoor swimming pools, while Shah's correlation underestimates the Evaporation rate of indoor swimming pools. It is clear, that literature equations can lead to significant errors in prediction of evaporation rates from indoor swimming pools, and as such should be used with caution.

Based on the measured results, a mathematical correlation of the water evaporation rate from an indoor swimming pool is fitted by application of the method of least squares. A general mathematical form for correlations of water evaporation rate from free water surface, based on literature, was used, where values of mathematical symbols A, B and N were determined by application of the least squares method. The general mathematical representation of the evaporation rates from the free water surface, with thorough analysis of the literature correlations, suitable for numerical simulations of the behavior of indoor swimming pools can be written as [21] (Fig. 5):

$$\dot{E} = \left(AV_a^N + B\right)\left(\mathrm{p_{sw}} - \varphi \mathrm{p_a}\right)/\mathrm{r} \tag{4}$$

Here, the evaporation rate is obtained as time dependent value, in kg/s. Factors A, B and N in the Eq. (4), should be determined based on empirical, i.e. measured results, where r is the latent heat of water evaporation. Based on the fitting of correlation curve with respect to measured results, the following correlation factors are determined to provide the best fit, i.e. least correlation error value, as written in Eq. (5) for the evaporation coefficient K with respect to Eq. (3):

$$K = 1.636585(0.000085643V_a + 0.66206089)^{1.1945}. \tag{5}$$

The results of the obtained evaporation rates in kg/h using the correlation in Eq. (5), are presented in Fig. 6 and 7. The Fig. 6 shows estimated evaporation rate for the temperature of air above the pool of 22 °C, with respect to relative humidity of Rh = 50% and Rh = 60%, and water temperature in the range from 26 °C to 29 °C. Figure 7 presents change of the estimated evaporation rate using Eq. (4), with respect to the change of partial differential pressure difference between the saturated air and air above the water surface.

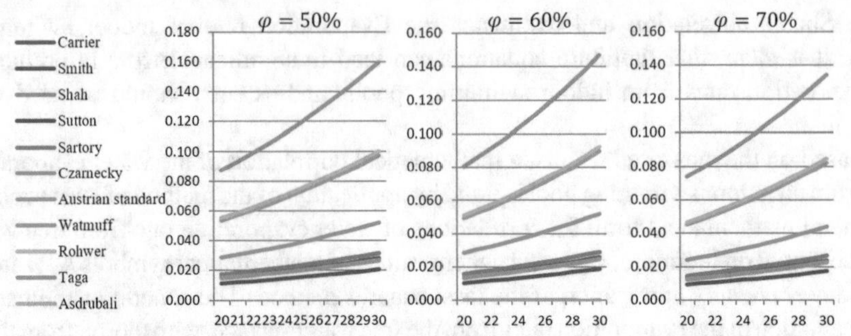

a) Evaporation rates obtained using evaporation coefficients from table 1. in kg/m^{2}h for air velocity of 0.05m/s, and relative humidity of 50%, 60% and 70%, respectfully.

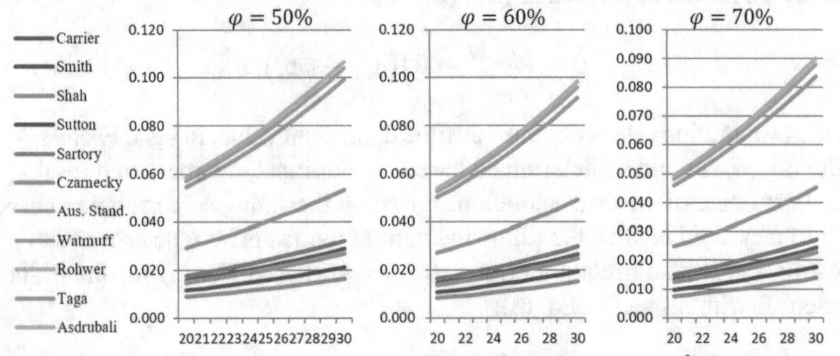

b) Evaporation rates obtained using evaporation coefficients from table 1. in kg/m^{2}h for air velocity of 0.08m/s and relative humidity of 50%, 60% and 70%, respectfully.

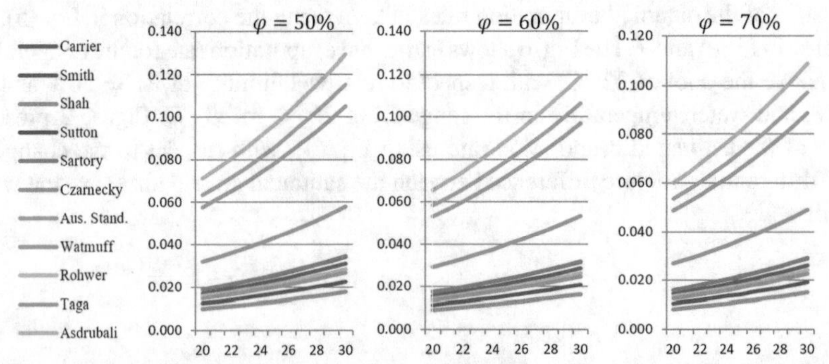

c) Evaporation rates obtained using evaporation coefficients from table 1. in kg/m^{2}h for air velocity of 0.17m/s and relative humidity of 50%, 60% and 70%, respectfully.

Fig. 5. Comparison of the evaporation rate results obtained for the correlation coefficients obtained with evaporation rate swimming pool correlations

The results obtained by applying the proposed mathematical correlation are compared to the actual measured results, for the values of the variables in the Eq. (4) obtained by measurements in the real object. The results obtained using the proposed correlation in Eq. (4), using the evaporation coefficient from Eq. (5). Show very good agreement

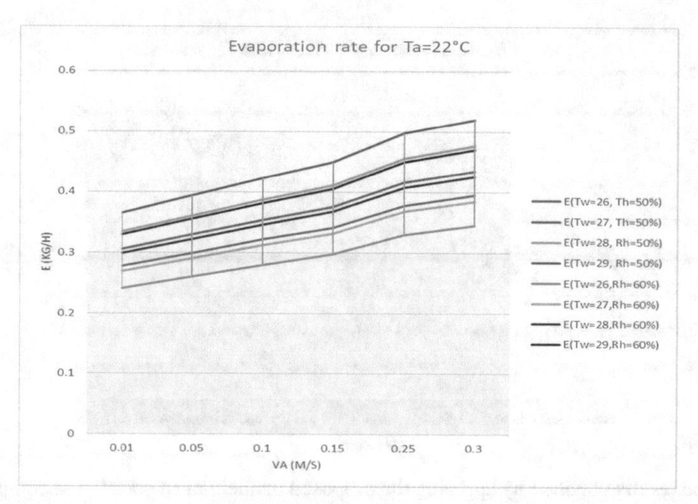

Fig. 6. Estimated pool water evaporation rates using the mathematical correlation obtained based on the measured results, with respect to relative humidity and pool water temperature, for indoor pool hall temperature of 22 °C

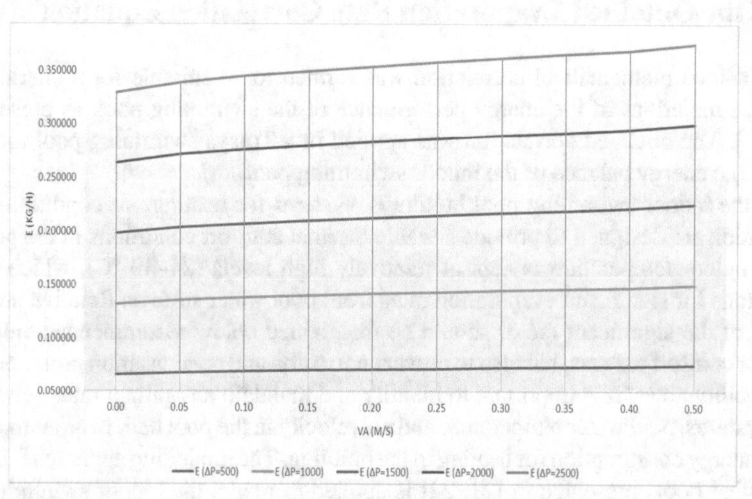

Fig. 7. Estimated evaporation rates in kg/h with respect to the change of partial differential pressure difference in Pa between the saturated air and air above the water surface.

with the measured results (Fig. 8). There is a significant disagreement with the measured results in one point, for the air velocity Va = 0.07 m/s. The significant difference could have occurred partially due to the normal imperfection of the mathematical approximation, and partially due to measurement error at the point of disagreement.

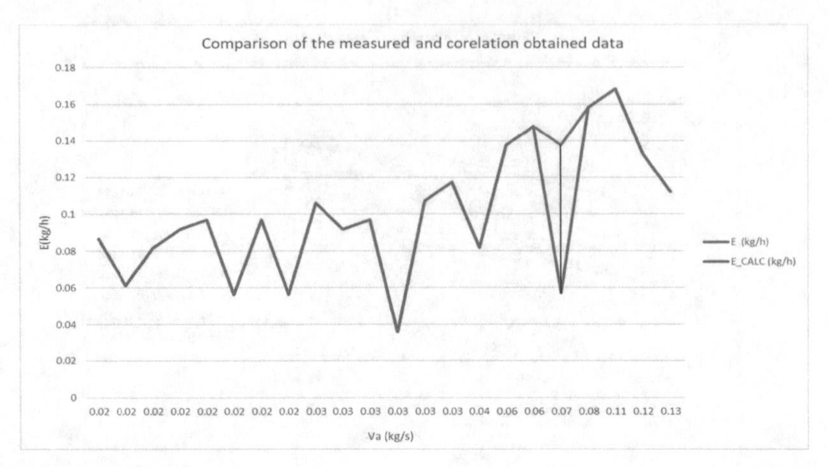

Fig. 8. The results obtained by applying the proposed mathematical correlation compared to the actual measured results

4 Energy Balance of the Indoor Swimming Pool with Application of the Obtained Evaporation Rate Correlation Equation

The obtained mathematical correlation was formed to be suitable for numerical quasistatic simulations of the energy performance of the swimming pool, as presented in [21, 22]. The obtained correlation was applied in a Trnsys swimming pool model, as part of the energy balance of the indoor swimming pool.

In the indoor swimming pool buildings, systems for heating, air conditioning and ventilation are designed to provide suitable thermal comfort conditions in the pool hall areas. Indoor temperature is kept at relatively high levels (24–30 °C), which creates conditions for significant evaporation rates from pool water surface. Relative humidity, as one of the significant factor should be maintained below recommended values not only for comfort reasons, but also to prevent corrosion and condensation problems in the pool facility areas. It is important to identify and maintain an optimal ratio between air temperatures, pool water temperature and air velocity in the pool hall, in order to achieve lower energy consumption for heating in the building. The modelling approach for indoor swimming pools presented in [21, 22] is applied to model the indoor swimming pool building analyzed in the paper. The pool water temperature change is calculated as [21, 22]:

$$\begin{aligned} \rho_w c_{pw} V_p \frac{dT}{d\tau} &= \dot{\underset{aux}{Q}} - \left(\dot{\underset{fw}{Q}} + Q_{evap} + Q_{conv} + Q_{rad} \right) \\ &= \underset{aux}{Q} - \underset{fw}{Q} - A_p \left(\dot{E}\, r + \alpha (T_w - T_{air}) + \varepsilon \sigma \left(T_w^4 - T_{wall}^4 \right) \right) \end{aligned} \tag{6}$$

Where: Evaporation heat losses Q_{evap} are a function of mass flow rate of evaporated water $\dot{E}$ and the latent heat of evaporation r; heat transfer by convection from the water surface Q_{conv} is a function of the convective heat transfer coefficient α, temperature of the pool water T_w and the indoor air temperature in the pool hall T_{air}; radiation losses Q_{rad} is the function of the emissivity ε[-], the Stefan-Boltzmann constant $\sigma = 5.67 \cdot 10^{-18}[\mathrm{Wm^{-2}K^{-4}}]$ and temperature of the indoor wall surface T_{wall}; Q_{fw} is the heat loss caused by supply of fresh water; and Q_{aux} is the heat gain from auxiliary heating equipment. The total water surface of the pool water area A_p is 1480 m^2. A balance model of an indoor swimming pool is presented in [22].

Evaporation heat losses are proportional to the flow of evaporated water from the water surface of the swimming pool:

$$q_E = \dot{E} \cdot r \tag{7}$$

Where $\dot{E}$ is the mass flow rate of evaporated water and *r* is the latent heat of evaporation. Heat transfer from convection from the water surface per unit surface area can be presented as:

$$q_{con} = \alpha(t_w - t_a) \tag{8}$$

Where α is the heat transfer coefficient, T_w is temperature of the pool water and T_a is the indoor air temperature in the pool hall. Convective heat transfer coefficient can be expressed as [22]:

$$\alpha = 2.8 + 3.0 \cdot V_a \tag{9}$$

Where V_a stands for air velocity above the water surface.

Radiation losses are calculated according to Stefan-Boltzmann law to sky for outdoor swimming pools are considered in [22]. The proposed model of the indoor swimming pool includes the heat transfer by radiation exchange with pool hall walls per unit of pool area [22]:

$$q_{rad} = \varepsilon \cdot \sigma \cdot \left(T_w^4 - T_{wall}^4\right) \tag{10}$$

Where ε is the emissivity average and σ is the Stefan-Boltzmann constant ($\sigma = 5.67 \cdot 10^{-8} \frac{\mathrm{W}}{\mathrm{m^2 \cdot K^4}}$) and T_{wall} is the temperature of the wall surface.

For the proposed swimming pool water it is assumed to have a constant water level in the pool. The mass of the evaporated water and waste water losses in the water treatment plant are compensated by fresh water supply system. Heat loss caused by supply of fresh water with lower temperature than that of the pool can be calculated by [22]:

$$\dot{Q}_{fw} = \dot{m_{fw}} \cdot c_{pw} \cdot \left(t_w - t_{fw}\right) \tag{11}$$

The heat losses should be compensated by heat gains in the pool area. Heat gains from auxiliary heating equipment to the swimming pool can be expressed as [22]:

$$\dot{Q}_{aux} = \dot{m_{aux}} \cdot c_{pw} \cdot (t_w - t_{aux}) \tag{12}$$

Where $\dot{m_{aux}}$ is the mass flow rate supplied by the auxiliary heating equipment, and t_{aux} is the temperature water supplied for auxiliary heating.

Hence, energy balance of the swimming pool can be written as [22]:

$$\dot{Q}_{aux} - \dot{Q}_{fw} - A_p \cdot (q_E + q_{con} + q_{rad}) = 0 \tag{13}$$

The simulation results in [22] indicate that the impact of radiation to the energy balance of the pool is negligible small.

The indoor swimming pool building facility was modeled as a multizone building in Trnsys software, with the envelope properties of the building described in [21, 22], thus determining the annual behavior of heating and cooling loads. The building was modelled as multi-zone in TRNSYS software [22]. Air velocity of 0.08 m/h in the swimming pool hall was assumed constant, determined based on the measured data. The simulation was done using the Meteonorm hourly weather data, which also included mains water temperature used for fresh water supply system.

The simulated results for the same ambient temperatures were compared with the measured results (Fig. 9, 10 and 11). The comparison of the results showed acceptable difference of results. The presented values of the simulated data in Fig. 9, 10 and 11 are rounded to values with two decimal places for easier comparison. Although a good agreement of the results of the simulation and the measured results on the real object is found, as presented in Fig. 9, 10 and 11, there is still slight error in the simulation results. This can be attributed to the complex mathematical model applied in Trnsys software, where, for the purpose of the simulation whose results re-presented in this paper, the heating and ventilation equipment was also modeled and all of this affects simulation result.

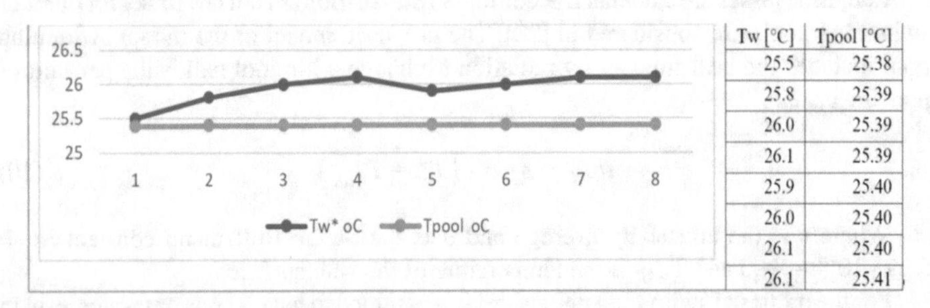

Tw [°C]	Tpool [°C]
25.5	25.38
25.8	25.39
26.0	25.39
26.1	25.39
25.9	25.40
26.0	25.40
26.1	25.40
26.1	25.41

Fig. 9. Measured pool water temperature (T_{pool}) and simulated pool water temperature (T_W)

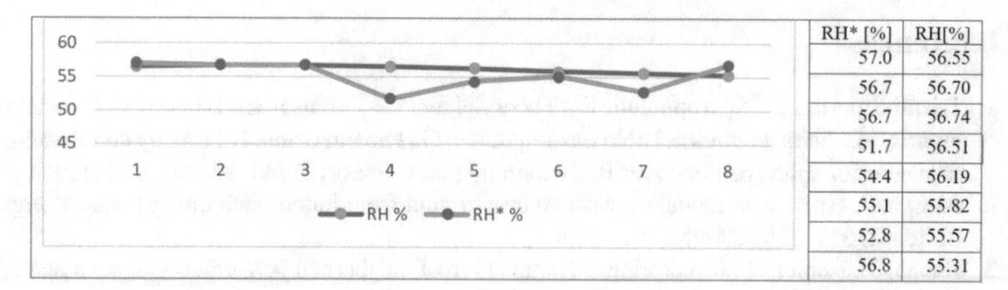

RH* [%]	RH[%]
57.0	56.55
56.7	56.70
56.7	56.74
51.7	56.51
54.4	56.19
55.1	55.82
52.8	55.57
56.8	55.31

Fig. 10. Measured relative humidity of the pool hall air (RH*) and the simulated relative humidity of the pool hall air RH

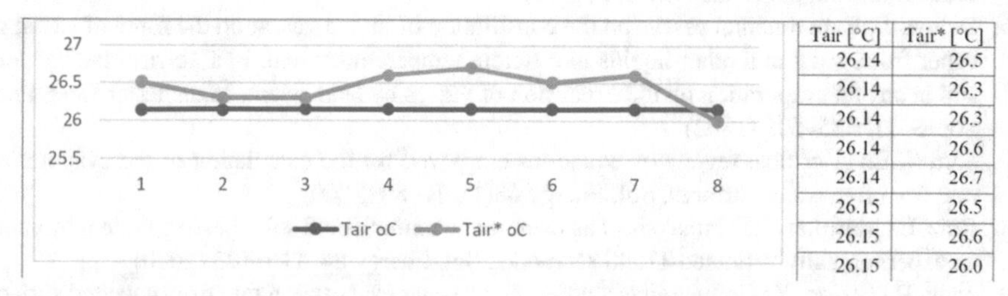

Tair [°C]	Tair* [°C]
26.14	26.5
26.14	26.3
26.14	26.3
26.14	26.6
26.14	26.7
26.15	26.5
26.15	26.6
26.15	26.0

Fig. 11. Simulated air temperature in the pool hall (T_{air}) compared to the measured temperature of pool hall air (T_{air}*)

5 Conclusion

This paper presents literature correlations for evaluation of the evaporation rates from free water surfaces, and applies them for estimation of this phenomenon in indoor swimming pools. Water evaporation from the water surface of an indoor swimming pool was also measured, as well as relevant parameters for its prediction. Measured results are presented in the paper and compared to the results calculated using available literature mathematical correlations for predicting evaporation rates from a water surface from the literature. It was found that most of the equations either overestimate or underestimate the evaporation rates, which implies extreme caution for their application for analysis of indoor swimming pools.

A new correlation should be fitted according to the measured data to ensure good prediction of evaporation rates for the purpose of predicting evaporation rates of indoor swimming pools. The measured data represents actual data found in a real indoor swimming pool building, while the measured air and water properties represent actual values found in operation of a real object. The domain covered by the measured parameters should be sufficient for creating a new correlation valid for typical conditions found in real indoor swimming pool buildings.

Acknowledgement. This research was financially supported by the Ministry of Education, Science and Technological Development of the Republic of Serbia.

References

1. Trianti-Stourna, E., Spyropoulou, K., Theofylaktos, C., Droutsa, K., Balaras, C.A., Santamouris, M., Asimakopoulos, D.N., Lazaropoulou, G., Papanikolaou, N.: Energy conservation strategies for sports centers: Part B. Swimming pools. Energy Build. **27**, 123–135 (1998)
2. Asdrubali, F.: A scale model to evaluate evaporation from indoor swimming pools. Energy Build. **41**, 311–319 (2009)
3. Hahne, E., Kübler, R.: Monitoring and simulation of the thermal performance of solar heated outdoor swimming pools. Sol. Energy **53**(1), 9–19 (1994)
4. ASHRAE HVAC Application Handbook, Atlanta, GA (1999)
5. Auer, T.: Assessment of an indoor or outdoor swimming pool, TRNSYS-TYPE 144, Transsolar, Energietechnik GMBH (1996)
6. Dalton, J.: Experimental essays on the constitution of mixed gases; on the force of steam or vapor from water and other liquids in different temperatures, both in a Torricellian vacuum and in air; on evaporation on the expansion of gasses by heat. Mem. Manchester Liter. Phil. Soc. **5–11**, 535–606 (1802)
7. Sartori, E.: A critical review on equations employed for the calculation of the evaporation rate from free water surfaces. Sol. Energy **68**(1), 77–89 (2000)
8. Ruiz, E., Martinez, P.J.: Analysis of an open-air swimming pool solar heating system by using an experimentally validated TRNSYS model. Sol. Energy **84**, 116–123 (2010)
9. Tang, R., Etzion, Y.: Comparative studies on the water evaporation rate from a wetted surface and that from a free water surface. Build. Environ. **39**, 77–86 (2004)
10. Smith, C.C., Lof, G., Jones, R.: Measurement and analysis of evaporation from an inactive outdoor swimming pool. Sol. Energy **53**(1), 3–7 (1994)
11. Mitrovic, N., Petrovic, A., Milosevic, M., Momcilovic, N., Miskovic, Z., Tasko, M., Popovic, P.: Experimental and numerical study of globe valve housing. Chem. Ind. **71**(3), 251–257 (2017)
12. Rajic, M., Banic, M., Zivkovic, D., Tomic, M., Mančić, M.: Construction optimization of hot water fire-tube boiler using thermomechanical finite element analysis. Therm. Sci. **22**(Suppl. 5), 1511–1523 (2018)
13. Milosevic, M., Mitrovic, N., Jovicic, R., Sedmak, A., Maneski, T., Petrovic, A., Aburuga, T.: Measurement of local tensile properties of welded joint using Digital Image Correlation method. Chem. Listy **106**, 485–488 (2012)
14. Todorovic, M., Zivkovic, D., Mancic, M., Ilic, G.: Application of energy and exergy analysis to increase efficiency of a hot water gas fired boiler. Chem. Ind. Chem. Eng. Q. **20**(4), 511–521 (2014)
15. Bowen, I.S.: The ratio of heat losses by conduction and by evaporation from any water surface. Phys. Rev. **27**, 779–787 (1926)
16. Heiselberg, L.Y.: CFD Simulations for Water Evaporation and Airflow Movement in Swimming Baths, Report for the project "Optimization of Ventilation System in Swimming Bath". Aalborg University, Denmark (2005)
17. Shah, M.: Prediction of evaporation from occupied indoor swimming pools. Energy Build. **35**, 707–713 (2003)
18. Shah, M.: Improved method for calculating evaporation from indoor water pools. Energy Build. **49**, 306–309 (2012)
19. Moghiman, M., Jodat, A.: Effect of air velocity on water evaporation rate in indoor swimming pools. Iran. J. Mech. Eng. **8**, 19–30 (2007)
20. Vinnichenko, N.A., Uvarov, A.V., Vetukov, D.A., Plaksina, Y.Y., Direct computation of evaporation rate at the surface of swimming pool, in Recent research in mechanics. In: Proceedings of the 2nd International Conference on Fluid Mechanics and Heat and Mass Transfer, Corfu island, Greece (2011)

21. Mančić, M.V., Živković, D., Đorđević, M.Lj., Jovanović, M.S., Rajić, M.N., Mitrović, D.M.: Techno-ecnomic optimization of configuration and capacity of a polygeneration system for the energy demands of a public swimming pool building. Therm. Sci. **22**(5), 1535–1549 (2018)
22. Mančić, M.V., Živković, D., Đorđević, M.Lj., Jovanović, M.S., Rajić, M.N., Mitrović, D.M.: Mathematical modelling and simulation of the thermal performance of a solar heated indoor swimming pool. Therm. Sci. **18**(3), 999–1010 (2014)

Total Fatigue Life Estimation of Aircraft Structural Components Under General Load Spectra

Katarina Maksimovic[1], Strain Posavljak[2], Mirko Maksimovic[3], Ivana Vasovic Maksimovic[4](✉), and Martina Balac[5]

[1] Secretariat for Utilities and Housing Services Water Management, Kraljice Marije Street 1, 11120 Belgrade, Serbia

[2] Faculty of Mechanical Engineering, University of Banja Luka, Vojvode Stepe Stepanovica 75, 78000 Banja Luka, Bosnia and Herzegovina

[3] Belgrade Waterworks and Severage, Kneza Milosa Street 27, Belgrade, Serbia

[4] Lola Institute, Kneza Višeslava Street 70a, Belgrade, Serbia
ivanavvasovic@gmail.com

[5] Faculty of Mechanical Engineering, University of Belgrade, Kraljice Marije street No 16, Belgrade, Serbia

Abstract. This work presents total fatigue life prediction methodology of aircraft structural components under general load spectrum. Here is presented an effective computation procedure, that combines the finite element method (FEM) and strain-life methods to predict fatigue crack initiation life and fatigue crack growth model based on the strain energy density (SED) method. To validate computation procedure in this paper has been experimental tested specimens with a central hole under load spectrum in form of blocks. Total fatigue life of these specimens, defined as sum of crack initiation and crack growth life, was experimentally determined. Crack initiation life was computed using the theory of low cycle fatigue. Computation of crack initiation life was realized using Palmgreen-Miner's linear rule of damage accumulation, applied on Morrow's curves of low cycle fatigue. Crack growth life was computed using strain energy density (SED) method. The same low cyclic material properties of quenched and tempered steel 13H11N2V2MF, used for crack initiation life computation, were used for crack growth life computation. Residual life estimation of cracked duraluminum aircraft wing skin/plate 2219-T851 under multiple overload/underload load spectrum was considered too. Presented computation results were compared with own and available experimentally obtained results.

Keywords: Aircraft structures · Fatigue life · SED · Crack initiation · Crack growth · FEM

1 Introduction

Many structural components are subjected to cyclic loading. In these components, fatigue damage is the prime factor in affecting structural integrity and service life. Fatigue

N. Mitrovic et al. (Eds.): CNNTech 2020, LNNS 153, pp. 394–412, 2021.
https://doi.org/10.1007/978-3-030-58362-0_23

damage is typically divided into three stages: crack initiation, crack propagation and final failure [1]. These three stages are important in determining the fatigue life of structural components. The fatigue life of structural elements can be generally divided into crack initiation and crack growth [2–5]. In this paper proposed computation procedures for total fatigue life estimation will be given.

Aircraft structural components and structural components of the other mechanical systems posses geometrical discontinuities with notches or notched structural components. As a rule, cracks appear at critical points which lie in the bases of notches.

Aircraft engine components are subjected to variable amplitude load conditions and usually they tend to experience fatigue damage [6, 7]. A reliable lifetime prediction is particularly important in the design, safety assessments and optimization of engineering materials and structures. Many fatigue damage accumulation theories have been put forward to predict the fatigue lives of structure components, such as linear damage rule, Grover-Manson theory and Corten-Dolan theory [8, 9].

Notches have important role in process of fatigue life analyses. Total fatigue life (TFL) of aircraft structural component is sum of crack initiation and crack growth life (CIL and CGL), and mainly depends of number, form, size, position and arrangement of its notches.

In this research two type materials are used; steel 13H11N2V2MF and duraluminum, 2219-T851. Low cyclic material properties for these materials are experimentally determined. For that purpose servo-hydraulic MTS system is used.

Here is necessary to estimate the life of structural elements, initial fatigue life and crack propagation life. This basically means estimating the total fatigue life. Precisely the subject of this research refers to the assessment of the total fatigue life of representative structural elements of aircraft structures under the load spectrum [6, 10–15]. Two types of materials are used here. The first is used for the structural elements of the aircraft engine and the second of the duraluminum used for the formwork of the aircraft wings. The conventional approach for estimating the total fatigue life uses low - cycle fatigue characteristics of the material for crack initialization life and dynamic characteristics of the material for crack growth analyzes [9–14].

CIL computation of aircraft structural components implies knowing: aircraft flight cycles, cyclic events in flight cycles, cyclic properties of material used or nominated for workmanship, stress-strain response at critical point or point of expected crack initiation and damages provoked by all cyclic events.

Stress-strain response at critical point (local stress-strain response) for all cyclic events and method of identification and counting of those events, have special importance. Local stress-strain response may be determined by strain gauge measurement, by the finite element method (FEM) [15–17] and by the methods that relate local stresses (σ_{loc}) and strains (ε_{loc}) to nominal values. Identification and counting of cyclic events in aircraft flight cycles may be carried out using rain flow counting method and range pair method [1].

The estimation of the fatigue life of structural elements until the occurrence of initial damage under the action of cyclic loading is in constant progress. Because of simplicity, Palmgreen-Miner's rule of linear damage accumulation is mostly in use [20–23].

For TFL computation, several investigators [5, 11–14] have combined the two approaches by combining the computed models for crack initiation and crack growth.

Primary attention in this research has been to develop an efficient and reliable computational method for structural components under cyclic load or under general load spectrum.

Methodologies of CIL and CGL computation of notched structural components are based on criterions of low cycle fatigue life (LCF) [1, 4] and strain energy density method (SED) [2, 15] as described in this paper. The results obtained by computation and experimentally obtained results are compared.

In this paper, the same low – cycle material properties of the material were used both for the assessment of the initial fatigue and for the crack growth. The use of the same low-cycle fatigue properties of the material makes the computation method of estimating the life more efficient on the one hand and more economical on the other hand. The use of the low-cycle fatigue properties of the material for the life to crack initialization and crack propagation makes this approach more efficient and more economical.

2 Crack Initiation Life of Specimens with Central Hole

The first stage of total fatigue life prediction is fatigue crack initiation. The procedure for determined this phase is based on strain-life method and cyclic strain- stress curves. Estimation of damage D, as damage at critical structural element point, can be carriad out using Palmgrem-Miner's rule of linear damage accumulation [1] defined as

$$D = \sum_{i=1}^{4} D_i = \sum_{i=1}^{4} \frac{N_i}{N_{fi}} \tag{1}$$

In above expression the damage provoked by i-th force cycle is marked with D_i. This damage presents the ratio of number of appearing N_i of i-th cycle within block of positive variable force and number N_{fi} of the same cycle that flat specimen made of steel 13H11N2V2MF in heat treatment state, can endure up to crack initiation. Numbers N_i are given in Table 1 and numbers N_{fi} were determined using Morrow's, Manson-Halford's and Smith-Watson-Topper's (SWT) curve [4] of low cycle fatigue. Morrow's curve [19] is the first equation of the next system

$$\begin{aligned} \frac{\Delta\varepsilon}{2} &= \frac{\sigma_f' - \sigma_{mi}}{E} N_f^b + \varepsilon_f' N_f^c \\ \frac{\Delta\varepsilon}{2} &= \frac{\Delta\varepsilon_i}{2} \qquad\qquad i = 1, 2, 3, 4 \end{aligned} \tag{2}$$

The first equation of system

$$\begin{aligned} \frac{\Delta\varepsilon}{2} &= \frac{\sigma_f' - \sigma_{mi}}{E} N_f^b + \left(\frac{\sigma_f' - \sigma_{mi}}{\sigma_f'}\right)^{\frac{c}{b}} \varepsilon_f' N_f^c \\ \frac{\Delta\varepsilon}{2} &= \frac{\Delta\varepsilon_i}{2} \qquad\qquad i = 1, 2, 3, 4 \end{aligned} \tag{3}$$

is Manson-Halford's curve [1] and the first equation of system

$$P_{SWT} = \sqrt{\sigma\max\frac{\Delta\varepsilon}{2}E} = \sqrt{\left(\sigma_f'\right)^2 (N_f)^{2b} + E\sigma_f'\varepsilon_f'(N_f)^{b+c}}$$

$$P_{SWT} = P_{SWT,i} = \sqrt{\sigma\max{}_{,i}\frac{\Delta\varepsilon_i}{2}E} \qquad i = 1, 2, 3, 4 \tag{4}$$

is Smith-Watson-Topper's curve [4] of low cycle fatigue, where PSWT presents Smith-Watson-Topper's perimeter in which σmax, i = σmi + Δσi/2. Stresses σmi in systems (2) and (3) are mean stresses.

Combining applied approaches of determining of nonlinear stress-strain response at critical specimen point, with different curves of low cycle fatigue, using Palmgren-Miner's rule, we obtained estimated values of damages D and crack initiation life (CIL = 1/D) expressed in blocks of positive variable force.

Fatigue life calculation of a varying load spectrum is a complex task. There are various approaches suggested to simplify the load spectrum into a simple major and minor cycle. One such approach is rainflow counting methodology.

The rainflow-counting algorithm is used in the analysis of fatigue data in order to reduce a spectrum of varying stress into an equivalent set of simple stress reversals. The method successively extracts the smaller interruption cycles from a sequence, which models the material memory effect seen with stress-strain hysteresis cycles. The rainflow method is compatible with the cycles obtained from examination of the stress-strain hysteresis cycles.

2.1 The Characteristics of Specimens

Primary attention in this investigation is focused to experimental total fatigue life estimation including crack initiation and crack growth lives. For that purpose three specimens denoted with SPC-1, SPC-2 and SPC-3 were made of steel 13H11N2V2MF. Structural elements (Specimen with central hole) shown in Fig. 1 are used for this purpose. Before final grinding, specimens were subjected to heat treatment (heating at 1000 °C,

Table 1. Cyclic properties of QT steel 13H11N2V2MF.

Low cyclic material properties of steel	Value
Modulus of elasticity, E [MPa]	229184.6
Cyclic strength coefficient, K′ [MPa]	1140
Cyclic strain hardening exponent, n′	0.0579
Fatigue strength coefficient, σ_f' [MPa]	1557.3
Fatigue strength exponent, b	−0.0851
Fatigue ductility coefficient, ε_f'	0.3175
Fatigue ductility exponent, c	−0.7214

oil quenching, tempering at 640 °C and air cooling). Steel 13H11N2V2MF, heat treated in the same manner, here is observed as quenched and tempered (QT) steel of tensile strength $R_m \approx 1000$ MPa. Low cyclic fatigue properties of this kind of steel are contained in Table 1 [11].

2.2 Experimental Determination Results of Crack Initiation, Crack Growth and Total Fatigue Life

Experimental determination results of CIL, CGL and TFL specimens with central hole were carried out using MTS servo-hydraulic system. The specimens were loaded by blocks of positive variable force. Time dimension of these blocks, according to Fig. 1, amounts 40 s.

The data of experimentally determined CI, CG and TF life expressed in blocks are given in Table 2.

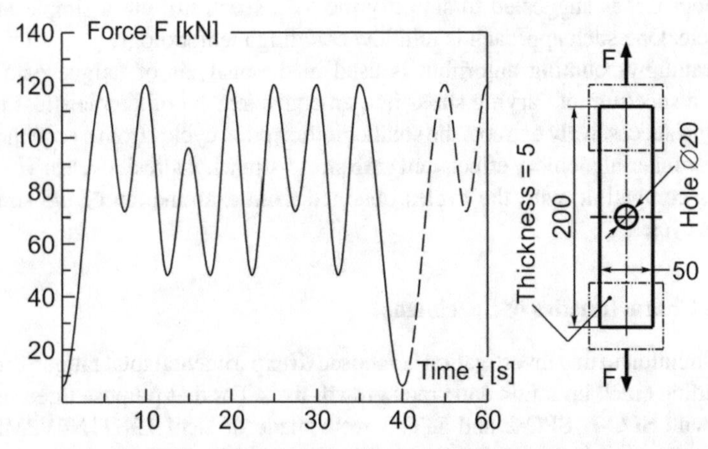

Fig. 1. Block of variable amplitude force.

Table 2. Experimentally determined CIL, CGL and TFL of flat specimens with central hole.

Specimen	CIL [Blocks]	CGL [Blocks]	TFL [Blocks]
SPC-1	4000	485	4485
SPC-2	3600	526	4126
SPC-3	4200	–	–

2.3 Converting of a Block of Variable Force

In this investigation variable amplitude load spectrum is used. For practical use this type spectra in life estimation is necessary to be converted. For that purpose rainflow counting method is used. The rainflow method is a method for counting fatigue cycles from a time history. The fatigue cycles are stress-reversals. The rainflow method allows the application of Minner's rule in order to assess the fatigue life of a structural element under complex loading.

For the purpose of CIL computation of flat specimens with central hole, block of variable force was satisfying modified and decomposed on simple X-Y-X force cycles (Fig. 2). Levels i and number N_i of these cycles in block are contained in Table 3.

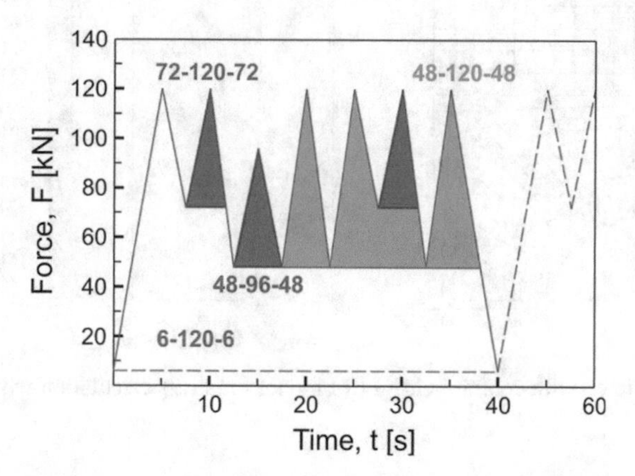

Fig. 2. Modified and decomposed block of variable force.

Table 3. X-Y-X force cycles.

Level i	X_i-Y_i-X_i force cycle [kN]	Number in block N_i
1	6-120-6	1
2	48-120-48	3
3	72-120-72	2
4	48-96-48	1

Decomposition of block of variable force was carried out using method of reservoir.

2.4 Linear Stress Response

At the beginning, flat specimens were observed as ideal elastic bodies. Stress response of these specimens provoked by maximal force $F_{max} = 120$ kN was obtained using FEM implemented in I-DEAS Master Series software [11]. Theirs symmetrical segment was separated and discretized by plane stress parabolic elements. In that manner FEM model for stress calculation was obtained. This model with principal stress σ_1 trajectories is shown in Fig. 3.

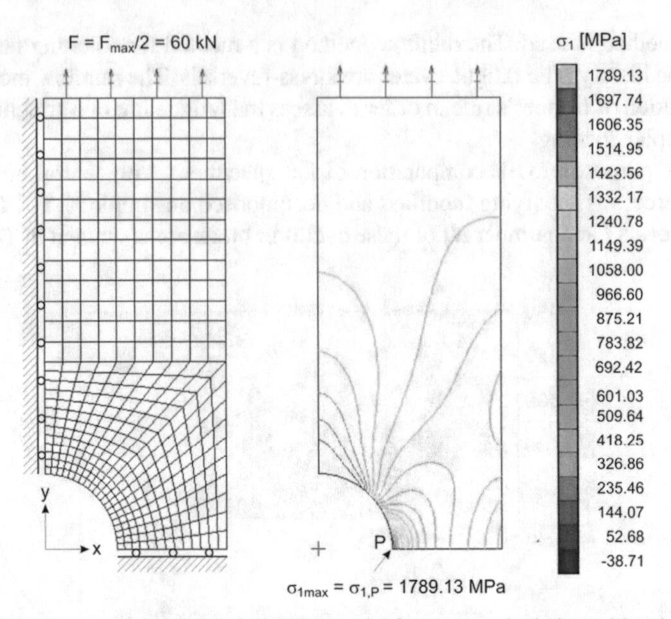

Fig. 3. Flat specimen with central hole - FEM model for stress calculation with principal stress σ_1 trajectories.

Principal stress σ_1 has maximum value at critical point P and amounts $\sigma_{1max} = \sigma_{1,P} = 1789.13$ MPa. That stress is over of the tensile strength $R_m \approx 1000$ MPa of QT steel 13H11N2V2MF which used for specimens with central hole.

Stress in area far from stress concentration area here is taken as nominal stress and its value $\sigma_n = 480$ MPa. This stress was obtained using simple expression

$$\sigma_n = \frac{F_{max}}{S} \tag{5}$$

where is with S marked surface of full specimen section ($S = 5 \cdot 50 = 250\,\text{mm}^2$).

Stress concentration factor $K_t = 3.727$ was obtained from relation

$$K_t = \frac{\sigma_{1,P}}{\sigma_n} \tag{6}$$

and it is applied for transformation of linear (ideal elastic) stress-strain response at critical point in real (nonlinear) stress-strain response, using approximate Sonsino's curve [12].

2.5 Nonlinear Stress-Strain Response at Critical Point

Real stress-strain response at critical point of flat specimen with central hole is nonlinear and it may be described by stabilized hysteresis loops [1, 12]. Stress-strain response provoked by block of variable force is shown in Fig. 4.

In Fig. 4 can be seen stabilized hysteresis loops assigned to all X_i-Y_i-X_i force cycles from Table 3.

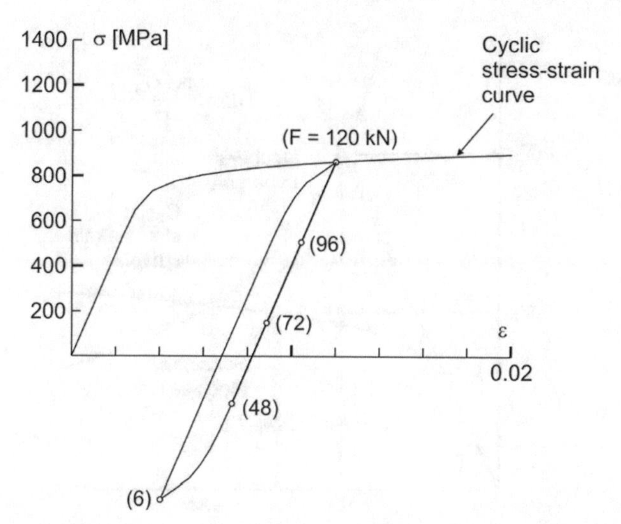

Fig. 4. Stress-strain response at critical point of flat specimen with central hole provoked by block of variable force.

Upper point as the first point of stress-strain response corresponding to force Fmax = 120 kN was determined with the help of system equations

$$\begin{aligned} \varepsilon &= \frac{1}{2}\left(\frac{K_t^2\sigma_{ni}^2}{\sigma E} + \frac{K_t\sigma_{ni}}{E}\right) \quad i = 1 \\ \varepsilon &= \frac{\sigma}{E} + \left(\frac{\sigma}{K'}\right)^{\frac{1}{n'}} \end{aligned} \tag{7}$$

Dimensions (ΔεxΔσ) of stabilized hysteresis loops were determined using the next system equations

$$\begin{aligned} \varepsilon &= \frac{1}{2}\left(\frac{K_t^2\Delta\sigma_{ni}^2}{\Delta\sigma E} + \frac{K_t\Delta\sigma_{ni}}{E}\right) \quad i = 1, 2, 3, 4 \\ \Delta\varepsilon &= \frac{\Delta\sigma}{E} + 2\left(\frac{\Delta\sigma}{2K'}\right)^{\frac{1}{n'}} \end{aligned} \tag{8}$$

The first equations in systems (7) and (8) are two forms of approximate Sonsino's curves derived on the base of Sonsino's modification of Neuber's rule (Fig. 5) [12, 13].

The second equations in systems (7) and (8) are equations of cyclic stress-strain curve and Masing's curve of QT steel 13H11N2V2MF. Mentioned systems were solved graphically using Drafting module of I-DEAS Master Series software. By special Fortran programs cyclic stress-strain curve, Masing's curve and approximate Sonsino's curves were copied in corresponding spline curves. Position of stabilized hysteresis loops were determined using memory of metals. Masing's curve was served for theirs modeling. Value for E, K′ and n′ was taken form Table 1. For stress concentration factor, above determined value $K_t = 3.727$, was taken. Values of nominal stresses σ_{ni} and nominal stress ranges $\Delta\sigma_{ni}$ were taken from Table 4 and can be determined using the next expressions:

$$\sigma_{ni} = \frac{Y_i \cdot 10^3}{A}$$

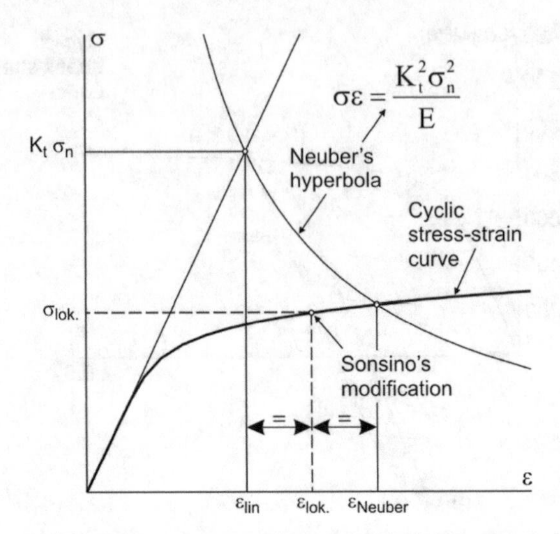

Fig. 5. Sonsino's modification of Neuber's rule.

Table 4. Nominal stresses σ_{ni} and ranges of those stresses $\Delta\sigma_{ni}$.

Level i	X_i-Y_i-X_i force cycle [kN]	σ_{ni} [MPa]	$\Delta\sigma_{ni}$ [MPa]
1	6-120-6	480	456
2	48-120-48	480	288
3	72-120-72	480	192
4	48-96-48	384	192

$$\Delta\sigma_{ni} = \frac{(Y_i - X_i) \cdot 10^3}{A} \tag{9}$$

Stress-strain response at critical point of specimens for 6-120-6 force cycle is presented in Fig. 6.

Complete numerical results of stress-strain response at critical point of specimen, described by stabilized hysteresis loops for all force cycles, are included in Table 5.

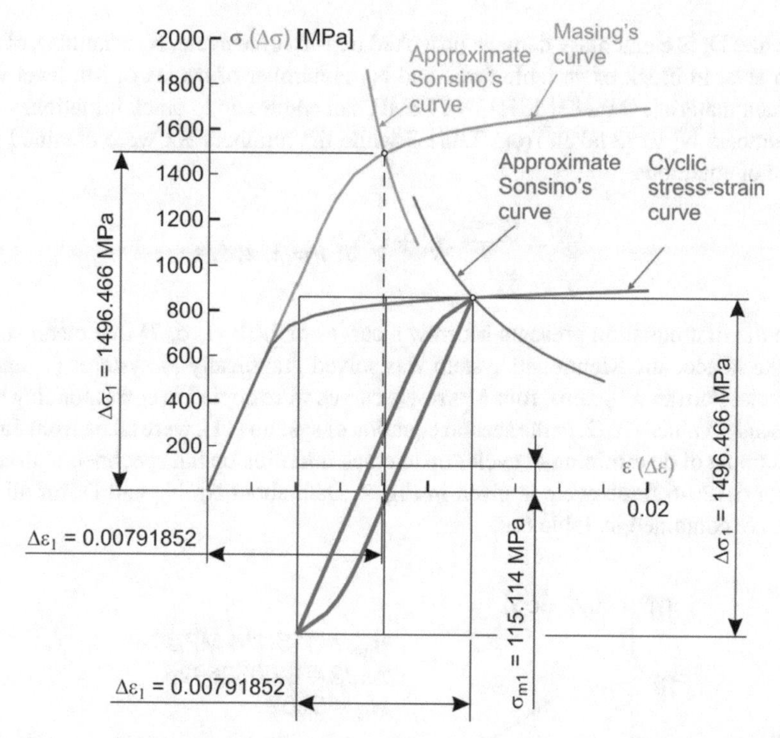

Fig. 6. Stress-strain response at critical point of flat specimen with central hole for 6-120-6 force cycle

Table 5. Numerical results of stress-strain response at critical point of flat specimen with central hole for all force cycles.

Level i	X_i-Y_i-X_i force cycle [kN]	σ_{mi} [MPa]	$\Delta\sigma_i$ [MPa]	$\Delta\varepsilon_i$	$\Delta\varepsilon_i/2$	$\sigma_{max,i}$ [MPa]
1	6-120-6	115.114	1496.466	0.00791852	0.003959260	863.347
2	48-120-48	326.998	1072.699	0.00468495	0.002342475	863.347
3	72-120-72	505.556	715.582	0.00312229	0.001561145	863.347
4	48-96-48	148.439	715.582	0.00312229	0.001561145	506.230

2.6 Damages and Crack Initiation Life Computation

Computation of damage D_B provoked by block of variable force was carried out using Palmgreen-Miner's rule of linear damage accumulation in the form

$$D_B = \sum_{i=1}^{4} D_i = \sum_{i=1}^{4} \frac{N_i}{N_{fi}} \tag{10}$$

where are: D_i is elementary damage provoked by i-th force cycle, N_i is number of cycle of i-th level in block of variable force and N_{fi} is number of cycles of i-th level which specimen material, QT steel 13H11N2V2MF, can endure up to crack initiation.

Numbers N_i were taken from Table 3 while the numbers N_{fi} were obtained using system of equations

$$\begin{aligned} \frac{\Delta\varepsilon}{2} &= \frac{\sigma_f' - \sigma_{mi}}{E} N_f^b + \varepsilon_f' N_f^c \quad i = 1, 2, 3, 4 \\ \frac{\Delta\varepsilon}{2} &= \frac{\Delta\varepsilon_i}{2} \end{aligned} \tag{11}$$

where the first equation presents Morrow's curves of LCF [1, 6, 7] that mean stresses σ_{mi} take in account. Mentioned system was solved graphically as systems (7) and (8). By special Fortran programs, four Morrow's curves were copied in corresponding spline curves also. Values $\Delta\varepsilon_i/2$, in the second equation of system (11), were taken from Table 5.

Example of determining of cycles up to crack initiation on flat specimen with central hole for 6-120-6 force cycle is given in Fig. 7. Data about N_i, N_{fi} and D_i for all force cycles are contained in Table 6.

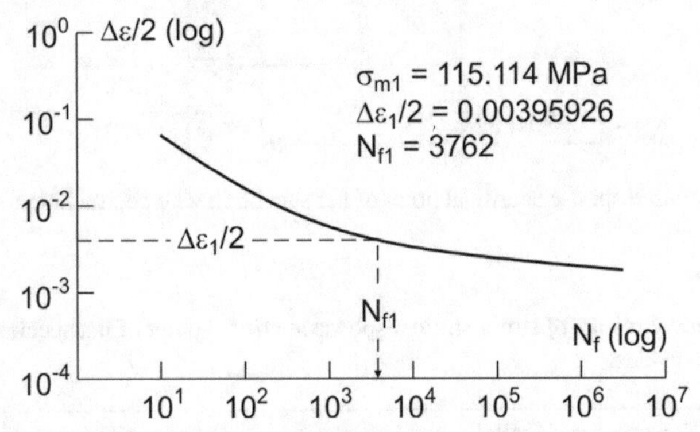

Fig. 7. Example of determining of cycles to CI on flat specimen with central hole for 6-120-6 force cycle

Table 6. Data about N_i, N_{fi} and D_i.

Level i	X_i-Y_i-X_i force cycle [kN]	Number in block N_i	N_{fi}	D_i
1	6-120-6	1	3762	0.000265816
2	48-120-48	3	38439	0.000078046
3	72-120-72	2	396753	0.000005041
4	48-96-48	1	10087673	0.000000099

Damage which one block of variable force provokes according to (10) has value $D_B = 0.000349002$. Reciprocal value of this damage, $1/D_B = 2865$ blocks, is equal to CIL of discussed flat specimens with central hole.

3 Residual Fatigue Life Prediction of Cracked Structural Element Using Strain Energy Density Method

To determine residual fatigue life of cracked structural element the strain energy density (SED) method is used. While predicting life of a structural element with initial damage it‘s necessary to establish the functional dependency between the crack growth gradient da/dN and stress intensity factor K_I.

3.1 Crack Growth Model Based on Theory of Strain Energy Density Method

In this section procedure for crack growth computation is established and it is based on SED theory. For that purpose structural element with central hole and initial crack, as shown in Fig. 1, is analyzed. Key value in process of crack growth analysis is stress intensity factor range ΔKI. Near stress intensity factor range ΔKI a threshold stress intensity factor range ΔKth exists that is value below which fatigue cracks do not growth. So here, we will formulate relation for crack growth rate da/dNp as a function of a stress intensity range ΔKI as well as a threshold stress intensity factor range ΔKth.

The first important parameter in fatigue crack analysis is the length of the process zone ahead of crack tip d^* (Fig. 8). That parameter can be considered as a constant by same authors [7, 9] and variable (as function of ΔK_I) by others [25, 26]. The length d^* can be expressed like function of ΔK_I and ΔK_{th} [26], like:

$$d^* = \frac{\Delta K_I^2 - \Delta K_{th}^2}{\pi E \sigma_y'} \tag{12}$$

where ΔK_{th} is range of threshold stress intensity factor and σ_y' is the cyclic yield stress. In order to define needed relation based on parameters ε_f', σ_y' and n', first we need to define relation for the cyclic plastic strain energy density ΔW_p in the units of Joule per cycle per unit volume [2, 18] as:

$$\Delta W_p = \left(\frac{1-n'}{1+n'}\right) \Delta\sigma_{eq} \Delta\varepsilon_{eq} \tag{13}$$

where $\Delta\sigma_{eq}$, $\Delta\varepsilon_{eq}$ are equivalent stress and strain along the crack line ($\theta = 0$), respectively, or

$$\Delta W_p = \left(\frac{1-n'}{1+n'}\right) \frac{\Delta K_I^2 \overline{\sigma_{eq}}(0; n') \, \overline{\varepsilon_{eq}}(0; n')}{E I_{n'} r} \tag{14}$$

where are ΔKI the range of stress intensity factors under mode I loading, or for the plate with central hole and crack ΔKI can be expressed as [15]:

$$\Delta K_I = \Delta\sigma \sqrt{\pi a} \left[1 + 2.365 \left(\frac{R}{R+a} \right)^{2.4} \right] \tag{15}$$

where a is crack length, $\Delta\sigma$ - stress range, R – radius of a central hole, and $I_{n'}$ is the non-dimensional parameter of exponent n'. For most metals the value of n' usually varies between 0.10 and 0.25, with an average value close to 0.15. Terms r and θ are the radial and angular positions, respectively, of any point from the crack tip, as shown in Fig. 8.

Further the $\overline{\sigma_{eq}}(0; n')$, $\overline{\varepsilon_{eq}}(0; n')$ are non-dimensional angular ($\theta = 0$) distribution functions for stress and strain, respectively.

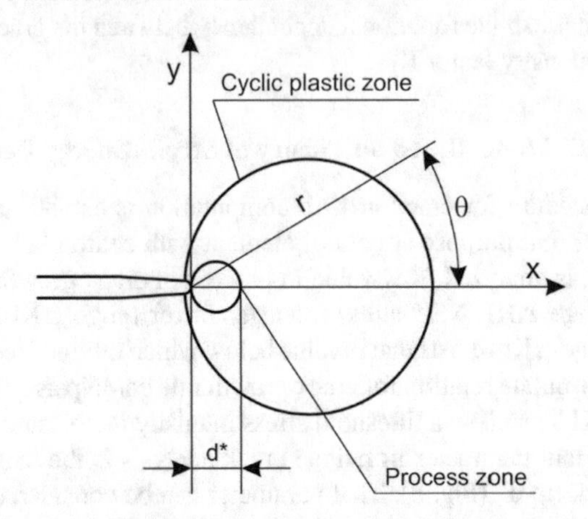

Fig. 8. A crack tip zones of damage

Equation (14) presents distribution of plastic strain energy density per cycle ahead of the crack tip. Unknown values are the angular distribution functions of equivalent stress and strain and parameter $I_{n'}$. The area near the crack tip is known as the process zone [15]. Next area is the region between cyclic plastic zone and process zone. The area where damage mainly accumulates is process zone.

Since we defined the length d^* it is possible to determine the plastic energy ω_p dissipated per cycle per unit growth. Needed relation for ω_p can be obtained by integrating the relation for the plastic strain energy density (16) and if r is substituted by d^*, or:

$$\omega_p = \int_0^{d^*} \left(\frac{1-n'}{1+n'}\right) \frac{\Delta K_I^2 \psi(n')}{E I_{n'} d^*} dr \tag{16}$$

after integrating,

$$\omega_p = \left(\frac{1-n'}{1+n'}\right) \frac{\Delta K_I^2 \psi}{E\, I_{n'}} \tag{17}$$

where $\psi = \left(\overline{\sigma_{ij}}(0;\ n')\ \overline{\varepsilon_{ij}}(0;\ n')\right)$ and it can be determined from experiments. It is known that external load increased from zero at the crack tip, it is blunted first and started to open when the stress intensity factor reached the threshold value K_{th}. Based on the previous statement it is possible instead of $\Delta K_I = K_{max} - K_{min}$ to formulate $\Delta K_I = K_{max} - K_{th}$, as:

$$\omega_p = \left(\frac{1-n'}{1+n'}\right) \frac{\psi(K_{max} - K_{th})}{E I_{n'}} \tag{18}$$

The plastic energy dissipated per cycle per unit growth can be formulated like function of energy absorbed till fracture W_p, as:

$$W_p \delta a = \omega_p \tag{19}$$

So relation for δa became:

$$\delta a = \frac{da}{dN_p} = \frac{(1-n')\psi}{4\,E\,I_{n'}\,\sigma_f'\varepsilon_f'}(K_{max} - K_{th})^2 \tag{20}$$

Equation (20) presents a special case for R = 0 ($R = K_{min}/K_{max}$). In that case, $K_{max} = \Delta K$, $K_{th} = \Delta K_{th}$, where is ΔK_{th} a material constant and it is sensitive to stress ratio R. An important aspect in the fatigue design of structural elements is the "design of safe cracks" on the basis of the crack growth threshold. Of much concern, however, is the effect of stress ratio on the fatigue threshold stress intensity range, ΔK_{th}. Regarding the R-effect on ΔK_{th}, many relations, mostly empirical, have been proposed, some of which are [14]:

$$\Delta K_{th} = K_{max}(1-R);\quad \Delta K_{th} = K_{th0}(1-R)^{\frac{1}{2}};\quad \Delta K_{th} = K_{th0}(1-R)^{\gamma};$$
$$\Delta K_{th} = K_{th0}\left(1-R^2\right);\quad \Delta K_{th} = K_{th0}\left(\frac{1-R}{1+R}\right)^{\frac{1}{2}} \tag{21}$$

where ΔK_{th0} is the range threshold stress intensity factor for the stress ratio R = 0 and γ is a material constant which varies from zero to unity. For most of materials, γ comes out to be 0.71 [13]. Despite the large difference value of proposed relations (21) between ΔK_{th} and R, substituting the value of K_{max} and K_{th} for a general stress ratio R, the fatigue crack growth relations are expressed as given:

$$\delta a = \frac{da}{dN_p} = \frac{(1-n')\psi}{4EI_{n'}\sigma_f'\varepsilon_f'}\left[\Delta K_I - \Delta K_{th0}(1-R)^{\gamma}\right]^2 \tag{22}$$

$$\delta a = \frac{da}{dN_p} = \frac{(1-n')\psi}{4EI_{n'}\sigma_f'\varepsilon_f'}\left[\Delta K_I - \Delta K_{th0}\left(\frac{1-R}{1+R}\right)^{\frac{1}{2}}\right]^2 \tag{23}$$

Equation (22) and (23) presents the low of crack growth based on strain energy density method. It's obvious that in these dependency cyclic characteristic of material from low-cycle fatigue domain are being used instead of dynamic parameters from more conventional laws for crack growth by Paris, Forman and others. It is clear from these equations that with an increase in stress ratio R, ΔK_{th} decreases and ΔK_I increases, increasing the fatigue crack growth rate but the influence is more propounce in stage I (near threshold region of da/dN_p vs ΔK_I plot) where ΔK_I and ΔK_{th} are comparable than in other regions of the plot. This is consistent with the experimental results [25, 26].

From fatigue crack growth relations (20) and (21) or (22) and (23) can be seen that they require only mechanical and fatigue properties E, σ'_f, ε'_f and n', which presents great advantage by application of this procedure.

3.2 Crack Growth Life Computation

To CGL computation Eq. (23) is used. Here is considered crack growth model of flat specimen with central hole with two cracks. For that purpose the stress intensity factor is defines by Eq. (15). In Fig. 9 are shown computation (prediction) results - relation between crack length *a* and number of blocks of variable force.

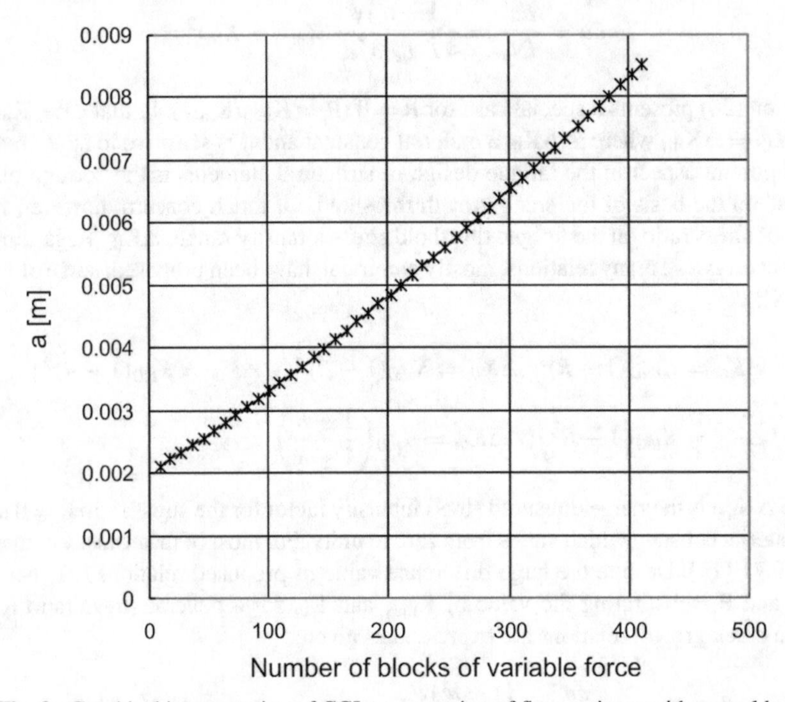

Fig. 9. Graphical interpretation of CGL computation of flat specimen with central hole

CGL obtained by computation, according to Fig. 10, amounts 412 blocks and that result has good correlation with experimentally obtained results (See Table 2).

3.3 Comparison of Results

For comparison of results of CIL, CGL and TFL (TFL = CIL + CGL) of discussed specimens, experimentally and by computation obtained, here is used histogram in Fig. 10.

Results in Fig. 10 shows good agreement between computation results with experiments for structural elements of specimens with central hole.

3.4 Residual Fatigue Life Estimation of Cracked Aircraft Skin/Plate Under Multiple Overload/Underload Load Spectrum

As already pointed out, the strain energy density method (SED) [2, 12, 15] was used for the analysis of crack growth, i.e. for determination of the residual fatigue life. In

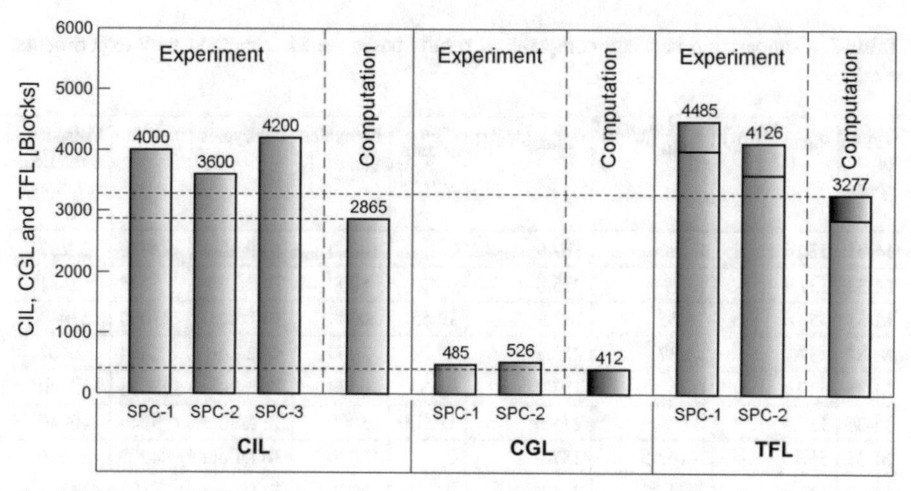

Fig. 10. Histogram of experimentally and by computation obtained results of CIL, CGL and TFL

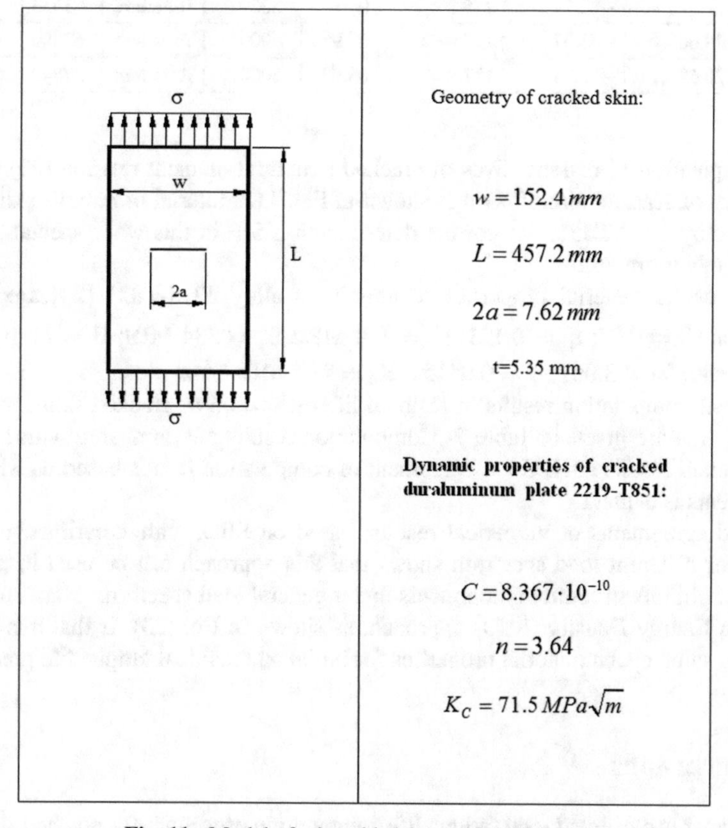

Fig. 11. Model of wing skin/plate with initial crack

Table 7. Comparation between computation results based on SED method with experimental results

Type of spec.	$\sigma^{I}_{\max}$ [MPa]	$\sigma^{I}_{\min}$ [MPa]	$\sigma^{II}_{\max}$ [MPa]	$\sigma^{II}_{\min}$ [MPa]	N_I cycles	N_{II} cycles	N_f^{exp}	Present numerical solution SED
M-31	55.16	0	137.9	0	10000	To failure	22430	22262
M-32	137.9	0	275.8	0	5000	To failure	5275	5131
M-33	55.16	16.55	137.9	16.55	10000	To failure	27000	27943
M-34	137.9	41.37	275.8	82.74	5000	to failure	6884	6462
M-35	55.16	0	137.9	96.53	10000	To failure	178858	145140
M-36	137.9	0	275.8	193.06	5000	To failure	9346	10940
M-37	55.16	−16.55	137.9	0	10000	To failure	24200	23076
M-38	137.9	−41.37	275.8	0	5000	To failure	5197	3325
M-39	0	−41.37	139	0	5000	To failure	19300	17862
M-40	0	−82.74	275.8	0	5000	To failure	5653	5309
M-41	−20.68	−41.37	137.9	68.95	5000	To failure	67400	69603
M-42	−20.68	−82.74	137.9	68.95	5000	To failure	57842	52857

this computation of residual lives of cracked structural element relation (23) is used. Geometry of cracked aircraft skin is shown in Fig. 11. Material of aircraft skin is aluminum alloy 2219-T851. For correct determination SIF in this work special singular finite elements are used.

Low cyclic material properties of aluminum alloy 2219-T851 [24] are : σ'_f = 613 MPa; $\varepsilon'_f = 0.35$; $n' = 0.121$; $k' = 710$ MPa; $S'_y = 334$ MPa; $E = 71\ 10^3$ MPa; $\Delta K_{th0} = 30$; $I_{n'} = 3.067$; $\psi = 0.95152$, $K_c = 71.5$ MPa $\sqrt{m}$.

Present computation results of residual life estimation of cracked skin for various load spectres are given in Table 7. Computation results are compared with available experimental results [24]. Good agreement of computation results based on SED with experiments is evident.

Good agreements of numerical results, based on SED, with experimental testing results for different load spectrum shows that this approach can be used in practical design of aircraft structural components under general load spectrum. Main advantage of Strain Energy Density (SED) approach, as shown in Eq. (23), is that this method using the same cyclic material properties for initial and residual fatigue life predictions [2, 16].

4 Conclusions

In this paper is presented total fatigue life computation procedure for notched structural components under general load spectrum. Computation procedure is based on combining strain-life approach for crack initiation and strain energy density (SED) method for crack growth.

The obtained results have been compared with available experimental data and good agreement has been achieved, therefore this procedure for total fatigue life prediction is significant for engineering applications.

Methodologies of CIL and CGL computation, developed for flat specimens with central hole and model of aircraft skin/plate with initial crack, in similar way, may be applied for notched structural components of different mechanical systems. This methodology is the most completive procedure because of using of the same cyclic material properties for crack initiation and crack propagation. An effective computation procedure is achieved, that combines the finite element method (FEM) and strain-life methods to predict fatigue crack initiation life and fatigue crack growth model based on the strain energy density (SED) method. The aim of this work is to predict total fatigue life of structural components using the same low cyclic material properties for crack initiation stage and for crack growth stage. This approach achieves efficient computation method for total fatigue life estimation. By using of the same material properties for crack initiation phase and for crack growth phase significantly simplifies total fatigue life estimation and does not require any additional material costs needed for experimental research except for those needed for fatigue crack initial prediction.

Acknowledgment. The authors would like to thank the Ministry of Education, Science and Technological Development of the Republic of Serbia for financial support.

References

1. Bannantine, J.A., Comer, J., Handrock, J.: Fundamentals of Material Fatigue Analysis. Prentice-Hall, Englewood Clifs (1990)
2. Boljanovic, S., Maksimovic, S., Carpinteri, A.: Residual strength evaluation under mixed mode loading. In: Proceedings Paper - ICMFF12 – 12 th International Conference on Multiaxial Fatigue and Fracture, vol. 300, br., str. (2019)
3. Dowling, N.E.: Mechanical Behavior of Materials, 2nd edn. Prentice Hall, Jersey (1999)
4. Smith, K.N., Watson, P., Topper, T.H.: A stress–strain functions for the fatigue of metals. J. Mater. **5**, 767–778 (1970)
5. Posavljak, S., Maksimović, S.: Increasing of fatigue resistance of aero engine disks. WSEAS Trans. Appl. Theor. Mech. **1**(2), 133–140 (2006)
6. Chang, J.B., Engle, R.M., Stolpestad, J.: Fatigue Crack Growth Behavior and Life Predictions for 2219-T851 Aluminum Subjected to Variable-Amplitude Loadings, ASTM STP 743 (1981)
7. Hong-Z, H., et al.: Fatigue life estimation of an aircraft engine under different load spectrums. Int. J. Turbo Jet-Engines **29**, 259–267 (2012)
8. Yin, Z.Y.: Manual for Design of Turbo Engine: Fascicule 18, Strength Analysis of Turbine Disk and Shaft. Aviation Industry Press, Beijing (2007)
9. Li, S.M.: The Fatigue and Reliability Design in Mechanics. Science Press, Beijing (2006)
10. Vasovic, I., Maksimovic, S., Maksimovic, K., Stupar, S., Bakic, G., Maksimovic, M.: Determination of stress intensity factors in low pressure turbine rotor discs. Math. Probl. Eng. **2014**, 9 pages. Article ID 304638. https://doi.org/10.1155/2014/304638
11. Posavljak, S.: Fatigue life investigation of aero engine rotating disks, Doctoral dissertation, Belgrade University, Faculty of Mechanical Engineering (2008). (in Serbian)
12. Sonsino, C.M.: Zur Bewertung des Schwingfestigkeitsverhaltens von Bauteilen mit Hilife örtlicher Beanspruchungen. Konstruktion **45**, 25–33 (1993)

13. Schijve, J.: Fatigue of Structures and Materials, 2nd edn. Springer, Berlin (2009)
14. Sehitoglu, H., Gall, K., García, A.M.: Recent advances in fatigue crack growth modeling. Int. J. Fract. **80**, 165–192 (1996)
15. Maksimovic, S., Vasovic, I., Maksimovic, M., Đuric, M.: Residual life estimation of damaged structural components using low-cycle fatigue properties. In: Third Serbian Congress Theoretical and Applied Mechanics, Vlasina Lake, pp. 605–617 (2011). ISBN 978-86-909973-3-6
16. Vasovic, I., Maksimovic, S., Maksimovic, K., Stupar, S., Maksimović, M., Bakic, G.: Fracture mechanics analysis of damaged turbine discs using finite element method. Therm. Sci. **18**(Suppl. 1), S107–S112 (2014)
17. Maksimovic, S., Kozic, M., Stetic-Kozic, S., Maksimovic, K., Vasovic, I., Maksimovic, M.: Determination of load distributions on main helicopter rotor blades and strength analysis of its structural components. J. Aerosp. Eng. **27,** br. 6 (2014)
18. Maksimovic, S., Posavljak, S., Maksimovic, K., Nikolic, V., Djurkovic, V.: Total fatigue life estimation of notched structural components using low cycle fatigue properties. J. Strain **47**, 341–349 (2010)
19. Balac, M., Grbovic, A., Petrovic, A., Popovic, V.: Fem analysis of pressure vessel with an investigation of crack growth on cylindrical surface. Eksploatacja i Niezawodnosc – Maint. Reliab. **20**(3), 378–386 2018. https://doi.org/10.17531/ein.2018.3.5
20. Bajić, D., Momčilović, N., Maneski, T., Balać, M., Kozak, D., Ćulafić, S.: Numerical and experimental determination of stress concentration factor for a pipe branches. Tech. Gaz. **24**(3), 687–692 (2017). ISSN 1848-6339. https://doi.org/10.17559/tv-20151126222916
21. Petrovic, A., Balac, M., Jovovic, A., Dedic, A.: Oblique nozzle loaded by the torque moment–stress state in the cylindrical shells on the pressure vessel. Proc. Inst. Part C J. Mech. Eng. Sci. **226**(3), 567–575 (2011). https://doi.org/10.1177/0954406211415907
22. Izumi, Y., Fine, M.E., Mura, T.: Energy consideration in fatigue crack propagation. Int. J. Fract. **17**(1), 15–25 (1981)
23. Vidanovic, N., Rasuo, B., Kastratovic, G., Grbovic, A., Puharic, M., Maksimovic, K.: Multidisciplinary shape optimization of missile fin configuration subject to aerodynamic heating. J. Spacecraft Rockets (2019). https://doi.org/10.2514/1.A34575
24. Geier, W.: Strength Behaviour of Fatigue Cracked Lugs. Royal Aircraft Establishment, LT 20057 (1980)
25. Pantelakis, Sp.G., Kermanidis, Th.B., Pavlou, D.G.: Fatigue crack growth retardation assesment of 2024-T3 and 6061-T6 aluminum specimens. Theor. Appl. Fract. Mech. **22**, 43–47 (1995)
26. Ellyin, F.: Fatigue damage, crack growth and life prediction. Fract. Mech. **48**(1), 9–15 (1997)

Study of Innovative Subsonic Ramjet

Nikola Davidović[1](✉), Predrag Miloš[2], Nenad Kolarević[3], Toni Ivanov[1], and Branislav Jojić[1]

[1] Faculty of Mechanical Engineering, Department of Aerospace Engineering, University of Belgrade, 11000 Belgrade, Serbia
nikola.davidovic@edepro.com
[2] EDePro Company, Kralja Milutina 33, 11000 Belgrade, Serbia
[3] Faculty of Mechanical Engineering, Department of Design in Mechanical Engineering, University of Belgrade, 11000 Belgrade, Serbia

Abstract. Subsonic ramjets, due to its simplicity, nowadays could be very attractive for expendable unmanned applications. However, due to very low efficiency at low Mach numbers, practically there isn't any important application at subsonic speed. In 50's subsonic ramjet was used for tip jet applications. Later in 70's there was a study of Fan Augmented Ramjet (FARJ) where a fan was in-tended to be powered by a piston engine, both in subsonic and supersonic regimes. Almost at the same time it was analyzed ejector augmented subsonic ramjet. Both of these studies did not find real application of such engines. Finally, there was an attempt for the use of an Electric Ducted Fan, used in RC models, with a combustor but exit fan pressure was almost at atmospheric pressure which results in very low efficiency.

This paper presents a study of propulsion system which consists of a pitot intake, electrically powered compressor/fan, combustor and nozzle (Electrically Fan Augmented Ram Jet, EFARJ). The electric motor connection to the compressor/fan is direct, without gearbox, which implies usage of a high-speed brushless motor. Such propulsion system is compared to pure ramjet and finally it is compared to performances of small expendable turbojet. Comparison is made regarding costs, weight and thrust/consumption. The study shows that the proposed propulsion system could be competitive for expendable, low speed and short duration applications.

Keywords: Ramjet · Electric propulsion · Subsonic · Turbojet

1 Introduction

Subsonic ramjet engines are not used in practice because of their low efficiencies. In the 50's there was an application of tip-jet powered helicopter with ramjet engine positioned at tip of the blade, where velocity was Mach number 0.7. Nowadays logic of unmanned vehicles for expendable use considers again subsonic ramjet application. Their logic is that, no matter specific fuel consumption is double than turbojet, for short duration of flight and small absolute consumption if we consume 5 or 10 L of fuel it doesn't change

N. Mitrovic et al. (Eds.): CNNTech 2020, LNNS 153, pp. 413–431, 2021.
https://doi.org/10.1007/978-3-030-58362-0_24

performances of vehicle but saves price due to ramjet simplicity. Again, it is through for very short flight duration and requires at least Mach number 0.7.

Few concepts were analyzed in subsonic ramjets. In order to overcome difficulties with operation at very low Mach numbers in reference [1] was analyzed ejector subsonic ramjet. Ejector is the component which increases pressure at low speeds. The possible source for that ejector was not deeply considered in that research.

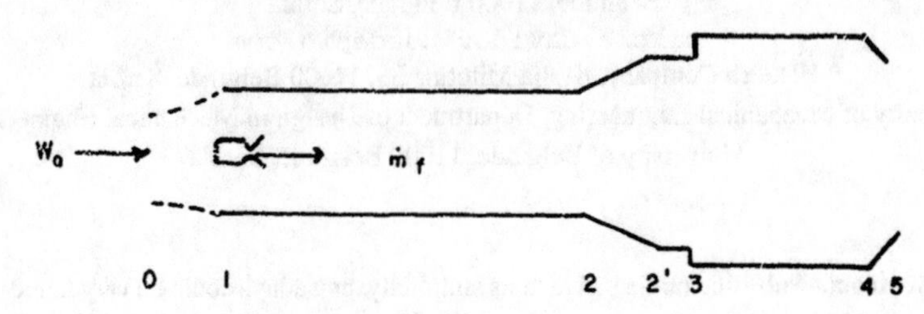

Fig. 1. Concept of ejector subsonic ramjet

Further, there was study of design combustion chamber for subsonic ramjet which would operate at very high altitudes, reference [2].

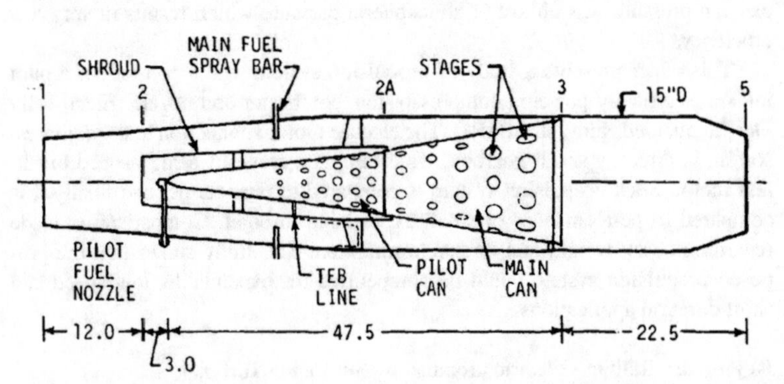

Fig. 2. Concept of subsonic ramjet for high altitudes

Most famous practical application of subsonic ramjet is tip-jet powered helicopter Hiller Hoe in 50's. Ramjet engines were positioned at the tip of the blades, rotating at Mach number around 0.7 as compromise between performance and stresses, and in such manner producing torque. Similar with performances but different type is application mentioned in references [3, 4] and [5] (Figs. 3 and 4)

Fig. 3. Hiller Hoe tip-jet helicopter

Fig. 4. Subsonic ramjet at tip of the helicopter blade

Here is proposed old idea but with new available components. In reference [6] it was proposed fan augmented ramjet concept, for subsonic and supersonic speeds. Fan was powered by piston engine and appropriate gear-box. In this work compressor/fan is directly connected to brushless electric motor saving weight and simplifying the system. There were investigations with similar systems but with "fan" in direct meaning, i.e. pressure after fan is very low.

Of course, basic idea is to increase pressure at low Mach numbers, say $M < 0.7$, and to make hybrid engine between ramjet and turbojet.

2 Model Description

Propulsion scheme of the model is presented at Fig. 1. Engine consists of Intake (inlet station 0, exit station 1), compressor (inlet station 1, exit station 2), combustion chamber (inlet station 2, exit station 3) and nozzle (inlet station 3, exit station 4). Compressor is powered directly by electric motor which is connected to battery. Fuel is injected into combustion chamber. Comparing to turbojet, there is no turbine which powers compressor, in this case electric motor serves as turbine. Because electric motor is in cold environment, exit temperature is not limited by turbine, as in case of turbojet. On other side, comparing to ramjet, this engine could work at zero speed because of compressor and also minimum Mach number of pure ramjet is shifted to lower values, say M = 0.4 (Fig. 5).

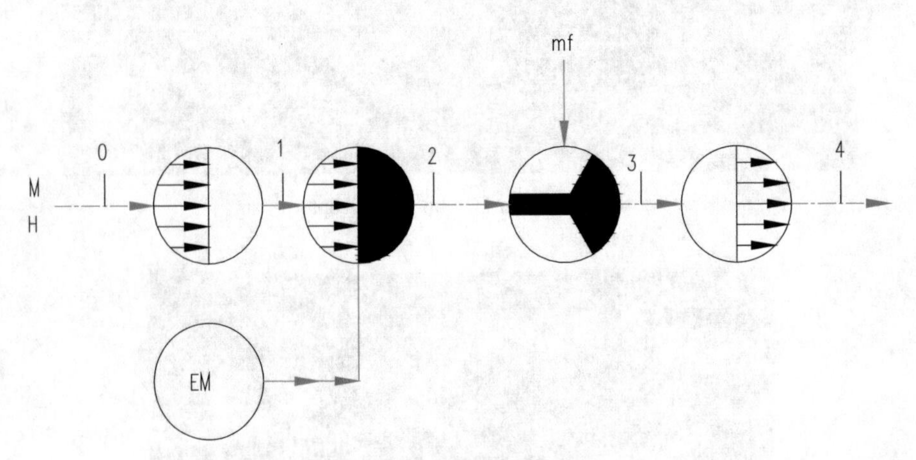

Fig. 5. Propulsion scheme

3 Mathematical Model

In order to perform cycle and performance analysis appropriate mathematical model was made. Also, some assumptions were made:

- Intake is pitot type, pressure recovery is constant and equal 0.98.
- Compressor characteristic is simplified to have constant efficiency of 0.8.
- Combustor is V-gutter type with constant pressure recovery both, due to hydraulic and heat addition losses, equal 0.95. Fuel type is kerosene Jet-A1.
- Nozzle is convergent type with coefficient of velocity equal 0.99 and discharge coefficient equal 0.99.
- Electric motor is brushless type, assumed to have the same rpm as compressor, for this certain analyses it is chosen Lenhner 3050, battery voltage 60 V, max rpm = 50000 and max power 30 kW (15 kW continuous)
- Compressor design, which is later proposed, is chosen to satisfy following conditions:

Thrust = 25daN, Mach number = 0.4, Altitude = 0, fuel flow rate <18 g/s, max diameter 155 mm, weight 5 kg. These data presents one existing system with turbojet engine, so idea is to propose replacement in same space and to have same fuel tank volume (Fig. 6).

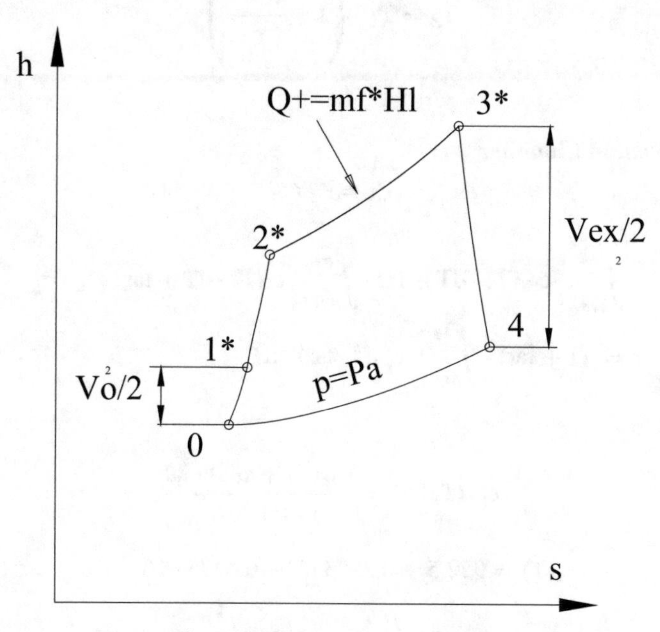

Fig. 6. Enthalpy entropy diagram

Engine cycle is the same as in the case of turbojet, the only difference is that electric motor powers compressor and there is no turbine. Enthalpy-entropy diagram is shown at Fig. 2. According to engine layout here are pressures and temperatures in characteristic stations. Equations for mathematical model refer to references [7] and [8].

3.1 Free Stream Conditions

$$P_o = p_o \cdot \left[1 + \frac{\kappa - 1}{2} \cdot M_o^2\right]^{\frac{\kappa}{\kappa - 1}} \tag{1}$$

$$T_o = t_o \cdot \left[1 + \frac{\kappa - 1}{2} \cdot M_o^2\right] \tag{2}$$

3.2 Intake

$$P_1 = P_o \cdot \sigma_i \tag{3}$$

$$T_1 = T_o \tag{4}$$

3.3 Compressor

$$\mathbf{P_2 = P_1 \cdot \pi_c} \tag{5}$$

$$T_2 = T_1 \cdot \left(1 + \frac{\pi_c^{\frac{\chi-1}{\kappa}}}{\eta_c}\right) \tag{6}$$

3.4 Combustion Chamber

$$P_3 = P_2 \cdot \sigma_p \tag{7}$$

$$\int_{298.15}^{T_2} c_{pa}(T) \cdot dT + far \cdot \int_{298.15}^{T_f} c_f(T) \cdot dT + far \cdot H_L \cdot \sigma_g$$
$$= (1 + far) \cdot \int_{298.15}^{T_3} c_{pp}(T, far) \cdot dT \tag{8}$$

While

$$c_{pp}(T, far) = \frac{c_{pa}(T) + a(T) \cdot far}{1 + far}$$

$$c_{pa}(T) = 920.5 + 0.2983 \cdot T - 6.4435 \cdot 10^{-5} \cdot T^2$$

$$a(T) = 1421.8 + 2.178 \cdot T - 4.43 \cdot 10^{-4} \cdot T^2$$

$$R_g = R \cdot \frac{1 + 1.0862 \cdot far}{1 + far}$$
$$\kappa_g = \frac{c_{pp}}{c_{pp} - R} \tag{9}$$

3.5 Nozzle

$$P_4 = P_3 \cdot \sigma_n \tag{10}$$

$$T_4 = T_3 \tag{11}$$

$$V_4 = V_{ex} = \varphi_{ex} \cdot \sqrt{2 \cdot c_{pp} \cdot T_4 \cdot \left(1 - \frac{1}{P_3/p_a^{\frac{\kappa_g - 1}{\kappa_g}}}\right)} \tag{12}$$

3.6 Electric Power Required

$$N_{el} = N_c = m_a \cdot c_{pa} \cdot T_1 \cdot \left(\frac{\pi_c^{\frac{\chi-1}{\kappa}} - 1}{\eta_c} \right) \tag{13}$$

$$N_{el} = I \cdot U \tag{14}$$

$$Cap = I \cdot \tau = \frac{m_a \cdot c_{pa} \cdot T_1 \cdot \left(\frac{\pi_c^{\frac{\chi-1}{\kappa}} - 1}{\eta_c} \right)}{U} \tag{15}$$

3.7 Performances

- Thrust

$$F = m_a \cdot [(1 + far) \cdot V_{ex} - V_o] + A_4 \cdot (p_4 - p_a) \tag{16}$$

- Specific thrust

$$F_{sp} = [(1 + far) \cdot V_{ex} - V_o] + \frac{A_4 \cdot (p_4 - p_a)}{m_a} \tag{17}$$

- Specific fuel consumption

$$C_{sp} = \frac{m_f}{F} = \frac{far}{F_{sp}} \tag{18}$$

- Specific battery capacity

$$Cap_{sp} = \frac{I \cdot \tau}{m_a} = \frac{c_{pa} \cdot T_1 \cdot \left(\frac{\pi_c^{\frac{\chi-1}{\kappa}} - 1}{\eta_c} \right)}{U} \tag{19}$$

4 Cycle Analysis

Specific thrust, specific fuel consumption and specific capacity of battery is analyzed as function of compressor pressure ratio, flight Mach number and fuel to air ratio. These results are shown at the Figs below (Figs. 7, 8, 9, 10, 11, 12, 13, 14 and 15).

In range of compressor ratio from 1.1 to 2, as expected, specific thrust and specific capacity of battery are increasing, while specific consumption is decreasing. Their optimum values are not in analyzed range. Again, that range was chosen to minimize required battery capacity and only to help ramjet in low Mach number region.

In range of Mach numbers from 0 to 0.8, specific thrust is almost constant up to Mach number 0.4 and later has the same trend as pure ramjet. It is obvious advantage in low Mach number region. The same story is for specific fuel consumption, almost constant

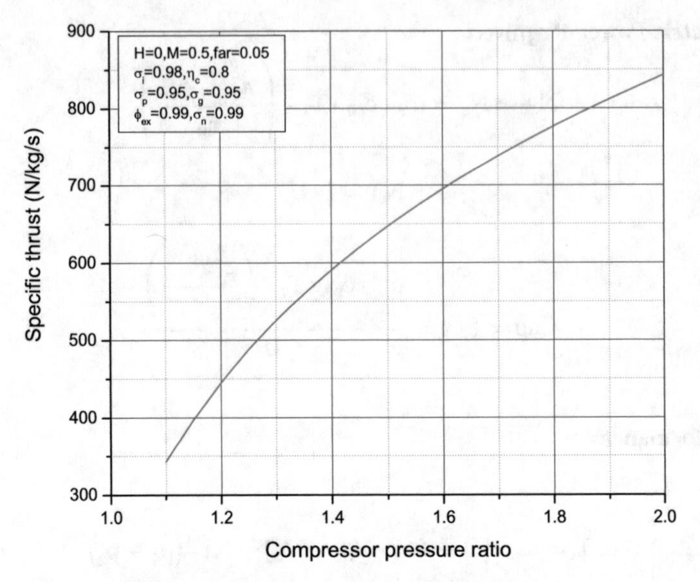

Fig. 7. Specific thrust vs. compressor pressure ratio

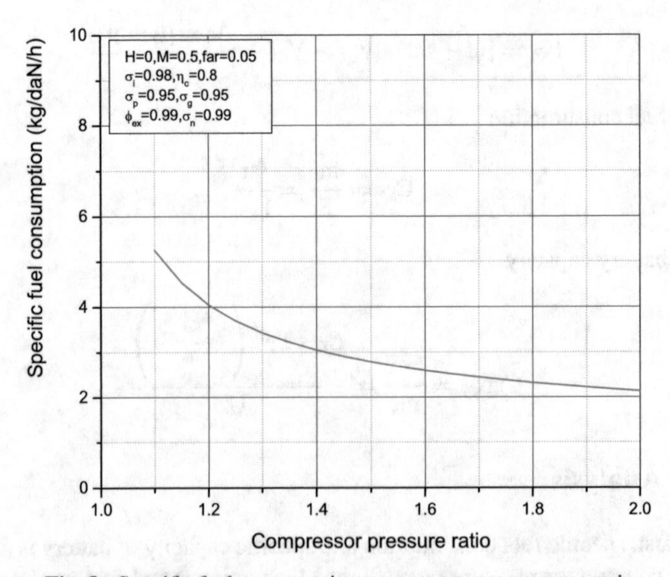

Fig. 8. Specific fuel consumption vs. compressor pressure ratio

up to Mach number 0.4 and later decreasing as pure ramjet. The values of specific thrust and specific fuel consumption are dictated with chosen fuel to air ratio, which is again chosen to compare thrust for existing turbojet systems, but their trend does not depend of fuel to air ratio. Specific capacity of battery practically don't depend on Mach number in such system.

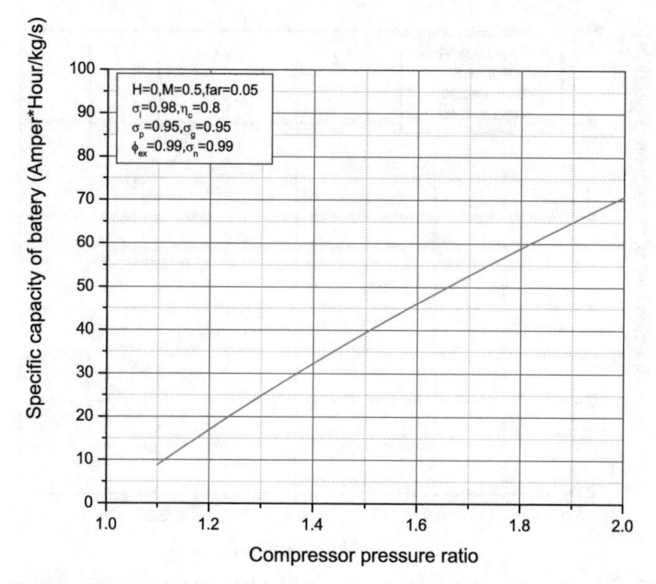

Fig. 9. Specific battery capacity vs. compressor pressure ratio

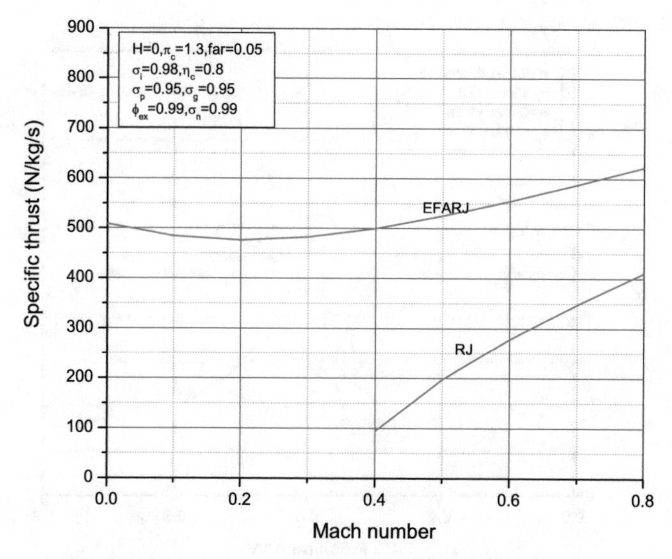

Fig. 10. Specific thrust of EFARJ and RJ vs. Mach number

Fuel to air ratio is increasing both thrust and specific fuel consumption, but it is shown that if lower values of specific fuel consumption are dominant that they are achievable in ranges of 0.015 to 0.025, typical for turbojets. In case or higher thrust required, fuel to air ratio could be throttled up to 0.06, where specific thrust increase is dominant comparing to increase in specific fuel consumption.

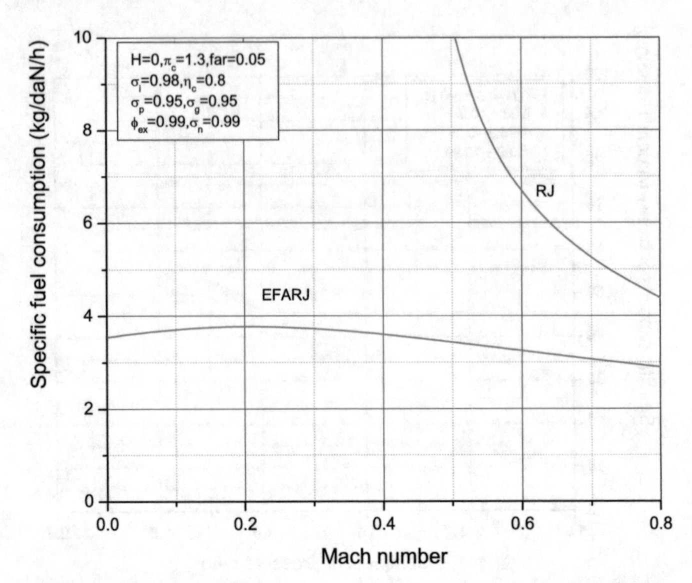

Fig. 11. Specific fuel consumption of EFARJ and RJ vs. Mach number

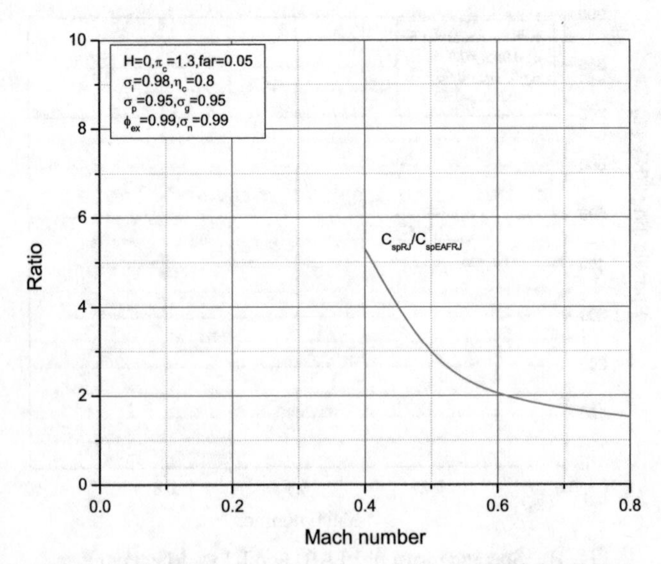

Fig. 12. Ratio of Specific fuel consumptions of RJ and EFARJ vs. Mach number

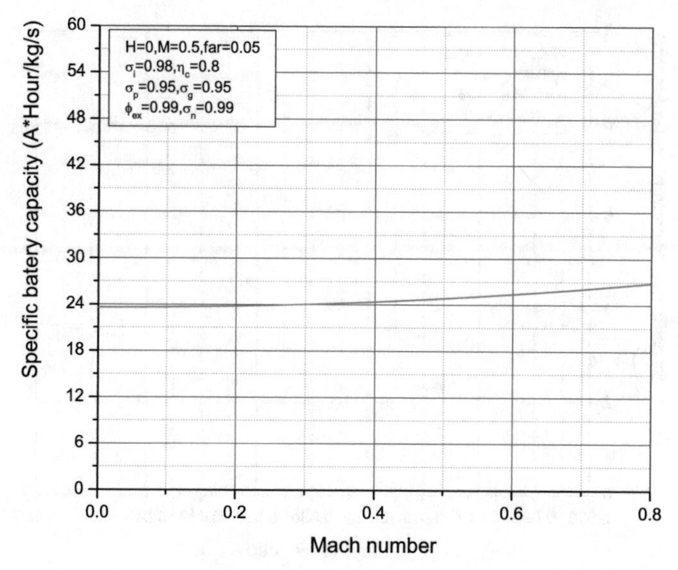

Fig. 13. Specific battery capacity of EFARJ vs. Mach number

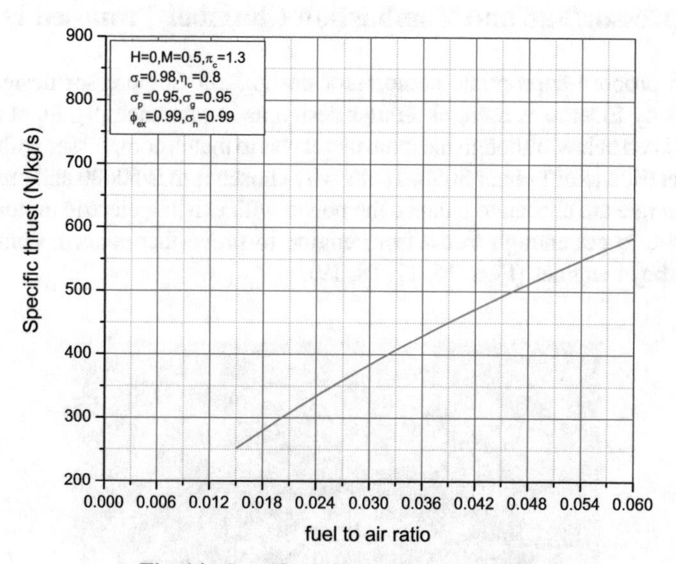

Fig. 14. Specific thrust vs. fuel to air ratio

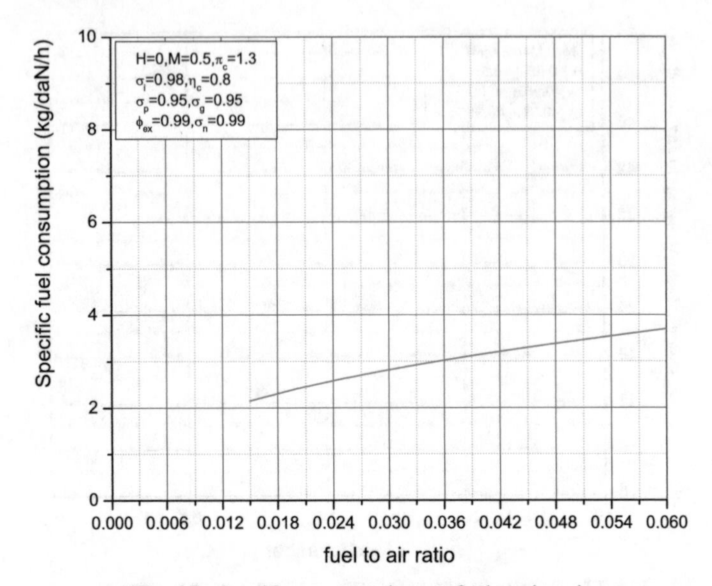

Fig. 15. Specific consumption vs. fuel to air ratio

5 Compressor/Fan and Combustion Chamber Proposed Design

In order to propose appropriate compressor design, 2D Compressor design program from company EDePro is used, as similar design as in reference [9]. Input and output values are listed below, although the main target was to match compressor with proposed, existing electric motor Lehner 3050. That's why chosen rpm is 50000 and pressure ratio and air flow rate are chosen to balance the power with existing electric motor, and from another side, to get enough thrust from engine to prove that concept comparable to existing turbojet engines (Figs. 16, 17, 18, 19).

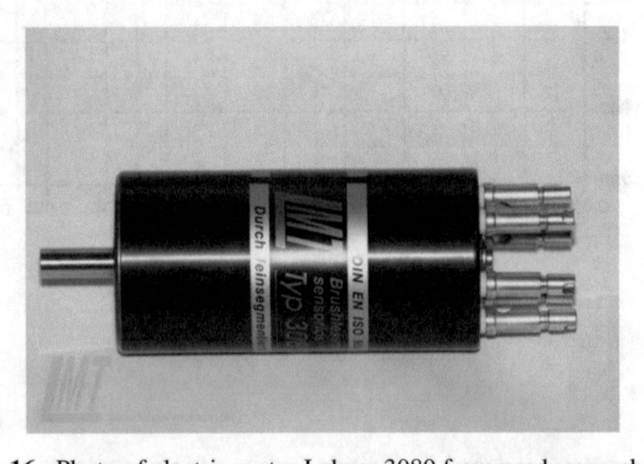

Fig. 16. Photo of electric motor Lehner 3080 from producer web site

Fig. 17. Similar compressor rotor designed and produced for small turbojet from company EDePro

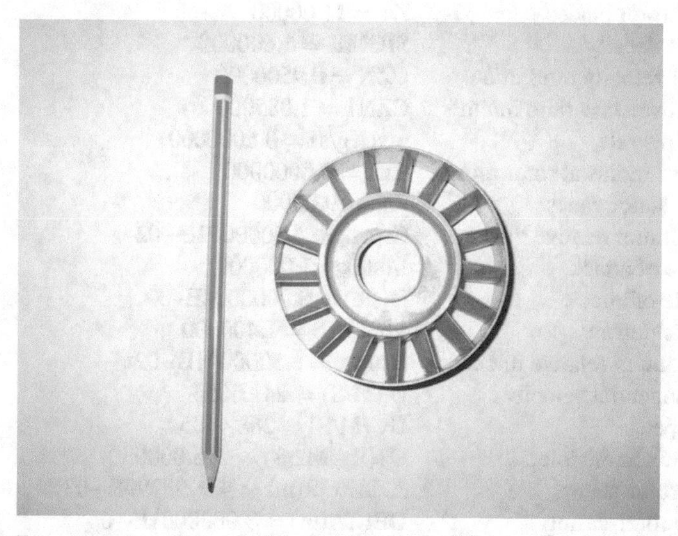

Fig. 18. Similar compressor stator designed and produced for small turbojet from company EDePro

Fig. 19. Proposed design for combustion chamber, similar design from company EDePro

Compressor Calculation at Average Diameter (2D Calculation)
Input Data For Calculation

Rotor inlet total pressure	Po(Pa) = 101300.0
Rotor inlet total temperature	To(K) = 288.0000
Flow inlet angle	AL1(rad) = 0.0000000E+00
Rotor geometrical turning angle	TETA(st.) = 25.00000
Stator geometrical turning angle	TETAS(st.) = 40.00000
Number of rotor blades	Zk = 11.00000
Rotor blade density	SIGMR = 1.600000
Rotor axial velocity ratio in/out	CZN = 0.9500000
Stator axial velocity ratio in/out	CZN1 = 1.085000
Air mass flow rate	mv(Kg/s) = 0.6000000
Inlet relative radius at rotor root	Rnk = 0.8000000
Number of stator vanes	Zs = 19.00000
Rotor maximum relative thickness	Crmax = 4.5000002E−02
Rotor angle of attack	Ir(ste) = 1.000000
Stator angle of attack	Is(ste) = 0.0000000E+00
Stator blade density	SIGMAS = 1.400000
Stator maximum relative thickness	Csmax = 8.5000001E−02
Average tangential velocity	U (M/S) = 241.6054
Rotor tip speed	Uv (M/S) = 268.4505
Air mass flux at the inlet	G1(Kg/M2/S) = 205.0000
Axial gap rotor-stator	AZAZOR(m) = 4.9999999E−03
Radial gap rotor-casing	DELTA(m) = 3.0000001E−04

Output Data
Rotor inlet

Absolute velocity	C1(m/s) = 199.1109
Axial velocity	Cz1(m/s) = 199.1109
Tangential velocity	Cu1(m/s) = 0.0000000E+00
Relative velocity	W1(m/s) = 313.0788

Apsolute Mach number	Mc1 = 0.6046843
Relative Mach number	Mw1 = 0.9555498
Static temperature	T1(K) = 268.3742
Static pressure	P1(Pa) = 79127.64
Density	r1(Kg/m^3) = 1.027320

Rotor geometry

Inlet flow angle	BET1(rad) = 0.8815224
Exit flow angle	BET2(rad) = 0.5368794
Mean line inlet angle	BET1L(rad) = 0.8640704
Mean line exit angle	BET2L(rad) = 0.4277702
Rotor blade width	SR(m) = 3.3658441E−02
Rotor blade chord	Br(m) = 4.2149648E−02
Rotor pitch	Tr(m) = 2.6343530E−02

Rotor outlet

Absolute velocity	C2(m/s) = 228.9676
Axial velocity	CZ2(m/s) = 189.1554
Tangential velocity	Cu2(m/s) = 129.0210
Relative velocity	W2(m/s) = 220.1250
Absolute Mach number	Mc2 = 0.6650378
Relative Mach number	Mw2 = 0.6393543
Static pressure	P2(Pa) = 105036.6
Total pressure	P2T(Pa) = 141311.5
Static temperature	T2(K) = 293.4022
Total temperature	T2T(K) = 319.3822
Density	Ro2 (Kg/m^3) = 1.247371
Air mass flux	Gr (Kg/m^2/s) = 235.9469
Static rotor pressure ratio	Pirtst = 1.327433

Rotor performances

Coefficient of losses	KSIR = 6.3837931E−02
Coefficient of pressure recovery	sigmr = 0.9716538
Degree of reactivity	ROK = 0.7949933
Diffusion factor	DFR = 0.4256847
Equivalent diffusion factor	DEQR = 1.735066
Polytropic efficiency	ETAR = 0.9166446
Politropic coefficient	Nr = 1.450760
Rotor pressure ratio	PIRT = 1.394980

Stator outlet

Absolute velocity	C3(m/s) = 205.2755
Axial velocity	CZ3(m/s) = 205.2336
Tangential velocity	Cu3(m/s) = 4.144831
Absolute Mach number	Mc3 = 0.5910886
Coefficient of stator losses	KSIS = 4.9786385E−02
Coefficient of pressure recovery	sigms = 0.9872198
Stator diffusion factor	DFS = 0.2982552
Stator equivalent diffusion factor	DEQS = 1.608179
Static temperature	T3(K) = 298.5021
Total temperature	T3T(K) = 319.3822
Static pressure	P3(Pa) = 110134.3
Total pressure	P3T(Pa) = 139505.5
Density	Ro3 (Kg/m3) = 1.285563
Air mass flux	G3(Kg/m2/s) = 263.8407

Stator geometry

Inlet flow angle	AL2 (rad) = 0.5986043
Mean line inlet angle	AL2L (rad) = 0.5986043
Mean line exit angle	AL3l (rad) = −9.9475980E−02
Flow exit angle	AL3 (rad) = 2.0192930E−02
Flow angle change	DAL3(rad) = 2.0192930E−02
Stotor vane width	Sst(m) = 2.0690640E−02
Stator vane chord	Bst(m) = 2.1352125E−02
Stator pitch	Ts(m) = 1.5251518E−02

Stage geometry

Tip radius	Rv(m) = 5.1270142E−02
Rotor root inlet relative radius	R1nk = 0.8000000
Rotor root exit relative radius	R2nk = 0.8289860
Stator root exit relative radius	R3nk = 0.8486965
Rotor blade height at inlet	ahl1(m) = 1.0254028E−02
Rotor blade height at exit	ahl2(m) = 8.7679131E−03
Stator vane height at exit	ahl3(m) = 7.7573503E−03

Stage performances

Stage pressure ratio	PIKT = 1.377152
Isentropic efficiency	ETAK = 0.8795804
Coefficient of secondary losses	ETAS = 0.8277239

Stage efficiency	eta = 0.7280497
Stage polytropic coefficient	Nk = 1.477529
Rotor absorbed power	Wk(W) = 22862.04

Combustion chamber designed is related to 6 V gutter-fingers positioned radially in 45° according to previous research in company EDePro.

6 Design Performances and Conclusion

Finally, in order to satisfy initial requirements following parameters were chosen:

- Air flow rate: $m_a = 0.6$ kg/s
- rpm = 50000
- Intake: $\sigma_i = 0.98$
- Compressor/Fan: $\pi_c = 1.38$, $\eta_c = 0.73$ (although it is at penalty of batteries not fuel!)
- Combustor: $\sigma_p = 0.95$, $\sigma_g = 0.95$, far = 0.03 (to benefit on efficiency and fuel consumption)
- Nozzle: convergent, $\sigma_n = 0.99$, $\varphi = 0.99$
- Electric motor (brushless): Lehner 3080, diameter 60 mm, weight 1 kg.

According to proposed ^{model} here are design performances:

- Thrust = 25.1daN at Mach number = 0.4
- Fuel flow rate = 18 g/s
- Specific fuel consumption $C_{sp} = 2.58$ kg/daN/h
- Exhaust temperature $T_3 = 1350$ K
- Required battery capacity for 3 min: 19.9 Ah (for 60 V). Expected to have weight around 1 kg and cost about 600 euros.
- Compressor diameter = 102.6 mm
- Combustor diameter (for inlet Mach number equal 0.1) = 131 mm

Final proposed design is shown at figure below. Layout is completely in real dimensions (Fig. 20).

- The numbers are showing that system can be packed in 150 mm, weight of dry system less than 4 kg with specific fuel consumption approximately higher for 70%, but for short mission that means that instead of 3 L of turbojet this concept will consume 5 L. Practically, it shows that proposed concept is comparable with turbojet at short durations. The system relies on existing brushless motor. If the electric motor is designed intentionally for this system the dimensions and weight could be further optimized. And finally, as package of electrical energy is becoming better and better, electric power in air breathing propulsion will be more and more present.

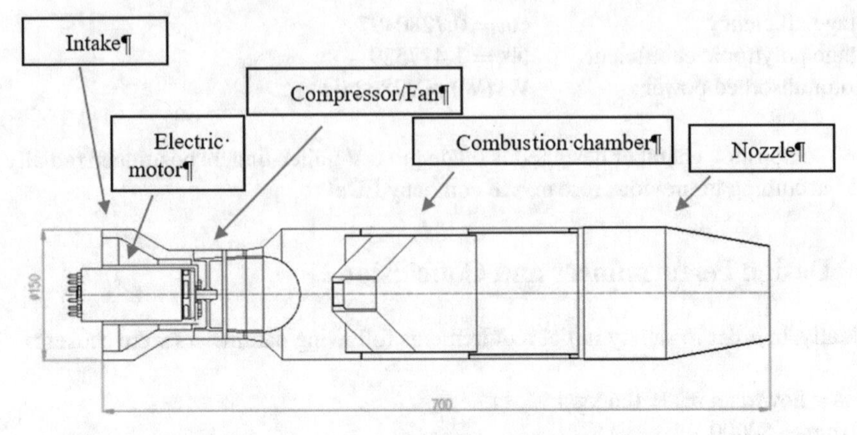

Fig. 20. Proposed design of electric subsonic ramjet according to calculation

Labels

A_i	Cross section i (m^2)
C_{sp}	Specific fuel consumption (kg/Ns)
C_{ap}	Battery capacity (As)
C_{apsp}	Specific battery capacity (As/kg/s)
c_{pa}	Air specific heat (J/kgK)
c_{pp}	Combustion products specific heat (J/kgK)
c_f	Liquid fuel specific heat (J/kgK)
F	Thrust (N)
F_{sp}	Specific thrust (N/kg/s)
h	Specific enthalpy (J/kgK)
H	Altitude (m)
H_L	Lower heating value (J/kg)
I	Electric current (A)
M_o	Free stream Mach number
m_a	Air flow rate (kg/s)
m_f	Fuel flow rate (kg/s)
N_c	Compressor absorbed power (W)
N_{el}	Power of electrical motor (W)
P_i	Total pressure of i-section (Pa)
P_a	Atmospheric pressure (Pa)
p_i	Static pressure of i-section (Pa)
far	Fuel to air ratio
R	Air gas constant (J/kgK)
R_g	Combustion products gas constant (J/kgK)
s	Entropy (J/kgK)
T_i	Total temperature of i-section (K)
T_f	Temperature of fuel (K)
t_i	Static temperature of i-section (K)

U Voltage (V)
V_o Free stream velocity (m/s)
V_{ex} Nozzle exit velocity (m/s)

Greek Letters

κ ratio of air specific heats
κ_g ratio of combustion products specific heats
η_c compressor efficiency
π_c compressor pressure ratio
τ time (s)
σ_i pressure recovery of intake
σ_g combustion chamber efficiency
σ_p combustion chamber pressure drop
σ_n pressure recovery of the nozzle
φ_{ex} exit velocity coefficient

References

1. Supp, E.W., Watson, A.K., Miller, H.J.: Subsonic performance potential of ramjets and ejector ramjets, Technical report AFAPL-TR-72-7, Wright Patersonn Air Force Base Ohio (1972)
2. Hornbeck, E.C.: Can type combustor design for a low cost subsonic ramjet, Technical report AFAPL-TR-75-71, Wright Patersonn Air Force Base Ohio (1975)
3. Davidović, N., Elmahmodi, A., Petković, S.: Tip-jet propulsion based on compressed air due to rotor blade rotation. In: Proceedings of Aerotech III, Kuala Lumpur, Malaysia (2009)
4. Kosanović, N., Davidović, N., Miloš, P., Jojić, B., Miloš, M.: Monitoring of engine parameters in tip jet helicopter tests. In: Symposium on Advances in Experimental Mechanics, Primosten, Croatia (2013)
5. Kolarević, N., Davidović, N., Miloš, P., Jojić, B., Miloš, M.: Experimental determination of light helicopter rotor lift characteristics with tip-jet propulsion system. In: Symposium on Advances in Experimental Mechanics, Primosten, Croatia, (2013)
6. Beans, E.W.: A Design study of fan augmented ramjet. In: ASME Proceedings, 75-GT-97 (1975)
7. Шляхтенко, С.М.: Теория и Расчет Воздушно-Реактивных Двигателей. Машиностроение, Москва (1987)
8. Davidović, N.: Metodologija proračuna karakteristikarealnog nabojno-mlaznog motora fiksne geometrije sa spoljašnjim sabijanjem, introductory lecture at the Faculty of Mechanical Engineering, University of Belgrade (2020)
9. Davidović, N., Miloš, P.: Profilisanje lopatice rotora transoničnog aksijalnog kompresora turbomlaznog motora– XXXIII Jupiter conference. Republic of Serbia, Zlatibor (2007)

A Geometric Theory for Robotic Manipulators Represented as Singular Control Systems

Ivan M. Buzurovic(✉)

Harvard Medical School, Harvard University, Boston, MA 02115, USA
ibuzurovic@bwh.harvard.edu

Abstract. In this article, dynamics of the constrained robotic manipulators have been investigated with the use of a geometric approach. The necessary and sufficient conditions for the existence and uniqueness of the solutions were derived for such systems. The proposed approach can be utilized for the dynamic investigations of the constrained robotic systems that are in contact with rigid frictionless surfaces in their working regime. The robotic manipulators were modeled as singular (semi-state, descriptor) systems so that dynamics originated from the environment interactions could be incorporated. The controllability condition of the system state implied that it was possible to steer the system from any initial to any final position within a predefined time window. The sufficient controllability conditions of such systems were based on the generalized Lyapunov equation that was derived while the geometric approach was used. The sufficient algebraic conditions were derived for controllability testing. The presented methodology could be used as a foundation for further investigations of the nonlinear time-variable and time-discrete singular systems.

Keywords: Robotic manipulators · Singular systems · Geometric approach

1 Introduction

The geometric approach is a mathematical technique developed to improve analysis and synthesis of linear multivariable systems. Many authors consider the notion of geometry in system theory to be mutual characteristics of the matrix pencils (*A*, *B*) or (*A*, *C*) for linear systems or (*A*, *E*) for singular systems, where A, B, C and D are regular system matrices in the state-space system representation, and *E* is the singular matrix of the system (1).

$$
\begin{aligned}
E\dot{x} &= Ax + Bu \\
x_i &= Cx + Du
\end{aligned}
\tag{1}
$$

Other authors view the geometric aspect as a study of characteristic subspaces of systems. The formal definition of the geometric approach (or aspect) to linear singular systems may be stated as: an approach to the study of singular systems, whose purpose is to

N. Mitrovic et al. (Eds.): CNNTech 2020, LNNS 153, pp. 432–446, 2021.
https://doi.org/10.1007/978-3-030-58362-0_25

determine and investigate characteristic subspaces that play a crucial role in the analysis of the matrix pencils (A, E). The geometric approach was first discussed in articles [1] and [2]. They discovered that the dynamic behavior of the time invariant linear control systems could be investigated by studying the characteristics of the system's invariant subspace of the matrices. As a result, system behavior could be predicted and the solution of many control problems could be tested by investigating these characteristics. The essence of this approach was the calculation and application of subspaces on computers using algorithms developed for that purpose. It was later shown in the literature that the geometric approach can be used to solve a variety of problems, such as by finding a control law for systems with feedback, observability problems, disturbance localization, design of the observers, control and tracking, robust control, etc. It can be concluded that the geometric approach is a mathematical concept developed in order to give greater insight into the most important characteristics of linear system dynamics. It is mostly represented in the state space domain and used to connect characteristics of single and multiple transfer systems. In the literature, the geometric approach did not always rely on algorithms and computer applications, but it should be kept in mind that the basic idea was to use computers when this approach first appeared.

In this article, we proposed a mathematical model of the robotic system, expressed as a singular system of differential equations. For some specific class of robotic systems, the mathematical model has a singular character due to a contact force that acts upon the system. In many applications, it is enough to consider the contact force as a disturbance to the system. Sometimes the unexpected range of the contact forces, due to their stochastic character, can significantly change or damage the contact surface, causing unacceptable alterations for some situations. Furthermore, a contact force which has an unknown value and characteristics could produce a compromised outcome. Therefore, it is necessary not only to measure the force, but to control it by obtaining adequate control algorithms which can keep the force within acceptable limits. The dynamical behavior of the described systems, represented as the singular system of differential equations, was investigated by many authors in [3–5]. The geometric approach to the robotic systems dynamic for linear non-singular systems was presented in [6] and partially in the article [7] based on the results of [8]. Several authors investigated the contact problem in robotic systems. In [9], the authors solved the problem of the practical stabilization of robots that are in contact with a dynamic environment. The goal of another article [10] was to shed light on the control problem of constrained robotic motion, from the viewpoint of the dynamical nature of the environment with which the robot was in contact. In [11] the authors presented the current state of the art of adaptive control of single rigid robotic manipulators in the constrained motion tasks. The same author in [12] investigated the problem of impendence control using a unified approach to contact tasks control in robotics. In article [13] the authors analyzed control problems for a specific mechanical system consisting of a rigid base body with an unactuated internal degree of freedom. The key assumptions were that the translational and rotational motions of the base body could be completely controlled by external forces and moments, while the internal degree of freedom was unactuated. This idea can apply to the system in the contact task. In [14] the model for the constrained robot dynamics, incorporating constraint uncertainties, was presented. The rigid constraining surfaces were represented by a set of algebraic functions.

In this article, we present a geometric theory for the robotic manipulators mathematically described as the singular systems with constraints. We investigated the necessary and sufficient conditions for the existence and uniqueness of the solutions for the robotic systems as well as the controllability conditions of the systems.

2 Geometric Conditions for System Transformation

It is assumed that the system studied here belongs to the class of time-continuous stationary linear singular systems, either with zero input matrix B:

$$E\dot{\mathbf{x}}(t) = A\mathbf{x}(t),\ \mathrm{x}(0) = \mathrm{x}_0, \tag{2}$$

$$\mathrm{x}_i(t) = C\mathrm{x}(t), \tag{3}$$

either in its standard form:

$$E\dot{\mathbf{x}}(t) = A\mathbf{x}(t) + B\mathbf{u}(t),\ \mathrm{x}(0) = \mathrm{x}_0, \tag{4}$$

with a matrix $E \in \mathrm{C}^{n\times n}$ obligatory singular and other matrices: $A \in \mathrm{C}^{n\times n}$, $B \in^{m\times n}$, $C \in \mathrm{C}^{p\times n}$, in the general case. In the majority of cases, the elements of the mentioned matrices will belong to the set of real numbers. It is known that the dynamics of robotic systems is represented by nonlinear equations. When a singular representation of the system is added, it can be observed that the linearization process can be particularly complex. The initial equation of the observed robotic system, developed using the Lagrange's energy approach, is:

$$M(q)\ddot{q} + H(q, \dot{q}) + G(q) = \tau + J'(q)f, \tag{5}$$

where the sign' denotes transposition, $q \in \Re^n$ are the generalized coordinates of the manipulator, $\dot{q}$ and $\ddot{q}$ are respectively velocity and acceleration in generalized coordinates; $M(q) \in \Re^{n\times n}$ is a matrix that represents the influence of the inertia of the manipulator, $H(q, \dot{q}) \in \Re^n$ is the vector of Coriolis and centrifugal forces; $G(q) \in \Re^n$ is the vector of gravitational forces acting upon the system. $\tau \in \Re^n$ is the vector of generalized forces acting on each segment, $f \in \Re^n$ is the contact force, a $J(q) = \partial h(q)/\partial q' \in \Re^{n\times n}$ is a Jacobian operator. The presence of an obstacle that a robotic system needs to overcome or act on will generally be represented by inequality

$$Py \geq d, \tag{6}$$

where $P \in \Re^{p\times d} id \in \Re^p$. It can be concluded that $Py > d$ for each $t \in [t_1, t_2]$, so then the dynamics of the observed system can be described by Eq. (5). Then the set of allowed positions of the system is limited, *i.e.* the constraint equality $P_i y = d_i$ as well as the associated inequality $P_i y \geq d_i$ are called the passive constraint, [7]. In the previous equation, the indices label the *i*-th row of defined matrices. On the other hand the physical constraint of a system is called active if $P_i y(t) \geq d_i$, for some $i \in 1, \ldots, p$. Based on [6], $P_i \dot{y} = 0$. Now the state of the system (5) can be defined as $x = [y^T, \dot{y}^T]^T$. If the system was linearized before joining the algebraic Eq. (6), we obtain a linear model of the differential equation of the second order of the form

$$M\ddot{y} + D\dot{y} + Ky = Lu. \tag{7}$$

Then the system (5), with constraint (6), can be transformed into the following form

$$\begin{aligned} \dot{\mathbf{x}}(t) = \hat{A}\mathbf{x}(t) + \hat{B}\mathbf{u}(t) &= \begin{bmatrix} 0 & I \\ -M^{-1}K & -M^{-1}D \end{bmatrix} x + \begin{bmatrix} 0 \\ M^{-1}L \end{bmatrix} u \\ \hat{C}\mathbf{x}(t) &= \begin{bmatrix} P & 0 \end{bmatrix} \mathbf{x}(t) \geq \hat{d} \end{aligned}. \tag{8}$$

So, the new system for further investigation becomes:

$$\dot{\mathbf{x}} = \hat{A}\mathbf{x} + \hat{B}\mathbf{u}. \tag{9}$$

with the constraint equation

$$\hat{C}\mathbf{x} \geq \hat{d}. \tag{10}$$

Taking into account the inequality of the contact surface, a combined expression for the observed system can be written, as

$$\begin{aligned} \dot{\mathbf{x}} &= \hat{A}\mathbf{x} + \hat{B}\mathbf{u} \\ 0 &\leq \hat{C}\mathbf{x} + d \end{aligned}, \tag{11}$$

with matrices of the system defined as: $\hat{A} \in \Re^{n\times n}$, $\hat{B} \in \Re^{n\times m}$, $\hat{C} \in \Re^{p\times n}$ and $d \in \Re^{p}$. Respecting the standard notation for the singular systems representation, Eq. (11) can be written as

$$E\dot{\mathbf{x}} = A\mathbf{x} + B\mathbf{u} + D, \tag{12}$$

where, in the case of constraints, in Eq. (10) the sign $\geq$ can be replaced with the sign $=$

$$E = \begin{bmatrix} I & 0 \\ 0 & 0 \end{bmatrix}, \quad A = \begin{bmatrix} \hat{A} & 0 \\ \hat{C} & 0 \end{bmatrix}, \quad B = \begin{bmatrix} \hat{B} & 0 \\ 0 & 0 \end{bmatrix}, \quad D = \begin{bmatrix} 0 \\ d \end{bmatrix}. \tag{13}$$

In order to prove the correctness of the previous transformation, the relation between the corresponding subspaces and mapping will be tested through the following theorem using a geometric approach.

Theorem 1: The transformation of system (7) with constraint Eq. (6) into system (12) is valid if and only if it is satisfied $\hat{C}\hat{B} = 0$, i.e. $N(\hat{B}) \subseteq R(\hat{C})$.

Proof: To show that the geometric condition $N(\hat{B}) \subseteq R(\hat{C})$ satisfied is sufficient to show the equivalence of the expression $\hat{C}\hat{B} = 0$ and $N(\hat{B}) \subseteq R(\hat{C})$. It is clear from the structure of the matrices $\hat{C}$ i $\hat{B}$ that their product is equal to zero. Based on the geometric characteristics of orthogonal invariant subspaces, it follows that $N(\hat{B}) \subseteq R(\hat{C})$. To make sure of that, the zero subspace of matrix $\hat{B}$ is calculated as $N(\hat{B}) = \{0\}$. As the domain of $\hat{C}$ also contains a zero vector, $R(\hat{C}) \supset \{0\}$, it can be concluded that the geometric condition for the transformation of the robotic system (7) with the equation of constraint (6) into the system (12), can be fulfilled if the condition $N(\hat{B}) \subseteq R(\hat{C})$. *q.e.d.*

Let $R(E)$ and $R(B)$ be the domains of the matrices E and B, respectively, given as a subspace $\mathrm{Y} = \Re^r$. Consider the following relation, which holds for the linear subspace V of Y for the robotic system given by Eq. (12)

$$\mathrm{Av} \subset \mathrm{Ev}. \tag{14}$$

Definition 1: The characteristic subspace of a pair (E, B) is the largest subspace V^* that satisfies relation (14).

For the observed system, a defined subspace exists because $\{0\}$ satisfies (14), and this relation is stable with respect to the addition of subspaces, so that V^* is the sum of subspaces satisfying the specified relation. In a special case, it cannot be ruled out that V^* may be trivial. This subspace is uniquely related to the solution of the equation

$$E\dot{\mathbf{x}}(t) = A\mathbf{x}(t). \tag{15}$$

and can be treated as a dynamic invariant. It can be shown that it consists of generalized eigenvectors of this system. The characteristic subspace of the pair (E, B) is responsible for the geometric description of the constraint. A practical interpretation of this subspace is that all the allowed initial conditions that provide contact (singularities) can be found in Eq. (6). In case the robotic system moves freely, then this subspace is an empty set. In the special case when a system has two input quantities, this is equivalent to the effect of disturbance acting upon the system

$$\dot{y}(t) = Ay(t) + Bu(t) + Cv(t). \tag{16}$$

where it is possible to perform the transformation using the matrix E, taking into account that $\mathrm{N}(E) = \mathrm{R}(C)$, a singular system is obtained where the effect of the disturbance is eliminated from the model due to the orthogonality of the subspace

$$E\dot{y}(t) = EAy(t) + EBu(t). \tag{17}$$

Bearing in mind Definition 1, it is valid

$$\mathrm{EAv} \subset \mathrm{Ev}, \tag{18}$$

which is equivalent to an expression

$$\mathrm{Av} \subset \mathrm{v} + R(\mathrm{C}). \tag{19}$$

It can be concluded that V is an A-invariant subspace modulo C and V^* is the largest A-invariant subspace modulo C.

Therefore, two geometric conditions of system transformation were presented here. The first is expressed by Theorem 1, and it gives the necessary condition for the transformation of the system into a singular system of differential equations. The constraint equation plays a key role here, because it depends on whether it is possible to perform this transformation. The second condition gives the geometric relation that must be satisfied, Eq. (19), in order for the effect of the disturbance on the system to be eliminated in the mathematical model. As a consequence, the system model acquired a singular character.

3 Characteristic Subspaces

After derivation of the geometrical condition for proper system linearization, a linearized system equivalent of the system (5) can be observed. Here, the linearization procedure is omitted, and it can be found in [15, 16]. A linearized system (20) is considered

$$E\dot{\mathbf{x}}(t) = A\mathbf{x}(t) + B\mathbf{u}(t) + d. \tag{20}$$

It can be noticed that the matrix E is singular, while the vector d represents perturbations that act upon the system. For a class of robotic systems in contact with the external environment, represented by singular differential equations, the matrices appearing in Eq. (20) are given by the following expressions

$$E = \begin{bmatrix} I & 0 & 0 \\ 0 & M(q_0) & 0 \\ 0 & 0 & 0 \end{bmatrix}, \quad A = \begin{bmatrix} 0 & I & 0 \\ \frac{\partial}{\partial q}(G - J^T D^T \lambda)|_0 & 0 & J^T D^T|_0 \\ DJ|_0 & 0 & 0 \end{bmatrix}. \tag{21}$$
$$B = \begin{bmatrix} 0 \\ I \\ 0 \end{bmatrix} \quad u = \delta\tau \quad d = \begin{bmatrix} 0 \\ \Delta\tau \\ 0 \end{bmatrix}$$

As all elements of matrices are physically responsible for a particular segment of dynamic behavior, the question arises as to what is their impact on the characteristic subspaces of the system (20).

Definition 2: Maximum subspace of the matrix triple (E, A, B) is the largest subspace W^* that satisfies

$$A\mathrm{W}^* + \mathrm{R}(B) = E\mathrm{W}^*. \tag{22}$$

Lemma 1: Subspace W^* can be calculated. It is a subspace of V^* and represents its boundary obtained in no more than n steps of the sequence W^k, according to the expressions:

$$\mathrm{W}^0 = \mathrm{V}^*, \mathrm{W}^{k+1} = E^{-1}(A\,\mathrm{W}^* + \mathrm{R}(B)) \cap \mathrm{V}^*. \tag{23}$$

Proof: Since $A\mathrm{V}^* + \mathrm{R}(B) \subset E\mathrm{V}^*$ is valid, the following is true

$$E\mathrm{W}^1 = A\mathrm{W}^0 + \mathrm{R}(B), \quad \mathrm{W}^1 \subset E\mathrm{V}^*. \tag{24}$$

Equation (24) holds for each step of the algorithm, allowing the index k to change. This sequence has a tendency to decrease, which means that it has a limit that satisfies (22). Such a subspace is stable with respect to the addition operation.

Definition 3: Minimum subspace of the matrix triple (E, A, B) is the smallest subspace W^* that satisfies (14) and $B = E\,\mathrm{V}\,\tilde{B}$ and contains N.

Theorem 2: Subspace W_* exists and is obtained as a limit value that converges in at most n steps by recursive formula (23), with initial value $\mathrm{W}_0 = \{0\}$. W_* represents the domain of the subspace of the state trajectory V of the system (16), where u and v are controls by position and by acceleration.

Proof: As Eq. (14) is not stable with respect to the intersection operation, the existence of the smallest subspace and other related conditions is not obvious. Consider the recurrent relation (23) with the initial condition $\mathrm{W}_* = \{0\}$.

$$\mathrm{W}_1 = E^{-1}(\mathrm{R}(B)) \cap \mathrm{V}^*. \tag{25}$$

As (26) holds, it follows that

$$E\mathrm{W}_1 = \mathrm{R}(B). \tag{26}$$

It can be observed that the sequence W_k is increasing and by induction satisfies all sequences of Eq. (24), as well as W^k, according to statement $N \in \mathrm{W}_k$, for each k. Therefore, the sequence has a limit value W_* that satisfies Eqs. (14), (16) and contains N. *q.e.d.*
According to Theorem 1, the range of values of the matrix V of the state trajectory space of Eq. (16) is precisely the space of the state trajectory $x(t)$, when the solution state starts from zero. According to the claim, that is exactly the limit value of the above expression.

W_* is the smallest subspace that satisfies (14) and (16) and contains N. It directly follows from the following Lemma.

Lemma 2: Any subspace W that satisfies (14) and (16) and contains N, contains the trajectory space of the system (20).

Proof: W is a subspace of V^* since it satisfies (14). Suppose that the matrix V is chosen in such a way that the matrix W, which generates W, is its submatrix.

$$V = [L\,W]. \tag{27}$$

Since $N \subset \mathrm{W}$, W can be chosen so that N is its submatrix. Further, the partitions of the matrix V can be selected, taking into account that $W = (I, N)$.

$$V = [L\,I\,N]. \tag{28}$$

where $M = [L\,I]$. Taking into account equations $\tilde{A}$: $A\mathrm{V} = E\ \mathrm{V}\ \tilde{A}$ and $\tilde{B}$: $B = E\ \mathrm{V}\ \tilde{B}$ it can be obtain

$$AM = EMA,\ AN = EMC,\ B = EMB, \tag{29}$$

so by introducing further partitions similar to the one introduced for the matrix M:

$$A_{11} = A_{13} = 0,\ B_1 = 0, \tag{30}$$

as can be seen from the structure of the linearized system, Eq. (21). This is the standard form of system description for which the system trajectory subspace is contained in W_*.

Theorem 3: The neutral space V_* exists and is identical to the space of the trajectory of the state for the matrix pair (A, C), for zero initial conditions.

Proof: The theorem is proved based on the previous condition when $B = 0$.

Let W be a submatrix of the matrix V, which defines the subspace W_* and let it be

$$W = [M \quad N]. \tag{31}$$

As mentioned earlier, there are matrices $\hat{A},\ \hat{B},\ \hat{C}$, so it is valid

$$A\hat{M} = E\hat{M}\hat{A},\ AN = E\hat{M}\hat{C},\ B = E\hat{M}\hat{B}. \tag{32}$$

If the system (20) starts from $x(0) \in W_*$, then the solution can always be represented by an expression

$$\dot{\hat{y}}(t) = \hat{A}y(t) + \hat{B}u(t) + \hat{C}v(t). \tag{33}$$

$$x(t) = \hat{M}\hat{y}(t) + Nv(t). \tag{34}$$

These equations are the minimal representation of the system (16), with v as the quantity that characterizes the unmeasured perturbations. The method described here was used in the next part to determine the solution of the singular system and according to the results, to analyze the stability of the system or stability at a finite time interval.

4 Solutions of Constrained Systems

This section covers existence and uniqueness of the solution when the contact force acts upon the system (20), i.e. when the control subspace S, responsible for position and velocity control, is extended by a reactive force control component.

4.1 Existence of the Solutions

The matrix structure (21) of system (19) can be denoted as:

$$J_D = \ \mathrm{AN}(E) = \left[0\ J^T D^T|_0\ 0\right]^T,\ I_3 = \mathrm{N}(E) = \left[0\ 0\ I\right]^T \tag{35}$$

where J_D represents the influence of the contact surface, over its gradient D, on the robotic system. Generally, the subscript "D" denotes a gradient. In this case, D denotes the gradient of the contact force that acts upon the system and it is not the same as a matrix in Eq. (21). N $(\cdot)$ is null space of the matrix $(\cdot)$ and R $(\cdot)$ is range of the matrix $(\cdot)$. x_0 represents the initial condition space matrix and I is identity matrix.

The system (20, 21) can be fully transformed to its state-space form:

$$\begin{aligned} E\dot{\mathbf{x}}(t) &= A\mathbf{x}(t) + B\mathbf{u}(t) \\ \mathbf{x}_i(t) &= C\mathbf{x}(t) + D\mathbf{u}(t) \end{aligned}. \tag{36}$$

The following *Theorems*, Lemma 3, and Corollary 1 are the original results that are obtained by using the geometric approach in the analysis of the system described by matrices (21).

Theorem 4: The solution of system (36), which comprises the reactive force control, on an arbitrary time interval exists for any control vector **u**(t) if and only if (37) and (38) are satisfied:

$$\mathrm{R}(B) \subset EV^* + J_D = EM + J_D. \tag{37}$$

$$x_0 \in V^* + I_3 = M + I_3. \tag{38}$$

Proof (necessary condition): Let us analyze arbitrary condition (39):

$$x(t) = z(t) + \varepsilon(t). \tag{39}$$

with

$$\varepsilon(t) \in I_3, z(t) \in \mathrm{z}, \tag{40}$$

where Z denotes the subspace that is mathematically defined in the following part. Choosing the proper constraints for the subspace Z, it is possible that $\mathrm{Z} \cap I_3 = \{0\}$; therefore, the proposed decomposition of the state space vector (39) is unique. Applying (39) to system (36) and neglecting disturbances to the system, the following equation can be written

$$E\dot{z}(t) = Az(t) + A\varepsilon(t) + B\mathbf{u}(t). \tag{41}$$

For given $\mathbf{u}(t)$ and $\mathbf{x}(t)$, $E\dot{z}(t)$ is uniquely defined. The direct consequence for the singular system is the following geometric distribution:

$$A\mathrm{z} \cap E\mathrm{z} + A\, I_3, \tag{42}$$

$$\mathrm{R}(B) \subset E\mathrm{z} + A\, I_3. \tag{43}$$

Further proof of the necessary condition relies on the following lemma.

Lemma 3: The maximum subspace which satisfies (42) and (43) is $\mathrm{V}^* + I_3$.

Proof: It can be shown that $\mathrm{V}^* + I_3$ satisfies (42) and (43). Let us assume that subspace Z satisfies the same conditions. In that case it can be written

$$\mathrm{z} = \mathrm{V} + I_3, \tag{44}$$

as well as

$$A(\mathrm{V} + I_3) \subset E\,\mathrm{V} + J_D. \tag{45}$$

Given relations (44) and (45), it can be stated that if $\forall a \in \mathrm{V}$, $\exists \tilde{a} \in \mathrm{V}$ and $b \in \mathrm{N}(E)$, the equation $Aa = E\tilde{a} + Ab$ is satisfied. Values $\tilde{a}$ and b could be chosen so that they are linearly related to a. Let matrix K generate $\mathrm{N}(E)$. Then a matrix exists with appropriate dimensions, so for every $a \in \mathrm{V}$, Eq. (34) is fulfilled:

$$Fa = E\tilde{a} + FKPa, \tag{46}$$

and also

$$F(I-KP)a = E\tilde{a} = E(I-KP)\tilde{a}. \tag{47}$$

Defining:

$$\overline{\mathrm{V}} = (I-KP)\mathrm{V}, \tag{48}$$

it can be stated:

$$\overline{\mathrm{V}} + \mathrm{N}(E) = \mathrm{V} + N(E) = \mathrm{Z}, \tag{49}$$

as well as

$$A\overline{\mathrm{V}} \subset E\overline{\mathrm{V}}. \tag{50}$$

The final conclusion that proves the necessary condition of Lemma 3 is given by Eq. (51):

$$\overline{\mathrm{V}} \subset \mathrm{V}^*,\ \mathrm{Z} \subset V^* + I_3. \tag{51}$$

Note: In order to obtain the unique solution $z(t)$, Z must be chosen to satisfy $\mathrm{Z} = \mathrm{M}$, where subspace M is a complement of I_3 in V^*, *q.e.d.*

Proof (sufficient condition): Let $\bar{K}$ be a matrix whose columns fully spans the null space of matrix E N(E) and let M be a matrix chosen as $AM = EMA$. Then,

$$x(t) = My(t) + \overline{K}w(t). \tag{52}$$

The conditions (37) and (38) imply the existence of matrices $\bar{B}$ and $\bar{P}$ and

$$B = EM\bar{B} + A\bar{K}\bar{P}. \tag{53}$$

Now it is possible to represent system (22) in its equivalent form,

$$EM\dot{y}(t) = EMAy(t) + A\bar{K}w(t) + EMN\bar{B}\mathbf{u}(t) + A\bar{K}\bar{P}\mathbf{u}(t). \tag{54}$$

and the solution of system (36), together with (52), as:

$$\dot{y}(t) = Ay(t) + \bar{B}\mathbf{u}(t). \tag{55}$$

$$x(t) = My(t) + \overline{KP}\mathbf{u}(t). \tag{56}$$

Here, note that since the constraints for matrix B are changed, it is not necessary to introduce the special relations between matrices B and $\bar{B}$ as well as B and $\bar{P}$. Thus, Eqs. (55) and (56) define the solution of system (36), which satisfies conditions (37) and (38), *q.e.d.*

4.2 Uniqueness of the Solutions

Theorem 5: The solution of system (36), which comprises the reactive force control, under conditions (37) and (38) is unique for any $\mathbf{u}(t)$, if and only if the matrix pencil (E, A) is a pencil of full column rank.

Proof: The question concerns under which conditions Eq. (54) has a unique solution. Denoting the derivatives of the values x and w as $\delta\dot{x}(t)$, $\delta w(t)$, the problem can be reformulated. Now it is necessary to find non-zero solutions of the following equation

$$EM\,\delta\dot{x}(t) = AK\,\delta w(t). \tag{57}$$

One of the possible solutions is zero if and only if the following is satisfied

$$\mathrm{N}(A) \cap \mathrm{N}(E) = \{0\} \tag{58}$$

and

$$EM \cap J_D = \{0\}. \tag{59}$$

Combining Eqs. (57) and (58) and using Eq. (35), it becomes clear that Eq. (58) is always fulfilled, because of the structure of matrices (35). This conclusion could be made based on injectivity of matrix EM. Consequently, for non-zero solutions, neither Eqs. (58) nor (59) should be fulfilled. Equation (59) in fact does not have to be fulfilled. It can be concluded that a non-zero element exists in the equation for $EM \cap J_D$.
When it is $\mathrm{V}^* = A^{-1}(EM)$, then

$$N = A^{-1}(EM) \cap \mathrm{N}(E). \tag{60}$$

It is now possible to verify the existence of an arbitrary linear operator A for two subspaces A and B and if the following statement is fulfilled

$$A\mathrm{A} \cap A\mathrm{B} = A\ [(\mathrm{A} + \mathrm{N}(A)) \cap \mathrm{B}]. \tag{61}$$

Applying statement (61) to Eq. (50) and applying $A^{-1}(EM) \subset \mathrm{N}(E)$, it can be shown:

$$AN = EM \cap J_D. \tag{62}$$

It can also be noted that $N(A) \subset \mathrm{V}^*$, and consequently that:

$$N \supset \mathrm{N}(E) \cap \mathrm{N}(A) \tag{63}$$

It can be concluded from Eqs. (62) and (63) that conditions (58) and (59) are not satisfied when subspace N is nontrivial. Consequently, the system is not column regular. On the other hand, if subspace N is nontrivial and if Eq. (58) is fulfilled, which is always true according to the previous discussion, then it can be stated that $N \subset N(E)$ holds and

$$N \cap \mathrm{N}(A) = \{0\}. \tag{64}$$

Given (64), AN has the same dimension as N, and so (62) shows that (59) is not fulfilled, *q.e.d.*

As in the previous part, in which solution uniqueness was discussed, a clear distinction between two types of the solution non-uniqueness can be made. In the case that (59) holds, but (58) does not, non-uniqueness of x includes only w and the statement is not fulfilled at any instant of time. The solutions $\mathbf{x}(t)$ are unique. In this case, non-uniqueness could be named *static*. Dynamic non-uniqueness is the product of the non-zero element in the statement $EM \cap J_D$.

Corollary 1: The regular matrix pencil $(sE - A)$ does not have any indefinite zeroes if and only if $\mathrm{N}(E) \cap A^{-1}(\mathrm{R}(E))$ exists. Let us assume that the regular matrix pencil $(sE - A)$ has indefinite zeroes. In this case, the static variables do not exist if and only if $A\mathrm{N}(E) \subset \mathrm{R}(E)$.

Proof: The first part of Corollary 1 can be proven using Theorem 4, because in that case, the matrix pencil $(sE - A)$ does not have any indefinite zeroes if and only if $\mathrm{W}_a^* = 0$, to i.e., $\mathrm{N}(E) \cap A^{-1}(\mathrm{R}(E)) = 0$. The second part is true because the matrix pencil $(sE - A)$ does not have indefinite zeroes and in that case $\mathrm{N}(E) \cap A^{-1}(\mathrm{R}(E)) \neq 0$ is fulfilled. Moreover, $\mathrm{N}(E) \cap A^{-1}(\mathrm{R}(E)) = \mathrm{N}(E)$ is true, or equivalently $\mathrm{N}(E) \subset A^{-1}(\mathrm{R}(E)) \Leftrightarrow J_D \subset \mathrm{R}(E)$, *q.e.d.*

5 Geometric Approach to Controllability

In this section we analyze how output zero (A, E, B) invariant subspace can be utilized to determine the possible geometric structure of a close-loop system. (A, E, B) invariant subspace carries information about an open system. The generalized Lyapunov Eq. (65) plays a central role in this part, where $S \subset \Re^n$ is the output zero (A, E, R(B)) invariant subspace of system (20).

$$\begin{bmatrix} A \\ C \end{bmatrix} \mathrm{S} \subset \begin{bmatrix} E \\ 0 \end{bmatrix} \mathrm{SF} - \begin{bmatrix} B \\ D \end{bmatrix} G. \tag{65}$$

The output zero subspace S satisfies Eq. (66)

$$\begin{bmatrix} A \\ C \end{bmatrix} \mathrm{S} \subset \begin{bmatrix} E \\ 0 \end{bmatrix} \mathrm{S} + \mathrm{R} \begin{bmatrix} B \\ D \end{bmatrix} \tag{66}$$

State-space feedback can be defined as

$$u = Kx \tag{67}$$

Combining (67) with system (20) it is obtained

$$\begin{aligned} E\dot{x} &= (A + BK)x \\ y &= (C + DK)x \end{aligned} \tag{68}$$

The following theorem shows how Eq. (65) can be used to determine the gain matrix K and provide the required dynamic properties of a closed system. Then, based on these conditions, the feedback control (67) is applied to system (20).

Theorem 6: S satisfies (66) if and only if there exists K such that, [17].

$$(A + BK)\mathrm{S} \subset E\mathrm{S} \tag{69}$$

$$(C + DK)\mathrm{S} = 0 \tag{70}$$

Proof: The existence of K satisfying (69) and (70) guarantees that (66) is fulfilled. Subspace S satisfies (66) if and only if (65) holds for some F and G. K is chosen as

$$K\mathrm{S} = G \tag{71}$$

Then conditions (64) and (67) of Theorem 6 become

$$(sE - (A + BK))\mathrm{S}(sI - F)^{-1} = E\mathrm{S} \tag{72}$$

$$(C + DK)\mathrm{S}(sI - F)^{-1} = 0 \tag{73}$$

Applying the results of Theorem 6 to a closed system (68), it can be noticed that S is an output zero *(A, E)* invariant subspace of a closed system, which satisfy (69) and (70), which should have been shown. *q.e.d.*

Corollary 2: Let S be given to satisfy (66) and K is defined as in (71), so that (72) and (73) hold. Then there is a unique system solution (68) for each $x(0) \in \mathrm{S}$, so $[sE - (A + BK)]$ is regular on S if and only if

$$\mathrm{N}(E) \cap \mathrm{S} = 0 \tag{74}$$

Proof: Relation (72) and (73) show the existence of a solution for each $x(0) \in \mathrm{S}$. This solution is independent of G and is unique if and only if there is a unique solution F of Eq. (65). For a given S and G there is a unique solution F of the expression

$$E\,S\,F = A\,\mathrm{S} + BG, \tag{75}$$

if and only if E S has full column rank or if $\mathrm{N}(E) \cap \mathrm{R}(S) = 0$, which is exactly expression (74).

According to expression (72) and (73) the quantity K from (71) which satisfies Lyapunov's Eq. (65) can be used to ensure the regularity of the system in space $\Re^n$. The relevant part of this equation can be written as

$$A\mathrm{S} - E\,S\,F = -BG \tag{76}$$

From this analysis, the characteristics of the observed system can be defined.

Definition 4: The system (20) is controllable if the matrix pencil

$$C(s) = \begin{bmatrix} sE - A & B \end{bmatrix} \tag{77}$$

has no finite or infinite zeros.

Theorem 7: Systems characterized by Eqs. (20) are controllable if and only if none of the matrix eigenvalues $C(s)$ are equal to zero. This condition is represented as

$$C(s) = \begin{bmatrix} sI & -I & 0 & 0 \\ \frac{\partial}{\partial q}(G - J^T D^T \lambda)|_0 & sM(q_0) & -J^T D^T|_0 & I \\ -DJ|_0 & 0 & 0 & 0 \end{bmatrix} \tag{78}$$

Proof: Taking into account expression (21) which defines the structure of a robotic system, replacing the corresponding quantities in (77), we obtain condition (78), *q.e.d.*

This condition defines the relationship between the dynamic structure of a robotic system that is in contact with the working environment and the geometric characteristic that determines the controllability of the system.

Definition 5: Systems (22) is reachable if (77) is fulfilled and if

$$rang\begin{bmatrix} E & B \end{bmatrix} = n \tag{79}$$

Theorem 8: System (20) is controllable if and only if condition (80) is fulfilled:

$$rang\begin{bmatrix} I & 0 & 0 & 0 \\ 0 & M(q_0) & 0 & I \\ 0 & 0 & 0 & 0 \end{bmatrix} = n \tag{80}$$

Proof: Similar to Theorem 7, using Definition 5.

It can be concluded, based on expression (80), that the controllability of the linearized system depends on the moment of inertia in the vicinity of the contact point. This means that the construction of the system itself can negatively affect the controllability. This result can serve as a check of the potential stability of the system at the time of design.

Definition 6: System (20) is controllable at infinity if (77) there are no zeros at infinity.

6 Conclusion

Robotic systems are typical examples in which external contact forces play an important role to the system dynamics. Mathematical modeling of these systems is challenging due to a variety of reasons. Mathematical models for these systems contain differential equations with the associate algebraic equation. The algebraic equations are mathematical description of physical constrains that influence to the system dynamics. Such systems are considered to be the singular system of differential equations in control system theory. In this article, system dynamics including controllability criteria, uniqueness of the solutions, and related characteristic subspaces for robotic systems with contact tasks, have been investigated. Contact forces were included in the stability and controllability conditions. The controllability conditions were investigated in order to achieve the desired dynamical system behavior. The main objective of this study was to find appropriate mathematical representations and solutions to the singular system in which kinematic constraints are imposed to the motion of the robotic system. The geometric approach was outlined and applied to these investigations.

References

1. Basile, G., Marro, G.: Controlled and conditioned invariant subspaces in linear system theory. J. Optim. Theory Appl. **3**(5), 306–315 (1969)
2. Wonham, W.M., Morse, A.S.: Decoupling and pole assignment in linear multivariable systems: a geometric approach. SIAM J. Control **8**(1), 1–18 (1970)
3. McClamroch, N.: Singular systems of differential equations as dynamic models for constrained robot systems. In: Proceedings of the 1986 IEEE International Conference on Robotics and Automation, vol. 3, pp. 21–28 (1986)
4. Huang, H.P.: The unified formulation of constrained robot systems. In: Proceedings of the 1988 IEEE International Conference on Robotics and Automations, vol. 3, pp. 24–29 (1988)
5. Mills, J.K., Liu, G.J.: Robotic manipulator impedance control of generalized contact force and position. In: Proceedings IROS'91. IEEE/RSJ International Workshop on Intelligent Robots and Systems, pp. 1103–1108 (1991)
6. Ten Dam, A.A., Dwarshuis, K.F., Willems, J.C.: The contact problem for linear continuous-time dynamical systems: a geometric approach. IEEE Trans. Autom. Control **42**(4), 458–472 (1997)
7. Mills, J.K., Goldenberg, A.A.: Force and position control of manipulators during constrained motion tasks. IEEE Trans. Robot. Autom. **5**(1), 30–46 (1989)
8. Cobb, D.: Descriptor variable systems and optimal state regulation. IEEE Trans. Autom. Control **28**(5), 601–611 (1983)
9. Stokic, D., Vukobratovic, M.: An efficient method for analysis of practical stability of robots interacting with dynamic environment. In: Proceedings of the 1997 IEEE/RSJ International Conference on Intelligent Robot and Systems. Innovative Robotics for Real-World Applications. IROS 1997, vol. 1, pp. 175–180(1997)
10. Vukobratović, M.K., Rodić, A.G., Ekalo, Y.: Impedance control as a particular case of the unified approach to the control of robots interacting with a dynamic known environment. J. Intell. Robot. Syst. **18**(2), 191–204 (1997)
11. Vukobratovic, M., Tuneski, A.: Adaptive control of single rigid robotic manipulators interacting with dynamic environment—an overview. J. Intell. Robot. Syst. **17**(1), 1–30 (1996)
12. Vukobratović, M.: How to control robots interacting with dynamic environment. J. Intell. Robot. Syst. **19**(2), 119–152 (1997)
13. McClamroch, N.H., Wang, D.: Feedback stabilization and tracking of constrained robots. In: 1987 American Control Conference, pp. 464–469 (1987)
14. Ho, Y.K., Wang, D., Soh, Y.C.: Modelling of constrained robot system with constraint uncertainties. J. Robot. Syst. **17**(1), 53–61 (2000)
15. Buzurovic, I.M., Debeljkovic, D.L.: A geometric approach to the investigation of the dynamics of constrained robotic systems. In: IEEE 8th International Symposium on Intelligent Systems and Informatics, pp. 133–138 (2010)
16. Buzurovic, I., Podder, T.K., Yu, Y.: Prediction control for brachytherapy robotic system. J. Robot., 1–11 (2010)
17. Ozcaldiran, K., Lewis, F.: A geometric approach to eigenstructure assignment for singular systems. IEEE Trans. Autom. Control **32**(7), 629–632 (1987)

Strain Measurement of Medical Textile Using 2D Digital Image Correlation Method

Nenad Mitrovic[1](✉), Aleksandra Mitrovic[2,3], and Mirjana Reljic[4,5]

[1] Faculty of Mechanical Engineering, Center for Optical Measurement, University of Belgrade, Belgrade, Serbia
nmitrovic@mas.bg.ac.rs

[2] Academy of Technical Vocational Studies, 11000 Belgrade, Serbia

[3] Faculty of Information Technology and Engineering, University Union "Nikola Tesla", Belgrade, Serbia

[4] Academy of Technical-Art Vocational Studies, 11000 Belgrade, Serbia

[5] CIS Institute, Belgrade, Serbia

Abstract. Medical textile plays an important role in the technical textiles sector as one of the most rapidly growing sectors in the technical textile market. The textile materials should have some adequate mechanical properties to be useful as medical textile. Tensile strength presents one of the basic mechanical properties used to describe textile specimens. Standardized tensile testing procedures on textile specimens were commonly used in the past. The aim of this paper was to measure in-plane strain field on the tensile medical textile specimen using 2D Digital Image Correlation method (2D-DIC). 2D-DIC is a non-contact optical method for accurate displacement and strain full-field measurement. In this study, two medical cotton textiles, with density 120 and 130 g/m^2, were used to create three specimens for each material. Each specimen was placed in the tensile testing machine and measured until the break. During the tensile testing, camera was automatically recording full-field displacement in X and Y directions. Textile 1 and Textile 2 showed significant differences in point distance values, despite the small deviation in densities (less than 10%). Mean value of the elongation for Textile 1 is more than a double than the elongation for Textile 2, although the difference for mean value of Maximum force if negligible. Also, it has been showed that 2D-DIC can play significant role for measurement in textile mechanical properties measurement.

Keywords: Medical textile · 2D Digital Image Correlation method · Tensile testing · Deformations

1 Introduction

The production of specific textile structures and products is essential focus of modern textile industry especially in the fields of medical textiles, protective textiles and smart textiles. Many researchers have been focused on exploring textile technologies in the

N. Mitrovic et al. (Eds.): CNNTech 2020, LNNS 153, pp. 447–464, 2021.
https://doi.org/10.1007/978-3-030-58362-0_26

medical area. This especially refers to medical devices, in which hollow fibers play most promising role. Textile materials used in medical devices are an important and growing part of the textile industry. Development of medical textiles is significant part of medical diagnosis and therapy since this group of materials present adequate interface between the human body and medical treatment, and thus their continued [1].

Group of medical textile materials include textile materials that are used for medical devices in health and hygiene applications. Thus, these types of materials have huge application referring to a group of products with different product performance and unit value. Medical textiles consist of four categories such as: non-implantable materials, implantable materials, healthcare and hygiene products, and extracorporeal devices. Different fibers are using for fabricating medical textiles for medical devices and they refer to non-implantable, implantable, and healthcare hygiene, respectively [1].

In medical and healthcare applications, textiles are unavoidable materials. It is very important that these materials have appropriate properties such as: biocompatibility, absorption/repellency, elasticity, specific tensile strength, good dimensional stability, stress tolerance, air permeability, water vapor resistance, good resistance to alkalis, acids and micro-organisms, free from contamination or impurities, etc. [2–5]. Progress with hygienic and antimicrobial functionalities is also one of the crucial areas of medical textiles.

Traditional applications of medical textiles envelop wipes, breathing masks, wound care products, protheses and orthoses, bedding and covers, diapers, ropes and belts, braces, etc. Examples of medical textiles used in extracorporeal medical devices include the use of hollow fibers and membranes (made from polyester, polypropylene, silicone, cotton and viscose) for production of bioartificial organs, such as the kidneys, liver and lungs.

1.1 Textile Material Strength, Loading and Failure Criteria

Appropriate strength and durability are important for any textile material, as material has to be strong enough to function properly. Strength is the main characteristic in the case of the performance testing on the fabric specimen, but for the items made of fabrics, the durability of that assembly adds another dimension to the problem. Considering this fact, it is not surprising that the industrial standards for quality assurance on fabrics and clothing mainly focus on strength and durability related issues. Solid engineering materials were initial focus of industrial standards related to strength and durability, as failure of those materials can lead to disastrous consequences: collapsing of buildings, pressure vessels, bridges, etc. and associated human casualties. So, the standardized tests, the theories behind them and the failure mechanisms have been developed with those materials in mind. Today's textile-related standard tests on strength and durability are based on the ones for solid materials. One of the most commonly used standardized tests is tensile testing on standardized specimens [6].

Simple tensile loading is one of the basic loading types, besides shear, bend, tear, and puncture/burst for sheet-like materials such as fabrics. In a simple tensile test, if a fabric is stretched to break, that maximal loading is called breaking load. Obviously, for the same material type, this breaking load value depends on the thickness of the material cross-section. Thus, the ultimate strength is defined, as well as the corresponding ultimate breaking strain, so this pair provides the strength indicator for the material under tensile load [6].

Since fibers are best in carrying tension, when a fibrous material starts to break, at the micro-level, fibers break almost exclusively due to extension (and in much rarer cases due to shear), regardless of the nature of the macro deformation mentioned above. The nature of the fabrics allows fibers to move from the loading. All of these lead to several scientific challenges including fracture behavior and failure criteria. For an isotropic material, its strength is identical, regardless of direction, however, strength direction for a woven fabric is much more complex. This happens due to different degrees of internal yarn re-orientation and movement when stretched in different directions of the fabric. Under a simple tensile test, the specimen failure occurs when the stress caused by the actual load reaches the stress limit (the strength) of the sample. Correlation of the actual stress with the maximum stress (strength) is straightforward in this case as they are both uniaxial [6].

Woven fabrics are well known for their property-direction dependence or property anisotropy. Anisotropy is responsible for many of the challenges in dealing with fibrous materials. As fibrous materials have bi-modular nature, the Young's modulus, as well as the entire stress–strain relationship is quite different in tension versus in compression. Woven fabrics are not only highly anisotropic, but dimensionally changeable and very susceptible to external loading. The important fabric properties critical to structural applications include tensile strengths, in-plane shear strengths, and normal compressive (in thickness direction) strength, as well as in-plane compressive strength, better known as the buckling strength.

Prediction of fabric strengths has its significance, both theoretically and practically, because, except for the uniaxial tensile strength, experimental measurement of all other strengths is exhausting and costs highly with no convenient test methods and instruments available. There are several approaches for strength prediction, but all of them are based on the experimental results. Optical methods are advisable for characterization of fabric properties due to difficulty in determination of deformations and displacements since this group of materials are tested under lower loading [7]. One of the established experimental and most often used optical method that enables additional datasets useful for further specimen analysis subjected to tensile loading is Digital Image Correlation (DIC) method [7–9].

1.2 Digital Image Correlation Method

The Digital Image Correlation (DIC) [10] method is a contactless optical method that overcomes limitations of conventional experimental methods (e.g., strain gauge) and enables full-field displacement and strain measurement. One experimental test enables acquisition of large datasets that replaces large number of strain gauges and significantly reduces experiment preparation time and costs. On the other hand, as numerical methods calculate full displacement and strain fields, model is easily verified by comparing to experimental results presented in the same way. Full strain field experimental measurement allows accurate determination of critical areas, i.e. areas with highest strain values, as well as principle stress directions that enables better theoretical analysis [11].

DIC method is usually used in a single-camera or a stereo-camera system, enabling two-dimensional (2D-DIC) and three-dimensional (3D-DIC) measurement results, respectively. For 2D-DIC, the movement of the planar object is assumed to occur in the object plane that is nominally parallel to the camera sensor plane [10]. For 3D-DIC, two movement restrictions occur: the object has to remain in focus during the experiment and its points of interest are imaged on two camera sensor planes. The 3D DIC method has high accuracy (up to 1 μm) and can be used in a variety of ways: testing of different materials [12–20], structure testing [11, 21, 22], model verification [23–26], fracture mechanics [27, 28], etc. 2D-DIC has some limitations when compared to 3D-DIC (out-of-plane movement should be avoided, object should be flat etc.), but also has some advantages – simple calibration, easy experimental setup, faster result computation etc.

In-plane displacement and strain measurements were the basics of early DIC application for solid mechanics measurements. In all cases, a nominally flat specimen was subjected to tensile testing. During the testing, it was assumed that the specimen deformed only in the planar specimen surface. So, three assumptions are usually employed when using 2D-DIC for in-plane measurement:

1. The specimen is nominally planar. This assumption is applicable, at least approximately, for initially flat specimens, with or without some geometric discontinuities (e.g., cracks, notches, complex cut-outs) or gradients in material properties without affecting this assumption.
2. The object plane is parallel to the camera sensor plane. In a general sense, it is assumed that a nominally planar object is subjected to a combination of in-plane tension, in-plane shear or in-plane biaxial loading so that the specimen deforms mainly within the original planar surface. When cracked or notched specimens are loaded, Poisson's effect in the crack tip region (which results in small amounts of out-of-plane motion) is assumed to be small relative to the applied in-plane deformations. When planar specimens with material property gradients are similarly loaded, the assumption remains.
3. The loaded specimen is deformed only in the original object plane.

The aim of this paper is to measure in-plane strain field on the tensile medical textile specimen using 2D Digital Image Correlation method (2D-DIC).

2 Experimental

Uniaxial tensile testing experiment of a planar textile specimen is performed according to procedure defined in the standard EN ISO 13943-1 [29]. EN ISO 13943-1 defines a procedure for determining the maximum force and equivalent elongation of textile fabrics using a strip method. The main application of the method is on woven textile fabrics, including fabrics that possess stretch characteristics imparted by the presence of an elastomeric fiber, mechanical, or chemical treatment [29].

In this study, two medical cotton textiles, Textile 1 and Textile 2, with density 120 and 130 g/m^2, respectively, were used to prepare three specimens for each group of the materials. As shown in Fig. 1, the strip specimens were made according to [29] with dimensions of 50 mm × 350 mm. Each specimen was placed in the tensile testing machine and measured until the break.

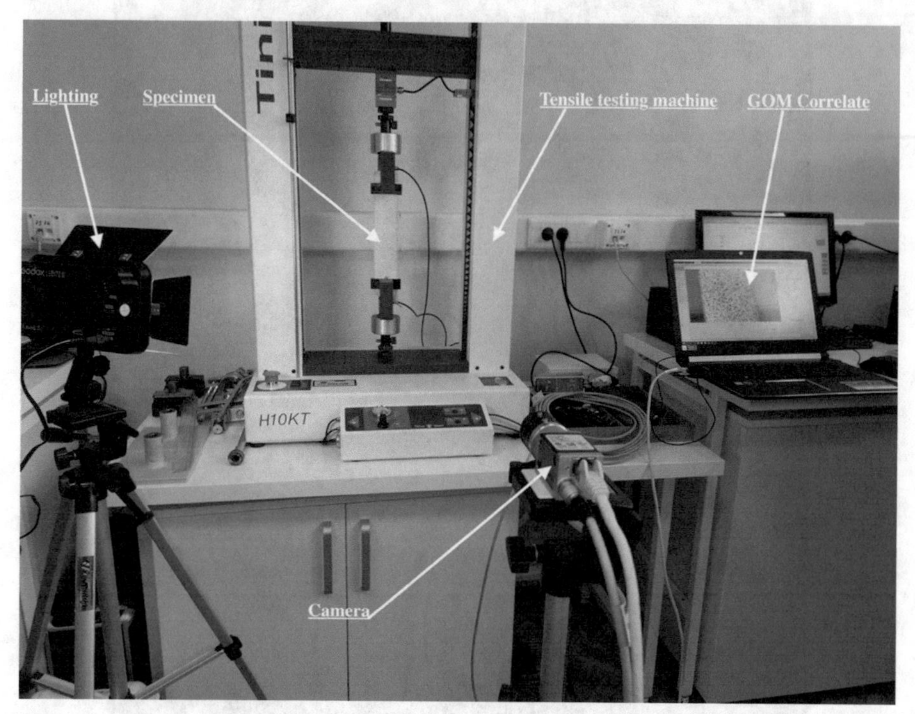

Fig. 1. Experimental setup for uniaxial loading of medical textile specimen. Camera is 1.38 m from the specimen.

All experiments were performed on tensile testing machine Tinius Olsen H10KT (Tinius Olsen, UK) with maximum loading of 10 kN. Pneumatic grips with flat plates were used to fix the specimens. Tensile loading was done using displacement control with the upper grip moving at a constant rate of 100 mm/min for all experiments. Pneumatic grips were positioned at the distance of 200 mm from each other. Preload for all the experiments was 2.0 N.

The surface structure is important for carrying out a measurement. The specimen's surface must have a pattern in order to clearly allocate the pixels in the camera images (facets). Thus, a pixel area in the reference image can be allocated to the corresponding pixel area in the target image. One surface of the specimen was coated with black paint dots to generate a random high-contrast stochastic pattern. All images were recorded using a 50 mm lens and acA-1920 Basler camera (Basler, Germany) with 2 MP camera resolutions. Continuous specimen lighting was provided by a LED lamp (Fig. 1). Figure 1 shows a photograph of the experimental optical setup. Figure 2 shows the region of interest of the strip specimen used for the surface strain measurement.

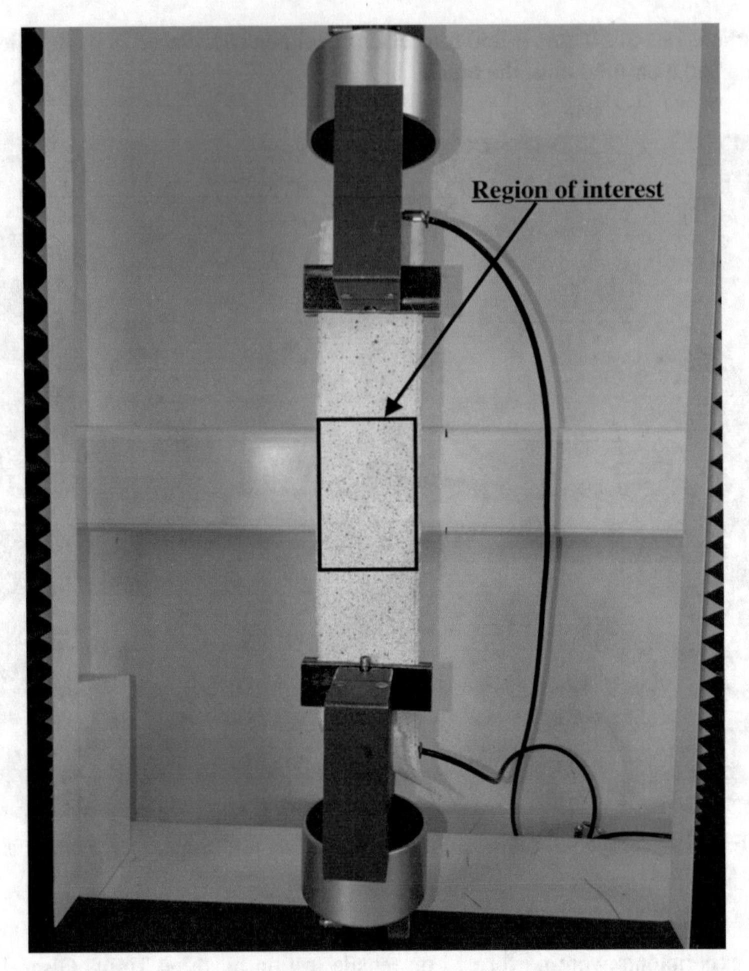

Fig. 2. Region of interest of the strip specimen

Images were recorded automatically for all experiments at a rate of 1 image per second. Tensile loading and image recording were started simultaneously. After the experiment, GOM Correlate software (GOM, Germany) was used for strain calculation.

3 Results and Discussion

As mentioned before, due to specific fiber structure, complex tensile deformations occur on the strip specimen under tensile loading. In the tensile load direction (y axis in this study), positive longitudinal deformation occurs while transverse deformation takes place simultaneously (x axis in this study). Longitudinal deformation is expressed as strain Epsilon Y and transverse contraction is expressed as Epsilon Y. The ratio of Epsilon X and Epsilon Y can be used for determining of Poisson's ratio, one of the fundamental properties of any material [7].

Epsilon X and Epsilon Y strain results are presented in this paper. Strain field was analyzed using one point (Point 1) and two orthogonal lines – Distance 1 and Distance 2, parallel to y and x axis, respectively. Experimental data is also presented graphically as function of time.

Representative images of Epsilon X and Epsilon Y strain field results for maximum axial force before the break are shown in Figs. 3 and 4 for Textile 1 and in Figs. 7 and 8 for Textile 2, respectively. Scale in % is given on ordinate. Value of 0 on the x axis (Figs. 5, 6, 9 and 10) represents the undeformed state of the Textile 1 and Textile 2 specimens, i.e. state before the tensile loading. At the beginning of each experiment (first 2 s), tensile loading values were small and shouldn't be taken for the further analysis.

For the representative sample of Textile 1 (Figs. 3, 4, 5 and 6), a total of 20 images were recorded before the specimen broke. As shown in Fig. 3, most of the Epsilon X strain field values are around −12% before the brake, with peak values around −15%. Most of the Epsilon Y strain field values (Fig. 4) are around 14% before the brake, with peak values around 15%.

Point 1 is placed in the center of the field of interest, as shown in Fig. 2. Three point 1 diagrams for Textile 1 are shown in Fig. 5, presenting time dependence of the Epsilon X, Epsilon Y and displacement Y values, respectively. After the initial loading, the change of the values in Fig. 5 is linear. Highest Epsilon X (−12.713%) and Epsilon Y (14.931%) values in Point 1 are, as expected, before the break, 19 s after the experiment has started.

Distance 1 is placed in the y directions and represents distance between two points positioned near the edges of the region of interest. Distance 2 also represents the distance between two points near the edges of the region of interest, but in the x direction. Distance diagrams for Textile 1 are shown in Fig. 6, presenting time dependence of the distance and deformation between two points, respectively. After the initial loading, the change of the values in Fig. 6 is close to linear. Highest values for Distance 1 (elongation of 5.177 mm and deformation of 14.44%) and Distance 2 (elongation of −5.797 mm and deformation of −12.595%) are, as expected, before the break, 19 s after the experiment has started.

For the representative sample of Textile 2 (Figs. 7, 8, 9 and 10), a total of 11 images were recorded before the specimen broke. As shown in Fig. 7, most of the Epsilon X strain field values are around −11% before the brake, with peak values around −18%. Most of the Epsilon Y strain field values (Fig. 8) are around 6% before the brake, with peak values around 8%.

Point 1 is placed in the center of the field of interest, as shown in Fig. 2. Three point 1 diagrams for Textile 2 are shown in Fig. 9, presenting time dependence of the Epsilon X, Epsilon Y and displacement Y values, respectively. After the initial loading, the change of the values in Fig. 9 is close to linear. Highest Epsilon X (−8.987%) and Epsilon Y (5.731%) values in Point 1 are, as expected, before the break, 19 s after the experiment has started.

Distance 1 is placed in the y directions and represents distance between two points positioned near the edges of the region of interest. Distance 2 also represents the distance between two points near the edges of the region of interest, but in the x direction. Distance diagrams for Textile 2 are shown in Fig. 10, presenting time dependence of the distance and deformation between two points, respectively. After the initial loading, the change

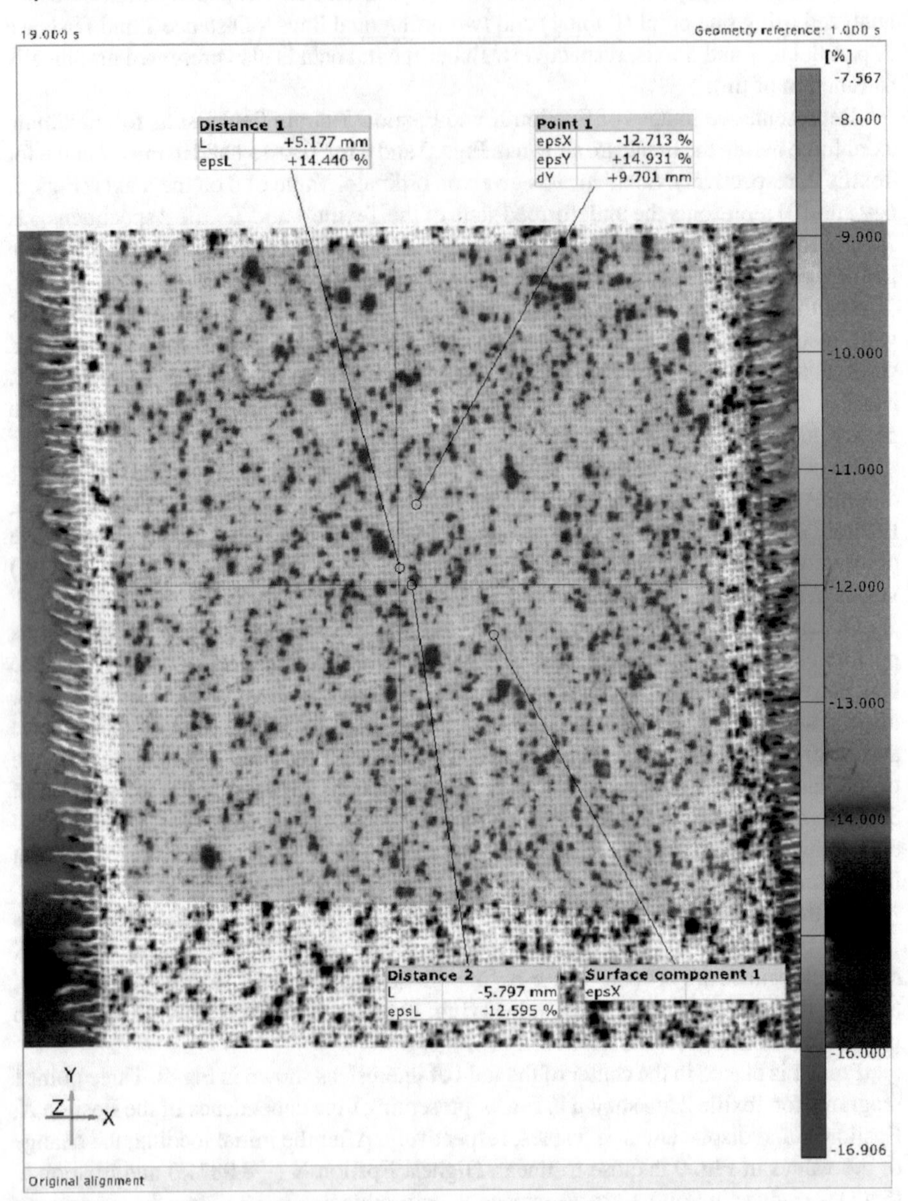

Fig. 3. Representative Epsilon X strain field of Textile 1

of the values in Fig. 10 is close to linear. Highest values for Distance 1 (elongation of 2.852 mm and deformation of 6.550%) and Distance 2 (elongation of −3.880 mm and deformation of −9.001%) are, as expected, before the break, 19 s after the experiment has started.

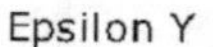

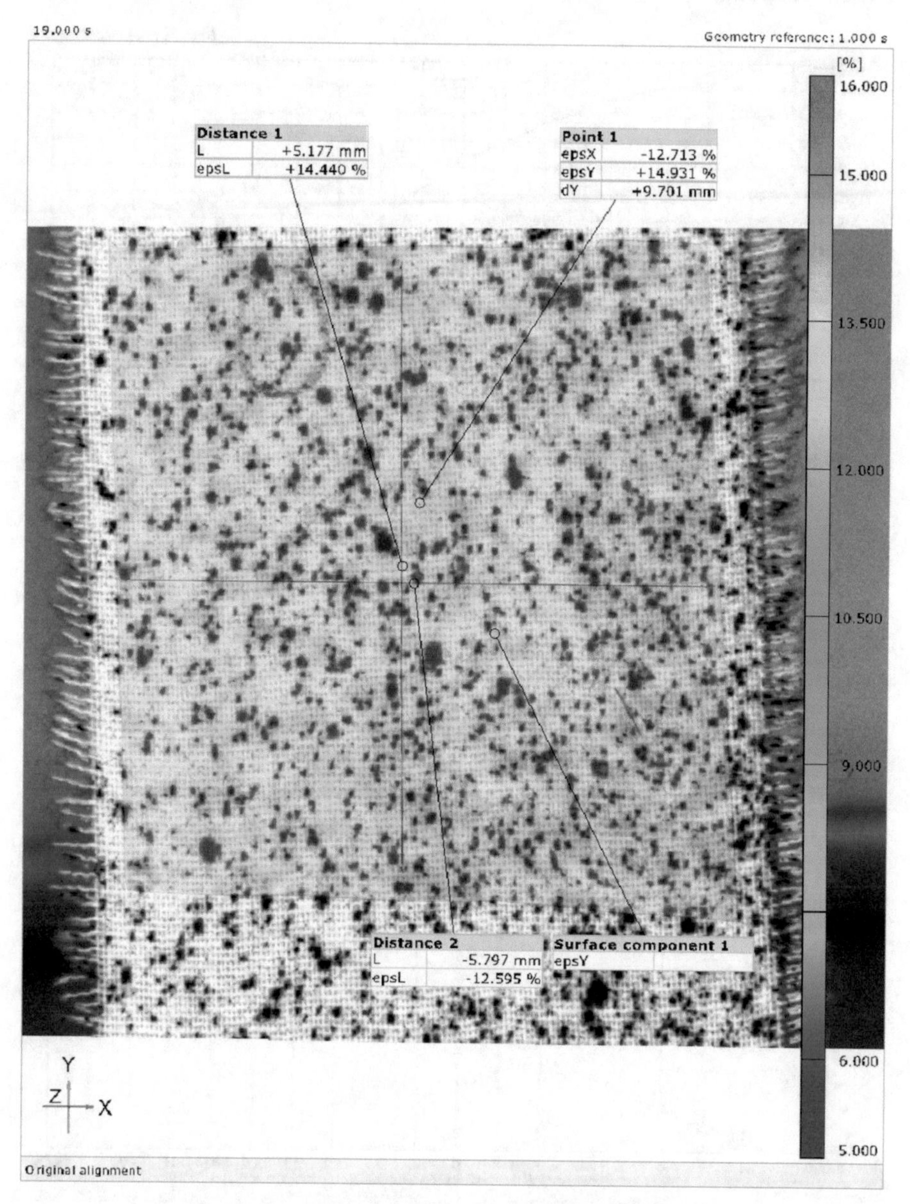

Fig. 4. Representative Epsilon Y strain field of Textile 1

Although there is a small difference between densities of Textile 1 and Textile 2 (less than 10%), difference between the strain values of two materials is much higher – values on the distance diagrams are nearly double between Textile 1 and Textile 2. Similar results were registered on the tensile testing machine (Table 1). Mean value of

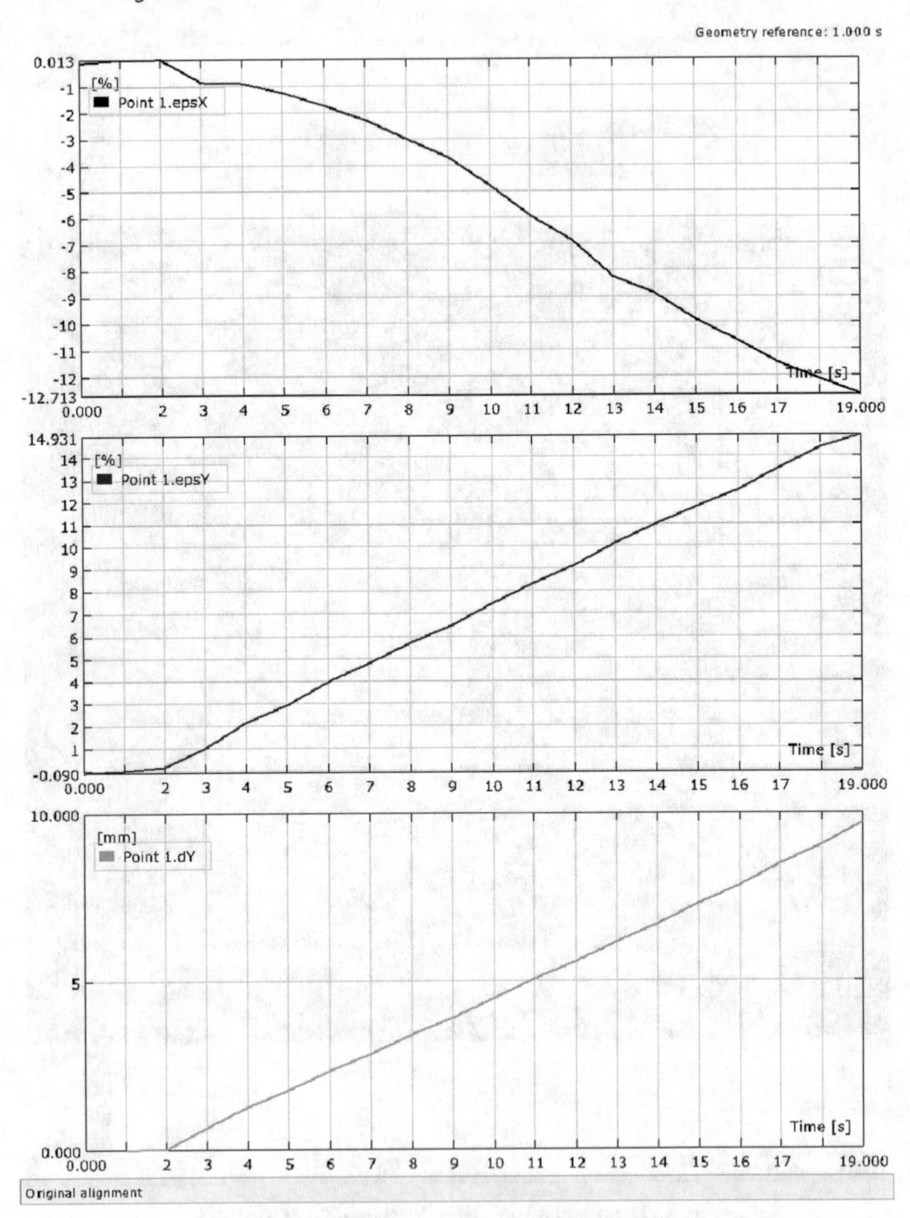

Fig. 5. Point 1 diagrams of Textile 1

the elongation for Textile 1 is more than a double than the elongation for Textile 2, although the difference for mean value of Maximum force is negligible. However, it is possible to get results that deviate from the real values due to incorrect adjustment and control of tension forces of material's ends during experiment.

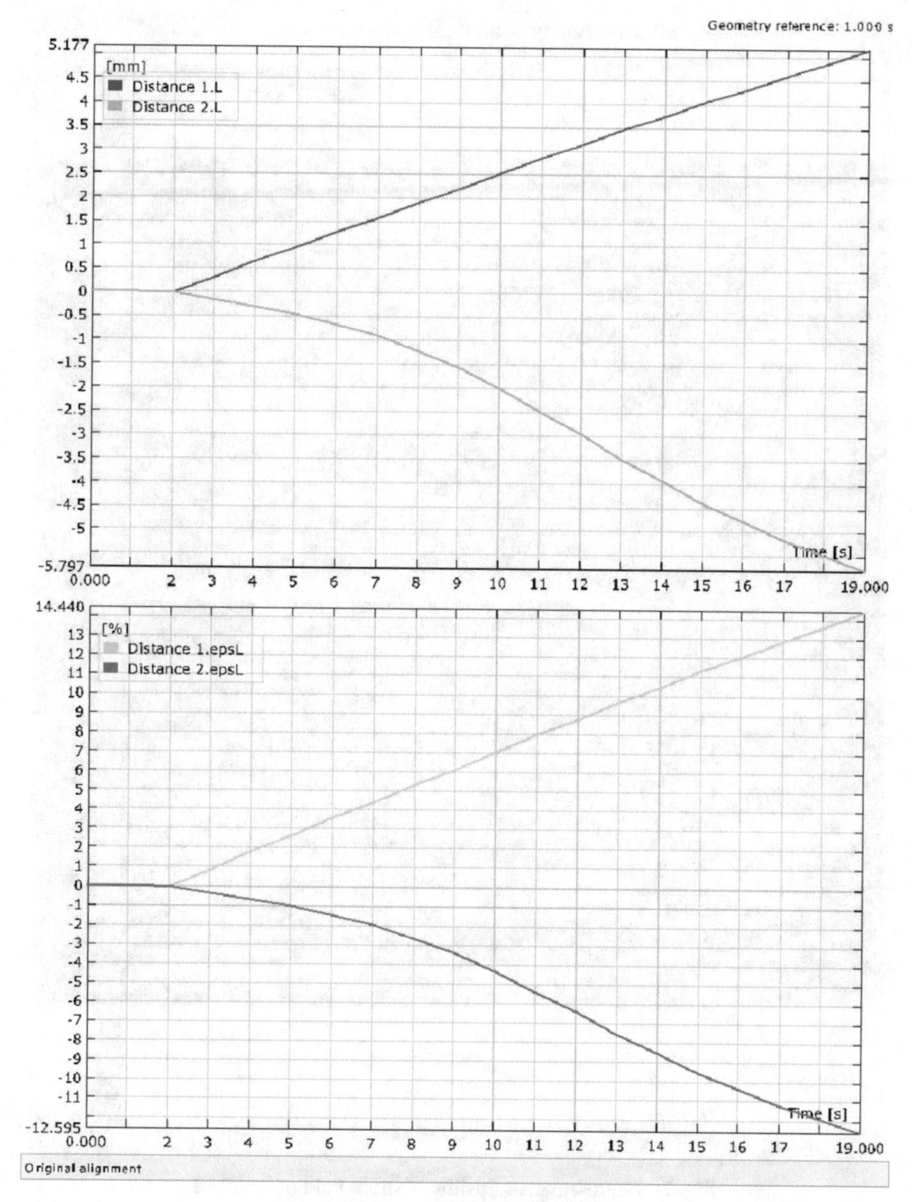

Fig. 6. Distance diagrams of Textile 1

2D DIC method has recently begun to be used for investigation of textile deformation behavior. Advantage of the digital image analysis is the use of fast and low-cost hardware and software for the digital image acquisition and processing. However, some limitations are related to the evaluation of spatial deformations and the specifications of the method.

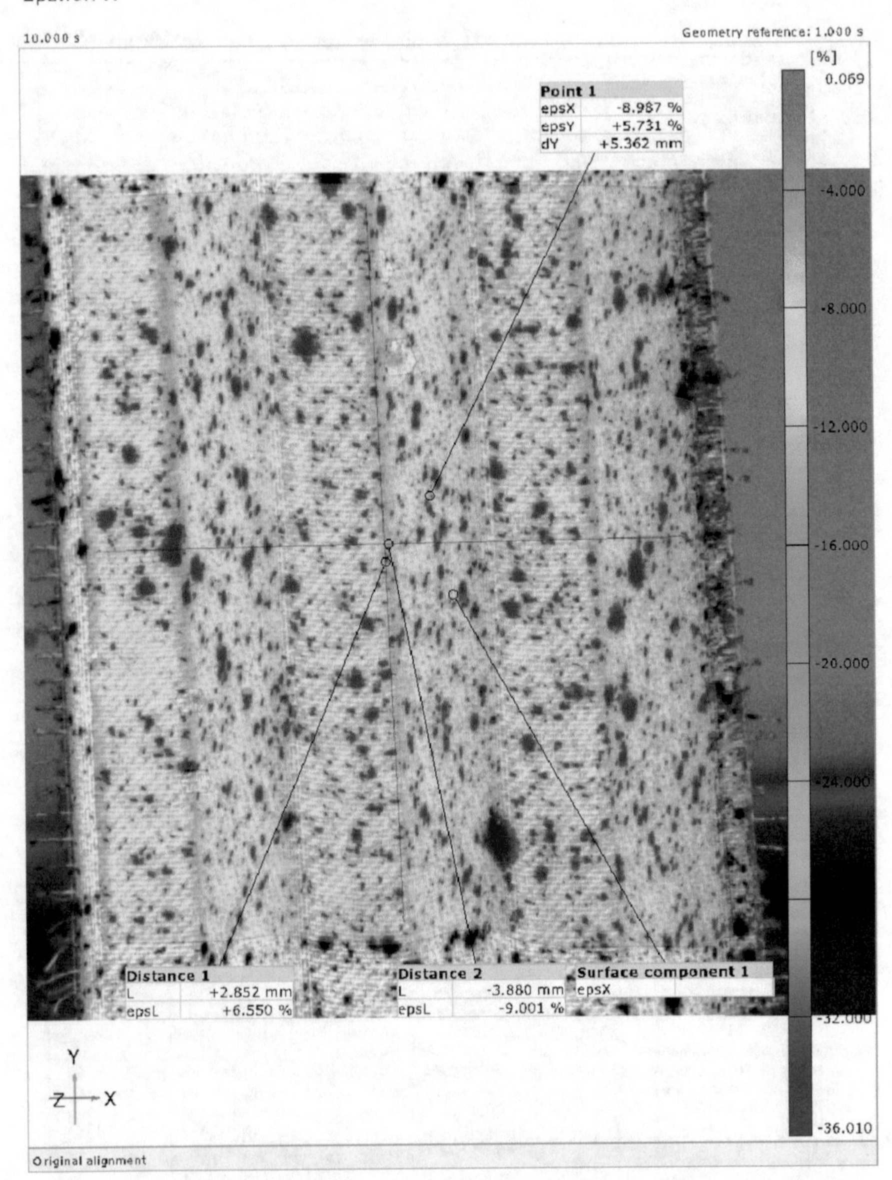

Fig. 7. Representative Epsilon X strain field of Textile 2

One of the most important parameters for digital image acquisition and image analysis errors present the digital camera characteristics, the distance between the object and camera, camera place and direction toward the object and illumination conditions [8].

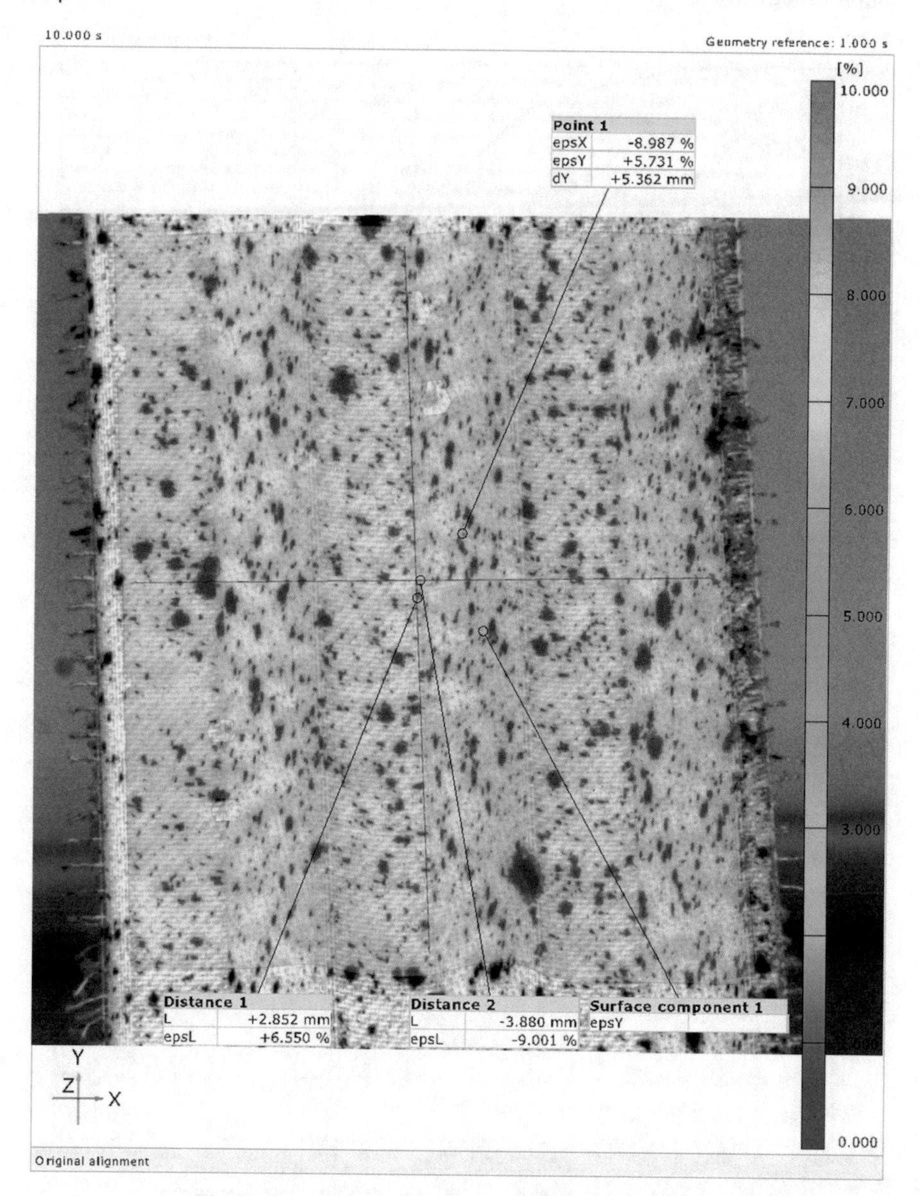

Fig. 8. Representative Epsilon Y strain field of Textile 2

As mentioned in the introduction, today's textile-related standard tests on strength and durability are based on the ones for solid materials, so some considerations related to textile fiber structures have to be mentioned. The main peculiarities of textile materials are as follows [6]:

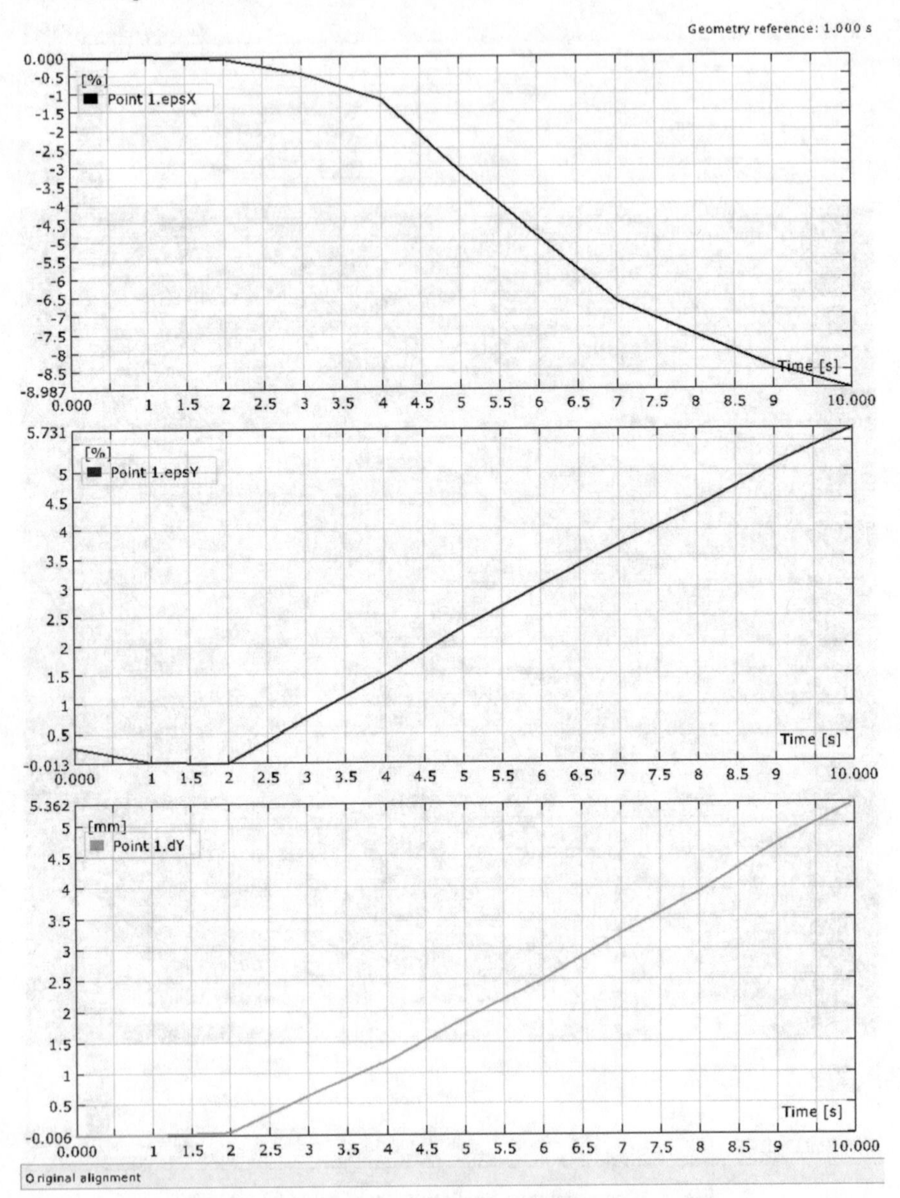

Fig. 9. Point 1 diagrams of Textile 2

- *The discrete nature of textiles.* Because of the porous and soft structure with hairy surfaces, it is difficult to measure the fabric dimensions in order to calculate the stress in the fabric. A much more convenient way is to calculate the fabric stress as force/yarn or its strength in force/tex where tex is the thickness of the yarn expressed in

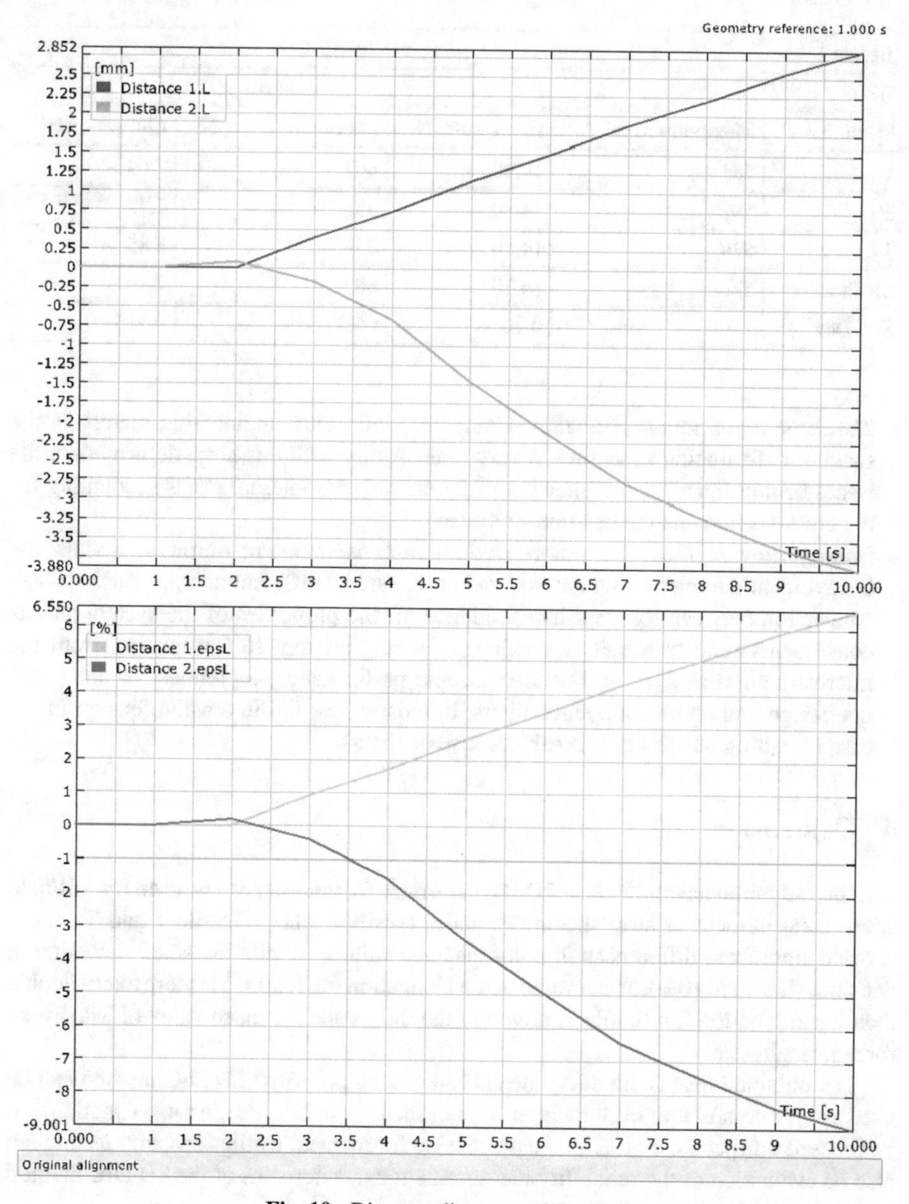

Fig. 10. Distance diagrams of Textile 2

the tex system. For the same reason, some of the analytical techniques in continuum mechanics become difficult or impossible to conduct. For instance, the vector and tensor tools, the internal force and stress resolution, and derivation of the principal eigen-stress components are unlikely to be applicable to fabrics.

Table 1. Maximum force and elongation recorded on tensile testing machine

Textile 1			Textile 2	
Warp				
Specimen	Maximum force, N	Elongation, %	Maximum force, N	Elongation, %
1	581	14.90	667	7.69
2	567	14.70	531	6.75
3	579	14.50	558	6.85
Mean value	576	14.70	582	7.10
Std. Dev.	7.40	0.20	66.3	0.515

- *The large deformation of textiles.* Compared with other engineering materials, the scale of deformation in textiles is very high. Along with this large deformation, the issues of non-linearity, the inter-yarn friction and true internal stress accounting for the cross-section change become significant.
- *Non-affinities between the macro- and micro-behaviors.* For fibrous materials, the behaviors at the micro- and macro-levels often are of different natures. Such a weak connection between, or even independence of, the properties of the system and its constituents renders a unique challenge for any attempt in formulating from the microstructural analysis to the macroscopic performance, a premise for any product design and application since fibers, in most cases, fail in tension (except in the case of cutting where fibers break because of shear).

4 Conclusion

This investigation has shown that 2D Digital Image Correlation can be used for full/field strain measurement of strip specimens under tensile loading. Textile 1 and Textile 2 showed significant differences in point distance values, despite the small deviation in densities (less than 10%). Mean value of the elongation for Textile 1 is more than a double than the elongation for Textile 2, although the difference for mean value of Maximum force is negligible.

The obtained results are adequate and precise so proposed 2D DIC method can be used as a successful tool in characterization of medical textile. Further investigation will be focused on specific digital image processing for medical textiles property testing but also on evaluation of the reliability measurement characteristics of the 2D DIC method in this field of textiles.

Acknowledgement. This research was supported by Ministry of Education, Science and Technological Development of Republic of Serbia under Project TR35031.

References

1. Anand, S., Kennedy, J., Miraftab, M., Rajendran, S. (eds.): Medical and Healthcare Textiles. Woodhead Publishing Limited, Sawston (2010)
2. Asanovic, K., Cerovic, D., Mihailovic, T., Kostic, M., Reljic, M.: Quality of clothing fabrics in terms of their comfort properties Quality of clothing fabrics in terms of their comfort properties. Indian J. Fibre Text. Res. **40**, 363–372 (2015)
3. Barbulov-Popov, D., Cirkovic, N., Stepanovic, J., Reljić, M.: The analysis of the parameters that influence the seam strength. Ind. Textil. **63**, 131–136 (2012)
4. Reljic, M., Stojiljkovic, S., Stepanovic, J., Lazic, B., Stojiljkovic, M.: Study of water vapor resistance of Co/PES fabrics properties during maintenance. In: Mitrovic, N., Milosevic, M., Mladenovic, G. (eds.) Experimental and Numerical Investigations in Materials Science and Engineering, pp. 72–83. Springer, Cham (2019)
5. Stepanovic, J., Cirkovic, N., Radivojevic, D., Reljic, M.: Defining the warp length required for weaving process. Ind. Textil. **63**, 227–231 (2012)
6. Wilusz, E. (ed.): Military Textiles. Woodhead Publishing Limited, Sawston (2008)
7. Hursa, A., Rolich, T., Ercegovic Razic, S.: Determining pseudo Poisson's ratio of woven fabric with a digital image correlation method. Text. Res. J. **79**, 1588–1598 (2009)
8. Meskuotiene, A., Dargiene, J., Domskiene, J.: Metrological performance of the digital image analysis method applied for investigation of textile deformation. Text. Res. J. **85**, 71–79 (2015)
9. Wijeratne, R.S., De Vita, R., Rittenhouse, J.A., Orler, E.B., Moore, R.B., Dillard, D.A.: Biaxial properties of individual bonds in thermomechanically bonded nonwoven fabrics. Text. Res. J. **89**, 698–710 (2019)
10. Sutton, M.A., Orteu, J.-J., Schreier, H.: Image Correlation for Shape, Motion and Deformation Measurements: Basic Concepts. Theory and Applications. Springer, Boston (2009)
11. Mitrovic, N., Petrovic, A., Milosevic, M., Momcilovic, N., Miskovic, Z., Maneski, T., Popovic, P.: Experimental and numerical study of globe valve housing. Hem. Ind. **71**, 251–257 (2017)
12. Mitrovic, A., Mitrovic, N., Tanasic, I., Miloševic, M., Antonovic, D.: Measurement of strain field in glass ionomer cement. Struct. Integr. Life. **19**, 143–147 (2019)
13. Kovačević, T., Rusmirović, J., Tomić, N., Mladenović, G., Milošević, M., Mitrović, N., Marinković, A.: Effects of oxidized/treated non-metallic fillers obtained from waste printed circuit boards on mechanical properties and shrinkage of unsaturated polyester-based composites. Polym. Compos. **40**, 1170–1186 (2019)
14. Manojlovic, D., Dramićanin, M.D., Milosevic, M., Zeković, I., Cvijović-Alagić, I., Mitrovic, N., Miletic, V.: Effects of a low-shrinkage methacrylate monomer and monoacylphosphine oxide photoinitiator on curing efficiency and mechanical properties of experimental resin-based composites. Mater. Sci. Eng. C **58**, 487–494 (2016)
15. Miletic, V., Manojlovic, D., Milosevic, M., Mitrovic, N., Stankovic, T., Maneski, T.: Analysis of local shrinkage patterns of self-adhering and flowable composites using 3D digital image correlation. Quintessence Int. (Berl) **42**, 797–804 (2011)
16. Miletic, V., Peric, D., Milosevic, M., Manojlovic, D., Mitrovic, N.: Local deformation fields and marginal integrity of sculptable bulk-fill, low-shrinkage and conventional composites. Dent. Mater. **32**, 1441–1451 (2016)
17. Mitrovic, A., Mitrovic, N., Maslarevic, A., Adzic, V., Popovic, D., Milosevic, M., Antonovic, D.: Thermal and mechanical characteristics of dual cure self-etching, self-adhesive resin based cement. In: Lecture Notes in Networks and Systems, pp. 3–15. Springer, Cham (2019)
18. Lezaja, M., Veljovic, D., Manojlovic, D., Milosevic, M., Mitrovic, N., Janackovic, D., Miletic, V.: Bond strength of restorative materials to hydroxyapatite inserts and dimensional changes of insert-containing restorations during polymerization. Dent. Mater. **31**, 171–181 (2015)

19. Mitrović, A., Antonović, D., Tanasić, I., Mitrović, N., Bakić, G., Popović, D., Milošević, M.: 3D digital image correlation analysis of the shrinkage strain in four dual cure composite cements. Biomed. Res. Int. **2019**, 7 (2019)
20. Tanasić, I., Mitrović, A., Mitrović, N., Šarac, D., Tihaček-šojić, L., Milić-Lemić, A., Milošević, M.: Analyzing strain in samples with all-ceramic systems using the digital image correlation technique. Srp. Arh. Celok. Lek. **2019**, 528–533 (2019)
21. Tihacek-Sojic, L., Milic Lemic, A., Tanasic, I., Mitrovic, N., Milosevic, M., Petrovic, A.: Compressive strains and displacement in a partially dentate lower jaw rehabilitated with two different treatment modalities. Gerodontology **29**, 851–857 (2012)
22. Tanasic, I., Milic-Lemic, A., Tihacek-Sojic, L., Stancic, I., Mitrovic, N.: Analysis of the compressive strain below the removable and fixed prosthesis in the posterior mandible using a digital image correlation method. Biomech. Model. Mechanobiol. **11**, 751–758 (2012)
23. Šarac, D., Atanasovska, I., Vulović, S., Mitrović, N., Tanasić, I.: Numerical study of the effect of dental implant inclination. J. Serbian Soc. Comput. Mech. **11**, 63–79 (2017)
24. Sarac, D., Mitrovic, N., Tanasic, I., Miskovic, Z., Tihacek-Sojic, L.: Experimental analysis of dental-implant load transfer in polymethyl-methacrylate blocks. Mater. Tehnol. **53**, 133–137 (2019)
25. Tanasić, I., Šarac, D., Mitrović, N., Tihaček-Šojić, L., Mišković, Ž., Milić-Lemić, A., Milošević, M.: Digital image correlation analysis of vertically loaded cylindrical Ti-implants with straight and angled abutments. Exp. Tech. **40**(4), 1227–1233 (2016)
26. Šarac, D., Mitrović, N.R., Tanasić, I.V., Tihaček-Šojić, L.: Experimental methodology for analysis of influence of dental implant design on load transfer. FME Trans. **46**, 266–271 (2018)
27. Milošević, M., Milošević, N., Sedmak, S., Tatić, U., Mitrović, N., Hloch, S., Jovičić, R.: Digital image correlation in analysis of stiffness in local zones of welded joints. Teh. Vjesn. Tech. Gaz. **23**, 19–24 (2016)
28. Sedmak, A., Milošević, M., Mitrović, N., Petrović, A., Maneski, T.: Digital image correlation in experimental mechanical analysis. Struct. Integr. Life. **12**, 39–42 (2012)
29. EN ISO 13934: Textiles—Tensile properties of fabrics—Part 1: Determination of maximum force and elongation at maximum force using the strip method

Author Index

N. Mitrovic et al. (Eds.): CNNTech 2020, LNNS 153, pp. 465–466, 2021.
https://doi.org/10.1007/978-3-030-58362-0

Printed in the United States
By Bookmasters

Printed in the United States
By Bookmasters